Exploitation Conservation Preservation

A Geographic Perspective on Natural Resource Use

FOURTH EDITION

SUSAN L. CUTTER
University of South Carolina

WILLIAM H. RENWICK
Miami University

John Wiley & Sons, Inc.

Acquisitions Editor	Ryan Flahive
Assistant Editor	Denise Powell
Marketing Manager	Kevin Molloy
Senior Production Editor	Patricia McFadden
Production Management Services	Charlotte Hyland
Senior Designer	Kevin Murphy
Photo Editor	Lisa Gee
Cover Photo	Lester Lefkowitz/TAXI/Getty Images

This book was set in Times Roman by UG / GGS Information Services, Inc. and printed and bound by Hamilton Printing Company. The cover was printed by Von Hoffmann, Inc.

This book is printed on acid free paper.

Copyright © 2004 John Wiley & Sons, Inc. All rights reserved.

No part of this publication may be reproduced, stored in a retrieval system or transmitted in any form or by any means, electronic, mechanical, photocopying, recording, scanning or otherwise, except as permitted under Sections 107 or 108 of the 1976 United States Copyright Act, without either the prior written permission of the Publisher, or authorization through payment of the appropriate per-copy fee to the Copyright Clearance Center, Inc. 222 Rosewood Drive, Danvers, MA 01923, (508)750-8400, fax (508)750-4470. Requests to the Publisher for permission should be addressed to the Permissions Department, John Wiley & Sons, Inc., 111 River Street, Hoboken, NJ 07030, (201)748-6011, fax (201)748-6008, E-Mail: PERMREQ@WILEY.COM. To order books or for customer service please call 1-800-CALL WILEY (225-5945).

ISBN 0-471-15225-0

Printed in the United States of America

10 9 8 7 6 5 4 3 2 1

PREFACE

Natural resource conservation has been an important college-level course for several decades, and many good texts have been written on the subject. Moreover, in the three-plus decades since the first Earth Day, students' interest in environmental issues has remained high. The textbooks most often used since the early 1970s have reflected the ideals of the recent environmental movement, with its concern for natural environmental processes, pollution control, the population explosion, and depletion of mineral and other resources.

The environmental movement of the 1960s and 1970s was one of idealism. Throughout the 1970s and 1980s those ideals became incorporated into many aspects of government policy, business practice, and the everyday concerns of the general population. In the 1990s we saw both a renewal of environmental concerns with a global focus and a maturation of our understanding of the interdependence of economic processes and environmental protection.

The terror attacks of September 11, 2001, have in many ways distracted the United States from environmental and resource issues. At the same time, the increased attention to global interactions and interdependency brought by those attacks and subsequent events have underlined the importance of environmental issues as one of many themes that bind the world together, for better or for worse. Oil is at the heart of U.S. involvement in the Persian Gulf area and the tensions it has produced. Water is one of the stressors aggravating the conflict between Israel and its neighbors, as well as a critical dimension to relations among several countries in that region. Agricultural subsidies in the United States and other countries continue to bedevil both efforts to increase food production in poor countries and negotiations over international trade. The reluctance of the United States to take serious action against global warming undermines its relationship with many other countries, especially in western Europe. These and other issues have increased in importance in much of the world, just as Americans are concentrating on terrorism and military security.

In this book, we integrate physical, economic, social, and political considerations into our examination of the major natural resource issues facing the world today. We take the view that none of these four factors alone determines the suitability of a resource for any particular use at any time. Rather, a dynamic interplay between these factors causes continuing changes in methods and rates of resource exploitation. The title *Exploitation, Conservation, Preservation* includes three value-laden and politically charged words that have been at the heart of the natural resources debate over the last century. The subtitle, *A Geographic Perspective on Natural Resource Use*, reflects the traditional use of geography, which integrates studies of physical and human phenomena to understand human use of the earth.

Although the authors share this approach to the subject, we have contrasting scientific and philosophical views. We have avoided, as much as possible, taking any one point of view. Instead, we have attempted to include a wide range of opinions and interpretations of natural resource issues, in the hope that this will provide both a balanced review and a basis for discussion. At the same time, no commentary on natural resources can be free of political content, and we recognize that this book must inevitably be influenced by its authors' personal views. We hope that students reading this book will learn to recognize and understand the political content of our discussions as well as others' presentations and arguments on these issues.

One way to cut through biases inherent in any discussion of natural resources is to ground such discussions in data that are as factual and objective as possible. We are fortunate that such data are readily available. In this edition we have used the most recent figures available, but we recognize that many of the data we use will be superseded by more recent information even as the book goes to press. We urge students and instructors to make use of the rapidly expanding data resources available via the Internet to inform their discussions whenever possible. These are invaluable instructional tools that have the potential to transform a course in environmental conservation from one that helps students to understand the issues to one in which they learn how to analyze and quantitatively evaluate the significance of resource patterns and trends. We have used a small fraction of these data to illustrate some of the more important topics covered in this text, but we encourage students and instructors using the book to exploit these data more fully, and especially to be aware of changing conditions over time.

In producing the fourth edition we have benefited not only from those who worked on previous editions, but also from several people who contributed specifically to this one. In particular, we would like to thank M. Victoria Berry (University of Georgia), Burrell Montz (Binghamton University), Donald A. Friend (Minnesota State University), James Hayes-Bohanan (Bridgewater State College), Leslie A. Duram (Southern Illinois University), Travis J. Ryan (Butler University), Wilbur Hugli (University of Western Florida), and Charles G. Manyara (Redford University).

We also thank our families: Langdon, Nathaniel, Megan, Debra, Sarah, Levi, Peg, and Oliver who continue to be understanding of the time pressures we face. The authors accept all responsibility for any errors, and we share credit with everyone who helped us for any praise this book may receive.

Contents

Preface

1 NATURAL RESOURCES: THOUGHTS, WORDS, AND DEEDS   1
What is a Natural Resource?   1
   Resource Cognition and Value   1
   Kinds of Resources   4
   Limits to Resource Classification   5
Conserving Resources: What Does It Mean?   5
ISSUE 1.1: Two Contrasting Views of Natural Resource Management   6
   Environmental Ethics   8
   Environmental Justice   8
   What Values Do You Bring to the Natural Resources Debate?   8
   Nature, Economics, and The Politics of Natural Resource Use   9
The Systems Approach   10
General Outline of the Book   11
References and Additional Reading   12
Study Questions   13

2 ECONOMICS OF NATURAL RESOURCES   14
Introduction   14
Economics and the Use of Resources   15
   Characteristics of Natural Resources   15
   Pricing Systems   17
   Economic Systems   17
   Supply and Demand   18
   Market Imperfections   18
Determining Resource Value: Quantifying the Intangibles   19
   Benefit-Cost Analysis   19
ISSUE 2.1: European Integration and the Environment: EEA and EIONET   20
   Quantifying Value   21
ISSUE 2.2: What Is the Value of a Human Life?   22
Management and Allocation of Resources   24
   Ownership   24
   Social Costs   27
   Economics of the Individual Firm   28
Business and the Environment: Recent Trends   29
   Diversification and Multinational Corporations   29
ISSUE 2.3: The Value of Nature   30
   The Greening of Business   32
   Deregulation   33
Conclusions   33
References and Additional Reading   34
Study Questions   35

3 ENVIRONMENTAL HISTORY, POLITICS, AND DECISION-MAKING   30
Introduction   36
Natural Resource Use: A Historical Perspective   36
Development of Natural Resource Policy   37
   U.S. Environmental Policy   37
   International Policy   45
ISSUE: 3.1: In Fairness to All: Agenda 21 and Environmental Equity   48
Current Natural Resource Policy   49

How Decisions Are Made   50
  Resource Decision-Making
    in the United States   50
  International Environmental
    Decision-Making   51
ISSUE: 3.2: Politics and the Arctic National
  Wildlife Refuge (ANWR)   52
The Decision-Making Process   54
  Organizations   54
  Strategies   55
  The Role of Public Interest   56
The "New" Environmental Politics   57
References and Additional Reading   58
Study Questions   60

4 ECOLOGIC PERSPECTIVES ON NATURAL RESOURCES   61

Earth's Resource Environments   61
  Bioregions   62
  Human Use of the Land   66
ISSUE 4.1: What Happens When the Geography
  Changes?   67
Energy Transfers and Material Flows   69
  Carbon Cycle   70
ISSUE 4.2: *Silent Spring* versus *Our Stolen
  Future*   71
  Nitrogen and Phosphorus   72
  Hydrologic Cycle   73
  Food Chains   74
  Carrying Capacity   76
The Scope of Human Impact   77
  The Extent of Environmental Pollution   77
  Human Impact on Biogeochemical
    Cycles   81
Ecological Concepts in Resource
  Management   82
  Any Given Environment Has Finite Carrying
    Capacity   82
  Be Aware of Limiting Factors   83
  Minimize Disruption by Mimicking
    Nature   83
  Close the Loops   84
Conclusions   85
References and Additional Reading   85
Study Questions   86

5 THE HUMAN POPULATION   87

A Brief History of Population Growth   88
Basic Demographics   89
  Birth, Death, and Fertility   90
ISSUE 5.1: AIDS and Population Growth
  in Africa   92
  Age Structure   95
  Migration   98
  Trends in Population Growth   99
The Distribution of Population and Population
  Growth   100
  Regional Disparities   101
  Increasing Urbanization   101
ISSUE 5.2: Megacities: The New Urban
  Demographic Transition   102
Population Control Strategies   102
  Socioeconomic Conditions
    and Fertility   103
  Contraception and Family Planning   103
Population Growth and Affluence   104
Conclusions   105
References and Additional Reading   105
Study Questions   106

6 AGRICULTURE AND FOOD PRODUCTION   107

Food Production Resources   108
  Crops   108
  Livestock   111
  The U.S. Agricultural Land Resource
    Base   112
  Modern American Agricultural
    Systems   115
Natural Resources for Agricultural
  Production   115
  Soil   116
  Water   116
ISSUE 6.1: Agriculture, CO_2, and Climate:
  The Only Certainty Is Change   118
  Fertilizers and Pesticides   120
  Seed   122
  Labor and Machines   124
ISSUE 6.2: The Digital Farmer   125
  Animals in the Food Production
    System   125

Environmental Impacts of Food
    Production   128
    Soil Erosion   128
    Rangeland Degradation   133
Agricultural Policy and Management   136
    Subsidies   136
ISSUE 6.3: Agricultural Subsidies,
    Trade, and Poverty in the Developing
    World   136
    Sustainable Agriculture   138
    Rangeland Management   138
Conclusions   140
References and Additional Reading   140
Study Questions   141

7 FORESTS   142
Forests as Multiple-Use Resources   142
Forests as Fiber Resources   144
    Principles of Sustainable Forestry   144
    Forest Management   144
ISSUE 7.1: Chipko: Grass-Roots
    Environmentalism or a Struggle
    for Economic Development?   146
    Forest Products Technology   148
Nonfiber Uses of Forest Resources   150
    Habitat   150
    Water Resources   151
ISSUE 7.2: The Pacific Lumber Saga   152
    Recreation   154
    Carbon Storage   154
    The Role of Fire   155
Deforestation and Reforestation:
    Three Examples   157
    The Amazon Forest   157
    The Siberian Forest   159
    The U.S. Forestland   160
Conclusion   166
References and Additional Reading   166
Study Questions   167

8 BIODIVERSITY AND HABITAT   168
The Value of Biodiversity   168
    Ecological Interactions   169
    Potential Resources   169
    The Inherent Value of Species   170
    The Pace and Processes of
        Extinction   172
    Causes of Biodiversity Loss   175
ISSUE 8.1: The Mass Extinction of Freshwater
    Mussels   178
Conservation of Biodiversity   180
    Species Protection   180
    Habitat Conservation   183
    The Endangered Species Act   186
ISSUE 8.2: Ecotourism: Loving Wild Places
    to Death   187
    The Convention on Biological
        Diversity   188
Conclusions   189
References and Additional Reading   190
Study Questions   192

9 MARINE RESOURCES: COMMON
    PROPERTY DILEMMAS   193
Introduction   193
The Marine Environment   193
    Physical Properties   193
ISSUE 9.1: Salmon in the Pacific
    Northwest   196
    Habitat and Biological
        Productivity   197
Fisheries   199
    Fisheries Production   199
    Fisheries in Distress   200
Minerals from the Seabed   200
    Energy Resources   200
ISSUE 9.2: Strip Mining the Oceans   201
    Deep-Seabed Minerals   202
Management of Marine Resources   203
    The Problem of Ownership   203
    The Law of the Sea Treaty   204
    Marine Pollution Problems   207
    Protecting Marine Ecosystems   209
    Example: Exploitation and Protection
        of Marine Mammals   212
Conclusions   214
References and Additional Reading   215
Study Questions   217

10 WATER QUANTITY AND QUALITY 218

Water Supply and Its Variability 218
   Spatial Variation in Surface Supply 219
   Temporal Variability 222
   Water Supplies and Storage 222
The Demand for Water 226
   Off-Stream Uses 227
ISSUE 10.1: Water Politics in the Western United States 228
   In-Stream Uses 230
Water Quality 232
   Major Water Pollutants and Their Sources 232
   Groundwater Pollution Problems 238
Water Pollution Control 239
   Wastewater Treatment 239
   Nonpoint Pollution Control 240
   Pollution Prevention 241
Quality, Quantity, and the Water-Supply Problem 241
   Relations Between Quality and Quantity 241
ISSUE 10.2: Water Pollution Legislation in the United States 242
ISSUE 10.3: Surf Your Watershed 244
   Water Quality in Developing Regions 244
Conclusions 246
References and Additional Reading 246
Study Questions 247

11 THE AIR RESOURCE AND URBAN AIR QUALITY 248

Introduction 248
Air Pollution Meteorology 248
   Composition and Structure of the Atmosphere 248
   Role of Meteorology and Topography 250
Major Pollutants 252
   Particulate Matter (PM) 252
   Sulfur Dioxide (SO_2) 252
   Nitrogen Oxides (NO_x) 253
   Carbon Monoxide (CO) 253
   Ozone (O_3) and Volatile Organic Compounds (VOCs) 253
   Lead (Pb) 253
Urban Air Pollution: The World's Megacities 254
   Monitoring Network 254
   Air-Quality Patterns 254
   Economic Develpment and Air Pollution 254
Urban Air Pollution in the United States 256
   Air Pollution Monitoring in the United States 256
ISSUE 11.1: On a Clear Day You Can See the Grand Canyon 259
   National Trends 259
   How Healthy Is the Air You Breathe? 262
ISSUE 11.2: Green Days, Red Days 263
   Air-Quality Control and Planning 264
ISSUE 11.3: Smog City, USA 264
Toxins in the Air 267
   Indoor Air Pollution 268
Conclusions 269
References and Additional Reading 269
Study Questions 270

12 REGIONAL AND GLOBAL ATMOSPHERIC CHANGE 271

Acid Deposition 271
   Formation and Emissions Sources 271
   Geographic Extent and Effects on the Environment 275
   Control and Management 277
Stratospheric Ozone Depletion 279
   Ozone-Depleting Chemicals 280
   The Ozone Hole Is Discovered 280
   Reducing ODCs: The Montreal Protocol 281
ISSUE 12.1: Black Market Freon 284
Global Climate Change 284
   The Greenhouse Effect 285
   Greenhouse Gases 285
   Impacts 287
ISSUE 12.2: The Costs of Global Warming 290
   Greenhouse Politics and Emissions Stabilization 290

Conclusions   292
References and Additional Reading   293
Study Questions   294

13 NONFUEL MINERALS   295
Introduction   295
Reserves and Resources   295
Availability of Major Minerals   298
   Geology of Mineral Deposits   298
   Variations in Reserves and Resources   299
   World Reserves and Resources   299
   U.S. Production and Consumption   301
Strategic Minerals and Stockpiling   302
Mining Impacts and Policy   304
   Environmental Considerations   304
   Social Impacts   305
ISSUE 13.1: The New Gold Rush: Prospecting Is Poison   306
   Nonfuel Minerals Policy   308
Conserving Minerals: Reuse, Recovery, Recycling   308
ISSUE 13.2: Living with Boom and Bust   309
ISSUE 13.3: Computers as Solid Waste   311
Conclusions   311
References and Additional Reading   312
Study Questions   313

14 ENERGY RESOURCES   314
Energy Use in the Industrial Age   314
   Wood, Coal and the Industrial Revolution   315
   Oil and the Internal Combustion Engine   315
   Energy Use in the Late Twentieth Century   316
Energy Sources   318
   Oil and Natural Gas   318
   Coal   323
   Other Fossil Fuels   327
   Nuclear Power   327
   Renewable Energy   331
ISSUE 14.1: The Legacy of Chernobyl   332
ISSUE 14.2: The Three Gorges Dam   336
   Energy Efficiency and Energy Conservation   340
Energy Futures   343
   High-Energy Options   343
   Low-Energy Options   344
   Energy Policies for the Future   345
ISSUE 14.3: Electric Energy Deregulation and the California Energy Crisis   346
References and Additional Reading   348
Study Questions   349

15 THE TRANSITION TO A GLOBAL SUSTAINABLE SOCIETY   350
Limits to Growth?   350
What Is Sustainable Development?   352
   Environmental Versus Economic Sustainability   352
   A Working Definition of Sustainability   353
How Does Sustainability Work?   353
   Waste Recycling   355
   Waste Reduction   356
   Design for Reuse and Recycling   358
   Changing Consumption Patterns   359
   Science and Technology for Sustainability   360
Tipping the Balance   360
   Individual Action   361
   Corporate Action   361
   Government Action   362
Looking Forward   366
References and Additional Reading   366
Study Questions   367

GLOSSARY   369

INDEX   382

Photo Credit List

Chapter 1
Figure 1.2: Steve Morgan/Liaison Agency, Inc./Getty Images. Figure 1.4: Grant Heilman Photography.

Chapter 2
Figure 2.3: Jonathon Nourok/Stone/Getty Images. Figure 2.4: Jacques Jangoux/Photo Researchers. Figure 2.7: Photo by William Renwick.

Chapter 3
Figure 3.1: Index Stock. Figure 3.2: W. A. Raymond/Corbis-Bettmann. Figure 3.3: Tim Barnwell/Stock, Boston/PNJ. Figure 3.4: Courtesy U.S. Department of the Interior, Bureau of Reclamation, B.D. Glaha.

Chapter 4
Figure 4.3: Nicholas Parfitt/Stone/Getty Images. Figure 4.4: Leland J. Prater/U.S. Forest Service. Figure 4.10: National Oceanic & Atmospheric Admin./MD. Figure 4.14: © Betty Press/Woodfin Camp & Associates.

Chapter 5
Figure 5.4: United Nations Photo Unit. Figure 5.7: Kontos Yannis/Gamma-Presse, Inc. Figure 5.12: J. Issac/United Nations Photo Unit.

Chapter 6
Figure 6.10a: Grant Heilman Photography. Figure 6.10b: Erwin Cole/U. S. Department of Agriculture. Figure 6.11: © Joe Munroe/Photo Researchers. Figure 6.12: Doug Wilson/U. S. Department of Agriculture. Figure 6.13: Grant Heilman Photography. Figure 6.14: © Billy E. Barnes/Stone/Getty Images. Figure 6.15: John McConnell/U. S. Department of Agriculture. Figure 6.17a: E. W. Cole/USDA/Soil Conservation Service. Figure 6.17b: U. S. Department of Agriculture. Figure 6.17c: Tim McCabe/U. S. Department of Agriculture. Figure 6.17d: Gene Alexander/USDA/Soil Conservation Service.

Chapter 7
Figure 7.4: Nature's Images, Inc./Photo Researchers. Figure 7.5: © AP/Wide World Photos. Figure 7.6: A. S. Sudhakaran/United Nations Photo Unit.

Chapter 8
Figure 8.3: © James H. Carmichael/The Image Bank/Getty Images. Figure 8.4: © Novosti/Liaison Agency, Inc./Getty Images. Figure 8.7: © Edward R. Degginger//Bruce Coleman, Inc./PNI. Figure 8.9: Tennessee Valley Authority.

Chapter 9
Figure 9.8: Corbis Sygma. Figure 9.9: © Dennis Capolango/Black Star.

Chapter 10
Figure 10.5: Department of Water Resource. Figure 10.11: © Kirk Condyles//Impact Visuals. Figure 10.12: Courtesy Dr. Diana Kashash, North Carolina State University. Figure 10.13: © Daniel S. Brody/Stock, Boston.

Chapter 11
Figure 11.1: © Anthony Suau/Liaison Agency, Inc./Getty Images. Figure 11.4: © Joel W. Rogers/Earth Images. Figure 11.8: © Grant Heilman/Grant Heilman Photography. Figure 11.11: © Gabe Palacio/Aurora/PNI.

Chapter 12
Figure 12.5: © Sigrid Estrada/Liaison Agency, Inc./Getty Images. Figure 12.6: Schmidt-Thomsen.

Chapter 13
Figure 13.5: Don Green/Kennecott Corporation. Figure 13.6: W. I. Hutchinson/U.S. Forest Service. Figure 13.7: © Paul Souders/Liaison Agency, Inc./Getty Images. Figure 13.9: Courtesy Reynolds Aluminum Recycling Company.

Chapter 14
Figure 14.11: U.S. Nuclear Regulatory Commission. Figure 14.14: State of New Mexico. Figure 14.15: Courtesy Pacific Gas & Electric Company. Figure 14.9: Grant Heilman Photography.

Chapter 15
Figure 15.4: Index Stock. Figure 15.5: (*right & left*): © AP/Wide World Photos.

CHAPTER 1

NATURAL RESOURCES: THOUGHTS, WORDS, AND DEEDS

WHAT IS A NATURAL RESOURCE?

Have you ever wondered what went into the manufacture of the pencil you are now using? A seed germinated and consumed soil nutrients, sprouted and was warmed by the sun, breathed the air, was watered by the rain, and grew into a beautiful straight tree. The tree was cut down. Perhaps it rode a river's current, was stacked in a lumberyard, and was sawn into small pieces. This wood was transported to a factory, where it was dried, polished, cut, drilled, inserted with graphite (which is made from coal), and painted. Then consider how the pencil made its way to you. It has been packaged attractively with appealing letters painted down its side, shipped via truck, and stored in a warehouse. Your pencil's active life will not end with you, for it may be used by other hands and minds if you lose or discard it.

Where are the natural resources in that description? *Resources* are things that have utility. *Natural resources* are those that are derived from the Earth and biosphere or atmosphere and that exist independently of human activity. The seed, tree, soil, air, water, sun, and river are all natural resources. They are out there, regardless of whether or not human beings choose to use them. They are the *"neutral stuff"* that makes up the world, but they become resources when we find utility in them (Hunker 1964).

Now, consider the role of human effort in the creation, sale, and use of that pencil. First, in addition to natural resources, nonnatural resources are needed, such as saws, labor, and the intelligence to create the pencil. But what motivates people to select and use some portions of the neutral stuff so that they become resources, while other things are neglected? It is here that we are able to isolate the subject matter of this book: the interactions between human beings and the environment or the neutral stuff. When geographers focus on natural resources, we are asking: What portions of the Earth's whole have people found of value? Why? How do these values arise? How do conflicts arise, and how are they resolved? Neutral stuff may exist outside of our use, and some may regard all things in nature to be of importance and therefore resources in some sense, but things acquire utility or value only within the context of politics, culture, and economics. Let us begin, then, to try to understand how and why resources emerge, are used, and ultimately are fought over.

Resource Cognition and Value

A resource does not exist without someone to use it. Resources are by their very nature human-centered. Different individuals or groups value resources differently. Let's look at the role of environmental cognition in the emergence of resource use.

Environmental cognition is the mental process of making sense out of the environment that surrounds us. To cognize, or think, about the environment leads to the formation of images and attitudes about the environment and its parts. Because we constantly think and react to the environment, our cognition of it is constantly changing on some level. Nonetheless, certain elements of environmental cognition will remain stable throughout our lives. Many factors influence our cognition of

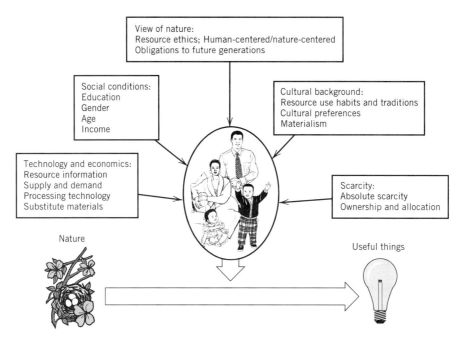

Figure 1.1 Factors involved in resource-use cognition include cultural evaluation, view of nature, social change, economic and technological factors, and resource scarcity.

resources and thus how they will be used. These factors can be grouped into five broad categories: (1) cultural background; (2) view of nature; (3) social conditions; (4) scarcity; and (5) technological and economic factors (Fig. 1.1).

With regard to the first category, there are many different cultures in the world, and each has a different system of values. What has value and meaning in one culture may be regarded as a nuisance in another. More to the point, the value and meaning assigned in one culture may be the complete opposite of the meaning and value of that resource in a different culture. Whaling provides a classic example. Native Americans, especially the Inuit, historically used whales as a source of food and the whale's fat as fuel. Later, the Inuit used whale bones in their arts and crafts, a usage that continues to the present. Today, most Americans appreciate the majesty and beauty of these marine mammals and value them, not as a consumable resource (food and fuel), but as an aesthetic one. Whale watching in California and New England draws thousands of people to view these migratory mammals in their natural habitat. Harvesting whales for food has led to protests against commercial whaling vessels, actions that garner world headlines and public sympathy (Fig. 1.2).

The mesquite, a deep-rooted drylands shrub, presents another example of cultural differences. Ranchers in West Texas feel the need to fight mesquite because they perceive that it dictates what will flourish and what will wither and die in the semiarid environment. Range grasses are shallow-rooted and do not compete well with mesquite, which thus deprives range animals of a source of food. The rancher has the same relationship with mesquite that Wile E. Coyote has with the Roadrunner: the rancher will try virtually anything to conquer mesquite. Yet, not too long ago, the Indians of the American Southwest lived quite harmoniously with mesquite. Mesquite was used for fuel and shade, while the bush's annual crop of highly nutritious beans was a staple resource. Even diapers were fashioned from the bark. Today mesquite is popular as a fuel for gourmet barbecues.

A society's view of itself relative to its natural environment is a second indicator of how it will ultimately use natural resources. On an idealized spectrum, different world views range from human domination and control of nature (technocentrism) to living in harmony with it (ecocentrism) (Pepper 1996). Of course, there is variation within any one group; not all members will agree

Figure 1.2 Society's view of nature. Nature can be viewed as a commodity or as a scenic wonder in need of preservation. These Greenpeace activists believe that killing whales is immoral and use dramatic actions such as this to call attention to their beliefs. In Norway, whales have been eaten for years, and many regard this as morally no different from eating any other animal.

on their view of nature. These underlying philosophical ideas form the basis for many of the modern environmental movements (Chapter 3).

Social conditions influence the value and use of resources. The composition of societies is constantly changing. People grow older, richer, and poorer, and the cultural makeup of societies changes. All of these factors, particularly ethnicity, gender, education, and income, influence how societies cognize and use resources. For example, higher-income households in the United States use more energy than do lower-income households. In colonial New England, lobsters were fed to indentured servants as a cheap food resource. It was not until the late nineteenth century and the influx of southern European immigrants, who regarded the lobster highly, that it became a valuable culinary delicacy.

Cognition of future resources is colored by historical and current use; cognitions also change over time. As a result, planning for future uses of natural resources must take account of these changes. Economists, politicians, and industrialists find it difficult to make accurate forecasts of future resource uses. We may overlook today a resource that will become invaluable in 20 years. Specifically, the solid waste we produce and discard today may be a source of raw materials in the future, and we may see mining reclamation projects in old landfills.

The fourth factor influencing natural resource cognition and use is resource scarcity. As a natural resource becomes scarce or is cognized as becoming scarce, its value may increase. This scarcity may be of two different types. *Absolute scarcity* occurs when the supplies of that resource are insufficient to meet present and future demand. The exhaustibility of all supplies and known reserves of some resources is possible, if improbable. The dwindling supply of certain land resources such as wilderness could conceivably lead to an absolute scarcity. *Relative scarcity* occurs when there are imbalances in the distribution of a resource rather than the insufficiency of the total supply. This imbalance can be either short-term or long-term. Climatic fluctuations resulting in floods, droughts, or frost routinely cause relative shortages of fresh produce. Open space was not considered a resource until it became relatively scarce in urban areas. Then it became something to be valued, protected, and incorporated into urban redevelopment plans. Relative scarcity also results when one group is able to control the ownership or distribution of resources at the expense of another group. In the energy crises of the early and mid-1970s, Americans were told by both environmental and industry experts that the supply of oil and gas was dwindling—and that it would be impossible to meet future demand because of the absolute scarcity of the resource. Yet today we see lower prices and a more-than-adequate supply, suggesting that relative scarcity was in fact the cause of the energy crisis.

Finally, the fifth set of factors that influence resource cognition and use are technological and economic, both of which are basic to understanding the role of scarcity. Technological factors relate to our knowledge and skills in exploiting resources. Groundwater is not a resource until it is made available by drilling a well and installing pumps or other means to bring it to the surface. Desert lands have little agricultural value unless we possess the technical capability to collect and distribute irrigation water, at which time they may become very valuable. Deuterium in the oceans is not at present a resource, except for its use in weapons. However, if we learn how to control the fusion reaction for energy production in the future, it may become a resource.

Economic factors combine technology and cognition, as reflected in our pricing system (Chapter 2). That is, the value or price of a good is determined by its physical characteristics as well as our ability and desire to exploit those characteristics. In a capitalist economy, a commodity will not be exploited unless it can be done at a profit. Therefore, as prices change, things become (or cease to be) resources. A deposit of iron ore in a remote location may be too expensive to exploit today, but if prices rise substantially it may become profitable to exploit and sell that ore; at that time it becomes a resource.

Rarely is the status of a resource determined by technological, cognitive, or economic factors alone; usually it is a combination of all three. The nuclear power industry is a good example. The development of fission reactors and related technology was necessary for uranium to become a valuable energy resource. But rapid expansion of nuclear-generating capacity depends on this energy source being economically competitive with other sources, such as coal and oil. Coal became costly to use, in part because of concerns about the negative environmental effects of global warming, air pollution, and mining. These concerns helped make nuclear power competitive. But the belief that nuclear power is unsafe necessitated modifications in plants that drove up the cost of nuclear power to the point where it is no longer economically attractive. In addition, many people, citing environmental and health fears, reject nuclear energy at any price. The interplay of these forces will continue to affect the selection of nuclear power relative to other energy sources for some time.

Kinds of Resources

There are various ways to classify resources. We can ask how renewable they are and who benefits from them. *Perpetual resources* (Fig. 1.3) are re-

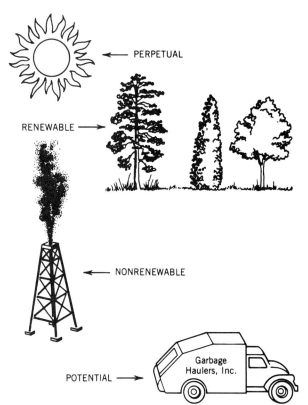

Figure 1.3 The four traditional resource classifications. In reality, a resource can shift from one category to another.

sources that will always exist in relatively constant supply regardless of how or whether we exploit them. Solar energy is a good example of a perpetual resource; it will continue to arrive at the Earth at a reasonably constant rate for the foreseeable future. In the past, the atmosphere and precipitation were regarded as perpetual resources. Recently, however, their quality and the absolute supply of rainfall in some locations have been questioned.

Resources that can be depleted in the short run but that replace themselves in the long run are called *renewable* or *flow resources*. Forests, most groundwater, and fisheries are good examples. Although they can be depleted by harvesting in excess of the replacement rate, if given sufficient time and the right conditions, natural processes will replace them. The key to maintaining the availability of renewable resources is keeping our rate of use at or below the rate of natural replacement.

Nonrenewable or *stock resources* exist in finite supply and are not being generated at a significant rate in comparison to our use of them. Once they are used up, that is the end of them. Most geologic resources, such as fossil fuels and mineral ores, are of this type, as is wilderness.

Finally, *potential resources* are not resources at present but may become resources in the future, depending on cognitive, technological, and economic developments. Their potential depends in part on decisions made about them today. If we make decisions that eliminate them from consideration (such as allowing a plant or animal species to become extinct), then there is no chance of our discovering a resource value in them. Some contemporary examples of potential resources that have recently come into use are solid waste as an alternative fuel or material source and wastewater that is treated and used in irrigation or for other purposes.

Limits to Resource Classification

Although these definitions are relatively clear, to a large extent the status of any resource as perpetual, renewable, or nonrenewable depends on the time scale in which we view it and on how we manage the resource. Even though rainfall on the global level is reasonably constant from year to year, in many areas the quality of that water has been changed by industrial and auto emissions that produce acid rain. On a longer time scale, there is evidence that we may be causing global climatic changes, resulting in increases or decreases in rainfall at the regional level, if not worldwide. Soil, generally regarded as a renewable resource, will recover some degree of its natural fertility if left fallow for a few years. But if accelerated erosion removes a substantial portion of the soil profile, the ability of that soil to support plants that restore nutrients and organic matter may be impaired. It may be centuries before the soil is again productive. That time period is probably too long to consider the soil renewable in human terms.

Similarly, groundwater is generally considered a renewable resource, but in many areas—particularly desert areas, where it is so important—the natural rate of recharge is very low, and in some cases there is presently little or no recharge. In these cases the groundwater is effectively a stock resource: once it is used, it is lost forever. For these reasons, the traditional definitions of resources tell us little about their true nature. In fact, they may be harmful, leading us to think that a renewable resource *will* always be available, regardless of how we exploit it. These classifications illustrate, however, that not all resources are equal to the demands put on them. They also indicate the importance of examining the detailed characteristics of resources and their ability to meet our needs under varying conditions.

CONSERVING RESOURCES: WHAT DOES IT MEAN?

Like Mom and apple pie, few politicians would ever admit to being opposed to the conservation of natural resources, but just as certainly, people disagree on what conservation means. Some believe that it means limited or no use of certain resources. A person with this point of view might maintain that any air pollution is unacceptable and that wilderness cannot be wilderness if there are any people in it. Others feel that conservation means efficient use. They argue that a resource should be used to produce the greatest possible human good. Resources are beneficial but only if they are used; disuse is seen as waste. Some of the history of the development of these two viewpoints in the United States is discussed in Chapter 3.

The disagreement, however, is even more complex than this. There are many definitions of "efficient," because few agree on what is truly beneficial. Is profit the highest benefit? Or is spiritual renewal the best use? If a beautiful valley is filled with four houses to the acre, each resident has a home and a quarter acre of land. Is this a more efficient and beneficial use than making the valley into a park, so that many more can enjoy it, albeit less often?

In addition, how much time should be considered for use of a resource? Should its beneficial use be spread over many years, in small amounts? Or should we gain all the benefits we can now and use other resources in the future? Do future generations have the same rights to a present resource as do people currently living, even though we can't tell whether they will actually want to use it? In some cases these questions can be answered in rational terms, but often they are philosophical or political in nature and may be considered in the context of *environmental ethics* and *environmental justice*.

ISSUE 1.1: TWO CONTRASTING VIEWS OF NATURAL RESOURCE MANAGEMENT

A Nature-Centered View
Consider the following statements:

1. Nature, including individual organisms, species, and ecosystems, has inherent value. Humans exist as one species among many, and like any species, we have the right to use nature to the extent necessary to maintain our existence. But because humans have intelligence and power that are much greater than those of any other species, we have a special obligation to use that intelligence and power wisely and not damage other organisms or their habitats unnecessarily.
2. Nature is the basis of the resource base we enjoy. While many materials we use are synthesized from inanimate substances, our most important resources—food, water, and oxygen—are produced by biogeochemical processes in nature. The function of these processes must therefore be allowed to continue unmodified by humans.
3. Humans are using resources at rates never before experienced on Earth, and these rates will probably double in the next 50 years. Human participation in Earth's biogeochemical cycles is globally significant and locally dominant. Most natural resource systems are fundamentally altered, and many are severely stressed. Growth in population and resource use over the next few decades will increase the degree to which natural resource systems are altered, and many resource systems will be stressed beyond the limits of utility.

These three statements sound reasonable enough and are not likely to offend many people. If we accept them as good environmental values, what resource management policies do they require? Clearly, they demand that we minimize our modifications of natural systems. Actions such as deforestation, soil erosion, pollution, and emissions that lead to global warming cause either permanent or long-term alteration of the environment, probably resulting in species extinctions or, at the very least, substantial alterations of natural systems.

These statements demand that our resource-use activities be completely sustainable, in the sense that everything we use from nature must be recycled and replaced at rates equal to our use. If we harvest the trees in a forest, we must allow them to regenerate, and we must do so in ways that do not prevent species that depend on mature forests from occupying them. If we use the soil to grow crops, we must do so in ways that do not deplete soil fertility. We should not use nonrenewable resources such as fossil fuels, for to do so deprives future generations of their use. We might mine metal ores, provided that we make it possible to supply future needs of those metals through the recycling of already-mined materials.

A Human-Centered View
What happens if we take another ethical position, one that is more human-centered? How does this position influence resource-use practices? Consider these statements:

1. Our primary concern should be to improve the quality of life for all humans. Quality of life depends on material goods, such as food, shelter, consumer goods, and good health, and intangibles, such as education, security, and aesthetics, including beauty in human and natural creations.
2. Human culture is so diverse that the value of intangibles cannot possibly be fixed but varies from indi-

Another central question relating to environmental ethics is whether nonhuman entities have rights, independent of those assigned by humans, and if so what those rights are. For example, many people believe that animals have a right to exist and that to deliberately kill them for any reason is immoral. Some may attach similar rights to plants. Or one might feel that it is acceptable to kill animals for human necessity, such as for food, but not for sport or convenience.

Some people may extend rights to nonliving entities, such as mountains or rivers. For example, some Native American groups believe that to plow the ground is to cut into the flesh of the Earth, while others would regard mining on a sacred mountain as an offense to the spirits of that mountain (Nash 1989; Stone 1993).

The dominant view in the United States today, at least as it is manifested in current resource-use practices, is quite different from all the foregoing.

vidual to individual, from time to time, and from one geographic location to another. Material goods, on the other hand, are universally important, and their value is measured in monetary terms with relative ease.

3. Throughout history, the single most important determinant of the material quality of life has been the ability of human societies to create goods from the natural resources that surround them. This ability is more dependent on human factors such as technology and wealth than on the inherent qualities of natural resources. For example, if we were in need of containers for cooking food, iron ore would be worthless without the knowledge of how to convert it to metal pans. With sufficient knowledge, however, we can make suitable cooking containers from a vast range of natural resources—not only other metals but also ceramics and many other materials.

This second group of statements, like the first one, contains few things that would be disputed by many. But you can probably see that they will lead us in a very different direction for resource use than those suggested by the first group.

Most significantly, by making quality of life for humans our primary concern, we immediately relegate the needs of other species and the habitats they occupy to a secondary role. This does not mean that we are driven to destroy other species; rather, in weighing a decision on how to manage a given environment, we need only concern ourselves with the value of that species to us and not with its inherent value.

For example, consider the tall-grass prairie ecosystem that once occupied the eastern Great Plains of the United States. Today this ecosystem is virtually nonexistent, except for a few remnant patches and some places where we are attempting to restore what we think were the original components of this ecosystem. We don't know how many species were lost through the destruction of this ecosystem, let alone what they were. In its place, however, we have farms that produce enormous amounts of food—the wheat, corn, and soybeans from which we make bread and which we feed to livestock to produce meat. These lands produce ample good-quality food for hundreds of millions of people. Which use of the land brings the greatest improvement in the quality of life? Tall-grass prairie or amber waves of grain?

What about the future? Should we be concerned about preserving specific environments or resources for future generations? While we should be concerned about the future, we cannot possibly anticipate society's needs and abilities more than a few generations into the future. If we look back to the 1890s, the internal combustion engine had only just been developed, use of electricity was in its infancy, and airplanes were still just dreams. Who could possibly have anticipated that plastics would replace steel in automobiles, that plywood and particle board could replace sawn lumber, or that telecommunications could replace physical transport of some commodities? Technological and social change will make resources that are insignificant today vital and today's precious commodities worthless, and these changes can happen very quickly. It would be foolishly arrogant of us to think that we can make decisions about resource needs 100 years in the future.

This view focuses on human needs and assigns rights only to humans. It considers plants, animals, and the nonliving parts of Earth to have value only to the extent assigned by humans. A bald eagle, for example, has value as our national symbol, and we prohibit killing it for that reason. Old Faithful geyser is unique and well known, and we would not tolerate turning the site into a parking lot. But we do not attach special values to "ordinary" natural things.

Environmental Ethics

Clearly, a wide range of views of nature and natural resources within human society have arisen. We have no intention of evaluating these different views or of making judgments about them—that is for the individual alone. As a society, we attempt to arrive at consensual value statements, usually through political processes, and our social expressions of values change through time.

As examples of the diversity of ethical views of natural resources that exist today and the implications of these views for natural resource policy, we can consider two different positions: a nature-centered view and a human-centered view (Issue 1.1). These two positions represent a strong contrast, but they are not necessarily extremes. Neither could be considered a majority view in the world today, but each includes elements that many people would agree with, at least to a limited extent. We present these two views solely as illustrations of the diversity of opinions that are brought to bear on the subject of this book.

Environmental Justice

Environmental justice is another important consideration in the use of natural resources (Mutz et al. 2001). Environmental justice examines how natural resource use benefits some stakeholders and communities, while burdening others. It also focuses on our collective obligation to future generations of humans to leave them with an environment that is at least as healthy, productive, and diverse as the one we have inherited. We must not make any assumptions about the desires, needs, or technological abilities of future generations that would limit their opportunity to lead satisfying and fruitful lives. For example, we must dispose of radioactive wastes in ways that will protect future generations from harm, but we must also ensure that economic resources (compensation for their stewardship) are available to our children and grandchildren to manage those wastes. Human-induced climate change is an especially severe violation of this principle, for it will likely have many far-reaching impacts on future natural resource availability, even with some cessation of emissions.

Finally, environmental justice looks at how policies and practices exacerbate social injustices and the problems of disadvantaged communities. Disparities in quality of life between rich and poor are large, and a significant portion of the Earth's human population has a quality of life that most people regard as unacceptably low. For the poorer portion of Earth's population, economic development would have a much greater impact on quality of life than preservation of natural resources. For example, we would be irresponsible to pass up the opportunity to significantly improve the quality of life for humans alive today, just to preserve a resource that might or might not be useful in the future. In the poorer nations of the world, millions of people (especially children) die each year from conditions such as diarrhea caused by untreated drinking water and respiratory ailments caused by indoor air pollution from wood-fired cookers (Cutter 1995). This suffering could be relieved through provision of safe drinking water supplies, sewage treatment plants, and electricity to power cooking stoves. If curtailing use of fossil fuels cost money that could otherwise be used to install water treatment systems, would it be right to do so just to keep the planet from being a few degrees warmer 100 years from now? If a large dam could be built that would generate hydroelectricity, would it be right not to build that dam for the sake of preserving habitats and species that would be replaced by the reservoir?

What Values Do You Bring to the Natural Resources Debate?

The values that we adopt individually and as a society are central to our decisions about resource management. As suggested by the title, this book will present many viewpoints regarding resource use along a spectrum from those who advocate full use (or exploitation) to those who would conserve (or balance efficient use with protection) to those who would preserve (or remove from use those resources in need of full protection). *Exploitation* is the complete or maximum use of a resource for individual profit or societal gain. *Conservation* is the wise utilization of a resource so that use is tempered by protection to enhance the resource's continued availability. *Preservation* is the nonuse of a resource by which it is fully protected and left unimpaired for future generations.

As we proceed through the book, we will examine a broad range of resource issues, each of which has many facets and many possible outcomes. You will find that in evaluating these issues you will need to know how you feel about such questions as the inherent value of nature and the extent of our obligations to future generations. Take some time to think about these questions as you go! For examples of others' ideas, see articles by Hardin (1974), Merchant (1992), Plant (1989), Warren (1990), Weiss (1990), and White (1967).

Nature, Economics, and the Politics of Natural Resource Use

We believe that a combination of natural, economic, and political factors determines resource use. This means that the study of natural resources must be interdisciplinary and integrative. Integrative approaches must also rely on increasingly specialized work on particular aspects of resource problems, and so we must be able to think in both specific and inclusive ways (Clawson 1986). Most modern views of natural resources involve systems thinking to achieve this integration, and that approach will be used here. To illustrate this view, consider the availability of a basic mineral commodity, such as a metal ore (Fig. 1.4). What determines how much of the mineral we will use?

Ores exist in rocks in a wide range of conditions and with a wide range of qualities. Thus, some deposits might be mined relatively cheaply, whereas others are costly to mine. There are also deposits that are as yet undiscovered but that might be found if one made a careful effort to do so (which is more likely if substantial profits are to be gained from mining the ore).

From the standpoint of a consumer of a mineral, availability depends on whether someone is able and willing to sell at a price the consumer is willing to pay. Willingness to sell usually depends on ability to do so profitably, which depends on a

Figure 1.4 Ore mine. The mining of iron ore in northern Minnesota provided the backbone of the regional economy for over a century.

combination of the costs of mining and processing and the market price. Economic factors affecting both costs of mining and consumer willingness to pay include general levels of wealth and economic activity and many more specific aspects of a national or world economy. Social goals and policies also come into play in that they affect both consumer preferences and environmental policies relating to mining activities, for example.

Figure 1.5 is a simplified representation of how all these factors interact in determining how much of a mineral is used. Clearly, being able to consider economics, geology, environmental regulation, marketing, and technology simultaneously has advantages. We also must rely on specialists who can analyze the details of individual components of that system and generalists who understand the totality of the system and its interconnectedness.

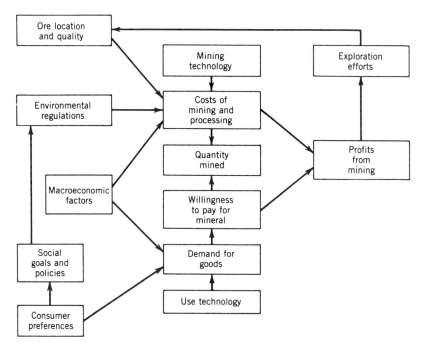

Figure 1.5 A schematic representation of the interaction between natural, economic, and political factors in determining mineral availability and use. Quantity mined is most immediately determined by mining costs and market price, but these are in turn affected by a wide range of economic, technological, and social factors.

THE SYSTEMS APPROACH

In the fourth century B.C., Aristotle stated that the whole is greater than the sum of its parts. This view, more fully developed over the centuries, argues that we should understand the entire world by examining all of it at once rather than looking at each of its constituent parts and then adding them up. During the twentieth century, this holistic view gained acceptance in many fields of study. It was formalized in the scientific community in the 1950s under the heading *general systems theory* (von Bertalanffy 1950).

Systems thinking is a way of viewing the world. The focus is on the comprehensive treatment of a whole by a simultaneous treatment of all parts. A systems approach not only examines the parts individually but also looks at how they interact both with each other and as part of the entire system. Geographers use the systems approach to make sense of natural and human systems and to better understand why the two interact as they do.

As we saw with the example of the pencil, natural resource use involves elements of both human and physical systems. Examples of natural systems are forest ecosystems, the hydrologic cycle, and atmospheric circulation. Human systems include technological, economic, and social systems. Figure 1.5 is a highly simplified illustration of these systems as they apply to mineral resources. The complexity and interrelatedness of human and natural systems make the systems approach particularly important. Natural resources cannot be viewed simply as parts of the physical environment or as commodities that are bought and sold. Instead, they must be considered in the context of the many natural and human factors that affect them and with concern for the potentially far-reaching impacts of changes in resource use.

The need for an integrated systems approach to natural resource use was made abundantly clear during the 1980s, when it appeared that "spaceship Earth" was rapidly deteriorating. Droughts plagued the northern Plains states, raw sewage and medical waste fouled the nation's beaches, the depletion of the ozone layer became fact, not conjecture, and unprecedented tropical deforestation continued. Dire predictions of eco-

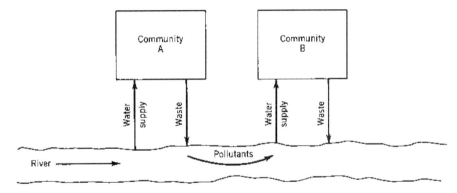

Figure 1.6 The external effects of one system on another. Communities A and B rely on the river as a source of drinking water and a receptacle for their waste. Unfortunately, Community B, because of its downstream location, must use the water after it has been polluted by Community A's waste.

logical collapse and environmental degradation echoed throughout the world. At the same time, it was clear that our science knowledge was insufficient to explain these processes. A new perspective, *Earth systems science,* emerged, which helps us understand the interactions between large-scale environmental changes and human societies. There are two primary tenets to this new approach to understanding environmental issues, including natural resource use. First, the Earth behaves as a single, self-regulating system (Lawton 2001). Second, human actions are so pervasive in their consequences, so complex and interactive at multiple scales from the local to the global, that they modify the very processes and components of the Earth's systems upon which humans depend (The Global Environmental Change Programme 2001).

Earth system science therefore requires integrative knowledge not only of the natural environment (knowledge acquired through earth science, biology, or physical geography) but also of human systems (expertise within the domain of the social sciences such as economics, political science, sociology, and human geography). To label only natural environmental processes as earth systems science misses the point! It is the understanding of the interaction between the natural world *and* human society that governs the modern approach to earth systems science, an approach we use in this book.

As we progress into the next century, environmental concerns will influence every aspect of our lives both here and abroad. The halcyon days of rapid exploitation are over and are being replaced by more conservation-oriented strategies that emphasize sustainable development, less resource use, and more awareness of the longer-term consequences of resource use (Fig. 1.6). Understanding the complexities of these human and natural systems and their interaction is essential as we fulfill our stewardship of the planet. You can make a difference, and we hope that this book will help you realize that.

General Outline of the Book

In this book, the analysis of natural resources and management policies has both physical and human foci. We stress the interrelations among the physical attributes of resources, their role in economic systems, and the political and social factors that govern decision-making about their use. We take the view that, even though resources can be classified as perpetual, renewable, nonrenewable, or potential at any given time, they are dynamic and subject to modification or redefinition. Human activity has as much effect on the nature of resources as do natural processes.

Part I focuses on the basic human and natural components of resource use. Chapter 2 provides an overview of the economics of natural resource use, including pricing systems, demand elasticities, externalities, and the relationship between economic growth and resource use. In Chapter 3 the decision-making processes governing resource use and the historical origins of current conservation philosophies in the United States are discussed. Chapter 4 provides a review of the ecological bases of natural resources. Chapter 5 examines the human population system.

Part II deals with specific resource issues. These issues include agriculture and food production (Chapter 6), forests (Chapter 7), biodiversity and habitat (Chapter 8), marine resources (Chapter 9), water resources (Chapter 10), urban air pollution (Chapter 11), regional and global air issues (Chapter 12), minerals (Chapter 13), energy resources (Chapter 14), and, last, a chapter on the transition to global sustainability (Chapter 15).

REFERENCES AND ADDITIONAL READING

Clawson, M. 1986. Integrative concepts in natural resource development and policy. In R. W. Kates and I. Burton, eds. *Geography, Resources, and Environment*. Vol. II, pp. 69–82. Chicago: University of Chicago Press.

Cutter, S. L., 1995. The forgotten casualties: women, children, and environmental change. *Global Environmental Change: Human and Policy Dimensions* 5(3):181–194.

Easterbrook, G. 1995. *A Moment on the Earth*. New York: Viking.

Ehrenfeld, D. 1978. *The Arrogance of Humanism*. London: Oxford University Press.

Ehrlich, P. and A. Ehrlich, 1991. *Healing the Planet*. Reading, MA: Addison-Wesley.

Engel, J. R. and J. G. Engel, eds. 1991. *Ethics of Environment and Development*. Tempe: University of Arizona Press.

Global Environmental Change Programmes. 2001. Earth systems science. *Environment* 43(8):21–27.

Gore, A. 1992. *Earth in the Balance: Ecology and the Human Spirit*. Boston: Houghton Mifflin.

Hardin, G. 1974. Living on a lifeboat. *Bioscience* 24:10.

_____. 1985. *Filters Against Folly*. New York: Basic Books.

Hunker, H. L., ed. 1964. *Erich W. Zimmermann's Introduction to World Resources*. New York: Harper & Row.

Kates, R. W. and I. Burton, eds. 1986. *Geography, Resources, and Environment*, Vols. I and II. Chicago: University of Chicago Press.

Kaufman, W. 1994. *No Turning Back: Dismantling the Fantasies of Environmental Thinking*. New York: Basic Books.

Lawton, J. 2001. Earth systems science. *Science* 292:1965.

McKibben, B. 1989. *The End of Nature*. New York: Random House.

Merchant, C. 1992. *Radical Ecology: The Search for a Livable World*. New York: Routledge.

Mutz, K .M., G. C. Bryner, and D. S. Kenney, eds. 2001. *Justice and Natural Resources*. Washington, D.C.: Island Press.

Nash, R. A. 1989. *The Rights of Nature: A History of Environmental Ethics*. Madison: University of Wisconsin Press.

Ophuls, W. and A. S. Boyan, Jr. 1992. *Ecology and the Politics of Scarcity Revisited*. New York: W. H. Freeman.

O'Riordan, T. 1986. Coping with environmental hazards. In R. W. Kates and I. Burton, eds. *Geography, Resources, and Environment*. Vol. II, pp. 272–309. Chicago: University of Chicago Press.

Pepper, D. 1996. *Modern Environmentalism*. New York: Routledge.

Plant, J., ed. 1989. *Healing the Wounds: The Promise of Ecofeminism*. Philadelphia: New Society Press.

Rolston, H. III. 1989. *Environmental Ethics: Duties to and Values in the Natural World*. Philadelphia: Temple University Press.

_____. 1994. *Conserving Natural Value*. New York: Columbia University Press.

Shabecoff, P. 2000. *Earth Rising*. Washington, D.C.: Island Press.

Smil, V. 1987. *Energy, Food Environment*. Oxford: Clarendon Press.

Southwick, C. H., ed. 1985. *Global Ecology*. Sunderland, MA: Sinauer Associates.

Stone, C. D. 1993. *The Gnat Is Older Than Man: Global Environment and Human Agenda*. Princeton: Princeton University Press.

von Bertalanffy, L. 1950. An outline of general systems theory. *Br J Philos Sci* 1(2):134–165.

Warren, K. 1990. The power and promise of ecological feminism. *Environmental Ethics* 12:125–147.

Weiss, E. B. 1990. Our rights and obligations to future generations for the environment. *Am J Int Law* 84:198–207.

White, L. 1967. The historical roots of our ecological crisis. *Science* 155:1203–1207.

Zimmerman, M. E., J. B. Callicott, G. Sessions, K. J. Warren, J. Clark, eds. 2000. *Environmental Philosophy: From Animal Rights to Radical Ecology.* 3rd Edition. Englewood Cliffs: Prentice-Hall.

For more information, consult our web page at *http://www.wiley.com/college/cutter*.

STUDY QUESTIONS

1. Pick an object you are familiar with—an item of clothing, a desk, whatever. List the natural resources that were probably used to make it. Which ones were renewable and which were nonrenewable? What portion of the value (retail price) of the object represents the cost of the raw natural resources, and what portion is attributable to human labor?

2. Do you eat meat? Do you wear leather? Define the extent to which you feel responsible for the welfare of the animals you use and those you don't (pets, for example). What are the moral bases of your decisions in this regard? In a few sentences, can you define the limits of your personal obligations to other species?

3. Examine the contents of your garbage container at home. Make a brief list of the major categories of materials present. For each of these, write down a potential use and what you think would have to change in your society in order for this potential resource to become useful and economically valuable (instead of having negative value as waste).

CHAPTER 2

ECONOMICS OF NATURAL RESOURCES

INTRODUCTION

Decisions on the exploitation, conservation, and preservation of natural resources are always made within the context of a particular culture, with its own economic system. This could be a centrally planned economy or an unregulated capitalist system based exclusively on the pressures of the marketplace. Most countries today have economies somewhere in between these two. No matter what the political or social system, a mechanism must exist for the exchange of goods and services. In most societies, this mechanism is price—the value society places on an item.

The price of a good or service is usually represented by its monetary equivalent. In some cases, however, a good or service can have less tangible value. For example, while many in the United States consider clean water and free-flowing streams valuable, it is difficult to place a monetary value on such a resource. Today, resource economists view natural resources differently from other commodities and are suggesting ways in which economies can include an accounting of the degradation or conservation of natural resources. In the past, a clean and healthy environment was seen by industry as "too expensive" and viewed more as a luxury than a necessity.

Economists have been debating the nature of economic systems and the relationship between economic growth and environmental quality for decades. The neoclassical view of economics as open systems unrestrained by environmental limits (either natural resources or residuals disposal) is being seriously questioned. In the 1960s, a lone voice in the economic wilderness dubbed this neoclassical view "cowboy economics." Boulding (1966) advocated replacing cowboy economics with a more sensible perspective in which economic systems are considered closed systems, with economic processes constrained by negative feedback effects. According to this "spaceship Earth" view of economic systems, every effort must be made to recycle materials, reduce wastes, conserve exhaustible energy resources, and seek more "limitless" energy sources such as solar. Kneese et al. (1971) refined this perspective in their *materials balance* principle. This approach suggests that wastes are pervasive in the economic system and thus require a production system in which inputs and outputs (wastes or pollution) are balanced. The primary contribution of the materials balance approach, as this was termed, was the recognition of the need for *residuals management*, in which the waste products of production and consumption are considered as well as the valued products themselves. Further refinements were made calling for steady-state economic systems (Daly 1996) with the objective to establish the lowest rate of throughput of energy and matter, not to maximize the output of goods and services. Georgescu-Roegen (1976) proposed a bioeconomic program based on the flow of solar energy and the minimal depletion of terrestrial matter (resources). Finally, the entropy concept (a measure of disorder in a system) was used to explain how economies decline as predicted by the second law of thermodynamics. Most environmental economists now favor a *steady-state* or *sustainable* view of economic systems, in which a system has equal inputs and outputs and resources are not significantly depleted over time.

These ideas have been brought together under the framework of *ecological economics*, which focuses on integrating economic and ecological views of human resource use, with the goal of working toward sustainability. In particular, ecological economists seek ways to quantify the value of natural resources in monetary terms so that natural and human-created resources are viewed in comparable terms. Ecological economics views the resources that we extract from nature as a form of capital, and just as traditional capitalist systems seek to accumulate and re-invest monetary capital, ecological economists argue that natural capital should be regenerated rather than depleted (Hawken et al. 1999; Daily and Ellison 2002; Brown 2001). We will examine issues of valuing nature later in this chapter.

William D. Ruckelshaus, first administrator of the U.S. Environmental Protection Agency (EPA), calls for the development of a "sustainability consciousness"—toward a way of living that does not destroy the environment but keeps it healthy for future use. Ruckelshaus (1989) states that such a consciousness requires the following beliefs:

- The human species is part of nature. Its existence depends on its ability to draw sustenance from a finite natural world; its continuance depends on its ability to abstain from destroying the natural systems that regenerate this world.
- Economic activity must account for the environmental costs of production.
- The maintenance of a livable global environment depends on the sustainable development of the entire human family.

This chapter provides a look at the evolution of thought on economics and environment, as they have shifted from more exploitative to more sustainable positions. Specifically, we examine the role of economics in natural resource management. Three questions provide the focus for our discussion.

1. What are the economic characteristics of natural resources? Natural resources are commodities regulated in part by supply, demand, and price, but several unique characteristics of natural resources alter the economic use we make of them.
2. How do we place a value on a natural resource? To operate within an economic system, a value or price of a resource must be determined. Yet, many values of a resource may not be reflected in price. Deciding on the value of a resource is the key to understanding how economic pressures influence both the use and the management of a resource.
3. How do economic forces influence the management of a resource? Short-term pricing mechanisms might dictate the use and management of a resource quite differently than pricing systems based on long-term social needs.

ECONOMICS AND THE USE OF RESOURCES

Characteristics of Natural Resources

Natural resources are the basic building blocks in the production system; they are raw materials. Because little of their value is derived from human inputs such as labor, they generally have a lower value per unit than other commodities. The value of a standing forest, for example, is rarely more than the cost of owning the land for a period of time, and usually it is much less. When the trees are cut they are somewhat more valuable, but milling and drying add much more value. By the time the wood is made into a house or a piece of furniture, the price of the standing tree accounts for a very small portion of the value of the finished product; most of the value was added after the tree was grown. In a few cases natural resources have a high value "in the ground," but in these instances it is the consumer that drives the price up, if demand is greater than the amount of the resource available.

A second important characteristic of natural resources is that over short periods of time supply is relatively inelastic. Most natural resources require substantial capital investment and planning to bring them into production. For example, suppose the demand for a particular metal increases significantly. To a certain extent, existing mines can step up activity by hiring more miners and buying more equipment. But a large increase in production will probably require that new mines be opened. This may entail geologic exploration, but even if the deposits are known and owned by mining companies, they must still build roads and other facilities to sink the initial mine shafts and build housing for labor, before extraction actually can begin. All of these activities take time, generally years, and in the short run the supply cannot keep up with increased demand. This means that the supply cannot be stretched quickly; in other words, supply is inelastic.

The inelastic nature of many resources can encourage wide fluctuations in price. For example, the prices of most mineral commodities are notoriously volatile (Fig. 2.1). Part of this volatility is inherent in the production and supply system of the resources themselves, and part of it is in the psychology of the market. During the energy crisis of the early 1970s, when imports from the Middle East were dramatically reduced, the price of gasoline at the pump more than doubled in a matter of months. In the years since then, gas prices have become very sensitive to real and apparent shortages in the supply. In 1989, for example, when the *Exxon Valdez* spilled its oil cargo along the Alaska shoreline, the price at the pump rose overnight in both the United States and Great Britain, even though the spill was virtually irrelevant to the supply of gasoline. Simultaneously, the intangible value of pristine coastline and healthy wildlife rose incalculably across the nation within a period of days. "Price gouging" was an issue again in 1990, when gas prices soared following Iraq's invasion of Kuwait, and in the days following the terrorist attack on September 11, 2001.

Coffee prices are another example. Good market prices for coffee (one of Colombia's primary export commodities) kept that country's external debt significantly lower than most of its other Latin American neighbors during the mid-1980s debt crisis in the region. Increased global production led to falling prices for coffee starting in 1987. In the early 1990s coffee prices rose sharply upon news of a bad harvest in Brazil. Each of these price fluctuations has worldwide repercussions for farmers, consumers, and the processors and handlers between.

Small changes in demand as well as supply can cause a dramatic rise or fall in price, yet there is little that the producers can do in the short run to ensure a steady, long-term trend in prices. To illustrate, bitter cold snaps occasionally hit the eastern United States. Consumer demand for fuel oil and propane supplies reduces existing supplies to low levels. Almost overnight, prices for fuel oil and propane may skyrocket to double and triple what they were earlier in the month. When warmer weather returns, consumer demand slackens, and home heating fuel prices once again fall in response to the lower demand.

Whenever possible, a relatively high degree of substitutability among raw materials is desirable. Not only can one metal be substituted for another in, say, an automobile, but recently plastics, fiberglass, carbon fibers, and other synthetic materials have begun to replace metals for many purposes. Beet sugar is a substitute for cane sugar; coal substitutes for natural gas. Although this substitutability contributes to stability for the makers of finished products, it often leads to considerable volatility in raw materials markets. The endless pattern of boom and bust cycles in one-employer mining towns is one of the sadder human consequences of this volatility.

For some natural resources, particularly minerals, supply is theoretically infinite, assuming we

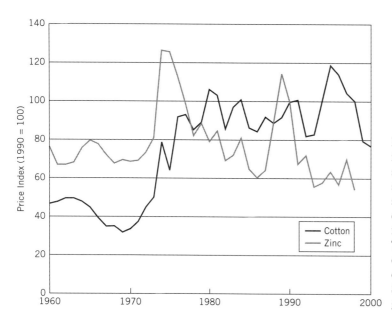

Figure 2.1 Recent variations in cotton and zinc prices demonstrate the volatility of natural resource commodity markets. The similarity in the patterns is coincidental. Cotton price data are from the New York Board of Trade (www.nyce.com). Zinc data are from the U.S. Geological Survey, Mineral Commodity Summaries (http://minerals.usgs.gov/minerals/pub/mcs).

are able (and willing) to pay a high enough price. Most metals exist in the earth's crust in much greater total quantities than we need at present and in the foreseeable future. The problem is a geographical one: only a limited number of locations have high enough concentrations or are close enough to the surface to allow the ores to be extracted at a profit. But as long as we are willing to pay a little more to obtain the resource, then we can dig a little deeper, or refine less concentrated ores, and still obtain the desired commodity. At some point, we find that it is cheaper to recycle used metals than to mine new ores, and it is at that point that we are able to supply much of our new requirements by recycling. When we also consider the substitutability of most substances, it seems unlikely that we will encounter a situation in which we run out of raw materials. On the other hand, the theoretical supply of energy may or may not be infinite, depending on what technologies are available to us, and activities such as mining and recycling may be limited by shortages of energy.

Pricing Systems

Natural resources are commodities, and we value them for their ability to provide the basic needs of life: food, clothing, shelter, and happiness. As commodities, they are exchanged among individuals, groups, and nations using some sort of pricing system as a medium of exchange. This pricing system can have a major impact on how resources are used. Resource price is dictated both by a society's determination of resource value (discussed in the next section) and by the economic system in use.

Economic Systems

Although there are three fundamental types of economic systems in the world—commercial, centrally planned, and subsistence—only commercial and centrally planned economies produce a surplus of goods. In a commercial economy, prices are set by the producers who sell goods and services. Producers are characterized by a profit motivation and pressure to produce at low cost. This motivation usually leads to specialization, thus allowing more efficient production. A producer will try to do one thing well instead of many things in a mediocre way: there is greater profit in using this approach. Profit and efficiency are balanced by market forces, where supply and demand govern the price and quantity of goods exchanged.

In other words, a producer can be efficient and offer a high-quality product, but once he or she enters the world market, certain economic forces are at work that are beyond the control of the individual company, and dreamed-of profits are not guaranteed. In a commercial economy, the use and allocation of natural resources are governed by many forces, especially market competition and profit maximization. Producers respond to these forces to protect their own best interests. Examples of commercial economies include those of the United States and Canada, European and most Latin American countries, India, Japan, and South Africa.

In a centrally planned economic system, the government controls the resources. Producers market goods and services to the central government, which in turn controls the supply and price according to its own objectives. These objectives can range from monetary gain to social and economic equality, goals that are not normally found in a commercial economy. China and Cuba are examples of economies that historically have been centrally planned, although they are changing and today are more a mixture of commercial and centrally planned economic systems. However, the failure of former socialist governments in eastern Europe and the former Soviet Union to consider environmental values in their planning, and the difficulty of recognizing amenity values in a materialist society, have contributed to the severe environmental problems evident there.

Subsistence economies based on pastoral herding or subsistence agriculture are found in very localized areas throughout the world such as Amazonia and portions of sub-Saharan Africa. In these places, there is no surplus of goods and therefore no need to implement a formal monetary or pricing system. Instead, goods are exchanged through an informal bartering system. Like centrally planned economies, true subsistence economies are becoming less common.

In surplus economies, a price is set on a resource, thus permitting exchange and use by society. Several different approaches are used to determine price. For example, Marxists follow the *labor theory* of value, which states that the price of a good is determined by the amount of labor required to produce it. An alternative approach is the *consumer theory* of value, which says that value is determined by how much a consumer is willing to pay. Still another view is the *cost theory*, which says that in a free-market system different producers will compete with each other by reducing prices until they equal the cost of production. Finally, the *production theory* emphasizes the inputs of some critical commodity, for example, energy. That is, the value of goods is a function of the energy required to produce them, and thus they are priced accordingly. Although each of these theories addresses at least one aspect of the value of resources, in the final analysis it is the interaction of supply and demand that determines both the price of a good and the quantity sold at that price. In other words, market forces are the dominant factor in pricing goods.

Supply and Demand

In a market economy, supply and demand are the primary factors determining price. The relations among supply, demand, and price tell us the amount of a good demanded by consumers and the amount the producers are willing to supply, depending on price. As price increases, the amount consumers are willing to buy decreases. Conversely, as price increases, the amount that producers are willing to sell increases. The intersection of these two functions determines the equilibrium price and quantity sold (Fig. 2.2). If there is a change in conditions, say, increasing scarcity of a commodity, then it will cost producers more to supply that commodity and they will either require a higher price to produce the same amount or supply less at the same price. In either case, the supply function shifts up and to the left, with the result that price increases and quantity sold decreases. Similarly, a change in the value consumers place on a commodity will change the demand function, with appropriate changes in price and quantity sold. This process assumes perfect competition, which exists only when no individual consumer or producer can exert influence on the market and when all producers and consumers have full information and access to the market. Obviously, perfect competition does not exist in the real world, so these assumptions are violated in many cases.

Market Imperfections

Unfair competition causes price to be determined by factors other than supply and demand, and this situation results in an imperfect market. A *monopoly* exists when a single buyer or seller dominates the market. Monopolistic or *oligopolistic competition* describes what happens when a few buyers or sellers either follow a price leader in fixing the price of a commodity or engage in discriminatory practices to set prices so as to maximize their own profits. Their purpose is to restrain competition by keeping prices artificially high.

There are many resource oligopolies in the world today, and these are normally referred to as cartels. A *cartel* is a consortium of commercial enterprises that work together to limit competition. These cartels can either be similar industries (such as refineries) or resource-rich exporting countries. In the latter case, these countries band together for economic advantage to fix the world prices of their commodity. The Organization of Petroleum Exporting Countries (OPEC) is a good example of a cartel. OPEC consists of 12 member nations: Algeria, Gabon, Indonesia, Iran, Iraq, Kuwait, Libya, Nigeria, Qatar, Saudi Arabia, United Arab Emirates, and Venezuela. OPEC nations supply around 40 percent of the world's supply of crude oil. Other examples of cartels include those controlling copper and bauxite.

Trade agreements and trade barriers also create unfair competition, leading to competitive advantages for some nations. The tensions between free trade and protective regulation are most pronounced in the environmental arena.

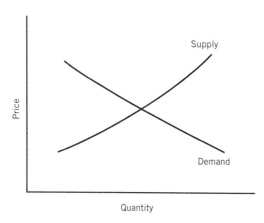

Figure 2.2 Classic supply, demand, and price relationships. As supply decreases and demand increases, prices also increase. Conversely, as supplies increase and demand decreases, prices also fall.

Protectionist producers want to restrict the importation of goods that can outcompete them in the marketplace because of cheaper labor costs or materials. Tariffs (which by definition distinguish between domestic and imported goods) increase the price of imports, often resulting in limiting the market through the use of quotas on the amount of goods allowed into the country. On the other hand, the trade community is worried about the use of trade barriers as a form of "eco-protectionism" used to restrict sales of goods produced in environmentally harmful ways in the exporting nation (Vogel 1995). The disparity in national environmental regulations is a continuing source of strain in the international trade community. However, trading alliances and economic integration such as those in the European Union and in North America are helping to improve the domestic environmental policies of less green trading partners (Issue 2.1).

DETERMINING RESOURCE VALUE: QUANTIFYING THE INTANGIBLES

Not all resources have a value that can be quantified and expressed in monetary terms. Clean air is one example; an unobstructed view of the landscape is another. Yet these resources are increasingly part of the picture in determining the use we make of our Earth. As mentioned in the chapter introduction, only in the past decade have economists begun to look for ways in which the environment and the marketplace can work together instead of being pitted against each other. Let us look at the historical development of the marketing of the environment.

The most widely used techniques to determine the value of intangibles such as clean air and scenic views are cost-effectiveness analysis and benefit-cost analysis. *Cost-effectiveness analysis* simply involves summing all the costs and monetary returns involved in a single plan to determine the expected return on investment. *Benefit-cost analysis*, on the other hand, compares all the costs and benefits of several different plans. It requires an understanding of the social context—the alternative values placed on a resource by different sectors of the society—within which the balancing of costs and benefits is made.

Cost-effectiveness analysis helps the decision-maker determine the least costly and most efficient strategy for carrying out a project or exploiting a resource once the decision has been made to proceed. Benefit-cost analysis is a tool for helping society make choices about the allocation of resources, whereas cost-effectiveness analysis aids the individual firm or agency in implementing those decisions. Benefit-cost analysis evaluates alternative uses of a resource within a particular social context, and it usually seeks to find the alternative that has the greatest ratio of total benefits to total costs.

Benefit-Cost Analysis

Because benefit-cost analysis historically was the building block of both environmental protection and exploitation, we will examine it in greater detail. When a shopper goes to the supermarket, he or she is constantly making choices about what and how much to buy. How long will that small jar of peanut butter last? Is it worth it to me to buy a larger jar at a lower unit price? Do I really need steak or should I settle for hamburger? Economic theory tells us that making such decisions usually involves thinking about the costs and benefits of each item and purchasing those items whose benefits appear to outweigh costs. Intangibles enter the picture even here, however, if the consumer considers the health aspects of cholesterol, fiber, and fats when choosing foods.

The balancing process observed in the supermarket also is found in natural resource economics. Decisions on whether or not to build a dam, invest in new timberland, or clean up a polluted river usually involve some sort of benefit-cost analysis. This analysis can range from the elaborate accounting used to justify large government water projects to simple judgments made by individual farmers, foresters, or fisherfolk. In either case, a value is placed on expected costs and benefits, and the two values are then compared within a particular time frame. Some sort of price must be placed on all the factors in the resource decision: the price of the resource in question and the price of improvements or costs of damages to resources and society expected to result from a project.

Benefit-cost analysis involves a number of factors, including price, interest rate, and time. Since the analysis frequently must take into account

ISSUE 2.1: EUROPEAN INTEGRATION AND THE ENVIRONMENT: EEA AND EIONET

Europe underwent rapid changes during the 1990s, and this situation can be expected to continue into the twenty-first century. It has become more prosperous than at any other time since before World War I, and its economy is more fully integrated today than it has ever been. Economic integration and prosperity are in large part attributable to the establishment of the European Economic Community, or European Union (EU) as it is known today.

The EU began in the 1950s as an economic association, with the aim of improving the standard of living in member states through elimination of internal trade barriers, while economically linking nations to reduce the likelihood of war. Since then, it has grown and become a political as well as an economic association. As economic cooperation has expanded, it has been necessary and desirable for the member nations to act in concert on other matters such as social and environmental policy. For example, social policy is often carried out through taxation and subsidies: taxing alcoholic beverages and tobacco or subsidizing rail systems. Social policies are thus linked to economic policies.

With the ratification of the Maastricht Treaty (1992), 15 countries became founding member states of the EU (Austria, Belgium, Denmark, Finland, France, Germany, Greece, Ireland, Italy, Luxembourg, Netherlands, Portugal, Spain, Sweden, and the United Kingdom). The EU is working toward a goal of an internal market with no barriers to the movement of goods, labor, or capital. This means the complete removal of any import restrictions, taxes, or regulations that could interfere with business across the international boundaries of the member nations. Many of the less contentious steps that must be taken—for example, elimination of import tariffs and standardization of product safety requirements and similar specifications—have already been achieved. More difficult matters, such as the complete elimination of barriers to international movements of labor, still remain. It is possible that a completely integrated and barrier-free internal market may not be achieved immediately, but most of the important barriers are likely to be significantly lowered, if not removed altogether. A unified currency (the euro), for example, has already been established.

In 1990 the EU established the European Environment Agency (EEA), whose purpose is to provide the EU with objective and comparable information for environmental protection, to inform the public about the state of the environment, and to provide the necessary technical and scientific support to achieve those goals. Currently, 29 countries are members of the EEA. One of the agency's first goals was to establish the European Environment Information and Observation Network (EIONET).

The primary function of the EEA is to monitor and assess the quality of the environment, pressures on it, and areas of environmental sensitivity. The EEA acts as a centralized clearinghouse for aggregated environmental data from all member states. As such, it is especially interested in transboundary and global implications of environmental quality, including the socioeconomic dimensions. Current priority areas include air quality/atmospheric emissions; water quality; status of soil and flora/fauna; land use and natural resources; waste management; noise emissions; hazardous substances; and coastal protection. The EIONET is the monitoring network for the EEA and consists of national monitoring centers from each of the member states. Thus, all data collection and management are the primary functions of EIONET.

But the environmental movement in Europe today is subject to a wide range of pressures. Many nations with diverse cultural and political systems are involved. The desire to improve living standards and stimulate trade provides the stimulus for economic cooperation, but because environmental and economic policies are closely intertwined, any improvement in economic conditions cannot be achieved without environmental improvement. It is too soon to evaluate the effectiveness of the European equivalent to our Environmental Protection Agency, but it is clearly a step in the right direction for building a cooperative and comprehensive environmental policy within the EU (European Environment Agency 2002).

nonmonetary variables, the values chosen for them have a major impact on the outcome of the analysis and the resulting decision on how the resource is to be used. For example, in the American Northwest, debate has been raging over what is of greater value: trees as lumber and as a source of employment or trees as a habitat for endangered species such as the spotted owl (Yaffee 1994).

One common approach to applying the result of a benefit-cost analysis is to determine the ratio by separately summing the costs and benefits over time, then dividing benefits by costs. A benefit cost (B:C) ratio greater than 1 indicates that benefits are greater than costs and the project should go forward. A B:C ratio can be very misleading when the analysis includes potential environmental impacts. Any project, for example, presumably has some environmental impacts, regardless of benefits or costs. These impacts should be applied to either the calculation of costs (by addition) or the calculation of benefits (by subtraction) and must be used consistently. Another problem is that the present value of future costs or benefits depends on the interest or discount rate used in the calculations (Turner et al. 1993). Small changes in interest rates can result in very large differences in project costs. Despite these problems, B:C ratios are frequently used to evaluate or justify public works projects such as river levees or shore protection structures. Increasingly, benefit-cost analysis is being used to evaluate environmental policies (Farrow and Toman 1999).

Frequently, the organization conducting a benefit-cost analysis has a vested interest in some particular outcome, leading to a bias in estimates of resource values that favor that outcome. This is especially true for estimates of amenity resources such as recreation or a scenic view. When several different pricing methods are available for incommensurable resources, it is tempting to choose the pricing method that gives the desired result—either high or low, depending on the wishes of the person doing the analysis. As a result, benefit-cost analyses are frequently biased and must be used with caution.

Quantifying Value

When faced with a decision about whether or not to permit lumbering in a National Forest, a resource manager seeks various types of information to include in an analysis of benefits and costs. Some of this information, like the value of standing timber, the cost of replanting trees after harvest, or the increased water yield as a result of cutting the forest, can be expressed in monetary terms. Other information is more speculative, such as the value of wildlife habitat lost and the resulting decline in the number of hunters and fisherfolk. These so-called incommensurables and intangibles can frustrate benefit-cost analysis, forcing the decision-maker to create an artificial or *shadow price* for that resource.

Incommensurables are effects (both benefits and costs) that cannot readily be translated into a monetary value or price. *Intangibles* are incommensurables that cannot be measured at all—they are truly outside the analysis. The trick in benefit-cost analysis is to separate the incommensurable effects from the intangible effects (see Issue 2.2). Although it is impossible to assign a value to intangibles, we can place a value or shadow price on incommensurables. In any assessment of the benefits and costs of resource exploitation, it is necessary to define the limitations of such methods in order to quantify these incommensurable effects.

Economists employ a broad range of techniques to establish the shadow price of incommensurable resources. Three of the most commonly used techniques are discussed here.

Willingness to pay is a method in which potential users of a resource are asked how much they are willing to pay for access to the resource. Conversely, one could ask how much society would be willing to pay the individual not to use the resource (avoidance costs). An example is a survey of beach users to determine how much they are willing to pay to use the beach or how much they would have to be paid not to go to the beach. The value of the latter could be used to estimate the value of recreation losses caused by an oil spill on the beach (Fig. 2.3). The former could be used to justify importing beach sand to keep an eroding beach the way it is. One specific method for measuring willingness to pay is the *contingent valuation method (CVM)*. This method employs a questionnaire to ask users (and often nonusers as a comparative group) a series of questions on their environmental preferences, for example, how much they would pay for that new park or how much they would be willing to pay to preserve their view of the coastline (Bateman and Willis 1999). Such a questionnaire often contains a number of possible scenarios, such as these in a CVM

ISSUE 2.2: WHAT IS THE VALUE OF A HUMAN LIFE?

Estimating the value of services provided to humans by an ecosystem (see Issue 2.3), in the form of materials produced, pollutants absorbed, or amenities provided, is a complex accounting task that involves a very large number of price estimates and assumptions but relatively few fundamental moral questions. The value of a human life, on the other hand, can be measured only in moral terms. We could never put a dollar figure on it. Or can we? Should we?

Like it or not, there may be good reasons to calculate the value of a human life. One reason is that our governments take actions to protect human life, through laws and regulations requiring seat belts in automobiles, controlling air pollution, maintaining safe drinking water, and the like. These measures cost money, and although one might think that any cost is justified when a human life is at stake, this is not necessarily true. The costs of life-saving measures are borne ultimately by consumers. These consumers' disposable income is reduced as a result.

It is a well-known fact that wealthy people tend to live longer than poor people. This is because the less money we have, the less likely we are to visit a doctor for a regular checkup, buy fresh vegetables instead of canned, take nutritional supplements, or perhaps move from a polluted area to a cleaner one. If a certain amount of money is taken out of the economy to pay for safety, then some people may suffer poorer health as a result because their income was reduced to pay for that safety. Inexpensive ways of saving lives (seat belts, for example) are cost-effective, but expensive ones may not be. One estimate is that a loss of $10 million from the U.S. economy results in one added death, due to people's reduced income. If this is true, we should not spend more than $10 million for safety per life saved—to do so would cost more lives than are saved (Passell 1995). In other words, we have just estimated the value of human life in the United States at about $10 million.

In the United States and other wealthy countries, most people have access to good food and health care, and the added cost of safety can be borne relatively easily without endangering health. Safety measures costing millions of dollars per life saved can be justified. In poor countries, however, many people suffer because they lack access to food and health care. In India, for example, average per capita income is around $400. A person earning this amount per year may not be able to afford clean drinking water and has a significant risk of sickness and death resulting from contaminated water. Providing such a person with enough money to buy clean drinking water or to buy clean fuel to boil water could easily save a life. In these situations, spending for safety is much harder to justify, as even a small loss of income could significantly increase the risk of death. In a poor country, it takes only a small amount of added income to save a life. In a rich country, saving lives costs much more. Put another way, it appears as though the value of a life in a rich country is greater than one in a poor country.

Does this valuation of life sound a little scary or at least morally questionable? Of course it does. But consider this: In the United States, insurance companies and courts routinely place monetary values on human life. Suppose a plane crashes and the airline is found to be at fault. The survivors of passengers killed in the crash can sue the airline for compensation for the loss of their loved ones. If the

study of the Monongahela River (Desvouges et al. 1987):

- Scenario 1: Keep the current river quality (suitable for boating only) and don't allow further decay.
- Scenario 2: Improve the water quality from the current levels (boating) to where fishing could take place.
- Scenario 3: Improve the water quality even more to make it swimmable.

Respondents in this study were asked how much they were willing to pay for each. The dollar value was then averaged across groups (users/nonusers) to produce a demand curve for water quality. Users were consistently higher in their willingness to pay in all three scenarios than nonusers ($45.30 for users and $14.20 for nonusers for Scenario 1 and $20.20 for users and $8.50 for nonusers for Sce-

person killed was, say, a 32-year-old brain surgeon with an annual income of $300,000 expecting to work to age 65, then the income lost to her family over the next 33 years is $10 million. If the person killed was a 75-year-old pensioner then there is no lost income. In fact, the death releases money that would have been used to sustain the pensioner. The court would likely award much more compensation to the survivors of the surgeon than to those of the pensioner. We thus estimate the value of a life in terms of the income an individual would earn. And, simply put, wealthy people's lives are worth more than those of poor people.

Individuals also place monetary values on their own lives on a routine basis, though usually not consciously. For example, suppose you are shopping for a new car. The model you want to buy is equipped with standard brakes, but an optional automatic braking system could be purchased for $500. Suppose that over the five-year period you expect to own the vehicle, the chance of you or one of your passengers dying in a crash is 1 in 1000 and that the automatic brakes would reduce that to 1 in 2000. Is the braking system worth the price to you? In the five-year period, your $500 investment would save 1/2000 of a life, giving a cost of $1 million per life saved. If you decide that the automatic braking system is too expensive, then you are implying that the value of your own and your passengers' lives is less than $1 million each. If, on the other hand, you decide that the braking system is worth the investment, then you are valuing a life at more than $1 million. This neglects other benefits you receive from the system, such as nonfatal crashes avoided and the peace of mind resulting from the added safety.

Note that a rich person would be more likely to buy the braking system than a poor one. Once again, rich people's lives seemingly are valued more highly than those of the poor.

In an era of global trade and manufacturing, with industrial activities and the pollution they create being shifted from wealthy countries to poor ones, these questions of the value of human life take an interesting twist. Suppose an industrial activity generates pollution that may shorten the lives of people living near the industry. Suppose that industry also generates income for those people. If this industry were located in a wealthy country, people would be unlikely to tolerate a significant amount of life-threatening pollution. But in a poor country, the added income from the industry might improve their lives to an extent that they would be better off with the pollution than without it. The number of lives saved by people being able to afford clean drinking water might be more than the number of lives lost to lung disease caused by the pollution.

If this is true, then would it be a good thing to move dirty or dangerous industries from rich countries to poor ones? Most people would respond emphatically, no. To do so is morally repugnant. We should find ways to increase the income of developing countries without also increasing pollution levels. We should never trade pollution for money, let alone trade one life for another. But the question does make us think a little bit about tradeoffs between environment and economic development. Perhaps it might lead one to think twice about condemning environmental degradation in poor countries, when that degradation is directly connected to economic development.

nario 3). Interestingly, the nonusers' willingness to pay is greater than zero, suggesting that the river has a public value at its current quality level. In other words, respondents did not want to see further deterioration in water quality and were willing to pay a considerable amount to achieve that goal.

A second technique is determining *proxy value* or the value of similar resources elsewhere.

This technique estimates the value of a day's hunting, for example, by summarizing the hunter's investment in supplies, time, and travel and dividing this figure by the number of days of hunting. This is an estimate of the value that would have resulted if the project or exploitation did not occur in the first place. Another example is the decline in commercial fish harvest due to the

Figure 2.3 Oil spills are an obvious and expensive form of pollution. We value coastal resources highly and demand cleanups after spills like this one at Huntington Beach, California. Usually when oil is spilled there is an easily identifiable party responsible, and the cost of the cleanup can be borne directly by those causing the pollution.

destruction of their habitat by pollution or a toxic spill.

The third technique is called *replacement cost*. This is simply the cost of replacing the resource that is being used, such as substitution of clean sand for polluted sand on a beach fouled by an offshore oil spill. Often there is no market for the replacement of extramarket goods, for the resources are not substitutable.

The accuracy varies when using these and other methods of estimating the shadow prices of resources. It is important to recognize the limitations of such techniques as well as their ability to provide the necessary data for a benefit-cost analysis. They are useful in placing a comparable value (price) on those resources that normally do not have one, but only within the limitations we have discussed.

MANAGEMENT AND ALLOCATION OF RESOURCES

Economic forces shape the price and utility of a resource (Fig. 2.4). For example, oil shale in western Colorado and Wyoming was valuable enough to extract and process when the world's oil price was high, but when that price dropped, oil shale became too expensive as a source of oil. Economic forces also effect who gets to use a resource and how it is managed. When the North American deserts were seen as valueless a century ago by European settlers, they became a place of exile for Native Americans who were displaced from more valuable lands elsewhere. Rich mineral deposits subsequently were found on many of these "valueless" tribal lands. In many cases, tribal corporations and private industry now have mutually beneficial business deals. Overall, the influence of economics on management decisions can best be classified into three categories. These are ownership of the resource, social costs, and the economics of the individual company or firm.

Ownership

Many natural resources are held in communal rather than private ownership. This is true both for resources that are government owned, such as National Forests and offshore mineral rights, and for resources for which no formal ownership is designated, such as air or the water in the oceans. Although these resources are commonly owned, that is, owned by everyone, they are exploited by private individuals and corporations for their own profit. This discrepancy between ownership and management responsibility causes problems.

Figure 2.4 This rainforest in Brazil is being cleared because its value to the owner is greater as pasture than as forest. It is being burned rather than harvested for wood because the cost of cutting, transportation, and marketing are greater than the market value of the wood that would be produced.

These types of resources are called *common property resources* because they are owned by everyone, even though they may be used by only a few who have the technical and economic means to do so.

This conflict was described by Garrett Hardin (1968) in his classic article entitled "The Tragedy of the Commons." Hardin argued that commonly owned resources nearly always are overexploited. The reason is simple: while the costs of exploitation are shared among all the owners of the resource, the benefits accrue to the individual, and so it is always in the individual's interest to increase exploitation, even to the point of overexploiting the resource. Hardin concluded that there must be some institutional arrangement that prohibits overuse and encourages conservation.

For example, if a group of people all dump sewage into the same body of water, they all suffer, but for only one person not to dump sewage would be foolish from a purely economic standpoint. This person would still suffer from everyone else's sewage and would not reap the full benefit of his or her own reduction in pollution. To manage such a resource, some governing body steps in, and through that body all agree to regulate their exploitation of the resource for mutual benefit.

In some cases this is relatively easy, and in others it is not. When large corporations are both exploiters and employers and thus politically powerful, or when enforcement of restrictions is difficult, then overuse and degradation of commonly held resources can result. For example, throughout the 1980s the United States avoided taking serious action to control emissions that contributed to acid rain, out of concern for the burdens these emission controls would place on steel mills and power plants that were basic components of the United States' midwestern economy. Only after intense pressure from both environmentalists and Canada, which received much of the acid rain caused by U.S. emissions, was significant action finally taken. The responsibility for management of these common property resources is a major issue in air, water, and ocean resources, as we will see in later chapters.

In the last few years, a new approach to the commons problem has emerged: How can government make it profitable to the individual to protect the commons? We can call this approach "the opportunity of the commons," in which government creates economic incentives that promise increased profits to a company that decreases its pollution output.

How might this work? In the case of acid rain, Stavins (1989) reports on a plan that would give pollution credits to companies that contribute to solving the acid rain problem. If a company lowered its emissions below the level required by law, it could make deals with the difference—to delay compliance with another regulation or to sell accumulated pollution-compliance points to a company that remains in violation of the law. This would earn a profit for the first company and a delay for the second. Pollution credits could thus be bought and sold among companies to allow the marketplace to decide the cheapest way to reduce smokestack emissions. But why restrict this plan to companies? Why shouldn't environmentalists be allowed to buy pollution rights (or credits) from publicly owned utilities and then retire the facility? In 1989 the going rate for the right to discharge a ton of sulfur dioxide per year in this case was between $1000 and $2000 per ton (Hershey 1989). By the late 1990s sulfur dioxide pollution credits were selling at between $150 and $200 per ton (Fig. 2.5), making them much more accessible to environmentalists. Each year a significant amount of pollution credits is purchased by environmental organizations, preventing some pollutants from being emitted (Solomon and Lee 2000). This program, discussed further in Chapter 11, is widely regarded as being a highly successful way to reduce pollution.

But can we trust the bottom line, a purely market-oriented approach to environmental protection? Even advocates for this approach warn that the market is limited as a tool for protection: while companies wheel and deal for points and profits, pollutants are still produced, and the commons continues to absorb them. Government remains a necessary tool to force industry to internalize the cost of pollution, but this command and control approach increas-

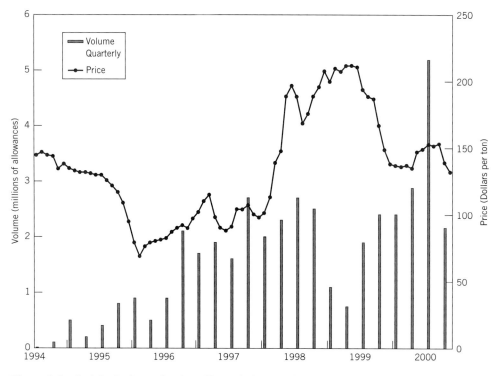

Figure 2.5 Activity in the market for sulfur emissions credits, 1994–2000. The price of these credits reflects: (1) what emitters (primarily electric utilities) are willing to pay for permission to emit sulfur; (2) what emitters would be willing to pay to avoid emissions by using cleaner technology; and (3) what society is willing to pay to prevent emissions. *Source:* U.S. Environmental Protection Agency, 2003. EPA's clean air market programs, www.epa.gov/airmarkets/trading/so2market/pricetbl.html.

ingly is becoming ineffective. Government has its own problems to address, for government-owned enterprises such as the Tennessee Valley Authority and nuclear weapons facilities are among the nation's most notorious and prolonged polluters and have voiced opposition to increased pollution abatement measures. How can industry be required to produce fewer pollutants? Let's look at economic concepts that, if made into law, would encourage industry to create fewer pollutants.

Social Costs

In commercial economies, both the producer and consumer bear the costs of production. These production systems are termed efficient if they maximize output (finished goods or services) per unit cost of production. Economists today recognize that not all production systems are truly efficient. Certain spillover effects from the production system may enter the environment and affect consumers disproportionately. These spillover effects are called *externalities* because they are commodities that change hands outside the marketplace, without a price being agreed upon by the principals. The unwanted byproducts of modern industrial processes—pollutants—that change hands as a side effect of production and consumption are often called environmental externalities. Such pollutants, or *residuals*, are transferred from one person or business to other people, but the transfer is not negotiated in a marketplace. Typically pollutants are discharged into one of the many common property resources—air, water, and public land—so they are transferred to society as a whole and are considered social costs (Fig. 2.6).

The problem with externalities is that they have clear environmental and social impacts. A private or government-operated producer wants to minimize costs and maximize profits, which ultimately forces the local community where the factory or plant is located to absorb the externality. Consider the example of a privately owned solid waste landfill (Fig. 2.7). The landfill owner willingly accepts very large amounts of pollution, in the form of other people's waste that is deposited on private land. This transfer is not an externality, however, because the landfill owner is compensated for accepting the waste, and the price is presumably negotiated to the mutual satisfaction of both the landfill owner and those transferring wastes to the landfill. On the other hand, residents who live near the landfill may notice unpleasant odors coming from it, and their property values might be lowered because of those odors. The transfer of unpleasant odors from the landfill owner to the nearby residents is not a market transaction and thus represents an environmental externality.

Social costs are those costs to society that not only involve externalities but also include the cost of producing the good in the first place. In most cases, natural resource use does not fully embrace this in the pricing structure, and we are thus left with the problem of managing and coping with these externalities or residuals.

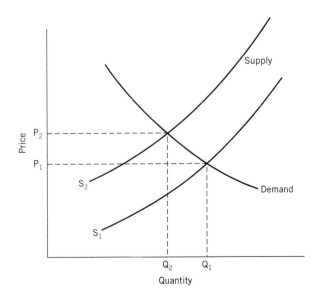

Figure 2.6 If social costs are considered in the production of a good, the supply function is shifted upward, from S_1 to S_2. This results in a decrease in the quantity sold, from Q_1 to Q_2, and an increase in price, from P_1 to P_2.

Figure 2.7 A privately owned landfill, near Cincinnati, Ohio. The owners of the landfill are compensated for accepting others' pollution in a government-regulated market transaction.

Residuals management is a major part of natural resource economics. Several techniques have been developed for understanding and managing residuals, and these usually involve economic or pricing mechanisms or government intervention in the form of pollution regulations. Residuals management is the term used to describe the first technique.

Residuals management (or materials/energy balance, as it is also known) is a process used to describe and quantify the inputs and outputs of a production system, including the residuals. The economic model that first described this process was developed by Kneese, Ayers, and d'Arge (1971). Residuals management advocates a steady-state production system in which inputs and outputs are balanced. The residuals, or waste products, are either fed back into the production system (recycled) or released into the environment. Specific techniques have been developed to manage residuals. A *residuals tax*, or effluent charge, was first proposed by Kneese and Bower (1971) as a method of controlling residuals. This can also be called the *polluter pays principle*. Producers of goods pay for the residuals they discharge, with taxes levied against firms in relation to how much pollution they release to the environment. A residuals tax forces the polluter to bear the true social costs of production and contamination of common property resources by internalizing the externalities. It would encourage the polluter to reduce the quantity of residuals, for the tax would be in direct proportion to the amount discharged. This fee would then be used either to clean up the environment or to compensate those consumers adversely affected by the pollution. In the early 1990s, a number of "green taxes" or more specifically "carbon taxes" were proposed to reduce greenhouse gas emissions in the U.S., but most of these failed for lack of popular support.

Another technique of residuals management is the *throughput tax*, or disposal charge. Producers of goods would be charged a materials fee that reflects the social costs of disposal of the residuals. This tax would be passed on to the consumer in the form of increased prices for the commodity. Although a throughput tax might provide an incentive to recycle and make products last longer, many of the disposal charges currently used, such as deposits on beverage containers, are discriminatory against consumers. These types of taxes influence the price of a commodity beyond the simple demand and supply accounting (Issue 2.3).

Economics of the Individual Firm

Microeconomics refers to the economics of the firm or individual company. In a commercial system, most resource exploitation is undertaken by the firm rather than by society as a whole, and so it is worthwhile to examine some microeconomic activities that influence natural resource decisions.

One problem affecting many resources is a tendency for firms to maximize the rate at which resources are used instead of spreading that use out over a long period of time. The reason for this practice is fairly simple. Firms earn profits by investing money (capital) in the physical plant needed for production (land, machinery, etc.) and in other production inputs (energy and labor) in order to produce an output that can be sold for more money than was invested. The producer's costs in making a given commodity can be divided into two categories, fixed costs and variable costs. *Fixed costs* are bills that must be paid regardless of how much of a product is made in a given time. Fixed costs are primarily the costs of owning the means of production (the physical plant). If the money needed to set up a plant was borrowed, then the interest expenses on that loan are fixed costs. *Variable costs* consist of labor, energy, transportation, and similar inputs to the production process itself, and these vary according to the rate of production. Variable costs are a relatively constant fraction of the selling price of the goods produced, although some levels of output are more efficient than others.

Let us assume that a given investment is required to open up a particular mine. Afterward, variable costs change in direct proportion to the rate of extraction of minerals. The faster the minerals are extracted, the greater the profit will be, for the fixed costs need to be borne for a shorter period of time. In practice, variable costs per unit output are low at high rates of production because of economies of scale. However, variable costs may increase at very high rates of production because of the need for greater capital and labor inputs to maintain high rates of production. The optimal rate of production is, of course, determined by a combination of fixed and variable costs. Fixed costs always make the optimal rate of output higher. As interest rates go up, the need to recover the initial investment in a short time period increases, hence the need to increase rates of production.

As a result of these pressures, demands are made on the firm to maximize the rate of production, regardless of whether it is good conservation policy in the long run. This generalization strictly applies only to a sole facility, such as a single mine or a single forest unit. A large corporation owning many mines varies its total output not by changing output at every mine simultaneously but by closing or opening mines. That is how it keeps its fixed costs as low as possible. But in most cases the extracting firm is pressured to maximize the rate of extraction to recover fixed costs, often at the expense of the environment.

Another important aspect of the firm relative to natural resources is the degree of liquidity of its assets, or how easily the firm can sell out if it needs to. Remember that the goal of any company is to turn a monetary investment into monetary return, and production of a particular commodity is simply a means to that end. An oil company consists of a group of people who have particular expertise in finding, extracting, and selling oil and who also own the equipment needed to do those things. If that oil company has an opportunity to invest its capital in a housing development or a soft drink bottling plant and receive a greater return on investment than it would in drilling for oil, then it will do so. Only the existing investment in oil-related equipment and experience prevent an oil company from taking its money elsewhere if the financial opportunities are more attractive in some other business.

BUSINESS AND THE ENVIRONMENT: RECENT TRENDS

Diversification and Multinational Corporations

One significant business trend in the 1970s and 1980s was toward *diversification* of large corporations. Tobacco companies bought soap companies; oil companies bought electronics companies; and steel companies bought oil companies. Among other things, this diversification served to protect large companies from unfavorable market conditions in particular sectors of the economy. It also served to weaken a company's commitment to the long-term stability of a particular enterprise or resource.

For example, recent decades witnessed a major increase in corporate ownership of farms. In the 1970s this increase was spurred by rapidly rising land prices that attracted speculative investments. In several areas of the United States, large diversified corporations became major holders of agricultural land. If they carry on a policy of maximizing return on investment, then they may adopt farming practices that lead to excessive soil erosion. In this instance, as in many others, the best interests of business are not the best interests of society as a whole.

ISSUE 2.3: THE VALUE OF NATURE

In the present era of neoliberal free-market economics, policymakers argue that solutions to environmental problems must be based on the market, a system of trading goods and services at prices that reflect society's valuation of them. This requires establishing prices for environmental goods and services such as biomass production and pollution dispersal that are currently traded but not in the marketplace. This is easier in some cases than in others. For relatively small areas and limited resources, it can be done fairly easily. But for large areas comprising a wide range of interdependent resources, the task is much more difficult, and the results are much less reliable.

In recent years, several attempts have been made to place price tags on the natural resources of large areas, including countries and even the entire planet. Because such estimates are necessarily crude, it is difficult to use the results to make decisions on specific resource exchanges as in a market. Rather, they are intended mostly as wakeup calls, showing the importance of natural resources' contributions to the wealth and prosperity of nations. And as natural resource accounting methods improve in the future, the potential for making real market-based decisions increases.

Natural resource valuations (like any financial balance sheet) take two fundamental forms: stocks (assets and debts) and flows (income and expenditures). Accountings of stocks, the total wealth of an individual, company, or country at any given time, are useful in determining the market price of an object if it is to be sold. This is not very meaningful in the case of, say, a country, unless the accounting is done repeatedly on a regular basis with consistent methods. For example, should the total wealth of a nation decline from one period to the next, then the resources or savings are being drawn down; the resource base is being mined, and future generations will have less to work with. Alternatively, accountings of flows show how much of a country's income comes from various sources, such as manufacturing, mining, and agriculture. When the productivity of natural systems is valued in this way, we can estimate the importance of natural processes in supplying humankind with our daily needs relative to our own labor. Recently, accountings of both types were attempted.

An accounting of the wealth of nations must recognize that such wealth takes many forms, and to a certain extent one form may be substituted for another. For example, meat can be obtained by relying almost entirely on natural resources, by raising cattle on natural grasslands. Meat also can be produced by feeding animals grain such as corn, using some natural resources but supplementing them with tractors, chemicals, and the farmer's skills. Or we can just use money to buy meat from someone else.

The World Bank calculated the wealth of nations by summing produced wealth (factories, goods, money), human and social capital (educated, healthy people capable of creating wealth through their labor), and natural capital (land, soil, water, forests, etc.). Calculations of natural resource value involved applying a particular value per hectare to each type of land, multiplying by the number of hectares of that land, and adding in wealth in the form of geologic resources such as oil or minerals. Not surprisingly, countries like

Historically, the large forest product (lumber and paper) companies were an exception to this tendency of corporations to be interested only in short-term returns. Although many of the forest products companies are diversified, they also own large tracts of land that require decades to produce harvestable trees. This creates an incentive for these companies to maintain a long-term, environmentally healthy commitment to their land, especially in those forest regions (southeastern United States, for example) with faster growing softwood species.

Another important trend of the last few decades was the formation and growth of *multinational corporations* (MNCs). These are companies that operate in several countries or own or collaborate with companies in several countries. They have the ability to shift resources, production, and marketing activities from one country to another, depending on where potential profits are greatest.

Japan and Germany have a large portion of their wealth in the form of human capital, while Russia and Canada have large amounts of natural capital. The total wealth of the world was estimated at $86,000 per person (20 percent, natural resources; 16 percent, produced wealth; and 64 percent, human capital). These measures allow us to evaluate the depletion of natural capital as an oil field is exploited, for example, in comparison to the increase in ability to generate wealth that may be represented by investing of oil revenues in industry or education (World Bank 1995).

Evaluating the amount of income produced by nature each year, as opposed to the amount of stored wealth, follows a similar approach: estimating the per-hectare amount of income generated from a particular ecosystem and multiplying by the total area of that ecosystem on earth's surface. In one study, 17 different kinds of services were evaluated, including such functions as gas and climate regulation, provision and regulation of water flows, soil formation, nutrient cycling, waste treatment, pollination, food production, and recreation (Costanza et al. 1997). For each of 16 biomes, estimates of the value of services provided were collected. Most of these estimates were based on contingent valuation—asking how much society would be willing to pay for a given amount of soil formation, for example, if we were able to increase such a process. The total value of these services per year was estimated at about $33 trillion, or a little less than double the gross world product of $18 trillion. Ecosystem services are not counted as part of gross product calculations, which only measure goods and services produced by humans.

Of the $33 trillion figure, the service with the largest value is soil formation, estimated at $17.1 trillion per year; recreation comes second at $3.0 trillion, and nutrient cycling is third at $2.2 trillion. Among biomes, coastal systems produce the most income—$12.6 trillion per year, followed by the open ocean ($8.4 trillion) and wetlands ($4.9 trillion) (Costanza et al. 1997).

Both the World Bank and Costanza studies offered are very rough approximations, and many uncertainties surrounded the actual values they estimated. But if we make the assumption that the two studies are in the right neighborhood, then some conclusions are possible. First, both studies make it clear that natural resources are very important to our well-being. Second, most of our stock of wealth is in human rather than natural form, whereas most of our income is derived from natural rather than human processes. In other words, nature produces a large amount of income from a relatively small amount of capital, while people store much more capital but produce a more modest amount of income from it. Perhaps the most critical issue, however, may not be measuring the value of nature. We all recognize, for example, that the biological productivity of the world's oceans is of enormous value or that air pollution increases costs for health care. The key question is how to get these costs internalized in our markets, so that our resource use recognizes these costs and damage to resources is minimized. The more we understand about the enormous value of the natural contributions to our well-being, the more likely we are to incorporate these values in our day-to-day management of resources.

They are generally large enough to have major control over markets in individual nations, if not at the world level. Their ability to move money and commodities internationally greatly limits the controls that individual governments have over them, and thus it is more difficult to force consideration of social costs in decision-making.

The environmental record and corporate role of MNCs in environmental management is subject to intense debate. Some advocates feel that the MNCs are industry innovators, and with advanced technology and an enhanced environmental ethos, they will make significant contributions to upgrade plants to meet environmental standards (Hoffman 2000; White 1999). The investment in pollution control (in the absence of strict environmental regulation) is based on minimizing the present and future investment costs of MNCs while offsetting any potential political or social conflict, including adverse public opinion (Roysten 1985). "Pollution prevention pays" is not only the anthem of environmentalists but also the rallying cry of many Fortune 500 companies (3M Corporation, DuPont), that cleaned manufacturing operations and saved money in the process.

Industry detractors believe that many MNCs deliberately seek pollution havens in developing countries in order to reduce their pollution control costs or their residuals costs. An example of this is multinational behavior and the disposal of hazardous wastes. New controls were established on the disposal of hazardous wastes in the United States and other wealthy nations. As a consequence of stricter regulations, which increased the costs of hazardous waste disposal, industrialized countries initially looked for other disposal options, including the transboundary shipment of these hazardous wastes. The U.S. EPA is required by law to fully inform waste recipients of what they are getting, both domestically and overseas. To prevent "toxic terrorism" (Deery Uva and Bloom 1989), that is, offering large sums of much-needed money to poor host countries to take the hazardous waste, the Basel Convention (1989) was enacted. The Basel Convention is a multilateral treaty adopted by 149 countries in an effort to manage the hazardous waste trade. The majority of the provisions are designed to ensure that hazardous wastes are controlled in an environmentally sound manner that protects the rights of both exporting and importing countries. As a result of these efforts, waste minimization (the reduction of waste at its source) is now a major issue for industry. It is also helping to focus attention on conservation and recycling efforts at the individual firm level. A global pollution consciousness is in the making, with the understanding that a pollutant produced on one side of the globe can affect life on the other side.

Multinational corporations have also had the effect of greatly increasing the degree of worldwide economic integration. Markets for resources are controlled by world supply and demand rather than at the national level. A shortage of a commodity in one country causes increases in prices in other countries.

The Greening of Business

Cleaning up the environment is a profit-making enterprise. The environmental control industry (solid waste companies, consulting companies, hazardous treatment firms, and so on) generates billions of dollars a year in revenues. The companies that are involved in cleaning up environmental pollution (oil spills, hazardous waste, solid waste, medical waste, asbestos, and so on) constitute one of the fastest-growing industrial sectors of the economy and will be growth industries as long as reducing environmental pollution is a government priority. Manufacturers and distributors of environmentally safe consumer products have increased their sales and clout in the marketplace and now constitute a more than $8 billion industry (Bogo 2000). Firms that specialize in environmental analysis and testing also will experience growth during the decade.

Investor interest in environmental stocks is also growing as environmentally concerned citizens put their money where their mouths are. Just as socially responsive investing was important during the 1980s (no military weapons contracts or business in South Africa), environmentally responsive investing was important during the 1990s and beyond. Several mutual funds specialize in environmental stocks. Consumers can even choose among a variety of different bank cards with the logo of their favorite environmental group emblazoned across their Visa or MasterCard, with a small percentage of the proceeds donated to the group.

As a result of enhanced consumer and stockholder concern for the environment, stricter environmental regulations, and the increasing costs of waste disposal (especially hazardous wastes), individual firms are seeking new ways to internalize their pollution costs. Pollution prevention strategies are one such mechanism gaining in importance in the industrial sector. It is more efficient (economically) to prevent pollution in the first place than to clean it up after the fact. The EPA's pollution prevention program assists industry in strategies that not only save money but clean up the environment as well.

Finally, business schools have joined the environmental bandwagon. Environmental management courses are now part of business curricula across the country at the behest of corporate sponsors who feel that students are totally ignorant about the business impact of environmental issues. Environmental issues have far-reaching implications, ranging from capital expenditures to cash flow to redesign of manufacturing processes.

In fact, corporate self-interest could become one of the primary forces encouraging environmental protection in the future (Levy and Newell 2000).

Deregulation

The 1980s and 1990s saw the rise of neoliberalism, a view that favored greater reliance on the private sector and free trade to promote economic development. One manifestation of this change was the trend toward deregulation of business, in both capitalist and centrally planned economies. Deregulation has enormous impacts on every aspect of society. Although in the United States the term *deregulation* is often mentioned in the context of industries such as airlines, banking, and telecommunications, natural resource use has been dramatically changed by deregulation activities as well. Let us consider some examples in the United States.

In the 1980s, early in the Reagan administration, the financial securities industry in the United States was deregulated. Deregulation unleashed a wave of corporate takeovers in which asset-rich companies were purchased and "restructured," usually involving the sale of valuable assets that generate relatively little income and concentration on the remaining profitable sectors of a business. This increased the pressure on natural resource companies to maximize short-term return on investments rather than managing for long-term income stability. For example, in 1985 the Pacific Lumber Company, holder of the largest remaining privately owned old-growth forests in the United States and a company known for practicing sustainable forestry, was purchased by a "corporate raider" who went deeply into debt to buy the company and subsequently raised rates of harvesting in order to increase income and pay off the debt (see Issue 7.2).

In 1996 the Freedom to Farm Bill was passed in the United States, deregulating U.S. agriculture, which continued a trend begun a decade earlier. The bill eliminated linkages between government subsidies and some important farm management decisions and increased the financial risk to farmers associated with variations in commodity prices. The full effects of this deregulation remain to be seen, but if deregulation continues, it may result in increased pressure to maximize short-term yields at the expense of long-term soil conservation (see Chapter 6).

Deregulation of the electric generation industry is under way. It will enable competition among different producers of electricity, instead of the regulated monopolies that have existed since the establishment of electric utilities. Deregulation is expected to open up opportunities for small generating companies to enter the business, possibly using unconventional technologies. It could therefore make it possible for photovoltaics, small-scale hydroelectric, or other renewable energy technologies to become commercially established. On the other hand, the volatility in electricity prices due to deregulation led to disastrous results in California during the summer and fall of 2000. Not only did some residents pay more than four times for their electricity, they were subjected to rolling blackouts and brownouts as the supply of electricity could not keep pace with demand. The economic losses were so great that many of the public utilities (which could not supply enough to meet the demand and thus had to purchase electricity from other providers at higher costs) incurred huge economic losses that ultimately required the state to bail them out.

As seen in these examples, deregulation has positive or negative effects on resource conservation. In some cases, regulation protected resources and encouraged conservation by allowing (or requiring) long-term goals and broader social needs to be considered by resource managers and users. In these cases, deregulation usually results in a focus on short-term monetary gains rather than on meeting the needs of a broader group of stakeholders. In other cases, where government regulation and subsidies encouraged waste or discouraged innovation, deregulation opens up opportunities for much more efficient and therefore conservation-oriented resource management.

CONCLUSIONS

The business of natural resource use is no different from any other business. It is governed by the same need to turn investment into profit as quickly as possible, and it is subject to the same vagaries of economics caused by fluctuating interest rates, inflation, and the ups and downs of business cycles. Although we often blame our government, big corporations, foreign governments, or natural calamities for problems related to natural resource supply or prices, in almost all instances the real causes of the problem can be traced to economic constraints on the businesses involved and the simple desire of companies to make as much profit as possible. In centrally planned economies, economic development was always favored over environmental protection. This was made exceedingly clear in past decades as eastern and central European countries had some of the worst air and water pollution in the world.

Natural resources are of fundamental importance to us all, because many of them are commonly owned. Decisions involving natural resources therefore are likely to have external costs and social effects that businesses normally do not consider. Government intervention is necessary to modify the management process, so that intangible resources, long-term needs, and social costs can be managed along with the commodities that move through our economic system.

REFERENCES AND ADDITIONAL READING

Abramowitz, J. 1997. Valuing nature's services. In L. R. Brown, ed. *State of the World, 1997.* New York: W. W. Norton.

Altman, D. 2002. Just how far can trading of emissions be extended? *The New York Times*, May 31, 2002: C1.

Bateman, I. and K. G. Willis. 1999. *Valuing Environmental Preferences: Theory and Practice of the Contingent Valuation Method in the US, EU, and Developing Countries.* Oxford: Oxford University Press.

Bogo, J. 2000. A greener bottom line. *E Magazine* XI (4) July–August: 26–33.

Brown, L. R. 2001. *Eco Economy: Building an Economy for the Earth.* New York: W.W. Norton.

Boulding, K. E. 1966. The economics of the coming spaceship Earth. In H. Jarrett, ed. *Environmental Quality in a Growing Economy.* Baltimore, MD: Johns Hopkins University Press.

Burger, J. and M. Gochfeld, 1998. The tragedy of the commons 30 years later. *Environment* 40(10):4–13, 26–27

Costanza, R., R. d'Arge, R. de Groot, S. Farber, M. Grasso, B. Hannon, K. Limburg, S. Naeem, R. O'Neill, J. Paruelo, R. Raskin, P. Sutton, and M. van den Belt. 1997. The value of the world's ecosystem services and natural capital. *Nature* 387:253–260.

Daly, H. E. 1973. *Towards a Steady State Economy.* San Francisco: W. H. Freeman.

_____. 1996. *Beyond Growth: The Economics of Sustainable Development.* Boston: Beacon Press.

Daily, G. C., T. Söderqvist, S. Aniyar, K. Arrow, P. Dasgupta, P. R. Ehrlich, C. Folke, A. Jansson, B. Jansson, N. Kautsky, S. Levin, J. Lubchenko, K. Mäler, D. Simpson, D. Starrett, D. Tilman, and B. Walker. 2002. The value of nature and the nature of value. *Science* 289:395–396.

Daily, G. C. and K. Ellison. 2002. *The New Economy of Nature.* Washington, D.C.: Island Press.

Deery Uva, M. and J. Bloom. 1989. Exporting pollution: The international waste trade. *Environment* 31(5):4–5, 43–44.

Desvouges, W. H., V. K. Smith, and A. Fisher. 1987. Option price estimates for water quality improvements: a contingent valuation study of the Monongahela River. *Journal of Environmental Economics and Management* 14:248–267.

European Environment Agency. 2002. www.eea.eu.int.

Farrow, S. and M. Toman, 1999. Using benefit-cost analysis to improve environmental regulations. *Environment* 41(2):12–15, 33–38.

Friends of the Earth. 1981. *Progress as if Survival Mattered.* San Francisco: Friends of the Earth.

Georgescu-Roegen, N. 1976. *Energy and Economic Myths.* New York: Pergamon Press.

Hardin, G. 1968. The tragedy of the commons. *Science* 162:1243–1248.

_____. 1972. *Exploring New Ethics for Survival: The Voyage of the Spaceship Beagle.* New York: Viking.

Hawken, P., A. Lovins, and L. Hunter Lovins. 1999. *Natural Capitalism: Creating the Next Industrial Revolution.* New York: Little Brown.

Hershey, R. D., Jr. 1989. New market is seen for "pollution rights." *The New York Times*, June 14, p. D1.

Hoffman, A. J. 2000. Integrating environmental and social issues into corporate practice. *Environment* 42(5):22–33.

Kneese, A. V., R. V. Ayres, and R. C. d'Arge. 1971. *Economics and the Environment: A Materials Balance Approach.* Baltimore, MD: Johns Hopkins University Press.

_____ and B. T. Bower, eds. 1971. *Environmental Quality Analysis: Theory and Method in the Social Sciences.* Baltimore, MD: Johns Hopkins University Press.

Krutilla, J. V. and A. Fisher. 1975. *The Economics of Natural Environments.* Baltimore, MD: Resources for the Future/Johns Hopkins University Press.

Levy, D. L. and P. Newell. 2000. Oceans apart? Business responses to global environmental issues in Europe and the United States. *Environment* 42(9):8–21.

MacNeill, J. 1989. Strategies for sustainable economic development. *Scientific American* 261(3):154–165.

O'Riordan, T. and R. K. Turner. 1983. *An Annotated Reader in Environmental Planning and Management*. Oxford: Pergamon Press.

Passell, P. 1992. Cheapest protection of nature may lie in taxes, not laws. *The New York Times*, November 24, p. C1.

_____. 1995. How much for a life? Try $3 million to $5 million. *The New York Times*, January 29, p. D3.

Pearce, D. W. and R. K. Turner. 1990. *Economics of Natural Resources and the Environment*. Baltimore, MD: Johns Hopkins University Press.

Portney, P. R. 1998. Counting the cost: the growing role of economics in environmental decision making. *Environment* 40(2):14–18, 36–38.

Rees, J. 1985. *Natural Resources: Allocation, Economics, and Policy*. London: Methuen.

Renner, M. 2000. *Working for the Environment: A Growing Source of Jobs*. Worldwatch Paper 152. Washington, D.C.: Worldwatch Institute.

Ripley, A. 2002. What is a life worth? *Time* 159(6):22–27.

Roysten, M. G. 1985. Local and multinational corporations: Reappraising environmental management. *Environment* 27(1):12–20, 39–43.

Ruckelshaus, W. D. 1989. Toward a sustainable world. *Scientific American* 261(3):166–175.

Solomon, B. D. and R. Lee. 2000. Emissions trading systems and environmental justice. *Environment* 42(8):32–45.

Stavins, R. N. 1989. Harnessing market forces to protect the environment. *Environment* 31(1): 5–7, 28–35.

_____, ed. 2000. *Economics of the Environment: Selected Readings*. 4th Ed. New York: W. W. Norton.

_____ and B. W. Whitehead. 1992. Dealing with pollution: market-based incentives for environmental protection. *Environment* 34(7):6–12.

Turner, R. K., D. Pearce, and I. Bateman. 1993. *Environmental Economics: An Elementary Introduction*. Baltimore, MD: Johns Hopkins University Press.

Vogel, D. 1995. *Trading Up: Consumer and Environmental Regulation in a Global Economy*. Cambridge, MA: Harvard University Press.

White, A. L. 1999. Sustainability and the accountable corporation: society's rising expectations of business. *Environment* 41(8):30–43.

World Bank. 1995. *Monitoring Environmental Progress: A Report on Work in Progress*. Washington, D.C.: World Bank.

World Resources Institute. 2003. *A Guide to World Resources 2002–2004: Decisions for the Earth: Balance, voice, and power*. New York: World Resources Institute.

Yaffee, S. L. 1994. *The Wisdom of the Spotted Owl*. Washington, D.C.: Island Press.

For more information, consult our web page at *http://www.wiley.com/college/cutter*.

STUDY QUESTIONS

1. When you buy a chicken at the grocery store, you pay a price that is divided (ultimately) among the grocery store, the chicken producer, the grain farmer, various transportation companies, and the many suppliers that provide each of these entities with materials they need in their businesses. At each of the steps along the way—grain farm, chicken farm, and grocery store—environmental externalities are generated. Make a list of the important environmental externalities generated from this process.

2. List the major environmental externalities associated with operating an automobile. Suggest three different ways these costs could be internalized.

3. Discuss why the externalities associated with operating automobiles are not likely to be eliminated in the United States any time soon.

4. Conduct a survey of your friends and classmates to determine how much they would be willing to pay to (1) preserve an acre of old-growth forest that would otherwise be harvested (a one-time investment) or (2) reduce sulfur emissions in the United States (or your country) by 50 percent (an annual fee). Multiply the result by the population of the country to estimate the value of these environmental improvements (assuming your friends and classmates are representative of the entire population).

5. Air pollution generated by a facility such as a coal-fired electric power plant constitutes an environmental externality that could be eliminated either by regulations (such as requiring the electric company to control its emissions under threat of a criminal penalty) or by a market-based system in which the costs of controlling emissions would be borne by those who benefit from generating it or those who would benefit from improved air quality. Discuss the advantages and disadvantages of each approach.

CHAPTER 3

ENVIRONMENTAL HISTORY, POLITICS, AND DECISION-MAKING

INTRODUCTION

Government policy is a key determinant of how natural resources are exploited and conserved. In the United States, the policies that control natural resource use were developed over three centuries by both governmental and private actions. Governmental policy is a product of the political process, constrained by history and precedent. The political process in turn is essentially one of confrontation and compromise among many different interests, both economic and ideological. International environmental policies are the outgrowth of cooperation between nations. Unfortunately, international policy often is difficult to produce because of wide disparities in legal, economic, and social systems, which exacerbate major differences between nations (Hempel 1996). Once these differences are ironed out, however, participation in an international treaty is only voluntary. Nations cannot be compelled to abide by treaties because of the concept of *national sovereignty* (Murphy 1994). In other words, a country has the right to look after its own interests first and foremost, and what happens within its territorial boundaries is that country's business. Thus, the enforcement of international treaties is even more doubtful because there are no international police, no international taxation mechanism to support enforcement, and you can't take a country to court without its permission. One last complicating issue for international environmental policy involves the transboundary nature of contaminants and the common property nature of many resources (Porter and Brown 1996).

This chapter examines the history of human impact on the environment worldwide and the development of natural resource policy, with a particular focus on the United States. The policy process is described, along with the various ideological and economic interest groups that are the major forces behind environmental management practices at both the national and global levels.

NATURAL RESOURCE USE: A HISTORICAL PERSPECTIVE

Human history can be viewed as a process of increasing ability to manipulate and alter usable aspects of the physical environment. For example, in the early stages of human evolution, at least 2 million years ago, the natural environment was largely unaffected by humans. Small numbers of protohumans in their hunting bands, using simple technology (bone, stone, and wood tools and hunting pits), were generally capable of competing with animal species. Like animals, protohumans were also at the mercy of climate and topography and did not have the technological skills to master the earth's more difficult climates. Thus, they were best able to utilize the food and shelter resources of open and coastal lands, locations that were far more vulnerable to natural hazards such as floods than to any alteration by people.

The first human tool to have a major environmental impact was fire. Early humans used it to drive animals into traps; when agriculture was developed between 10,000 and 7000 B.C., fire was used to clear land for crops and to create grazing areas for livestock. Fire is the only example in which the capacity of modern technology to alter the environment is matched by that of the pretechnological humans. The deliberate use of fire introduced three types of environmental effects: (1) it was widespread, affecting a large area; (2) it was a repetitive process and could cover the same areas at frequent intervals; and (3) it was highly selective in its effects on animal and plant species, having a negative effect on some while encouraging those with rapid powers of recovery or resistance to fire (Pyne 1997). The environmental result was to improve the yield of certain species for human use and to modify the vegetation cover. These early effects were confined largely to tropical, subtropical, and temperate forests and grasslands, as well as some wetlands.

At least 10,000 years ago the human race had spread to all continents except Antarctica. With the shift from hunting and gathering to agriculture, human culture developed more sophisticated food production tools for planting, harvesting, and transporting. Also, in drier areas in the Middle East and later elsewhere, large-scale irrigation works were built. The sedentary life of the agriculturalist went hand in hand with the development of cities. These two developments, agriculture and urbanization, led for the first time to a substantial change in land use, from natural to human-made forms of productivity, in the form of fields, streets, homes, and irrigation ditches. The development of cities led to large-scale environmental disruption and change because of the concentration of large populations and the wide areas in which land was cultivated, grazed, cleared of trees, and subjected to erosion to support the urban population. In addition, through the domestication of plants and animals, people were able to direct the energy and nutrients of an ecosystem to produce more of certain foods than the environment would naturally. This in turn permitted the growth of human populations beyond the limits set by their pre-agricultural patterns. Thus, agriculture raised the *carrying capacity* of the earth to support human beings.

After about 1000 B.C., humans began to move freely around the world, and rulers began to dominate large regions from a distance. Settlements and their impacts were no longer necessarily small in scale or localized in effect. The era of European colonialism that began in the fifteenth century A.D. placed the environments and resources of far-distant lands under their control. These colonial powers were interested in removing and using resources, with little regard for environmental consequences either abroad or close to home. The advent of industrialization led to a global-scale use of fuel and mineral resources that altered or destroyed local and regional ecosystems, perhaps ultimately affecting global climate and other environmental patterns (Simmons 1996; Turner et al. 1990).

The last three millennia, and particularly the last 500 years, saw a transformation in the kinds and scales of natural resource use in the world (Turner and Butzer 1992). Early societies depended primarily on locally available resources with relatively little trade, whereas now most of the goods we consume come from quite far away. Resource-use systems have become complex, with a wide variety of goods utilized in everyday life. This increasing complexity isolates us somewhat from the basic raw materials provided by the environment and makes us more dependent on human systems of resource manipulation and distribution. There clearly are innumerable ways of making a living in the world today. No single commodity or geographic area is indispensable, and resource management has become a task of selecting which resource utilization techniques are most appropriate for our needs at any given time.

DEVELOPMENT OF NATURAL RESOURCE POLICY

Natural resource policies are established at a variety of levels—local, state, national, regional, and international. As you might expect, the decision-making, mechanisms, and impacts of these policies vary not only between different states in the United States but also from country to country.

U.S. Environmental Policy
The history of natural resource policy in the United States from the seventeenth century to the present can be divided into different phases, but these do not have distinct beginning and ending dates and often overlap. The following is a summary of some of the major actions and events that form the basis of much of contemporary U.S. conservation philosophy and policy.

Phase 1: Exploitation and Expansion (1600–1870) When the early European settlers arrived in North America, they found a vast continent with natural resources in apparently limitless supply, compared to the more urbanized and developed Europe. Their goal was to establish stable and profitable colonies. To accomplish this goal, the European landholders who controlled settlement promoted population growth and resource extraction to maximize their security and prosperity. The colonial economy was, by design, based on exporting raw materials to industrial Europe, with agriculture for domestic food production. The enormous land area of North America was the primary resource for this economic development, and exploitation of its natural resources was the means to the desired end.

The forests that covered most of the eastern third of the continent were seen partly as a resource and partly as a nuisance. Wood was needed for fuel and construction purposes, but the vast amount of forest compared to productive agricultural land meant that timber cutting was a low-value land use. The forests, then, were cleared as rapidly as possible to make room for agriculture. In addition, the prevailing aesthetic attitudes toward forests were different. Forests were seen as unproductive, undesirable, and dangerous, whereas agricultural land was productive, attractive, and secure (Fig. 3.1). Except in a very few cases, regulations limiting the clearing of forests were unknown, as was the notion of natural resource conservation. This exploitative attitude toward the land prevailed for about the first 250 years of European occupation of North America, until the middle or late 1800s (Cronon 1984). The growth of an industrial economy in the nineteenth century had little effect on this attitude, except perhaps to increase the demands of urban populations for food, timber, and, later, coal. Forests were first culled of the most valuable trees, and later the remaining timber was generally clear-cut and often burned. Agriculture in many areas was largely cash cropping of a very few crop types, and except for liming soils in some areas, little was done to maintain, let alone enhance, soil fertility. As a result, soil erosion was rapid, and declines in fertility forced abandonment of land after only a few years, particularly in the southeastern United States (Dilsaver and Colten 1992).

As the nation expanded westward, political as well as economic goals required rapid settlement

Figure 3.1 A mixed landscape of forest and farms has replaced the near-continuous forest that covered most of the eastern United States when Europeans began migrating here in the seventeenth century. Clearing the forests was viewed by individuals and their governments as an improvement of the resource base.

and development of the Great Plains. With each major territorial expansion from the Louisiana Purchase in 1803 to the Alaska Purchase in 1867 (the annexation of Texas excepted), the federal government acquired possession of vast acreages. In the early nineteenth century, much government land was sold to provide income to the fledgling republic as well as to promote settlement. Several laws were passed in the mid-1800s to promote settlement, largely by transferring government-owned lands to private ownership, either for free or at a nominal cost. The most notable among these laws were the *Homestead Act* of 1862, the *Railroad Acts* of the 1850s and 1860s, the *Timber Culture Act* of 1873, and the *Mining Act* of 1872.

The Homestead Act and the Railroad Acts were specifically designed to encourage settlement, especially in the Great Plains. The Homestead Act gave any qualified settler 160 acres free, and the Railroad Acts granted large rights of way to the railroad companies to finance construction of transcontinental and other rail lines that would further accelerate settlement of the West. Most of the land granted to the railroads was sold to other private interests, but substantial acreages remain in railroad ownership today, particularly in California. The Timber Culture and Mining Acts granted free access to forests and minerals to anyone willing to exploit them. There were widespread abuses of these privileges, which resulted in land companies and speculators acquiring vast acreages at nominal expense. Although these laws were successful in stimulating settlement and economic development, in many cases they encouraged excessive exploitation by artificially depressing the price of resources. Environmental degradation usually followed, as with the forests of the upper Midwest (Michigan, Wisconsin, and Minnesota), in which much timber was lost to wasteful logging practices and fires and soil was lost to accelerated erosion.

The primary themes of this era included resource exploitation for economic prosperity and land transfers from public to private ownership. In fact, this era is best characterized by the massive land transfers from federal to private ownership, be it individuals, developers, or selected industries such as the railroads.

Near the end of this phase, the practice of promoting exploitation of resources for economic growth was limited somewhat by the growth of the conservation movement. Exploitative policies continued into the twentieth century, however, with legislation such as the Reclamation Act of 1902, which provided for the development of water at public expense for crop irrigation in the arid West. Today, natural resource exploitation for economic prosperity is still the basis of government management of mineral resources such as coal and oil, as well as being an important consideration for other resources such as water and rangelands.

Phase 2: Early Warnings and a Conservation Ethic (1840–1910)

As the westward expansion continued, Americans escalated their efforts to exploit the environment for their own needs, and with improved technology, settlers had a much easier job of "taming the land." For example, the mechanization of farming enhanced the settlement of the Great Plains. McCormick's grain reaper allowed the timely harvest of wheat. Steel plows developed by John Deere helped to break up the prairie soil for cultivation (Fig. 3.2). The cotton gin provided a mechanical means to sort lint from the cotton, and barbed wire allowed farmers and ranchers to demarcate property. As a result of many of these mechanical inventions, wildlife populations were particularly hard hit, as they were displaced from their ecosystem, were outcompeted by the domesticated animals (cows and sheep), or succumbed to harsh winters and the rifle.

During the time when settlement rapidly was advancing westward with the government's stimulation and encouragement, a few individuals were suggesting that the exploitation of resources was too rapid and too destructive. In general, these persons were intellectuals and academics who did not enjoy popular audiences for their criticisms; thus, the effects of their writings were limited at the time. Eventually, however, their warnings were heard by decision-makers in government, and this led to a new concern for conserving and preserving resources.

Figure 3.2 An early tractor plowing prairie soil for wheat production in Oregon about 1890. The development of equipment such as this encouraged specialization in agriculture and strengthened links between agricultural and industrial sections of the economy.

Among the early American writers advocating wilderness preservation were Ralph Waldo Emerson and Henry David Thoreau, who argued on philosophical grounds in the 1840s and 1850s against continued destruction of natural areas by logging and similar activities. George Perkins Marsh's *Man and Nature, or Physical Geography as Modified by Human Action*, published in 1864, was perhaps more influential in the conservation versus exploitation debate. Marsh was both a public servant and a scientist, which led him to advocate government action to protect natural resources. Although a native of Vermont, Marsh traveled widely in the Mediterranean lands, areas long damaged by overgrazing. He saw a parallel between the Mediterranean situation and the damage done by sheep grazing in the Green Mountains in his home state. In *Man and Nature*, Marsh argued that humans should attempt to live in harmony rather than in competition with nature (Fig. 3.3). More important, Marsh argued that natural resources were far from inexhaustible. This book was widely read and had considerable influence on Carl Schurz, who later became Secretary of the Interior under President Rutherford Hayes in 1877.

Phase 2 included a series of developments in the late nineteenth century, when many of the basic doctrines of the government's natural resource conservation policy were established. The primary tenet was that land resources should be managed for long-term rather than short-term benefits to the general population. This phase, dominated by concern for forest resources and to a lesser extent wilderness preservation, began in the 1860s and was marked by the first significant government action aimed at restricting exploitation of natural resources. One important governmental action was the formation, in 1862, of the Department of Agriculture's Land Grant College System, which was designed to help improve the management and productivity of agriculture and forestry through improved education. The establishment of the Cooperative Extension Service by the Smith-Lever Act of 1914 also helped to improve conservation education by linking the local farmer to agricultural experts in the state universities.

By the late nineteenth century, the forests of much of the eastern United States either were entirely cut over or were rapidly disappearing. Thus, it is not surprising that the forest resource was the first focus of the emerging conservation efforts in the 1870s. Carl Schurz launched an attack on corrupt and wasteful practices in timber harvesting on federal lands and brought the severity of the problem to the public eye. In 1872 the Adirondack Forest Reserve Act halted the sale of state forest lands, an action that eventually led to the creation of the Adirondack Forest Preserve (now Adirondack State Park). The most significant development during this period was the passage, in 1891, of a rider on a public lands bill that gave the President the authority to set aside forested lands by proclamation, thus reserving them from timber cutting. President Benjamin Harrison quickly began withdrawing land from timber cutting, and in 1897 additional reservations by President Grover Cleveland brought the total forest reserves to about 40 million acres (16 million hectares [ha]). The fed-

Figure 3.3 Formal gardens like these in Manteo, North Carolina, show our love for nature's forms but, more important, our domination over them.

eral government had established what would later become the National Forest System, but at the time it had no real management policy for these lands. In 1898 Gifford Pinchot was appointed as the first Chief Forester. Pinchot was trained as a forester in Europe, where the field was well established. He brought with him knowledge of the scientific basis for land management, in particular the notion of *sustained yield* forestry. The principle of sustained yield management of renewable resources has now been firmly incorporated into all aspects of official federal policy, although there is some debate as to whether the principle is truly followed in practice.

In 1901 Theodore Roosevelt became president, and his administration represented the culmination of this phase of America's natural resource history. Roosevelt was an adventurer and outdoorsman and thus had a personal appreciation for the values of undeveloped land, particularly the still untouched wilderness areas of the western United States. Pinchot was one of his key advisers, and with the forester's advice, Roosevelt added large acreages to the nation's forest reserves. In 1905 the United States Forest Service was established with Gifford Pinchot as its first chief, and "forest reserves" became National Forests. By the end of Roosevelt's administration, these reserves totaled 172 million acres (70 million ha). Later, large acreages were added to the National Forests in the eastern United States after the passage of the Weeks Act in 1911, which provided for federal acquisition of tax-delinquent cutover lands.

Theodore Roosevelt was also instrumental in expanding what would later become the National Park System. Yellowstone was reserved as a National Park in 1872, and several other parks were created during this period. Roosevelt protected the Grand Canyon from development by invoking the Antiquities Act. Passed in 1906, this act primarily was intended to allow the President to preserve national historic sites such as buildings and battlefields. Roosevelt, however, used it to create the Grand Canyon National Monument, which later became a National Park. Some 90 years later, in 1996, President Clinton used this same act to preserve thousands of acres of wilderness in Utah. Finally, near the end of his presidency, Roosevelt sponsored the first White House Conference on Conservation, further bringing the issue to public attention and concern.

Another important figure during this period was John Muir, who founded the Sierra Club in 1892. Muir was a strong preservationist and wilderness advocate, whose primary area of interest was the Sierra Nevada Range of California. The Yosemite region was one of his favorite spots, and he led the battle to protect the area from damage by sheep grazing by establishing what would later become Yosemite National Park.

One of the most significant battles of Muir's life was fought over the preservation of Hetch-Hetchy Valley. This valley is adjacent to Yosemite Valley and was very similar in scenic beauty. However, Hetch-Hetchy was a convenient source of water for the growing city of San Francisco and an excellent dam site. Muir fought hard to prevent the damming of the Tuolumne River but eventually lost in a battle with a former ally in the conservation movement, Gifford Pinchot (Miller 2001). Although Pinchot was a conservationist, he believed in conservation as a means of maintaining the productive capacity of natural resources. To prevent development was contrary to the notion that resources could be used for the general benefit of the population, and Pinchot opposed Muir in the debate over Hetch-Hetchy. In the end, the development interests prevailed, and today the valley is a reservoir providing water and electricity to the cities of northern California. Almost a century later, there is occasional talk of draining Hetch-Hetchy and restoring the valley to its original state.

The Hetch-Hetchy controversy made clear the distinction between conservationists, who encourage careful husbanding of resources yet do not condemn their use, and preservationists, who would stop all use or development on the basis that some areas and resources are too valuable to be used. This second phase ushered in major achievements in resource conservation and saw the establishment of the principles of both sustained yield management and preservation of outstanding natural features for future generations. This period also witnessed the emergence of two of the major ideological camps, the preservationists and the conservationists, which still dominate debates over natural resources today.

Phase 3: Conservation for Economic Recovery (1930–1940) The Great Depression of 1929–1941 and Franklin Roosevelt's (FDR's) New Deal of the 1930s had more impact on all aspects of modern domestic policy than events in any other period in American history, and natural resource use was no exception. The Depression provided the impetus for massive programs aimed at relief, recovery, and prevention of similar problems in the future. Most of the major programs of this period were primarily economic rather than conservation-oriented in emphasis. The Civilian Conservation Corps, for example, did not represent a major new policy but rather was a makework program that put many of the unemployed to work on conservation projects, principally planting trees and maintaining or constructing park facilities. In contrast, two major agencies established by the New Deal, the Tennessee Valley Authority (TVA) and the Soil Conservation Service (SCS), were aimed at correcting problems that, if not major contributors to the Depression, were very much worsened by it.

The Appalachian region of the Southeast had been economically depressed for a long time and was among the areas hardest hit by the Depression. The forests were largely cut over, farms were not competitive with those in the Midwest, soil erosion and flooding were particularly severe, and no significant industrial employment was available. The TVA was the first major effort to address this wide range of problems in an integrated regional resource management and economic development program. The major elements of the program were the construction of dams for hydroelectric power generation and flood control, with the generated power used to support new industries, particularly fertilizer and later munitions production. Forests were replanted to control erosion, and many smaller erosion control measures were instituted, in part to protect the newly created reservoirs from sedimentation. Today the TVA is mostly an electric power–generating authority, but it has an important legacy in natural resources. It represents the recognition that good natural resource management and economic vitality are interdependent and that to ensure long-term economic stability, both must be undertaken together.

In addition to dam construction in the Tennessee Valley, many large dams were completed in the arid western states, including the Hoover Dam on the Colorado River (Fig. 3.4) and several dams in the Columbia River basin. These were seen as important government investments in agriculture and electric power generation, which would help revitalize agriculture and provide new sources of energy for industry.

The agricultural expansion in the Midwest and Great Plains during the late nineteenth and early twentieth centuries took advantage of the naturally fertile soils of that region, and farming was successful without significant inputs of fertilizers or other means to maintain soil fertility. Severe soil erosion was widespread, but it took the economic collapse of the 1930s and the ensuing dust

Figure 3.4 Hoover Dam. This 1938 Bureau of Reclamation photo shows the multipurpose dam, which spans the Colorado River at the Nevada–Arizona border. Hoover Dam and its reservoir, Lake Mead, provide flood protection, water storage, hydroelectric power, and recreation. These benefits typify the goals of the governmental public works projects during the 1930s and 1940s.

bowl conditions in portions of the Midwest and Great Plains to focus attention on the problem. Several dry years on land that was marginal for farming, combined with economic hardship brought on by low farm prices, led to severe wind erosion in Oklahoma, Colorado, and nearby areas, forcing thousands off the land (Worster 1979).

The Soil Erosion Service, created in 1933, was established in response to these problems. Hugh Hammond Bennett became the first director of the newly renamed Soil Conservation Service (now known as the Natural Resources Conservation Service) in 1935. He led an extensive research effort to determine the causes of soil erosion and the means to prevent it. This resulted in the development and implementation of many important soil conservation techniques, which yielded dramatic reductions in soil erosion in much of the nation. The Agricultural Adjustment Administration, forerunner to the Agricultural Stabilization and Conservation Service (now known as the Farm Service Agency), was established to provide payments to farmers who reduced crop acreage. This not only reduced farm surpluses but also helped support prices and reduced the rate of soil erosion. Some of the subsidy programs established in the New Deal continued through the 1990s but are being phased out (see Issue 6.3).

Another significant piece of legislation of this period was the *Taylor Grazing Act* of 1934, which established a system of fees for grazing on federal lands, with limitations on the numbers of animals that could be grazed. This law was a partial response to the widespread accelerated erosion caused by overgrazing. The act also closed most of the public lands to homesteading, effectively ending the large-scale transfer of public lands to private ownership that had begun in 1862. Today, these public lands are administered by the Bureau of Land Management. The Natural Resources Planning Board was another milestone of resource management in the FDR years and was a major step toward establishing long-term comprehensive natural resources planning.

In summary, the FDR years saw important advances in federal resource management and conservation activities. Most of the new programs were conceived as a result of the Depression and were designed to alleviate the problems of the time as well as prevent future mismanagement of resources. The need for careful management of renewable resources, particularly soil and water, was recognized, and the close relation between economic and resource problems became clear.

Phase 4: The Environmental Movement (1962–1980) The years 1940 to 1960 witnessed few new developments in conservation policy. The 1940s were dominated by war, and the economic recovery and ensuing prosperity of the 1950s diverted attention from natural resources issues. However, there was considerable progress in soil, water, and forest conservation, expanding on the achievements made under FDR. The major federal actions of this period were largely in the area of recreational activities, with the expansion of the National Parks and similar recreational areas in response to increased use by the American public.

By the 1960s, attention became focused on the quality of life available to Americans, and natural resources became more broadly defined. Two significant books published in 1962 and 1963 called attention to this expanded view of natural resources and signaled the beginning of a new era in which environmental pollution was recognized as a major threat to natural resources and the quality of life.

One of these books was *The Quiet Crisis* (1963), by Stewart Udall, Secretary of the Interior under President John F. Kennedy. In this work Udall presented much of the history of natural resource use in the United States, particularly focusing on the destruction of natural environments and wildlife. He called for renewed attention to the human effects on the environment, echoing many of the sentiments of G. P. Marsh 100 years earlier. The other book was Rachel Carson's *Silent Spring* (1962), which described the effects of pesticides on the ecosystem and predicted drastic environmental consequences of continued pollution.

Throughout the 1960s, a popular movement for pollution control grew, led largely by scientists, such as Barry Commoner and Paul Ehrlich, student activists, and a few government officials such as Stewart Udall. Many influential authors argued that the environment was damaged severely already and that urgent action was needed to restore its health and prevent further damage to both natural and managed ecosystems. A major focus of the movement was the disparity between a limited resource base on "spaceship Earth" and a rapidly growing world population that already faced severe shortages of food and raw materials. A long list of laws was passed in the late 1960s and early 1970s aimed at reducing pollution, preserving wilderness and endangered species, and promoting ecological considerations in resource development. Some of the more important of these were the Wilderness Act of 1964, the Clean Air Act of 1963 and its amendments of 1970 and 1977, the Federal Water Pollution Control Act of 1964 and its amendments of 1972, the Coastal Zone Management Act of 1972, the Endangered Species Act of 1973, and the National Environmental Policy Act of 1970 (NEPA). The laws relating to specific resource problems such as air and water pollution were the most important in terms of improving environmental quality, and they are discussed in more detail later in this and other chapters.

The National Environmental Policy Act (NEPA) represents the first comprehensive statement of United States environmental policy and is illustrative of the character of this phase of American natural resources history. Section 101 of NEPA contains a statement of the federal government's environmental responsibilities. These are to

1. fulfill the responsibilities of each generation as trustee of the environment for succeeding generations;
2. assure for all Americans safe, healthful, productive, and aesthetically and culturally pleasing surroundings;
3. attain the widest range of beneficial uses of the environment without degradation, risk to health or safety, or other undesirable and unintended consequences;
4. preserve important historic, cultural, and natural aspects of our national heritage, and maintain, wherever possible, an environment which supports diversity and variety of individual choice;
5. achieve a balance between population and resource use which will permit high standards of living and a wide sharing of life's amenities; and
6. enhance the quality of renewable resources and approach the maximum attainable recycling of depletable resources (CEQ 1980:426–427).

These are lofty goals, but they reflect the idealism of the time as well as the far-reaching concerns of the environmental movement. They emphasize quality of life, preservation, or maintenance rather than exploitation and the concern with a limited and finite resource base supporting a rapidly growing population. NEPA also established the requirement for environmental impact statements in order to ensure compliance with its policies.

By the late 1970s a complex set of laws, regulations, and procedures was in place, along with a bureaucracy to administer them. The mass of environmental legislation generated in the preceding decade was being translated into everyday action, and the energy crises of the mid-1970s re-emphasized the need for resource conservation. Substantial improvements in environmental quality were made, particularly in the areas of air and water pollution. With an upsurge of public concern about the effects of pollution and toxic chemicals on health, the Toxic Substances Control Act and the Resource Conservation and Recovery Act were signed into law in 1975 and 1976, respectively.

Phase 5: Pragmatism and Risk Reduction (1976–1988) The next phase focused attention on improving human health by controlling toxic substances in the environment. The Comprehensive Environmental Response, Compensation, and Liability Act (Superfund) was signed into law in 1980. Its purpose was to reduce the toxicity, mobility, and volume of hazardous wastes and to clean up existing hazardous waste sites. The Superfund Amendments and Reauthorization Act (SARA), passed in 1986, further clarified the goals of the Superfund program to enhance the long-term prevention of health effects through waste reduction and better treatment and incineration of wastes. Land disposal of hazardous wastes was not considered a viable policy option. Risk reduction as environmental policy placed the Environmental Protection Agency in the leading role. However, scientific uncertainty regarding the nature of toxic risk, coupled with the scale and complexity of abandoned waste sites and over 5000 known toxic chemicals, resulted in very little cleanup and standard setting during this phase. Inadequate funding for enforcement and cleanup, compounded by a lack of agency and governmental priorities, exacerbated the situation.

At the same time, public debate shifted away from the rather abstract issues of ecological stability and environmental quality toward economic problems. With a downturn in the national and world economy, some began to see the costs of improving environmental quality as contributing to economic problems, and others saw environmental initiatives as simply too expensive for the benefits derived. When President Reagan took office in 1981, he rode a tide of political conservatism that turned away from the idealism of the 1960s and focused more on stimulating economic development. Public lands policy shifted from federal management and conservation to state or private control of resources and exploitation to improve supplies of raw materials, especially energy. Federally owned coal, which had not been sold during earlier administrations because of an oversupply of minable coal, was once again made available to the industry. In its rush to divest the federal government of its holdings, the Interior Department sold coal leasing rights in many areas at prices below market value. Pollution abatement efforts by the federal government were reduced in favor of state and local control over these policies. Attention also was turned toward reducing the costs of pollution control to industry. Clearly, resource conservation had entered a new phase that considered the economic aspects of resource decisions along with the ecological goals established in the 1960s and 1970s.

Phase 6: The New Environmental Consciousness: Global and Local Responsibility (1988–2000)

During the late 1980s, a new awareness of global environmental issues reemerged. The spaceship Earth philosophy, which had been popular 20 years earlier, fell out of favor in the intervening years as predicted catastrophes (widespread famines, climate change, and species extinctions) failed to materialize, and environmentalists who voiced these fears were regarded as alarmist or, worse, catastrophists (Gore 1992; McKibben 1989)! But a number of dramatic events in the mid-1980s, including the nuclear accident at Chernobyl and the "discovery" of the Antarctic ozone "hole," made people aware of the global scale of environmental problems. They were no longer problems of the smelly smokestack in town, but a real source of threat to the planet. Environmental protection was fast becoming an international concern, one requiring cooperation between developed and developing nations. A key event in this phase was the United Nations Conference on Environment and Development, discussed later in this chapter.

One of the fundamental tensions facing U.S. leaders in this era, as in many times before, was between isolationism and internationalism. Environmental issues, most notably the debate over global warming and the Kyoto Protocol, were center stage in this debate. George W. Bush campaigned for the presidency in 2000 from a position that was relatively isolationist, arguing that the U.S. should not be involved in mediating disputes in other countries and rejecting the Kyoto Protocol as damaging to the U.S. economy.

Phase 7: Retrenchment (2000–)

The election of President George W. Bush may mark a new phase in the evolution of natural resources policy. Following the September 11, 2001, terrorist attacks, the U.S. called upon many other countries to join in the fight against terrorism and received widespread support. At the same time, many of the countries that supported the U.S. on this issue are in favor of the Kyoto Protocol and resent the fact that the U.S. is the world's largest contributor to global warming, yet has rejected calls to control its emissions. The contrast between U.S. environmental policies and attitudes and those of other regions (especially western Europe) became increasingly stark. It remains to be seen whether economic, political, and military cooperation between the U.S. and other countries will extend to environmental issues in the next few years.

Domestically, natural resource policies are in a period of retrenchment from the gains of the previous decade. The Bush-Cheney agenda for energy and timber resources is to open up more public lands for exploitation, including Alaska (oil development) and the Pacific Northwest (timber harvesting). The balance between public good and private enterprise appears to have shifted more toward private gain, with a subsequent backlash against environmentalists (McCarthy 2001a, b; Goldstein and Cooper 2002).

International Policy

At the international level, the evolution of environmental policy took a different path (Kamieniecki 1993). Prior to World War II, a few international environmental agreements existed, as did a number of international institutions. Most of the early agreements concerned migratory birds and marine life (fur seals, whales). Institution building during this time included the well-known World Meteorological Organization, founded in 1873 (to coordinate weather observations), and the International Council of Scientific Unions, founded in 1919 (to coordinate international science).

The United Nations System, 1944–1972

International environmental policy and institutions got their real start with the founding of the

United Nations in 1945. As this international organization evolved, more specialized programs and subagencies were founded. The International Union for the Conservation of Nature and Natural Resources (IUCN) (1948) included both nation-state and nongovernmental members. Others operated under the auspices of the United Nations, including the International Maritime Organization and the World Health Organization (WHO).

Also established during this period were two important financial institutions—the International Monetary Fund and the World Bank. The World Bank's mission is to provide financial assistance to developing countries in order to help improve economic conditions. Initially, the World Bank was the leading financier of many of the economic development projects that accelerated environmental decline in the developing world, such as large dams. Since 1985, the World Bank has funded environmental restoration projects to minimize resource degradation and pollution of habitats. Finally, the Food and Agriculture Organization (FAO) was established in 1945 to improve nutrition and the standard of living among the world's rural residents. Some of the more important global institutions are listed in Table 3.1.

While this period reflected substantial growth of the United Nations' infrastructure, it also ushered in the era of environmental diplomacy and the negotiation of environmental treaties (Susskind 1994). The enhanced protection of whales was one of the first big success stories for international environmental policy. The 1946 International Convention for the Regulation of Whaling established the International Whaling Commission to preserve and manage these marine mammals, which is still in effect today (see Chapter 9). Similarly, the conservation of transboundary resources and the global commons were of utmost concern and were reflected in a number of important treaties on plant protection (1951), preventing oil pollution of the seas (1954), fishing resources (1958), and pristine Antarctica (1959, 1964). The most significant treaty during this time, however, was the 1963 Nuclear Test Ban treaty, which prohibited the testing of nuclear weapons in the air, under water, or in outer space (CIESIN 1997; Tolba et al. 1972).

The 1972 Stockholm Conference Most of the United Nations' work is done through conferences and the meetings that prepare for them. This

Table 3.1 Global Institutions with Environmental Interests

Institution	Date Established	Mandate
Food and Agriculture Organization (FAO)	1945	Raise nutritional levels and standards of living, improve food production, better conditions of rural populations
General Agreement on Tariffs and Trade (GATT)	1946	Expand international trade and remove trade/tariff barriers
International Atomic Energy Agency (IAEA)	1957	Monitor nuclear programs for peaceful uses
International Maritime Organization (IMO)	1958	Responsible for sea safety, international shipping, and protection of the marine environment from shipping
UN Educational, Scientific, and Cultural Organization (UNESCO)	1945	Promote peace and security through education, science, and culture
World Health Organization (WHO)	1948	Improve the health of the world's population
UN Children's Fund (UNICEF)	1946	Provide assistance for children's health and welfare services
UN Development Program (UNDP)	1966	Assist developing countries in social and economic development programs
UN Environment Program (UNEP)	1972	Coordinate and stimulate environmental action; monitor global environmental trends

Source: Tolba et al. 1992.

is a slow, deliberative process that allows all member nations to participate in the drafting of the policy (Table 3.2). Once the treaties are drafted, they are presented to member states for signature and ratification.

One such conference that many people point to as the turning point in the international environmental arena was the 1972 United Nations Conference on the Human Environment, held in Stockholm. This conference produced divergent opinions among member nations. The industrialized world was concerned about pollution from increasing industrial activity and the need to preserve natural resources, while the developing nations argued that poverty and the inefficiency of resource use caused by underdevelopment were the primary issues. Despite the often rancorous debate, the Stockholm conference was a defining moment for international environmental policy. It led to the establishment of the United Nations Environment Programme (UNEP), which collects and disseminates environmental data. UNEP now plays a leading role in developing and implementing international environmental policy.

In 1987 the United Nations World Commission on Environment and Development published its influential report *Our Common Future* (WCED 1987), linking resource degradation to unsustainable development practices. Ushering in a new emphasis on sustainability, the Bruntland Report (named for the Norwegian Prime Minister Gro Harlem Brundtland, who chaired the Commission) fostered new recognition among the world's countries of the interconnectedness of human development, resource decline, and pollution.

The Earth Summit and the World Summit on Sustainable Development Marking the twentieth anniversary of its Stockholm Conference on the Human Environment (1972), the United Nations held its Earth Summit in Rio de Janeiro in June 1992. The United Nations Conference on Environment and Development (UNCED) was designed to forge a new path in international cooperation on environmental protection and sustainability. The United States sent a large contingent led by the newly elected Vice President, Al Gore. New agreements (on climate change and biodiversity) were signed at the Earth Summit. In addition, a new intergovernmental infrastructure was created to monitor progress on Agenda 21, the series of goals and principles adopted at the conference (Haas et al. 1992; Jordan 1994; Parson et al. 1992) (Issue 3.1). These achievements indicated a new and perhaps unprecedented period of international cooperation to

Table 3.2 Selected Major International Treaties/Conventions

Date	Place	Treaty
1959	Washington, DC	Antarctic Treaty and Convention
1963	Moscow	Nuclear Test Ban Treaty
1971	Ramsar	Convention on Wetlands of International Importance
1972	London	Ocean Dumping Convention
1972	London	Biological Weapons Convention
1973	Washington, DC	Convention on International Trade in Endangered Species (CITES)
1978	London	Protocol on the Prevention of Pollution from Ships (MARPOL)
1979	Bonn	Convention on Conservation of Migratory Species of Wild Animals
1982	Montego Bay	Convention on the Law of the Sea (UNCLOS)
1985	Vienna	Vienna Convention for the Protection of the Ozone Layer
1987	Montreal	Montreal Protocols on Substances that Deplete the Ozone Layer
1989	Basel	Basel Convention on the Control of Transboundary Movements of Hazardous Wastes and Their Disposal
1992	Rio de Janeiro	Framework Convention on Climate Change (FCCC)
1992	Rio de Janeiro	Convention on Biological Diversity
1997	Kyoto	Kyoto Protocol to the Framework Convention on Climate Change

ISSUE 3.1: IN FAIRNESS TO ALL: AGENDA 21 AND ENVIRONMENTAL EQUITY

One of the greatest obstacles in international environmental diplomacy is the sharp divide between industrialized and developing nations. This North/South conflict, as many call it, has led to many divisions among nations, often causing them to fall along the lines of haves and have nots. As part of the Earth Summit, an attempt was made to reduce these sharp economic divisions among nations. Through the application of equity, the Earth Summit tried to convince nations of the need for sustainability—not just in the developing world but among the most industrialized nations as well.

Agenda 21 provides the blueprint for achieving sustainability. But it is much more than that. It also provides a framework for implementing a code of conduct or fairness among nations as they strive to meet the ambitious goals of the Earth Summit.

Equity has many forms. For example, *social equity* refers to the differences in social, economic, and political forces in the consumption of resources and the ability to degrade environments. The transboundary shipment of hazardous wastes from industrialized countries in Europe to poorer nations in Africa is one example of social inequity; closer to home, dumping your garbage in another poor community is another example. *Generational equity* is perhaps the most important form of *environmental equity*. Generational equity embodies the concept of fairness to future generations. By this we mean that our generation has a responsibility to our children to leave the environment in the same or better condition than what was left to us. This type of generational stewardship is most often seen in the preservation of wild areas (Biosphere Reserves, National Parks) and more recently in biodiversity treaties. The last form of equity is *procedural*, which applies to the ways in which laws and treaties are applied in a fair way.

Agenda 21 has 27 specific principles, and close to half of these directly address some form of equity. For example, Principle 3 states the right of all people to a standard of development that meets the environmental needs of present and future generations (generational equity). Principle 22 recognizes the rights of indigenous people and communities, while Principle 20 encourages the full participation of women in environmental management (social equity). Finally, Principle 11 proposes that environmental standards reflect the context within which they apply (procedural) and argues that universal codes might affect some nations more than others.

It remains to be seen whether these equity goals will be achieved, for it will require not only a rethinking of international relations but a potential redistribution of wealth as well. In other words, Earth's resources must be shared for the benefit of all, not just a select few (United Nations 1992).

---

facilitate the sustainable use of the earth's natural resources. It remains to be seen how long this new awareness will last or how important it will be in changing policies and behavior. It is clear, however, that unprecedented popular support has developed for the establishment of an international agency to solve environmental problems (Dunlap et al. 1992). In many nations, residents are keenly aware of the linkage between environmental problems and their own health and welfare.

The years following Rio were a mix of successes and failures (Flavin 1997). While the link between environment and development has been recognized, the goal of an environmentally sustainable global economy has not been achieved. Carbon emissions are climbing, biodiversity is declining, world population is increasing, and millions more people are living in poverty. Progress on the ambitious Agenda 21 has been slower than many expected. The United Nations World Summit on Sustainable Development (WSSD), held in Johannesburg, South Africa, in 2002, re-energized the commitment to improving the human condition and environmental protection. Five areas were targeted: water and sanitation, energy, agricultural productivity, biodiversity and ecosystem management, and health (Annan 2002). The redirection of economic growth to emcompass environmental sustainability, a tangible commitment from developed nations, and an understanding of the root causes of environmental degradation will be required to achieve the goals established at the

World Summit on Sustainable Development (Speth 2002). Despite widespread pessimism on actually achieving a sustainable future, grassroots activism has increased through which people are implementing practical plans for solving local environmental problems. This "grass-roots" spirit throughout the world is giving new meaning to the old slogan "Think globally, act locally."

Current Natural Resource Policy

The history of policy development for natural resource use reviewed in the previous section shows that at one time or another government decision-making motivated many important goals, many of which are embodied in present policy. These goals can be grouped into four general categories: to promote economic development; to conserve resources for the future; to protect public health; and to preserve important natural features.

Clearly, the most frequent motivation for government actions with respect to natural resources has been to promote economic development. This began with the land divestitures of the eighteenth and nineteenth centuries and continues today in the management of our national forests, offshore oil resources, rivers, and grazing land. The construction of major dams and reservoirs on rivers for hydroelectric power generation, irrigation, or flood control addresses this goal. Economic development was the motivation when the United States increased mineral lease sales in the 1980s, with development rights being sold at prices below market value to stimulate production. It is also the primary justification for one of the basic tenets of public land management policy in the United States, that of multiple use. The concept of multiple use was incorporated into the U.S. Multiple Use and Sustained Yield Act of 1960 and was restated in NEPA, but it originated much earlier.

In encouraging economic development, the companion to the multiple use concept is the idea of *sustained yield*, which aids in achieving the second goal, to conserve resources for future generations. This is the fundamental principle of renewable resource management established by Pinchot in forest management, but it applies equally to the mission of the U.S. Natural Resources Conservation Service, the U.S. Fish and Wildlife Service, and indeed every agency that manages natural resources. More recently, sustainability has been the approach used to reconcile economic development goals within environmental limits (Chapter 15).

The third goal, to promote public health, is the basis for most pollution control legislation. Many of the early American laws regulating potential health hazards in the environment were enacted at the state level, and major federal actions in this area did not appear until the late nineteenth and early twentieth centuries. Today most of the water and air quality standards established by governments are based on health criteria and risk assessments. At the international level, environmental health is a key objective of the World Health Organization. Treaties limiting emissions of ozone-destroying chemicals are motivated primarily by health concerns.

The fourth major goal of natural resource policy is to preserve significant natural features that are valuable for aesthetic or scientific reasons, if not for economic ones. This is the aim of the extensive legislation enacted regarding wilderness preservation and protection of endangered species, and it is the principal mission of the National Park Service and the United Nations' Biosphere Reserves (Chapter 8). This goal also forms the basis of some water-quality criteria, and it is considered one of the uses of public lands in multiple-use planning.

Many natural resource policies combine these different goals. A U.S. example is water pollution control, which not only protects the public health but also provides recreational, aesthetic, and economic benefits to fisheries. All the agencies involved in managing resources address these multiple goals in devising their plans, and this combination of purposes also plays an important role in creating political coalitions to enact new laws. Together they form the basis for resource management decisions.

After decades of polarization and politicization, environmental issues now have broad political support in the United States and most other industrial nations (Paarlberg 1996). Conservation and environmental issues are part of the national and international agenda in such diverse areas as national security, international trade, and population policies. International concerns over acid rain, global warming, and tropical deforestation, to name but a few, fueled the international debate on how to best manage these resources. Domestic

Figure 3.5 Celebrating Earth Day. Despite three decades of environmental concerns and activism, we still have a long way to go.

concerns such as clean air and toxic emissions still provide a focus for the annual American Earth Day observances (April 22), a gentle reminder not of how far we've progressed but of how much farther we need to go to protect the environment (Fig. 3.5).

How Decisions Are Made

Resource Decision-Making in the United States

Several different groups are involved in any decision over the use of natural resources: resource managers, social agents, and interest groups. Membership in these groups is not constant, for any individual may shift from one role to another as the decision-making process unfolds (Fig. 3.6).

A resource manager is the individual or agency that is in immediate contact with the resource and has a direct stake in how that resource is used or misused. Examples of resource managers include an individual farmer concerned with soil erosion, a forest ranger charged with managing a particular national forest, or the Secretary of the Interior, who manages the resources under his or her jurisdiction—parks, public lands, and so on.

Resource managers are subject to outside influences or social agents, also called *stakeholders*. These range from the forest rangers' superiors in the Department of Agriculture to the United States President, who oversees the Secretary of the Interior. These social agents provide technical expertise and direction to the individual managers in the field. The goals, objectives, and responsibilities of the social agents are broader than those of the resource manager. Social agents are thus influenced by interest groups who have a stake in how a resource is eventually used. Special interests range from timber companies seeking access to a national forest to supermarket chains buying farm produce to environmental groups seeking to preserve a piece of nature.

Conflicts between participants inevitably lead to disagreement over management policies. One example is whether to manage forests for water

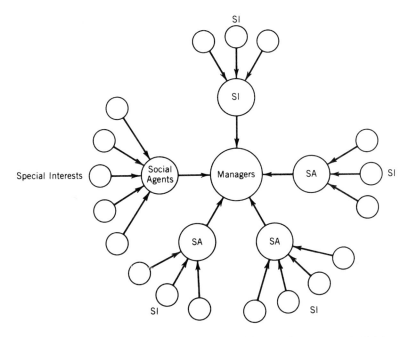

Figure 3.6 Participants in the resource-use decision-making process. Individuals may shift roles, depending on the resource issue under consideration, becoming a manager in one instance, a social agent in another, or even a special interest in yet a third.

yield, timber harvesting, or species habitat. These disagreements are further complicated by goal-oriented and mitigation-oriented management strategies. Some governmental bodies are charged with the management of a resource (U.S. Fish and Wildlife Service or the National Forest Service), that is, they are goal-directed; others are charged with regulation and protection of environmental quality (U.S. EPA), or mitigation-oriented. Often the two conflict, even when they are in the same agency (Table 3.3). Decision-making then becomes very difficult and usually involves conflict, cooperation, and compromise between the resource manager, social agents, and interest groups. In addition, conflicting environmental ideologies complicate the matter further (Pepper 1996).

Ultimately, federal decisions are made by Congress and then implemented by the appropriate agency such as the Environmental Protection Agency or the Department of the Interior (Issue 3.2). Passing an environmental bill in Congress is fraught with partisanship and compromise, and often the finished bill (which is based on the compromise between the House and Senate versions) is not as strict or as comprehensive as environmentalists would like.

International Environmental Decision-Making

As mentioned previously, environmental decision-making at the international level is conducted through conferences and conventions. However, some regional organizations also have environmental policy interests. Often these organizations also serve as economic or strategic alliances, such as the European Union (EU) and the Organization of American States (OAS). The most successful of these regional environmental efforts is the United Nations' Regional Seas Program. Member nations that border many of these important seas, such as the Caribbean or Mediterranean, participate in multinational treaties designed to protect the sea from the overexploitation of its biotic resources and to reduce land-based pollution originating from the bordering nations. Relatively few conflicts arise inasmuch as every regional nation has a vested interest in protecting this shared common property resource. In many instances, regional alliances are more effective than global treaties in

ISSUE 3.2: POLITICS AND THE ARCTIC NATIONAL WILDLIFE REFUGE (ANWR)

At the national level, the fundamental decision-making process in the U.S. is the congressional legislative process. Laws are initially drafted by individual legislators or, more commonly, groups of legislators and their staffs. They are discussed before the various congressional committees that have responsibility for particular areas of government policy. These committees modify the proposals and, if approved, forward them to the full Senate or House of Representatives for amendment and approval. Usually after much debate and revision, they are passed by both houses of Congress and then forwarded to the President for signing into law. At each stage of the process, resource managers, social agents, and interest groups make their positions known, and their opinions are either incorporated or ignored in the proposed legislation. Each law is unique, and each is subjected to different forces depending on the course the process takes from initial proposal to final enactment. An example of the politics of environmental legislation is the Alaska National Interest Lands Conservation Act (ANILCA), passed in 1980. This act, better known as the Alaska Lands Bill, is a good example because the battles over it were particularly intense and involved many different actors. Few environmental laws in recent decades have been as controversial as this one was and continues to be. It also provides the historical context for understanding the ongoing debates over oil exploration in the Arctic National Wildlife Refuge (ANWR).

Alaska has an area of about 375 million acres (152 million ha), and virtually all of this land was federally owned when Alaska became a state in 1959. The terms of the Statehood Act, however, required that 104 million acres (42 million ha) eventually be turned over to the state. To preserve America's last unspoiled wilderness area, President Eisenhower created the ANWR in 1960, preserving more than 9 million acres in northern Alaska. In 1971 the Alaska Native Claims Act was passed, which paved the way for construction of the Alaska pipeline by providing a settlement of the land claims of the native peoples. This act called for 44 million acres (18 million ha) to be turned over to the natives. But before these lands could be transferred, the federal government had to decide, by the end of 1978, which lands it would retain in federal ownership, and of these which would be preserved as wilderness and which would be open to development. The Alaska lands issue was thus a classic battle of preservation versus development, and the stakes were high: spectacular and unique natural areas containing potentially very valuable mineral and timber resources. The battle took much longer than was expected, and to prevent development of some areas while Congress debated, President Carter proclaimed about 44 million acres (18 million ha) as national monuments and doubled the size of the ANWR.

The House of Representatives was the first to take up the Alaska lands issue, in 1977. Morris Udall, a leading environmentalist in Congress, introduced a bill that would place nearly 170 million acres (69 million ha) in the "four systems": the national parks, wildlife refuges, forests, and wild and scenic rivers. In contrast, a proposal introduced by Alaska Representative Don Young would place only 25 million acres (10 million ha) in the four systems and another 57 million acres (23 million ha) in a joint state-federal management area to be managed for multiple uses. The battle lines were drawn, and the special interests went to work. One of the most effective of these was the Alaska Coalition, a group of conservation organizations that banded together to press for a preservation-oriented bill. On the other side were the state of Alaska, which wanted as much development potential as possible, and industry groups such as the American Mining Congress and the Western Oil and Gas Association. In the numerous negotiations the various proposals were modified, and eventually they were narrowed to two: the Udall bill and another that would have preserved much less land. Public sentiment for preservation was mobilized by the Alaska Coalition, which produced and distributed literature and films depicting the spectacular wilderness. In the end, this sentiment was very important, and the Udall bill passed by a wide margin.

Once the House had passed its bill, the Senate began deliberations on its own versions of the bill. In general the Senate was less conservation-minded than the House, although there was a powerful group of senators who supported a bill very similar to the House-passed bill. But the bill that finally emerged from committee in the Senate was rather different, including substantial reductions in areas designated as wilderness and more access for development in other areas. This was far from acceptable to the two Alaskan Senators, Ted Stevens and Mike Gravel, who vowed to filibuster to prevent passage of a bill that would not meet the desires of the state. The debate became heated, and at one point the Senate went into closed session after a shouting match between Stevens and Colorado Senator Gary Hart. Eventually, Stevens and Gravel's attempts at a filibuster failed as the Senate voted to cut off debate. The Senate version finally passed and placed 104 million acres (42 million ha) in the four systems, substantially less than the 127 million acres (51 million ha) in the House bill. Wilderness designation was given to 67 million acres (27 million ha) in the House bill but only 57 million acres (23 million ha) in the Senate version, and mineral exploration was permitted in some wildlife refuges.

After the Senate passed its bill, the House again took up the issue under threats of a filibuster if it failed to agree to the Senate version. Finally, it did accept the Senate version of the bill, and in late 1980 the bill was signed by President Carter. It was most certainly a compromise but not an entirely happy one. Morris Udall said that he got most of what he wanted but that some provisions were still unacceptable to him, which he hoped would be modified in the next Congress. Alaska Representative Don Young was pleased that the bill did permit more mineral exploration than the original House bill, but he, too, said he wanted to change things in the next Congress to permit even more exploration and development.

Throughout the debate on the Alaska Lands Bill, the changing strengths and fortunes of the actors could be seen. Conservationists were buoyed by support from the Carter administration, particularly Interior Secretary Cecil Andrus. In the midst of the debate, the Alaska legislature was repealing income taxes and rebating millions of dollars to its citizens as a result of accumulating oil revenues, actions that earned no extra sympathy for their demands for resource development. President Carter and most environmental groups hailed the bill as his administration's most significant environmental achievement. Mike Gravel was defeated in a primary election during the height of the debate, and one of the major issues was his inability to force the Senate to recognize Alaska's interests. At the same time, industry won some crucial battles, and most of the valuable mineral and timber resources are now open to development.

The congressional process is one of compromise, and the most effective means for compromise is to make sure everyone gets something they want. Each actor and each interest group is given at least token recognition of its interests in an attempt to gain support for the final outcome. The particular characteristics of that outcome—which interests get more of what they want and which get less—are dependent on the strengths of their various power bases. In the case of the Alaska Lands Bill, the environmentalists had much more popular support than their opponents, and the result was a bill that is generally regarded as a significant achievement for conservation.

Two decades later, the controversy continues, only now it revolves around oil development in the Arctic National Wildlife Refuge. During the Clinton administration, Congress wanted to open up 1.5 million acres along the coastal plain for oil drilling and development. The bill was passed but vetoed by President Clinton. The Bush-Cheney administration again is pushing to open this tract of land, called Area 1002, to commercial oil exploration and development as part of their energy policy. There is considerable scientific debate on both the amount of oil in ANWR (4.2–11.8 billion barrels, according to most estimates), its reduction on U.S. foreign dependency (less than 2%), and its effect on caribou herds and other wildlife (under debate). American public opinion is against drilling, but the oil industry believes that the reserves will help reduce foreign oil bills (and, not surprisingly, put money in the pockets of American oil companies). In 2002, Congress voted to block oil and gas drilling in ANWR, thus defeating a large component of the Bush-Cheney energy plan. The debate, which began almost a half century ago, will no doubt continue (Gibbs 2001; McCarthy 2001a; Kerr 2002; Kaiser 2002).

curtailing transboundary pollutants and the degradation of common property resources.

THE DECISION-MAKING PROCESS

The legislative process and diplomacy are not the only ways environmental law is made. For example, over 50 U.S. federal entities are involved in natural resources policy and decision-making (Table 3.3). Many times overlapping jurisdictions, different goals, and antagonistic staffs result in interagency squabbling over the management of specific resources. There are also intra-agency conflicts between temporary political appointees who head the agencies and their professional civil servant staffs. How these agencies go about making decisions and implementing policy is crucial to our understanding of natural resources management, especially in the United States.

Nongovernmental organizations (NGOs) are not-for-profit entities that operate at national and international levels. Increasingly, NGOs are becoming a considerable force in environmental decision-making. In the environmental area, their primary purpose is to improve the human condition and manage resources in a sustainable way. They apply pressure to government to act responsibly. In this way, they act like both special interests and social agents. Most NGOs were started in the 1980s, but a few were operating much earlier. Examples of NGOs include the Worldwide Fund for Nature, Catholic Relief Services, Rainforest Action Network, Greenpeace, and the Sierra Club.

Organizations

There are few differences between how organizations make decisions and how you do. Their decisions differ from yours only in scale and complexity. As we saw in the previous chapter, private industry makes decisions based on the "bottom line." Governmental and other public organizations are less motivated by profit margins. Instead, their decisions usually are made in response to human needs and require government efforts for implementation. Government agencies are influ-

Table 3.3 Federal Agencies with Major Responsibility for Environmental Policy or Management

Department of the Interior	Department of Defense
National Park Service	Army Corps of Engineers
Bureau of Land Management	Departments of Army, Navy, and Air Force
Fish and Wildlife Service	Department of Transportation
Biological Survey	Department of Labor
Geological Survey	Occupational Safety and Health Administration
Minerals Management Service	Department of Health and Human Services
Bureau of Reclamation	Food and Drug Administration
Surface Mining Reclamation and Enforcement	Environmental Protection Agency
Department of Agriculture	Executive Office of the President
Forest Service	Council on Environmental Quality
Natural Resources Conservation Service	Federal Emergency Management Agency
Farm Service Agency	Other Independent Agencies
Department of Commerce	Tennessee Valley Authority
National Oceanographic and Atmospheric Administration	Bonneville Power Commission
	Water Resources Council
United States Coast Guard	National Science Foundation/National Research Council/National Academy of Science
Department of Energy	
Federal Energy Regulatory Commission	Nuclear Regulatory Commission
National Center for Appropriate Technology	Great Lakes Basin Commission
Nuclear Waste Policy Act Project Office	

enced by the opinions of others, such as lobby groups or political action committees. In addition, these decisions can be influenced by the motivations and political philosophy that underlie the decision-makers' choices. Finally, decisions often are avoided because they are painful in terms of conflict between the governmental entity and the other groups or individuals. This results in nondecision, which in fact becomes a form of decision-making. Non-decision-making is more pervasive in the United States than most people realize (O'Riordan 1981).

In theory, responsible, objective decisions are possible, but in practice many factors bias both decision-makers and their conclusions, resulting in less-than-perfect decisions. One of these factors is the constraint imposed by organizational tradition—we have always done things this way, and there is a tradition to maintain. Moreover, there are constraints imposed by bureaucratic procedures, such as the endless arguments between regional and home offices or between divisions of the same organization. Conflicts between regional offices of the EPA and Washington headquarters are well known. Some constraints on decisions also are imposed by the demands of the executive role. A decision-maker may feel that she cannot show friendliness to subordinates, for it might be construed as a sign of weakness and would hamper negotiations with a lobbyist or other interest groups.

Perhaps one of the most important constraints is the lack of objective standards for assessing alternative outcomes, which can force the decision-maker to be sympathetic to social and political pressures and special interests. Decision-makers often rely on stereotypes, such as believing that the information of uneducated people is always unreliable, resulting in biased decisions. Bias can be introduced by an individual decision-maker's cognition of his or her role and intuitive assessment of the likelihood of the success or failure of the chosen course of action. Last, decisions often are made with insufficient or imperfect information, particularly in the case of environmental contaminants.

Strategies

Given that we live in an imperfect world with many complexities, it is surprising that we are able to make sound environmental decisions at all. Decision-making in natural resources management is divided into three general categories: satisficing, incrementalism, and crisis management. *Satisficing* is the consideration of two policy alternatives at a time, which are examined sequentially and compared to one another. The best choice is then selected from these two. The goal of the satisficing approach is to look for the course of action or alternative that is just good enough and meets a minimal set of requirements. This type of approach is cost-effective because the full range of alternatives is not researched, which would be too costly in time and money; thus, the collective resources of the decision-maker or agency are used more efficiently. A negative aspect of this strategy is the limited range of alternatives from which to select the best choice. Satisficing is an appealing approach to managers because it is simple, and it is used in many other areas besides resource-use decision-making.

Incrementalism is used when the problem or resource issue is not clearly defined or when there are conflicting goals, values, or objectives. Incremental decisions are made by "muddling through" as they come across an administrator's desk. Administrators may not know what is wanted, but they do know what should be avoided. As a result, incrementalism is not used to set broad policy guidelines, as is the satisficing strategy, but rather to alleviate the shortcomings in the present policy in its day-to-day administration. This approach is regularly used to cope with the bureaucratic politics that often result in compromising and shifting coalitions. Incrementally made decisions are often disjointed and seemingly contradictory and reflect minute changes in policy.

The third strategy, *crisis management*, is the approach most commonly used in government today. Crisis management is the response to an issue once it becomes a critical problem. It begins with a seat-of-the-pants planning effort to come to grips with the looming impact of the problem, and policy is then determined on a piecemeal basis to deal with the immediate problem at hand. There may be little consideration of long-term effects in the rush to get something done quickly. For example, when it was realized that certain industries contributed to local air pollution, regulations were put into effect in the 1970s that required higher smokestacks so that the pollutants would not afflict nearby communities. In the long run, this decision led to larger negative impacts, as these airborne pollutants contributed to the acid rain problem, leading to major deforestation and water pollution problems hundreds of miles from the smokestacks. Thus, with crisis management choices, there is no time for a discussion of larger policy questions. All decisions must be made immediately and implemented as quickly as possible, with very little time to discuss all the alternatives or the implications of new rules and regulations. Unfortunately, many of our environmental regulatory agencies routinely operate in this fashion.

You might think that the cumulative effect of all these imperfections in the decision-making process prevents good decisions from ever being made. Some might agree. In fact, with most decisions of this kind there is a wide range of opinion on how problems should be approached, and in most cases only a portion of the population is satisfied with the result. That, of course, is the nature of the political process. But the important thing to recognize is that the push and pull of politics go on at many levels of decision-making—not just at election time. The administrator and enforcer are just as susceptible to the forces that sway decisions as is the legislator up for re-election.

The Role of Public Interest

The public interest can be defined in many ways because there are many different "publics." At the international level, for example, who constitutes "the public" and what is the public interest? At the

Table 3.4 Selected Public Interest and Environmental Groups

Advocacy	Litigation
Clean Water Action Project	Environmental Defense
Common Cause	Natural Resources Defense Council
Cousteau Society	The Public Citizen
Defenders of Wildlife	Public Interest Research Group
Ducks Unlimited	Research/Education
Earth First!	Center for Marine Conservation
Earth Island Institute	Center for Science in the Public Interest
Environmental Action Foundation	Center for the Study of Responsive Law
Friends of the Earth	Center for Research on Endangered Plastic Pink Flamingos
Fund for Animals	
Fund for Renewable Energy and the Environment	Center for Health, Environment and Justice
Greenpeace	Conservation Foundation
Izaak Walton League	Environmental Law Institute
League of Conservation Voters	Institute for Local Self Reliance
National Audubon Society	Population Reference Bureau
National Wildlife Federation	Resources for the Future
The Nature Conservancy	World Resources Institute
Physicians for Social Responsibility	WorldWatch
Planned Parenthood	Industry
Rainforest Action Network	American Petroleum Institute
Sierra Club	American Water Well Association
Trout Unlimited	Atomic Industrial Forum
Union of Concerned Scientists	Chemical Manufacturers Association
The Wilderness Society	Edison Electric Institute
World Wildlife Fund	Keep America Beautiful
Zero Population Growth	National Solid Waste Management Association

national level, perhaps these are easier questions to address, largely because of our penchant for opinion polls.

There are many types of public interest groups, each with a particular cause and management style. Some groups are politically or socially conservative and work with lobbyists. Other groups are more radical and often take their message to the forest, oceans, or streets, wherever they can command media attention. Table 3.4 lists some of the different types of public interest environmental groups that are active today. This is not an exhaustive list but is provided simply to illustrate the diversity and abundance of different "public interests." It is important to understand that environmental groups make decisions just like any other organization and are subject to the same pressures and stresses. Just because it is the Sierra Club or the Natural Resources Defense Council does not mean that the decisions are perfect and unbiased. These groups operate just like any other organization or governmental agency.

Public opinion regarding environmental issues has always been strong, yet often this support does not translate into electoral power. In the past decade, in particular, the public has endorsed stronger environmental laws even if improved environmental quality means higher prices and costs (Dunlap 1987; Dunlap and Mertig 1992). Polls confirm the notion that public support for environmental protection remains high not only in the United States but in other countries as well (Dunlap et al. 1993).

Despite this overwhelming support, environmental issues still do not decide national elections. At the national level, this concern is often tapped by environmental activist groups that are able to mobilize public support and increase their membership and ultimate lobbying positions. During the early Reagan years, membership in many environmental organizations rose markedly, largely in response to Reagan's anti-environmental policies (Gottlieb 1994). The environment remained a hot political topic in the 1990s with the elections of President Clinton and Vice President Gore (a strong environmentalist). Even Congress was supportive until the environmental backlash of 1994. The attempt to reduce environmental regulations as part of Newt Gingrich's Contract with America was severely denounced by the American people, who have maintained their pro-environment opinion. As a result, the Republican Congress backed off and "rediscovered" environmental issues by the end of that congressional session. From the mid-1990s onward, however, environmental problems took a back seat to more pressing national problems such as the economy, health care, and terrorism (Kaufman 1994).

THE "NEW" ENVIRONMENTAL POLITICS

Two additional factors currently influence environmental decision-making. These include the increased role of private industry as an environmental innovator, and the shift from top-down policies to more localized innovations.

In many countries, especially the United States, private businesses are the major environmental policy innovators. Take, for example, the ISO 14000 program that is a standardized compliance auditing system. ISO 14000 sets up a standardized reporting system so that companies can see whether their customers or vendors comply with a given country's environmental laws. ISO 14000 also sets up periodic third-party audits of vendors to monitor such compliance. For example, Levi Strauss, the blue jeans manufacturer, buys denim from a variety of vendors, some in the United States, others in Singapore, Taiwan, and so forth. If one of these vendors is contaminating the environment in its stone washing (bleaching) process, for example, Levi Strauss (the parent organization), not the local vendor, is sued. In an effort to control product and environmental liability, then, many private industries have agreed to ISO 14000. While motivated by reducing their own liability, private industries actually have helped local manufacturers comply with local environmental regulations, or else the big companies simply take their business elsewhere. Obviously, this provides a strong incentive to follow the letter of the environmental law, be it here in the United States or abroad. The ISO14000 program and its forerunner, ISO 9000, provide the framework for industry to integrate aspects of process design and management into everyday business practice.

The second factor is a shift from the top-down legislative approach, in which Congress passes national laws and states adhere to them, to a more bottom-up one in which individual states set policies that then are nationalized. Compromise politics is especially important when governmental power is divided between two political parties. Nonetheless, one of the most significant legacies of the Reagan administration was to shift environmental responsibility from the federal government to state and local governments. The result was a fundamental transformation in how laws are made in the United States. Beset by inactivity at the national level, state legislators devised their own laws to tackle pollution issues within their state but not exclusive to them. For example, in 1987 legislators in Sacramento, California, insisted that refiners change the mix of ingredients in gasoline to inhibit evaporation since these vapor fumes were a major source of hydrocarbons in Southern California. New York, New Jersey, and the six states in New England followed suit a year later. By 1989 the EPA announced a major national program for controlling evaporation of unburned fuel. California is clearly the policy innovator for clean air legislation, for it has the worst air quality in the nation and hence the greatest need to clean it up. Another innovative state is New Jersey, which experienced the garbage crisis before anyone else. In response to the mounds of solid waste generated daily by its residents, the state developed a comprehensive recycling master plan that mandates the recycling of 25 percent of the municipal solid waste stream. This program has become a model for the rest of the nation. California continued this innovator role in 2002, passing a law intended to limit carbon dioxide emissions from automobiles.

Why are these interstate problems being solved at the state rather than the national level? Some point to the fact that many states have such severe pollution problems (e.g., smog in Los Angeles) that they cannot wait for Washington to act. As a consequence, if California acts and doesn't fall flat on its face, other states may adopt California's programs. In addition, aggressive national lobbyists don't often frequent state houses, which means that the likelihood of passing a controversial piece of legislation (meaning one that industry does not favor) is greater at the state than the national level. Finally, some argue that state legislators are more in tune with political change and thus more responsive to local environmental concerns. This immediately translates voter support for environmental concerns into political action. The trend at the national level is less clear.

As we have tried to convey in this chapter, natural resource policy is determined by choices and compromise. There is no right or wrong policy, nor is there a good or bad one. To render these subjective evaluations depends on your own perspective and role in the decision-making process.

REFERENCES AND ADDITIONAL READING

Annan, K. A. 2002. Toward a sustainable future. *Environment* 44(7):10–15.

Carson, R. L. 1962. *Silent Spring*. Boston: Houghton Mifflin.

Commoner, B. 1966. *Science and Survival*. New York: Viking Press.

Consortium on International Earth Science Information Network (CIESIN). 1997. Web page for their international treaties (ENTRI): http://www.ciesin.org/entri.

Council on Environmental Quality (CEQ). 1980. *Environmental Quality: The Eleventh Annual Report*. Washington, D.C.: U.S. Government Printing Office.

_____. 1986. *Environmental Quality: The Seventeenth Annual Report*. Washington, D.C.: U.S. Government Printing Office.

Cronon, W. 1984. *Changes in the Land: Indians, Colonists and the Ecology of New England*. New York: Hill & Wang.

Diamond, J. 1999. *Guns, Germs, and Steel: The Fates of Human Societies*. New York: Norton.

Dilsaver, L. M. and C. E. Colten, eds. 1992. *The American Environment*. Lanham, MD: Rowman & Littlefield.

Dunlap, R. 1987. Polls, pollution, and politics revisited: Public opinion on the environment in the Reagan era. *Environment* 29(6):6–11, 32–37.

_____ and A. G. Mertig, eds. 1992. *American Environmentalism: The U.S. Environmental Movement 1970–1990*. New York: Taylor & Francis.

_____, G. H. Gallup, Jr., and A. M. Gallup, 1993. Of global concern: Results of the health of the planet survey. *Environment* 35(9):6–15, 33–39.

Easterbrook, G. 1995. *A Moment on the Earth: The Coming Age of Environmental Optimism*. New York: Viking Press.

_____. 2000. Green surprise? How Bush or Gore, as President, might pull a "Nixon goes to China" on environmental issues. *The Atlantic Monthly*, September:17–24.

Flavin, C. 1997. The legacy of Rio. In L. R. Brown et al., eds., *State of the World 1987*. Washington, D.C.: Worldwatch Institute, pp. 3–22.

Gibbs, W. W. 2001. The arctic oil and wildlife refuge. *Scientific American* 284(5):63–69.

Goldstein, A. and M. Cooper. 2002. How green is the White House? *Time*, April 29, 2002:30–33.

Gore, A. 1992. *Earth in the Balance: Ecology and the Human Spirit.* Boston: Houghton Mifflin.

Gottlieb, R. 1994. *Forcing the Spring.* Washington, D.C.: Island Press.

Goudie, A. 2000. *The Human Impact on the Natural Environment.* Cambridge, MA: MIT Press.

Haas, P. M., M. A. Levy, and E. A. Parson. 1992. How should we judge UNCED's success? *Environment* 34(8):6–11, 26–33.

Harrington, W., R. D. Morgenstern, and P. Nelson. 1999. Predicting the costs of environmental regulations. *Environment* 41(7).10–11, 40–44.

Hempel, L. C. 1996. *Environmental Governance: The Global Challenge.* Washington, D.C.: Island Press.

Hughes, J. D. 2001. *An Environmental History of the World: Humankind's Changing Role in the Community of Life.* New York: Routledge.

Huth, H. 1957. *Nature and the American: Three Centuries of Changing Attitudes.* Berkeley: University of California Press.

Jordan, A. 1994. Paying the incremental costs of global environmental protection: The evolving role of GEF. *Environment* 36(6):12–20, 31–36.

Kaiser, J. 2002. Caribou study fuels debate on drilling in Arctic Refuge. *Science* 296:444–445.

Kamieniecki, S., ed. 1993. *Environmental Politics in the International Arena: Movements, Parties, Organizations, and Policy.* Albany, NY: SUNY Press.

Kaufman, W. 1994. *No Turning Back: Dismantling the Fantasies of Environmental Thinking.* New York: Basic Books.

Kennedy, D. and R. W. Sant. 2000. A global environmental agenda for the United States: Issues for the new U.S. administration. *Environment* 42(10):20–30.

Kerr, R. A. 2002. A modest drop in a big bucket. *Science* 296:444.

Kline, B. 2000. *First Along the River.* 2nd Ed. San Francisco: Acada Books.

Leopold, A. 1949. *A Sand County Almanac and Sketches Here and There.* New York: Oxford University Press.

Lomborg, J. 2001. *The Skeptical Environmentalist.* Cambridge: Cambridge University Press.

McCarthy, T. 2001a. War over Arctic oil. *Time*, February 19:24–29.

———. 2001b. High noon in the west. *Time*, July 16:18–32.

McKibben, B. 1989. *The End of Nature.* New York: Random House.

Marsh, G. P. 1864. *Man and Nature, or Physical Geography as Modified by Human Action.* New York: Scribner.

Miller, C. 2001. *Gifford Pinchot and the Making of Modern Environmentalism.* Washington, D.C.: Island Press.

Murphy, A. 1994. International law and the sovereign state: Challenges to the status quo. In G. J. Demko and W. B. Wood, eds. *Reordering the World: Geopolitical Perspectives on the 21st Century.* Boulder, CO: Westview Press, pp. 209–224.

Najam, A., J. M. Poling, N. Yamagishi, D. G. Straub, J. Sarno, S. M. DeRitter, and E. M. Kim. 2002. From Rio to Johannesburg: progress and prospects. *Environment* 44(7):26–38.

Nash, R. 1982. *Wilderness and the American Mind.* Rev. ed. New Haven: Yale University Press.

O'Riordan, T. 1981. *Environmentalism.* London: Pion.

Paarlberg, R. L. 1996. A domestic dispute: Clinton, Congress, and international environmental policy. *Environment* 38(8):16–20, 28–33.

Parson, E. A., P. M. Haas, and M. A. Levy. 1992. A summary of the major documents signed at the Earth Summit and Global Forum. *Environment* 34(8):12–15, 34–36.

Pepper, D. 1996. *Modern Environmentalism.* New York: Routledge.

Porter, G. and J. W. Brown. 1996. *Global Environmental Politics.* 2nd Ed. Boulder, CO: Westview Press.

Pyne, S. J. 1997. *World Fire: The Culture of Fire on Earth.* Seattle: University of Washington Press.

Rennie, J. 2002. Misleading math about the earth. *Scientific American* 286(1):61–71.

Scheberle, D. 1998. Partners in policymaking: forging effective federal-state relations. *Environment* 40(10):10–20, 28–30.

Shabecoff, P. 2000. *Earth Rising: American Environmentalism in the 21st Century.* Washington, D.C.: Island Press.

Simmons, I. G. 1996. *Changing the Face of the Earth: Culture, Environment, and History.* Oxford: Basil Blackwell.

Speth, J. G. 2002. A new green regime: attacking the root causes of global environmental deterioration. *Environment* 44(7):16–25.

Susskind, L. E. 1994. *Environmental Diplomacy: Negotiating More Effective Global Agreements.* New York: Oxford University Press.

Swem, T. and R. Cahn. 1983. The politics of parks in Alaska. *Ambio* 12:14–19.

Tolba, M. K. et al. 1992. *The World Environment 1972–1992.* New York: Chapman & Hall.

Turner, B. L. and K. W. Butzer. 1992. The Columbian encounter and land use change. *Environment* 34(8):16–20, 37–44.

———, W. C. Clark, R. W. Kates, J. F. Richards, J. T. Matthews, and W. B. Myers, eds. 1990. *The Earth as Transformed by Human Action.* Cambridge: Cambridge University Press.

Udall, S. 1963. *The Quiet Crisis.* New York: Holt, Rinehart.

United Nations. 1992. *Agenda 21: Programme of Action for Sustainable Development.* New York: United Nations.

Vale, T. R., ed. 1986. *Progress Against Growth: Daniel B. Luten on the American Landscape.* New York: Guilford Press.

Wood, W. B., G. J. Demko, and P. Motson. 1989. Ecopolitics in the global greenhouse. *Environment* 31(7):12–17, 32–34.

World Commission on Environment and Development (WCED). 1987. *Our Common Future.* Oxford: Oxford University Press.

Worster, D. E. 1979. *Dust Bowl: The Southern Plains in the 1930s.* Oxford: Oxford University Press.

———. 1993. *The Wealth of Nature: Environmental History and the Ecological Imagination.* New York: Oxford University Press.

For more information consult our web page at *http://www.wiley.com/college/cutter*.

Study Questions

1. The phrase "Think globally, act locally" has taken on new meaning in the past decade. What kinds of actions can you take in your local area that might ultimately lead to some reduction in global consumption or pollution?
2. What are the major differences between international environmental policies and those developed for individual countries?
3. How have the following laws/treaties/conventions shaped natural resource policies over the last two decades?
    (a) NEPA
    (b) Homestead Act
    (c) Nuclear Test Ban Treaty
    (d) Stockholm Conference
    (e) Superfund
4. Go and visit your local city or town government meeting when officials are holding discussions on an environmental issue. How do they make decisions? How does it compare to what goes on nationally? How is their decision-making process different from yours?
5. Who are the important stakeholders in your area who influence environmental policy? In what ways is this influence manifested?

CHAPTER 4

ECOLOGIC PERSPECTIVES ON NATURAL RESOURCES

Natural resources, by definition, are produced by natural processes. Some resources such as metal ores or fossil fuels were produced by processes operating millions of years ago and in human time scales are essentially fixed. But many of the natural resources that we depend on today are continually being produced and replenished. Resource use that is truly sustainable in the long run must be based on these renewable resources. An ecological perspective on natural resources focuses on the role of natural processes in sustaining human activity rather than on human controlled processes.

An ecological view of resource management is also a nature-centered view. As discussed in Chapter 1, this view holds that:

- Nature has inherent value.
- Natural processes provide us with many, if not all, of the resources on which we depend.
- We have an obligation to future generations of humans to leave them with an environment that is at least as healthy, productive, and diverse as the one we have inherited.

If natural processes are the basis of resource use, and if we are to preserve the integrity of these processes to achieve sustainability, then we must learn to use resources in such a way that the basic structure of natural systems remains unaltered. This requires considerable knowledge of these natural systems. Although some may argue that we can never adequately understand nature in all its complexity, we can benefit from a careful description and analysis of nature.

This chapter introduces the basic concepts for an ecological perspective on natural resources, which derives from the scientific study of natural processes operating on Earth's surface. We first summarize the physical landscapes that exist on the planet, their extent, and our use of them. We then describe a few of the processes that function in these environments to create and maintain the resources we use. Finally, we will examine some of the ecological consequences of the enormous burden humans are placing on Earth's natural resource systems.

EARTH'S RESOURCE ENVIRONMENTS

The heart of the study of *ecology* is the interrelationship between animals and plants and the living and nonliving components and processes that make up their environment. This is why the study of ecological systems is so basic to the conservation of natural resources. Without an understanding of how natural systems work, we cannot begin to conserve, manage, and protect them.

An *ecosystem* is a system in which matter and energy are exchanged among organisms and with the larger environment. The ecosystem receives inputs of energy and material, which are stored, are utilized, or flow through the ecosystem and leave as outputs. Interaction between the physical environment of the Los Angeles basin and the human patterns imposed on it produces a complex picture of inputs and outputs (Fig. 4.1). In this urban ecosystem, inputs include raw materials and other imports, migration of new residents, relatively clean Pacific air, water supplies, and electric power. Within greater Los Angeles, products are

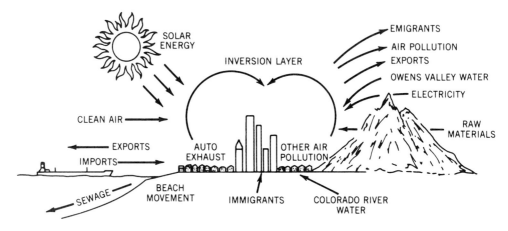

Figure 4.1 An urban ecosystem. This simplified and idealized model of the Los Angeles urban ecosystem shows some of the physical, social, and economic inputs and outputs.

created for export, such as air pollution, and emigrants move on in search of other opportunities. This diagram is a simplification and presents only a few of the major elements that make up the input-output model of an urban ecosystem.

An ecosystem can encompass a large or small geographic area and usually consists of several organisms whose needs and requirements are complementary, so that the available resources are used in a stable and nondepleting fashion. It is difficult to define the precise boundaries of an ecosystem, for organisms are often participants in more than one. *Ecotones* or transitional zones are the more flexible alternative to outlining a sharp boundary. Within a single ecosystem, researchers tend to separate the plants and animals into *communities*, which consist of groups of interacting local populations. Finally, the ecologist can examine the individual organism's function in this larger system.

Although much of the research on ecosystems has focused on small scales such as individual communities or relatively homogeneous ecological regions, there is growing recognition that the entire Earth may be viewed as a single environmental system. Events occurring in one area have such far-reaching impacts as to be essentially global in nature. For example, deforestation in the Amazon basin and other tropical areas is a significant contributor to the worldwide increase in atmospheric carbon dioxide content, which influences global climate changes. Fluctuations in ocean circulation in the eastern Pacific were recognized as a factor in the decline of the Peruvian anchovy fishery in the 1960s, but now we recognize that these events play a major role in weather patterns throughout the Earth. The explosion at the nuclear reactor at Chernobyl in the Ukraine in 1986 spewed a plume of radioactive pollutants that spread across most of Europe and was detected at monitoring stations around the world.

The integrated nature of the world environmental system is the basis of the *Gaia hypothesis*, a controversial but stimulating view of the Earth (Lovelock 1979). The Gaia hypothesis suggests that the Earth functions as a unified, self-regulating ecological system that has internal feedback mechanisms not unlike those of a single organism. For example, variability in carbon dioxide content creates changes in the rates of plant photosynthesis, which in turn remove carbon dioxide from the atmosphere at differing rates as well.

Bioregions

One way to look at the world's land resources is in terms of their general ecological characteristics. A region's resources are closely tied to the basic characteristics of its ecosystems, which themselves are related to the prevailing geologic, topographic, soil, and climatic environment. Ecological regions, or *bioregions*, serve as a valuable framework for environmental management decisions—one that is based primarily on natural resources rather than their use by humans, although human factors are combined with natural ones in classifying resource areas. The explanation for delimiting bioregions is that they form a rational basis for resource assessment and management decisions (Ricketts et al. 1999; Abell et al. 2000). Such regions have a certain amount of internal homogeneity, and presumably similar resource policies should be applied within them, whereas different ecological areas may demand different policies. Unfortunately, political boundaries usually have little correspondence to natural ones, and it is often more expedient to base resource management strategies on the political map than on the ecological map. But at somewhat smaller scales, within nations or cooperating groups of nations, bioregions offer considerable potential for guiding the management process.

Ecosystems also are useful to the researcher because they are large enough to be a reasonably representative slice of the environment, yet not so large as to be unmanageable. Ecosystems are themselves ultimately part of larger systems. At the broadest level is the *biosphere*, the worldwide envelope of organic and inorganic substances within which all life functions. Envision the millions of tiny and large ecosystems that make up the biosphere, and try to comprehend the awesome complexity of the interactions and "meshing of gears" that developed over billions of years so that this global system can function smoothly.

The *biome* concept helps us make sense of the patterns of interaction between plants, animals, and the physical environment. A biome is a major ecological unit within which plant and animal communities are broadly similar, both in their general characteristics and in their relations to each other and to the physical environment. Because biomes are defined on the basis of organism-environment interactions, it is within each of the world's major biomes that the researcher makes sense of individual ecosystems.

Figure 4.2 is a map of the major world biomes. Looking first at the equator, we see that the equatorial or tropical rainforest is an exuberant response to a rainfall schedule that is year-round and frequent, with little variation in day length or seasonal change. The resulting vegetation cover is a complex array of broad-leaved evergreens, trees that constantly shed some leaves but are never bare. A larger diversity of species is found in the rainforest than anywhere else. For example, an acre of forest in the northeastern United States might sustain five to ten tree species, whereas an acre of tropical rainforest might yield several times that many. In addition, tropical species often are unique to only a very small area of rainforest, unlike their northern counterparts, which are found over a very wide geographic area. Tropical rainforests straddle the equator, with the largest areas found in the Amazon basin (Brazil), in Indonesia and the Southeast Asian peninsula, and in Africa's Congo basin.

South and north from the equator we find a seasonally drier climate and a correspondingly less heavily vegetated biome called *savanna*, which occurs near latitude 25 degrees north and south, notably in Africa, South America, and Southeast Asia. Most savanna is located in the tropical wet and dry climate zone and is characterized by heavy summer rainfall and an almost completely dry winter season. The characteristic vegetation varies from open woodland with grass cover to open grassland with scattered deciduous trees (Fig. 4.3). Researchers generally agree that much of Africa's savanna is derived not from natural processes but from the human use of fire. If this is true, savanna is perhaps the oldest of human-shaped landscapes.

The dry climates that produce the desert biome are found in two locations worldwide: in the subtropical latitudes as the result of high-pressure zones and in the midlatitudes in continental interiors far from ocean moisture. Deserts vary from the cartoon image of bare rock and blowing sand dunes where no rainfall is recorded year after year to areas with shrubs and 100 to 300 mm (4–12 inches) of rain a year. *Potential evapotranspiration*, or the amount of water that would be evaporated or transpired if it were available, is much higher, leading to a water deficit. Desert vegetation consists of plants with special structural adaptations that enable them to store moisture, to retain it under waxy leaves, and to search for it via long root systems. These species complete their brief life cycles during the short rainy season.

The subtropical deserts are the largest on Earth and include the Sahara Desert, which stretches across Africa eastward to join the Arabian, Iranian, Afghani, and Pakistani deserts. Other subtropical deserts are found in the southwestern United States, northern Mexico, Australia, Chile, Peru, and southern Africa. Continental deserts are found in the interiors of Northern Hemisphere continents: between the Caspian and Aral seas in Central Asia

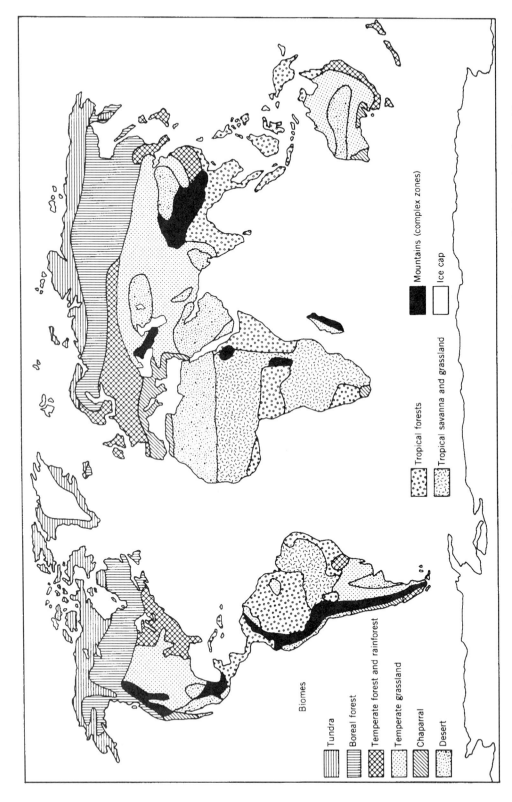

Figure 4.2 The world's major biomes. The broad geographical distribution illustrates the diversity of ecosystems found throughout the world.

Figure 4.3 Tropical savanna in the Serengeti Plains of Tanzania. This landscape has been influenced by humans for many thousands of years, but today human exploitation is being limited so that the remaining wild animal populations can be preserved.

in Mongolia (the Gobi Desert), and between the Rocky Mountains and the Sierra-Cascade ranges in the United States.

The Mediterranean biome is named after the region that stretches around the Mediterranean Sea, characterized by a cool, moist winter and a hot, dry summer. Typical vegetation consists of a thorny, glossy, and sometimes impenetrable mass of fire-prone species called *chaparral*. Mediterranean climate and chaparral are found in other coastal locations between 30 and 45 degrees latitude north and south of the equator, including coastal Southern California, Chile, South America, and parts of southern Australia.

The midlatitude *grasslands* are found at 30 degrees north and south latitude in semiarid interior areas. With not quite enough moisture to support trees and shrubs, these fire-prone grasslands once stretched from Texas to Alberta and Saskatchewan in North America, before being put to the plow in the nineteenth and twentieth centuries. Ecologists sometimes half-jokingly refer to the ecosystem that replaced the grasslands in the central United States as the corn-hog biome. The grasslands of South America in Argentina and Uruguay also have been converted to agricultural development. The grassland steppes of Asia extend from the Ukraine east as far as Manchuria in China.

The vast midlatitude or *temperate forests* (Fig. 4.4) once stretched from 30 to 50 degrees north and south latitude across the eastern United States, much of northwestern Europe, eastern China, Japan, and small areas of South America, Australia, and New Zealand. With the colonization and population growth of the last several centuries, much of this deciduous (in the United States) and mixed deciduous/evergreen (in Europe) woodland cover was removed. The characteristic climate of this region is cold winters and warm summers, with average annual precipitation generally equal to or greater than potential evapotranspiration.

The *boreal*, or northern coniferous forest, biome is located between 50 and 60 degrees north latitude (there being no significant land mass at this latitude south of the equator). In contrast to the diversity of the tropical rainforest, the number of species in this biome is extremely limited. These woodlands stretch in a belt across Alaska and Canada and as far south as the northern portions of Michigan, New York, and New England in North America. In Europe this belt continues across Scandinavia and Russia. The vegetation cover is dominated by fir, spruce, pine, and larch, which have thick needles and bark to withstand the cold. In the far north, trees only 1 meter high may be 100 or more years old.

At this tree line boundary the *tundra* biome begins, generally poleward of 60 degrees north and south latitude. The average temperature of the growing season does not exceed 50 degrees for more than a few weeks, which prohibits tree growth. Much of this area is underlain by *permafrost*, or permanently frozen ground. With low precipitation, the tundra is often called a frozen desert; in its brief thaw, like the warm deserts, a low and colorful mat of shrubs, mosses, lichens,

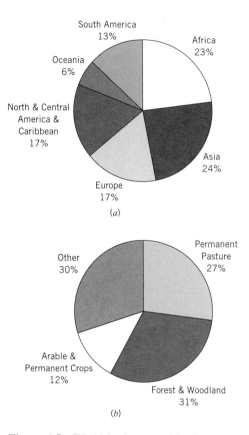

Figure 4.4 A stand of old-growth broadleaf deciduous forest in the Nicolet National Forest of Wisconsin. Much of this biome, especially in North America and Europe, has been either replaced by agriculture or significantly altered by harvesting in the last few hundred years. Old-growth stands like the one in this photo are rare.

Figure 4.5 World land area and land use, excluding Antarctica. (a) Land areas of the continents. (b) Use of the world's land area. *Source:* FAO, 2002.

and grasses temporarily springs up. The tundra region contains substantial storages of carbon, which may be released if climate warms substantially.

The preceding descriptions present a highly simplified picture of a complex patterning of biomes. There are numerous exceptions to the commonly occurring biomes, caused by microclimate, soil variability, and human impact. The main point to note is that vegetation and climate have interacted over millions of years, resulting in adaptive vegetation patterns. We should remember that substantial portions of these biomes have been altered by people already, and that is the cause of much discussion and dissent among those who would conserve and those who would develop and further alter the landscape.

Human Use of the Land

The land area of the world (excluding Antarctica) totals about 32 billion acres (13 billion hectares) (Fig. 4.5). The best data on world land use are provided by the Food and Agriculture Organization (FAO) of the United Nations. The FAO classifies land as *arable* (land that is capable of producing crops through cultivated agriculture), permanent pasture (land that is suitable and available for grazing animals, regardless of the dominant vegetation type), forest and woodland, and other (including parks, urban areas, roads, deserts, and other nonagricultural areas). On the basis of this classification, about 12 percent of the world's land is used for arable agriculture and another 27 percent for permanent pasture. Thirty-one percent of the world is in forest and woodland, while 30 percent is in other uses. The continents vary considerably in the proportions of land in these uses. For example, Oceania, Africa, and South America are only about 6 percent arable, whereas Asia, North/Central America, and Europe have between 10 and 20 percent arable land (UN Food and Agricultural Organization 2002). Europe's arable land increased from 5 to 26 percent between 1990 and 1994, not because of agricultural land conversion but because of the reclassification of Russia. This statistical artifact (see Issue 4.1) is complicating our understanding of longer-term trends at the continental scale.

Although the overall portion of land in each of these categories at the world and continental scales is relatively stable, locally changes are occurring. During the past decade, minor but significant changes took place: the amount of arable land increased by about 1.3 percent worldwide. The largest increases were in Africa, South America, Asia, and Oceania. The increases in arable land were caused by expansion of cultivation into areas that were previously forested to meet the food demands of a rising population (see Chapter 6). In the wealthier parts of the world (North America and Europe), the amount of arable land decreased. In much of the world, especially Africa, South America, and Asia, forest cover is declining while permanent pasture is increasing. This is a result of forests being cleared for grazing or arable agriculture. Estimates of the total amount of change vary, but generally forest cover decreased about 1 to 4

ISSUE 4.1: WHAT HAPPENS WHEN THE GEOGRAPHY CHANGES?

The rise of independent republics in the aftermath of the fall of the communist empire profoundly altered world affairs. The Soviet Union no longer exists; in its place we have 15 independent countries. Czechoslovakia is now Slovakia and the Czech Republic. When the United Nations was formed in 1945, there were 51 member states; in 1989 there were 157; and today there are 190.

What do these geopolitical changes have to do with conservation of natural resources? The simple answer is *everything*! Most of the comparative statistics on population, economics, and the environment are reported by individual country. In 1989 the total land area of the Soviet Union was 2190 hectares. Because the Soviet Union no longer existed as a separate political entity in 1990, the statistics on land area were divided among its successor states, of which Russia is the largest (1699 million hectares). Although the total amount of land hasn't changed at all, the way we politically carve it up and report data on it has.

This change in how we report statistical data also affects continental comparisons. Both the United Nations and the World Resources Institute report individual country and continental averages. The Soviet Union spanned two continents, so which continent gets allocated what land? In the statistical reports, Russia, Belarus, Estonia, Latvia, Lithuania, Moldova, and the Ukraine are categorized as being in Europe, whereas Armenia, Azerbaijan, Georgia, Kazakhstan, Kyrgyz Republic, Tajikistan, Turkmenistan, and Uzbekistan are listed as being in Asia.

One of our biggest headaches in this regard involves comparing countries or continents over some period of time. For example, if we try to examine changes in arable land from 1980 to 1995 (see Figure 4.5), the task becomes very difficult. In order to compare the 1995 and 1980 data, we must sum all the individual country-level data and generalize to the old geography. If we parcel out the 1980 data to individual countries, the original data may not be available. While we may already be convinced that geography matters in natural resources and environmental issues, perhaps now we can appreciate just how important it really is.

In addition to the impacts of these geographic changes on statistical trends, the economic collapse of the Soviet Union and eastern Europe had far-reaching consequences for the world resource trends. For example, the steady increase in total world carbon dioxide emissions was halted temporarily in 1989, and total emissions remained relatively steady between 1989 and 1994. Emissions were increasing in most of the world, but the dramatic drop in coal combustion that took place in Russia and eastern Europe was enough that in some years world totals actually declined. Similarly, a dramatic decrease in grain production in the former Soviet Union in the early 1990s was enough to significantly reduce world grain supplies. We will discuss these trends in more detail in later chapters of this book.

percent during the 1980s and around 2 percent during the 1990s (UNEP 2002). The fact that woodlands are being cleared to make way for cultivation indicates that agriculture is being forced onto poorer and poorer land, with concomitant increases in problems of erosion, unfavorable moisture conditions, poor fertility, and so on. The amount of land in "other" uses changed little overall, but locally significant changes occurred as a result of urban development, land degradation, and the establishment of large areas for parks and wildlife reserves.

The total land area of the United States is 2264 million acres (957 million hectares). This consists of about 20 percent cropland, 25 percent pasture and range, 30 percent forest, and 25 percent other land. This land-use pattern has undergone many changes since the Europeans arrived in North America. Obviously, little cropland existed prior to European settlement, and there was substantially more forest. Most of the forest clearance took place during the eighteenth and nineteenth centuries, and forest area reached a minimum in the late 1800s or early 1900s—earlier along the eastern seaboard and later in the central and western states. As settlement and agriculture spread westward from the early 1600s to the 1920s, forestlands in the east were cleared and tall-grass prairie was plowed under. Land that wasn't cleared for farming or pasture was cleared for other purposes, usually to provide fuel and building materials. But as the productivity of agriculture grew in the nineteenth century, eastern lands less suitable for farming were abandoned and gradually reverted to woodland, meaning that much of the eastern United States today now is reforested. But the tall-grass prairie of the eastern plains and most of the forest of the central states were replaced by farms; thus, what was once oak forest or grassland is now fields of corn, soybeans, and wheat.

In addition to this dramatic land-use change, there were equally drastic changes covering smaller areas such as natural preserves, cities, and other developed areas. Urban lands now occupy about 7 percent of the country. Reservoirs also have taken substantial areas. The amount of land in urban uses is increasing at about 2.5 percent per year (NRI 2002). Rural areas are also being committed to new uses, most importantly as parks and wildlife areas. National Parks occupy over 3 percent of the land area of the United States, and national wildlife refuges and designated wilderness areas each cover an additional 8 percent of the nation's land (FWS 2002; Wilderness Information Network 2002; NPS 2002).

About a third of the land in the United States is publicly owned. In areas of the eastern United States that were territories in the original British colonies, the British government granted land to various companies and private individuals or the states granted land as payment for military service. Almost all the land in the east has been in private hands since the late eighteenth century. But most of the land west of the Mississippi River (excluding Texas) was acquired by the U.S. government by treaty or purchase. Through homesteading, grants to railroads, and other means, most of the more productive land came under private ownership especially during the second half of the nineteenth century. But semiarid and arid areas and high mountain areas that were not desirable for agriculture remained unclaimed into the early twentieth century. Many of the mountainous areas were reserved as National Forests, and so most federally owned land consists of these arid and mountainous areas in the western United States. Almost 30 percent of U.S. land is in federal ownership, another 10 percent is state-owned or owned by Native Americans, and the remaining 60 percent is in private hands (BLM 1999; U.S. Bureau of the Census 2002).

Land-use decisions are probably the most important single issue in environmental management, for land use essentially defines what kinds of resources we exploit and the extent of environmental impact associated with our activities. If a parcel of land is used for agriculture, then not only does this contribute to our total food production but also it means that runoff from that land contains certain byproducts of agriculture (sediment, nutrients, and pesticides) in non–point source water pollution. The use of a land parcel for one purpose has implications for other parcels: urban growth increases demand for recreational lands nearby and the need for watershed protection in other areas to supply a growing population with drinking water.

In most of the world, land-use decisions are made through some process of balancing competing interests. In market economies, allocation of land among various uses is generally determined by the relative income generated at any given time. Agricultural land near cities is sold for urban development because those uses generate greater profits for the landowner than agriculture. Similarly, farmers shift land among crops, pasture, and other uses, depending on production costs and the return on investment for various commodities.

Governments play a role in this process. Public interests are served by the allocation of land to transportation, parks, water resources, defense, and similar common needs. In most countries, patterns as well as types of land uses are regulated. In the United States this regulation is accomplished by zoning laws, usually at the local level, which are generally designed to separate incompatible uses, such as industry and residences. Land-use regulations also help ensure adequate open space, such as through the designation of greenbelt areas around cities, and environmental protection, such as through restrictions on development in wetland areas or prohibition of waste disposal on floodplains.

As populations grow and land becomes more scarce, competition for land resources and the intensity of debates over land-use decisions inevitably increase. Many examples of such problems exist, ranging from the restriction of oil exploration from wildlife refuges (Issue 3.2) to local debates over where to locate a sewage treatment plant.

Energy Transfers and Material Flows

Energy in ecosystems (and human systems) is ultimately derived from the sun. This energy passes through a series of storages via many paths, before finally being returned to space as radiant energy. Two fundamental laws govern all energy transfers: the first and second laws of thermodynamics.

The *first law of thermodynamics* is the *law of conservation of energy*, but it also governs conversion of matter into energy. It simply states that in any energy transfer, the total amount of energy is unchanged; energy is neither created nor destroyed. Another way of saying this is, "You can't get more out than you put in." The second law is called the *law of entropy*, and it is a little less obvious than the first. It states that any time energy is converted from one form to another, the conversion is inefficient. Energy is always converted to a less concentrated form or dissipated as heat. Another way of saying this is, "Not only is it impossible to get more out than you put in, but you can't even break even." Entropy is a measure of the degree of organization present. Greater entropy means greater disorganization or randomness. The following two examples illustrate these concepts.

The first example is the conversion of solar energy to food energy by *photosynthesis*, producing all the food needed to feed humans and other organisms. A leaf on a plant is exposed to sunlight, and this stimulates a chemical reaction in which carbon dioxide and water are converted to carbohydrates, with oxygen given off as a byproduct. In the process, however, the leaf must be heated, causing a loss of energy by radiation or convection from the leaf surface. In addition, water must be moved through the leaf stem to deliver nutrients to the leaf and is also evaporated by the leaf as a cooling mechanism. This water loss involves a conversion of water from liquid to vapor form, which requires energy. Finally, plants also must respire. Respiration is a process in which food energy is converted to heat energy, which is then dissipated. When all these things are considered together in an energy budget for a single plant or for an entire plant community, only a very small fraction of the incoming solar energy is converted to food energy or biomass, generally about 1 percent or less. The rest of the energy is reflected, reradiated, or used in the conversion of liquid water to water vapor.

A coal-fired electric-generating plant provides a second example. Coal is formed by the chemical modification of formerly living matter, mostly plants. The energy released initially was stored by plant photosynthesis at some time in the geologic past. When the coal is burned, some heat is lost in the smokestack, but most of the heat is used to convert water to steam. The steam then drives turbines, which drive generators, which in turn produce electricity. The steam is cooled in the process but not enough to condense it, although it must be returned to the boiler as water. To do this it must be cooled, usually by dissipating the heat in the nearest river or other body of water. Heat is also produced by the generator and by friction in moving parts, and this must be dissipated as well. In the end, only about 35 percent of all the heat stored in the coal finally is converted to electric energy.

The rest is dissipated as heat, either in the stack gases or in the steam condensing system. This dissipation of energy is an example of entropy.

Just as energy flows through an ecosystem, materials necessary for life—carbon, oxygen, nitrogen, potassium, and water—cycle through the system as well. The paths these substances take are called *biogeochemical cycles*. Some biogeochemical cycles are dominated by large storages in the atmosphere; the nitrogen cycle is a good example of this type. Others are dominated by terrestrial storages, usually in rocks and sediment. The phosphorus, potassium, sulfur, and calcium cycles are examples of this latter type. Although the cycles have different chemical and biological processes regulating them, their patterns generally are similar.

These biogeochemical cycles are of enormous importance in regulating natural processes and strongly affect the viability of natural resources. They provide the major means for resource renewal following harvesting. For example, when a forest is logged, the forest ecosystem is altered drastically. Removal of trees causes a reduction in evapotranspiration, with a corresponding increase in water moving both through and over the soil. This increased water movement, together with the decay of plant matter left behind (stumps and smaller branches and leaves), contributes to increased removal of nutrients from the area, some that dissolve in water and others that are part of soil particles. Were it not for weathering, nitrogen fixation, and the additions of nutrients from other processes over time to replace these losses, the soil would be unable to support the regrowth of the forest for future harvest. Similarly, grazing, cultivation, and water resource development depend on replacement of substances by these natural cycles. In addition, the timing of these cycles places significant constraints on human use in the same way that finite, or stock, resources impose constraints. Clearly, "renewable" resources also have an element of finiteness.

The operation of biogeochemical cycles also has implications for activities that disturb some portion of the cycle, by either removing or introducing substances. The use of nitrogen fertilizers in agriculture causes a substantial increase in the inorganic nitrogen content of the soil. This increase benefits crop plants, but it also leads to greater leaching of nitrogen by water and hence greater nitrogen concentrations in rivers and lakes draining agricultural areas. This added nitrogen causes serious pollution problems in some areas, for it not only modifies aquatic ecosystems to trigger algal blooms but also poses a danger to humans if it is consumed in drinking water. By modifying the nitrogen inputs to the soil, we also change conditions farther downstream, often with undesirable consequences. The total amount of biologically available nitrogen introduced into the environment is enormous, exceeding total natural fixation of nitrogen (Vitousek 1994). A second example relates to the use of pesticides in the environment. An insecticide is intended to act on only one small component of an ecosystem—the population of some particular insect species—but the food chain carries both food and unwanted substances to other organisms, oftentimes with unforeseen and damaging effects (Issue 4.2).

Thus, biogeochemical cycles serve as conduits for energy and matter from one part of the environment to another. They also cause the effects of human activities to extend beyond the immediate area of impact. For these reasons, we have become increasingly aware of the interrelatedness of natural resources, particularly renewable resources, and of the need to understand these phenomena completely in order to manage these resources properly. Next we discuss four of the important biogeochemical cycles: the nitrogen, phosphorus, carbon, and hydrologic cycles.

Carbon Cycle

One of the most important biogeochemical cycles is the carbon cycle, which utilizes large storages in the atmosphere, in living organisms, and in rocks. Carbon dioxide in the atmosphere enters plant leaves and through photosynthesis is incorporated in living matter to form a basic part of starches, sugars, and other foods. As it passes through the food chain, carbon dioxide is returned to the atmosphere by respiration of consumers and decomposers. But significant amounts of organic matter accumulate in soils, marshes, and lake bottoms, and this organic matter is largely carbon. Some of it may be oxidized from time to time, but most is semipermanently stored in sediment. In addition, large amounts of carbon exist in the oceans, both in living organisms and as dissolved carbon dioxide. Carbon is continually deposited on the ocean floor in sediment, which over time becomes sedimentary rocks. Limestone, which is primarily cal-

ISSUE 4.2: *SILENT SPRING* VERSUS *OUR STOLEN FUTURE*

When *Silent Spring* was first published more than 30 years ago, it warned of the dangers of synthetic pesticides, especially DDT. In this book, first serialized in *The New Yorker* magazine, Rachel Carson eloquently explained how these persistent chemicals damaged the environment and how they were accumulated in species, including humans. Both industry and government criticized and tried to discredit the book by attacking its author, claiming she was not a scientist but just an extremely emotional woman (Hynes 1989). Despite her master's degree in marine biology from The Johns Hopkins University, Carson was not taken seriously until the President's Science Advisory Committee produced its own report a year later, confirming her findings. *Silent Spring* has sold millions of copies worldwide and laid the foundation for governmental action in protecting the environment. It is truly a classic in the environmental field.

Our Stolen Future (www.ourstolenfuture.org) picks up where Rachel Carson left off. Written by two scientists and an environmental journalist (Theo Colburn, Dianne Dumanoski, and John Peterson Myers), the 1996 book chronicles the effect of synthetic chemicals on reproductive systems. Providing evidence from laboratory experiments, wildlife studies, and human epidemiological data, the book illustrates the role of these hormone-disrupting chemicals and how they are affecting the reproductive and growth-regulating systems of many species. Worldwide, human male sperm counts have dropped by nearly 50 percent, and nearly one-quarter of American couples have infertility problems. Frogs are disappearing in the upper Midwest, and defective sexual organs have been found among alligators in Florida. These problems are not specific to the United States but have been found in other industrialized countries as well.

In response to public concern about these reproductive toxins, or endocrine disrupters as they are more formally known, the Environmental Protection Agency (EPA) established a large research initiative to establish the linkage between these synthetic substances and reproductive health. Unfortunately, we still don't quite know which specific chemicals pose the greatest risks to endocrine functions. Nor do we know how widespread they might be in the environment. However, as a precautionary measure, both the Food Quality Protection Act and the Safe Drinking Water Act passed in 1996 mandated screening and testing for endocrine disrupters, a program that the EPA began in 1999. The EPA screening program is now prioritizing more than 87,000 chemicals on the basis of their danger to humans, fish, and wildlife and conducting more detailed hazard assessments. While it's too soon to judge whether our future has indeed been stolen, it's not too late to ask that environmental toxic releases be reduced for the benefit of all species.

---

cium carbonate ($CaCO_3$), is formed in this way, as are fossil fuels. Carbon in rocks reenters the atmosphere through combustion of fossil fuels. In the past few hundred years, humans removed and burned much more carbon from terrestrial storages of coal, oil, and natural gas than was returned in that time, and the atmospheric concentration of carbon dioxide has increased accordingly (see Chapter 12). In addition, clearing of forests and depletion of soil organic matter by poor land management practices reduced these storages of carbon and contributed to the increase of atmospheric carbon dioxide. Regrowth of forests, on the other hand, removes carbon from the atmosphere. In some parts of the world today there is a net loss of carbon from the biosphere to the atmosphere, while in others the biosphere is taking more carbon out of the atmosphere than is released. These exchanges are an important way in which human activity alters the carbon cycle.

Measurements of the movements of carbon between human systems, the atmosphere, the biosphere, and the oceans and of the changes in storage of carbon in various systems allow an accounting of the relative contributions of each of these flows to the total *carbon budget* (Fig. 4.6). On an average annual basis, about 6.3 billion tons of carbon are emitted to the atmosphere

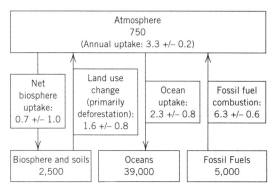

Figure 4.6 The carbon cycle. Carbon is stored in the form of carbon dioxide in the atmosphere, in plants, and in the decayed remains of organic organisms found in rocks and oceans. Fossil fuel combustion has altered the normal carbon cycle by removing carbon from terrestrial storage, burning it, and returning it to the atmosphere faster than it can be placed into storage (unit measures are billion tons).

by fossil fuel combustion and cement production. Of this, about 3.3 billion tons accumulate in the atmosphere, and 2.3 billion tons enter the oceans. Carbon is also delivered to the atmosphere by changes in land use, especially deforestation and reforestation. These flows are estimated at 1.6 and 0.7 billion tons, respectively. The geographic distribution of these flows is of critical importance to international agreements relating to global warming, as will be discussed in Chapter 12.

Nitrogen and Phosphorus

Nitrogen comprises about 80 percent of the Earth's atmosphere, and most of the Earth's nitrogen is in the atmosphere at any given time (Fig. 4.7). Nitrogen is also an essential nutrient and a fundamental component of many proteins. In its gaseous form, nitrogen cannot be used directly by most organisms. In order for it to be available to living matter it must be fixed or incorporated in chemical substances such as ammonia, nitrates, or organic compounds that plants are able to use. Some nitrogen is added directly to the soil as impurities in rainfall, but the much more important mechanism is the action of nitrogen-fixing bacteria, some of which live in association with plant roots. These bacteria are able to extract nitrogen directly from the air. Some plants, such as legumes, have symbiotic or mutually beneficial relationships with particular nitrogen-fixing bacteria. Many other plants also accommodate nitrogen-fixing bacteria, and some nitrogen fixers are not dependent on plant roots at all. Once nitrogen is incorporated into organic matter, it follows much the same route as energy in the food chain, passing from producer to consumer and ultimately to decomposer. Decomposers return nitrogen to the soil in mineral forms such as ammonia that are again available to plants. In addition, nitrifying bacteria convert nitrogen from ammonia to nitrates and eventually to the gaseous forms N_2O, NO, and N_2, which are returned to the atmosphere. Finally, some nitrogen is leached from the soil or

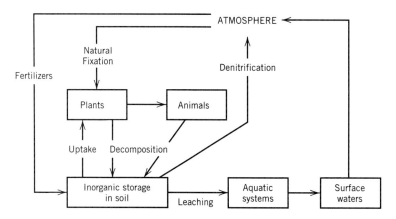

Figure 4.7 The nitrogen cycle. The atmosphere provides the primary storage of nitrogen, which must be converted from its gaseous state by nitrogen-fixing bacteria before it can be used by plants.

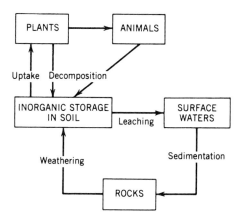

Figure 4.8 The phosphorus cycle. Phosphorus naturally enters the environment through the weathering of rocks. Phosphorus forms used by humans include phosphate fertilizers and detergents.

phosphate rocks must derive their phosphorus from trace amounts contained in rainfall. Once in the soil, phosphorus travels through the food chain and is returned to the soil by decomposers. Considerable amounts of phosphorus are leached or eroded from the soil, however, and this phosphorus eventually accumulates in the sea, where it is concentrated in the bones of fish. As the fish die and their bodies decay, phosphorus is deposited on the ocean floor and eventually incorporated in sedimentary rocks. Fish-eating birds excrete large amounts of phosphorus, and their dung, or guano, which accumulates on rocks or offshore islands where seabirds roost, is an important source of phosphate fertilizer. Fish bones from processing plants also are used as fertilizer in some parts of the world. In the United States, most of our phosphate fertilizer is derived from the mining of phosphate-rich rocks.

incorporated in runoff and makes its way into groundwater, rivers, lakes, and the sea, where it is ultimately returned to the atmosphere.

The phosphorus cycle (Fig. 4.8) is a good example of a biogeochemical cycle that is dominated by terrestrial rather than atmospheric storage. Phosphorus, an essential nutrient, is found primarily in rocks and enters the soil by the weathering of those rocks. But many rocks contain little phosphorus, and areas underlain by non-

Hydrologic Cycle

Another example of an important environmental cycle is the hydrologic cycle (Fig. 4.9), which is not a biogeochemical cycle in the same way as the others just discussed but is more of a regulator of energy flows and nutrient cycling. The hydrologic cycle is the set of pathways that water takes as it passes from atmosphere to Earth and back (Fig. 4.10). It is regulated primarily by climate, but the terrestrial components of the cycle—rivers, lakes,

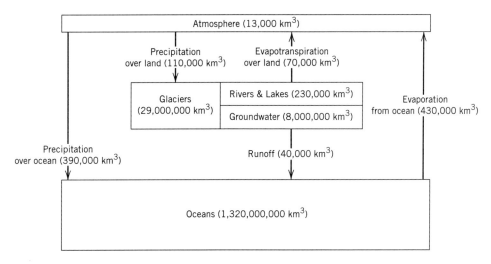

Figure 4.9 The hydrologic cycle. Water flows in the environment play a major role in regulating material and energy cycles.

Figure 4.10 Precipitation. Moisture held in storage in the atmosphere precipitates out as rain, illustrated by this thunderstorm. The water is used by plants, runs off into surface waters, or percolates through the soil for storage as groundwater.

soil, and groundwater—are also regulated by the characteristics of surficial materials and by topography. Analysis of water budgets, which quantify various components of the hydrologic cycle, is essential for water management.

Beginning with the atmosphere, water is delivered to the Earth's surface by precipitation. Rain strikes the leaves of plants, and some remains there and is evaporated, but most reaches the soil surface. Once on the soil, the water evaporates, soaks into the soil, or runs off. Several factors determine how much water soaks in and how much runs off, but the primary controls are the rate at which rain falls, or precipitation intensity, and the ability of the soil to soak up water, or its *infiltration capacity*. These factors are of critical importance in controlling soil erosion and are discussed later in that context. Water that runs off the soil surface or through the regolith enters stream channels and becomes surface water. Through gravity, surface waters flow to the ocean via rivers, lakes, swamps, and so on. Depending on climatic factors such as atmospheric humidity and temperature, varying amounts of surface water are lost by evaporation. Water is temporarily stored in the soil, where it becomes available to plants. As they use water, it is returned to the atmosphere by evapotranspiration from their leaves. Water that is not used by plants percolates into the ground, where it eventually reaches a level below which the pores in the rocks are saturated, known as the *water table*. Water in this saturated zone is called *groundwater* and flows by gravity. Over long periods of time it may return to the surface in valleys and become surface water. The velocity of flow of groundwater is considerably less than that of surface water, and depending on subsurface characteristics, very large amounts of water may be stored there. Groundwater flow is primarily responsible for maintaining river flow during dry periods between rains. Eventually, most water is returned to the atmosphere, either by evapotranspiration by plants or by evaporation from surface waters, particularly the oceans. Some water is stored for long periods of time, such as groundwater, ice caps, and isolated deep water bodies. Water in such long-term storage is removed, for all practical purposes, from the hydrologic cycle.

Food Chains

Ecosystems consist of all the living organisms in a defined geographic area, together with all the physical entities (soil, water, dead organic matter, and so on) with which they interact. As such, they are exceedingly complex, and the energy and material transfers within them are difficult to quantify. As one type of organism in an ecosystem consumes another, a pattern of energy flow through the ecosystem is set up, called a *food chain* (Fig. 4.11). Some food chains are simple—for example, when a plant is consumed by a rabbit, which is consumed by a fox. Such a simple chain is usually part of a more complex *food web* in which several animals and plants are dependent on one another. Many organisms eat more than one kind of food and in turn can be appealing to several other species.

Energy is transferred from one *trophic level* to another within the chain or web. Terrestrial green plants are producers, as they convert solar energy to food energy at the first trophic level. Consumers are classified as primary consumers, which feed on producers at the second trophic level; secondary consumers, which feed on primary consumers at the third trophic level; and so forth. In addition there are decomposers, which feed on

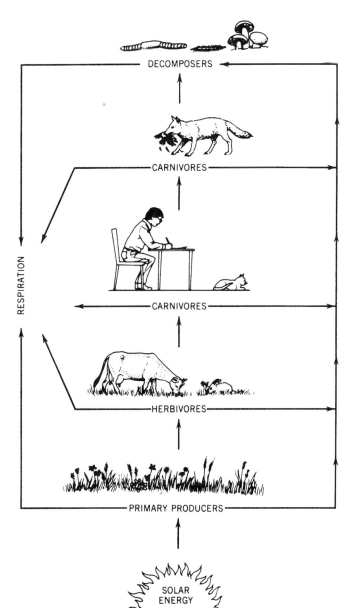

Figure 4.11 Food chain. Herbivores, or primary consumers, eat producers and are then consumed by carnivores. Sometimes a food chain can support additional trophic levels of consumers, before decomposers take their turn.

dead organic matter and return nutrients to the soil or water, where they become available to producers. There may be fourth- or fifth-level consumers, but rarely do we find more than five levels in an ecosystem, for the energy produced at the first level was consumed at intermediate levels. Human beings are able to take advantage of the food energy at different levels because we consume energy in the form of both plants and animals. There is some debate over whether or not we are wasteful of the world's food energy because our diets are high in animal products. It is generally more efficient in terms of energy production and consumption to obtain our food directly at the first trophic level.

Figure 4.11 illustrates a relatively simple food chain. Notice that at each step in the system, energy is either stored as biomass (the living and dead organic matter in an ecosystem) or used in respiration. Most of the energy consumed at any given level is used in organism metabolism, with only a small percentage stored, as biomass, and available for the next higher level to consume.

Carrying Capacity

No matter how complex or simple the ecosystem, its component organisms always are working to reproduce themselves and to find adequate food. Of course the number of organisms cannot exceed the amount of food available to them for very long, or the stability of the ecosystem is threatened. For an ecosystem to sustain itself, population size and food supply must be balanced over the long run, although there can be short-term fluctuations. As a result, we find an intricate relationship between the size of populations of the different species in an ecosystem and their competitive or complementary food needs relative to other populations in that ecosystem. These relationships change over time, as the ecosystem's population dynamics shift in response to internal and external changes. *Carrying capacity* is the number of individuals of a given species that is sustained by an ecosystem over a long period of time without environmental degradation.

With ample food, sufficient living space, good health, and no predators, a species population could grow at its *biotic potential*. This is the maximum rate of population growth resulting if all females breed as often as possible, with all individuals surviving past their reproductive period. It is obvious that a species breeding at an exponential rate of increase would soon outstrip the available food supply for it and other species in its ecosystem. Just as obvious is that various types of environmental resistance, such as exhaustion of the food supply, adverse weather, and disease, ensure the population is kept at a level far below its biotic potential. In systems terms, *environmental resistance* is a good example of negative feedback. For example, as a population grows, its food consumption increases and food becomes more difficult to find. As food becomes scarce, competition for food intensifies and survival of organisms to reproductive age is less likely, and thus population growth is reduced.

Although these inhibitions on population growth prevent exponential rates of increase, there are several fairly predictable growth patterns for populations within ecosystems (Fig. 4.12). An S-shaped growth curve describes a population with only a small difference between the rates of

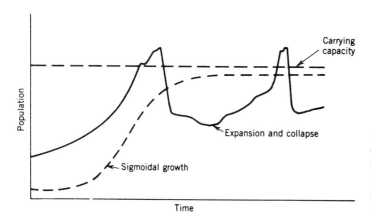

Figure 4.12 Population patterns in ecosystems. Populations can stabilize below carrying capacity. If the carrying capacity is exceeded, however, populations collapse and dieback can occur.

growth (birth and immigration) and decline, as a result of disease, predation, uncertain food supply, and other forms of resistance. Over time, however, the population may increase more rapidly as long as there is enough food and other necessities. Eventually, the resources demanded by a population exceed those available for its use, and environmental resistances such as disease and malnutrition put a damper on the rate of increase.

The near-level portion of the S-shaped curve suggests a zero growth rate in which births plus immigrations equal deaths plus emigrations. In fact, this is an equilibrium situation in which biotic potential equals environmental resistance. The carrying capacity for this species in this ecosystem has been reached, which is the maximum number of organisms of one species that can be supported in the given environmental setting. Extinction of a species occurs when a population fluctuates dramatically, dropping so low that the species cannot reproduce quickly enough to remain in competition with others for available resources. These factors provide the biological basis for renewal resources management.

THE SCOPE OF HUMAN IMPACT

Scientists have been aware of the human impact on the environment for a century or more (Marsh 1864). Agriculture and urbanization generated the most dramatic transformations of the natural landscape (Goudie and Viles 1997; Turner et al. 1990). While often regionally specific, the global patterns of agriculture and urban changes permanently have altered natural ecosystems in all parts of the world.

In the 1960s, spurred by books such as *Silent Spring* (Carson 1962), environmental scientists became more interested in local and regional impacts, such as pollutants in a river or acid deposition in a lake. This localized research demonstrated that many systems were profoundly altered by such pollutants. In the 1970s, more extensive global-scale data collection networks were assembled, so that by the 1980s sufficient data were amassed to allow rough estimates of human impacts on a global scale, especially in the area of climate change and its global and regional impacts. Finally, in the 1990s, Colburn et al. (1996) realerted the public to the dangers of toxic substances in the environment and their links to reproductive problems in species, including humans (Issue 4.2). This section examines some of the results of the recent research on pollution impacts, focusing on toxic substances. Other forms of water and air pollution are discussed in Chapters 10, 11, and 12.

The Extent of Environmental Pollution

One of the most significant human impacts in recent decades was the creation and release of a wide variety of chemical substances into the environment, some of which have significant adverse effects on natural organisms or on humans. Ecological processes are very important in the study of toxic substances, both because ecosystems are affected by toxic substances and because ecological processes, like the biogeochemical cycles discussed earlier, are important in distributing toxic substances through ecosystems. Problems associated with toxic substances are discussed throughout this book in relation to various resource issues. The following sections provide an introduction to some of the general problems and characteristics of toxic substances.

Toxic Substances Divergent definitions of toxic substances are used in different circumstances. We consider a *toxic substance* to be any material that is harmful to humans, plants, or animals at very low concentrations. In practice, most of the toxic substances we examine are hazardous to humans and other animals, and by low concentrations we usually mean parts per million or less per liter of water, air, or soil. Thousands of toxic substances are of concern, and the lists of of them change frequently as new information on toxicity or environmental concentrations becomes available. Toxic substances are derived from a multitude of human and natural sources and move along many different environmental pathways. In the following sections, we provide a glimpse of the diversity of toxic substances and their sources.

Manufacturing activities are an important source of toxic substances in the environment, and in the United States polluters are required to report toxic substance releases through a system known as the Toxic Release Inventory (TRI). The reported releases and transfers of toxic substances totaled about 7.77 billion pounds (3.52

Table 4.1 Toxic Releases Reported Under the Toxic Release Inventory, 1999

	Thousands of Pounds					
Source	Total Air Emissions	Surface Water Discharges	Underground Injection	Landfills	Other Land Disposal	Transfers to Off-Site Disposal
Manufacturing	1,175,055	253,592	199,548	12,440	311,227	374,648
Metal mining	4,453	447	35,092	0	3,934,846	2,179
Coal mining	1,772	235	144	0	9608	0
Electric generation	841,920	4,510	5	1,299	258,121	57,958
Chemical wholesale distributors	1,318	3	0	0	1	649
Petroleum terminals and bulk storage	4,044	44	0	<1	15	166
Treatment, storage & disposal, and solvent recovery facilities	803	51	22,861	206,756	13,707	43,825

Source: U.S. Environmental Protection Agency, 2000.

billion kg) in 1999 (USEPA 2000; Graham and Miller 2001). The majority of these releases are emitted into the air and account for nearly half of the total. Table 4.1 summarizes the major release sources and destinations. In general these releases have declined since the TRI was established (Fig. 4.13).

Many toxic substances are released to the environment deliberately for beneficial purposes. Pesticides are an obvious example. In 1997 more than 1.0 billion pounds (454 million kg) of pesticides were applied to agricultural crops (CEQ 1998). Another beneficial use is the application of tributyl tin as an anti-fouling agent in marine paints, which is used to keep algae off the bottom of boats.

Lead is an important toxic substance that received much attention in recent years. It is toxic to

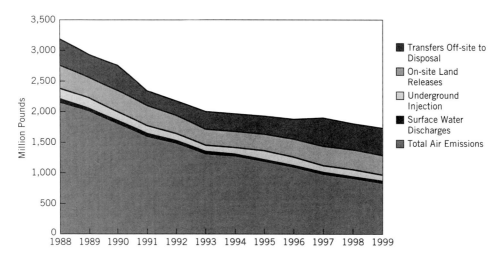

Figure 4.13 Trends in Toxic Release Inventory releases, 1988–1999. The types of releases and entities releasing them reported in the TRI have changed over the years. This graph shows releases only for those industries that have been reporting since 1988. Releases have steadily declined since then. *Source*: U.S. Environmental Protection Agency, 2000.

humans and was used as an additive in motor fuels for decades. In the early 1970s increasing concern over air pollution led to the requirement that cars be equipped with catalytic converters to reduce hydrocarbon emissions. Leaded fuel damages these converters, and so lead-free gasoline was introduced, with the added benefit of reduced lead emissions. Today all gasoline sold in the United States is unleaded. As a result, lead concentrations in the air and water decreased markedly in the 1980s and beyond (see Chapter 11).

Many toxins have natural origins besides their human-caused presence in the environment. For example, most toxic metals are present in rocks and soils and are released naturally to water and biota through rock weathering and water flow in soils. Many plants produce toxic organic chemicals, of which some are released directly by plants and others are produced by processes of biochemical decay.

Under most circumstances, these natural sources of toxic substances do not present major problems, either because the concentrations involved are low or because organisms have adapted to them. However, in some circumstances human activities aggravate or are aggravated by these natural sources. Metals are a good example. Several metals are natural constituents of water but usually in relatively low concentrations. In some cases, however, human sources of these metals cause the total concentrations to rise to toxic levels, creating or increasing hazards to humans and other organisms. Another possibility is that toxins are present in relatively harmless forms naturally, but human actions release these toxins to the environment. An example would be acid precipitation, which may result in a decrease in soil pH and thus cause metals that were in insoluble forms to be released and taken up by plants or enter surface water. Natural substances also combine with pollutants of human origin to produce new toxic substance problems. For example, a group of substances called trihalomethanes is produced in drinking water from the combination of natural organic substances (such as the products of plant decay in streams) and chlorine added to water to reduce bacterial concentrations.

The fact that some toxic substances have both natural and human origins makes management and control efforts particularly difficult. It is possible to control some human sources but not all, and sometimes we eliminate human sources and still find the pollutant present because of natural sources. If a pollutant is found in the environment, it is often difficult to determine how much of it is natural and how much is human in origin. These uncertainties often are used to political advantage by those attempting to point a finger at polluting industries, for example, or by those trying to avoid pollution control efforts.

Once pollutants are introduced into the environment, they migrate within it and their concentrations change as a result. The processes of pollutant movement in the environment are critical in understanding and evaluating their impacts on resources. In the following sections, we will examine three different mechanisms by which the concentrations of pollutants change through time: pollutant decay, bioaccumulation, and physical transport from one place to another.

Pollutant Decay Some pollutants decay gradually in the environment and become less toxic over time. This is true for compounds such as pesticides but not for elements such as metals. *Biochemical decay* may be accomplished in organisms that ingest pollutants, or it may take place as a result of exposure to water, sunlight, and other substances in the environment. Some pollutants do not break down rapidly in the environment, and it is this stability that often causes them to become a problem. Persistent pesticides, so named because they are relatively stable in the environment, are an example. Many compounds of carbon, hydrogen, and chlorine (chlorinated hydrocarbons) are resistant to decay. Persistent pesticides such as DDT and chlordane are examples of chlorinated hydrocarbons that were restricted from use in the United States because of their long-term danger to organisms and ecosystems.

Other pollutants decay quite rapidly, and their concentrations decrease through time. Most of the pesticides in use today quickly decay and thus have relatively short-term impacts. As a rule, rapid decay is advantageous, although it sometimes requires that pesticides be applied in higher concentrations to overcome the effects of rapid decay.

Bioaccumulation and Biomagnification
Some pollutants are accumulated selectively by organisms. This phenomenon is called *bioaccumulation* and is particularly important for metals

and some organic compounds. Plants and animals absorb pollutants through their food, through skin contact, or through the air. If the pollutant has a tendency to bind with other substances in the organism, then it accumulates over time. Usually, the pollutant lodges in a particular part of the organism; for example, iodine tends to accumulate in the thyroid gland. As collection continues, the concentration of the pollutant in an individual organism rises, sometimes to toxic levels.

Biomagnification is a process whereby the concentration of a substance in animal tissues is increased step by step through a food chain. As a rule of thumb, it takes about 10 kg of food to make 1 kg of tissue in the consumer. If a substance has a tendency to remain in animal tissues rather than be metabolized or excreted, then its concentration in those tissues is increased potentially by as much as 10 times for every step in the food chain. When a food chain has many steps, the ultimate concentration of the toxin in the top predator is quite high.

Many of the persistent pesticides accumulate in fatty tissues of animals and tend to be concentrated in the food chain. Although concentrations in most waterways now generally range from not detectable to parts per billion (ppb), fish and bird tissues frequently contain DDT and other pesticides at the parts per million (ppm) level. Average concentrations of DDT and its metabolites in human tissues in the United States were about 8 ppm in 1970, 5 ppm in 1976, and less than 1 ppm in 1990 (CEQ 1994). Other hazardous substances that bioaccumulate include polychlorinated biphenyls (PCBs) and strontium-90, which tends to concentrate in bone tissues. Persistent pesticides generally are found in higher concentrations in aquatic sediment than in the water and remain in animal and human tissues for a long time.

Pollutant Transport Because toxic substances are dangerous at relatively low concentrations, transfers of small amounts from one place to another may be of significance. Many processes are involved, including transport in surface water and groundwater; in air, dust, and sediment; and through organisms. Often a pollutant deposited in one medium turns up in another. For example, organic compounds in wastes disposed of in a landfill enter groundwater, flow into surface water, are taken up by fish, and are passed through the food chain. Likewise, soil in a mine-tailings pile containing high concentrations of metals may be eroded by wind and transported considerable distances before being deposited.

Management problems become more complex as a result of the low concentrations and the mobility of toxic substances from one medium to another. Not the least of these problems is the expense of sampling and analyzing water, soil, or air for concentrations of a wide range of substances at the parts per million or parts per billion level. Laboratory analyses for some substances cost in excess of $1000 each, in addition to sampling costs. But more troublesome is the lack of development of regulatory mechanisms that simultaneously govern several different media. For example, a substance may be relatively benign when found in soil but be very hazardous in water. Regulations governing disposal of wastes on land must include some means to evaluate the likelihood that pollutants enter water and the concentrations in which they may occur. Such regulations become extremely complex and are both difficult and costly to enforce.

Uncertainty in Toxic Substances Management Managing toxic substance pollution presents many problems, but perhaps the most difficult one is that decisions often are made with less than adequate information. In the first place, a number of substances (at least several hundred) are toxic to humans or other organisms. For many of these we have little information on the exact level or dose at which toxicity occurs and even less information on their concentrations in the environment. Assessment of the risk to humans associated with exposure to a substance becomes one of extrapolation from very limited information, and the conclusions we draw are shaky (Cutter 1993; National Research Council 1994, 1996). On the one hand, we seek to eliminate any chance of adverse health impacts associated with a pollutant. However, we cannot do so with any certainty unless we reduce exposures to very low levels of the toxin, often at very high (and weakly justified) costs.

Careful and painstaking research by ecologists provides a strong argument for thoughtful management of resources. Ecosystems are so complex that one can never assume that an action affects only the immediate location. The vast body of ecological research in the past few decades showed that the interconnectedness of ecosystems leads to more far-reaching and often unexpected effects. Disruption in natural systems is likely to occur and lead to further degradation of these connected systems or possibly to a domino effect in which distantly related systems and organisms may suffer.

We must always remember that the ecological viewpoint is open to political uses, as are all scientific approaches. Ecosystems are both real and abstract (Kormondy 1969). The abstract idea of interrelatedness is used by individuals or organizations eager to protect some special interest from change or development. They argue that because everything is interrelated, the act of damming a certain stream eventually will lead to the collapse of civilization or life on Earth. Ecology as a serious field of study makes no such sweeping guarantees or generalities. In the context of resource management, it needs to be used in a sober and factually based manner.

Human Impact on Biogeochemical Cycles

One of the startling conclusions of global-scale studies of biogeochemical cycles is that human impact is sufficiently large to be easily recognizable at the global scale (Vitousek et al. 1997; Smil 1997a, b). Even though 71 percent of the Earth's surface is water and human population is concentrated in a small portion of the land area of the planet, the quantities of basic substances—carbon, nitrogen, water, and so on—processed by humans is a significant part of the global total. This is especially obvious in the carbon cycle, described earlier, but is no less significant for other biogeochemical cycles. Here we describe two examples in more detail: the hydrologic cycle and the food chain.

Hydrologic Cycle Less than 3 percent of the world's water is fresh, but most of this is found in glaciers. The fresh water that is accessible to humans is found in lakes, rivers, and the soil. This amounts to a total of perhaps 200,000 km^3. Groundwater is more plentiful (perhaps as much as 8 million km^3), but this is a finite quantity that cannot be drawn down indefinitely because its recharge is based on surplus surface water. Surplus surface water is derived from net precipitation: the difference between precipitation and evapotranspiration. Worldwide, this net precipitation on land areas totals about 40,000 km^3. This is the renewable supply of water that is available for river flow and for replacing depleted soil and groundwater.

Humans withdraw water from lakes, rivers, and wells and use it for irrigation, industrial and commercial purposes, consumption by themselves and domestic animals, and other purposes (Fig. 4.14). The total amount of water withdrawn from the hydrologic cycle worldwide is a little over 3000 km^3,

Figure 4.14 Irrigation of crops, as is being carried out by these women in The Gambia, accounts for most of the water withdrawn for human use worldwide.

or about 8 percent of the total supply. Of the water that is withdrawn, some is evaporated in the atmosphere, especially from irrigated lands. The remainder is returned to rivers and lakes after use, usually with waste products added. The total withdrawals for agricultural uses are estimated to be about 2880 km^3 per year, with 65 percent of this, or 1870 km^3, evapotranspired to the atmosphere by irrigated crops, and another 415 km^3 evaporated from other uses, mostly reservoirs in arid and semiarid areas (Postel et al. 1996). Thus, the average runoff from the continents to the oceans is decreased by human activities by 6 percent, or 2285 km^3. Most of this depletion occurs in the drier parts of the populated world, such as Asia and western North America, and in these regions the net depletion of runoff is much higher—virtually 100 percent for some basins.

The Food Chain Photosynthesis by green plants is the basic process that creates food for virtually all living organisms. *Net primary production* (NPP) is the net amount of biomass created by the plants in an ecosystem, after respiration by those plants is deducted. The world's biomes differ greatly in the amount of NPP that occurs in them. Wetlands and tropical rainforests are the most productive environments, sometimes producing 2000 to 4000 grams of biomass (dry weight) per square meter of area per year. At the other end of the spectrum are deserts and polar regions, where NPP is only a few grams or tens of grams per square meter per year. The total amount of NPP that occurs in the world includes all the food that is available to support life—human and otherwise. This is estimated to be about 220 petagrams (1 Pg = 10^{18}g) per year (Vitousek et al. 1986).

How much of this primary production is consumed by humans? Vitousek et al. (1986) estimated that humans consume about 0.8 Pg from cultivated land directly, but an additional 6.4 Pg are consumed by domestic animals, harvested as wood from forests, or caught in the world's fisheries, for a total of 7.2 Pg, or 3.2 percent of total global NPP, consumed or used directly by humans and their domestic animals. They extended their calculations to include that portion of NPP that is either controlled by humans or lost due to poor land management. Productivity controlled by humans would include the waste portions of crop plants or trees burned in shifting cultivation. Poor land management includes the decreased productivity of farmlands due to soil erosion and declining soil fertility. When these additional portions of global NPP are included, Vitousek and others estimate that humans directly consume, control, or waste about 39 percent of terrestrial NPP and about 25 percent of global NPP. Subsequent work has emphasized uncertainties associated with these estimates, but the approximate magnitude of human appropriation of NPP is certainly large (Rojstaczer et al. 2001).

The global-scale estimates described in the preceding paragraphs are very rough, but they provide an indication of the magnitude of the human impact on Earth's biological processes. What they show is that human participation in biogeochemical cycling, whether the hydrologic cycle, the food chain, the carbon cycle, or other cycles, accounts for a significant portion of the total amounts of materials processed. We can no longer consider humans to be just one of many species on the planet, consuming a portion of its output but otherwise not altering it. We are not merely nibbling at the edge of the natural world. We are major, if not the dominant, processors of material, intentionally controlling or inadvertently altering many ecological systems. At the local scale, in those environments most intensely used by humans, we have altered natural processes to the extent that entire biomes are destroyed and replaced by new, human-made and human-controlled systems.

ECOLOGICAL CONCEPTS IN RESOURCE MANAGEMENT

If humans are now the dominant players in the world's ecological systems, then we need to act accordingly. Whether we are concerned for the health and longevity of all species or just our own, we need to manage the Earth in ways that maintain its ability to support life. In this section, we describe four principles of environmental management that derive from an ecological view of natural resources.

Any Given Environment Has Finite Carrying Capacity

Earlier in this chapter, we discussed the concept of carrying capacity. This concept originally was developed for application to natural and human-managed systems and used to determine how many organisms could be supported in a given area. The concept is extended to human use of a given system as a way to suggest the extent to which we can use that system without damaging it. For example, we might calculate how much sewage a river can absorb, breaking it down and removing the waste products, without significant degradation of water quality. In this case, "significant" is defined in terms of whether sufficient dissolved oxygen is maintained to support a diverse aquatic community.

The problem with applying the carrying capacity concept to human use of the Earth arises when we ask how many humans can be supported by a given resource base. To do this we must make assumptions about what level of resource consumption constitutes an adequate level of human support and what level of environmental degradation is acceptable. For rabbits or elephants, we usually define support as meaning a level of food supply and habitat that allows the population to reproduce itself, without worrying too much about whether the animals are hungry or well-fed, uncomfortable or happy. But with humans we recognize that reproduction alone doesn't translate immediately into an acceptable quality of life. If people are poor and die young because of inadequate food supply, then we might argue that the carrying capacity of a system was exceeded even though the population can still reproduce itself.

Nonetheless, the idea of a finite carrying capacity for any given system is useful, provided we can agree on a definition of acceptable environmental quality and quality of human life. If, for example, we can agree that a forest must be maintained with a sufficient number of large trees to support those species that depend on mature forests for their survival, then we can agree on what level of harvest would be sustainable in that system. Questions such as these are at the core of the debate over maintaining habitat for biodiversity preservation (see Chapter 8).

Be Aware of Limiting Factors

One of the most important principles relating populations to ecosystem characteristics is the *principle of limiting factors*. Usually, many factors (nutrients, physical site characteristics, etc.) are necessary for an organism to exist, and the availability of these factors varies from site to site, with some factors plentiful and others rare. An organism requires different environmental conditions in varying degrees or in different amounts, and there is no reason to assume that every condition is available in exactly the amount required. Small changes in limiting factors have profound effects; similar changes in nonlimiting factors may have little or no effect.

Plant nutrients provide a simple example. A particular plant requires sunlight, water, a stable substrate, and a variety of nutrients from the soil. These nutrients include nitrogen, phosphorus, potassium, magnesium, copper, zinc, and many other elements. They are available in plentiful supply except one, say, phosphorus. The plant's rate of growth is restricted by the lack of phosphorus, even though there is more than enough of the other nutrients. In this situation, nitrogen fertilizer applied to the soil does nothing to help the plant, but phosphorus fertilizer will be very effective. Phosphorus is thus said to be the limiting factor in this case.

Nutrients are not the only factor that is limiting. Sunlight, carbon dioxide, frost, or any other environmental characteristic that an organism requires limits its growth. If substances are present in excessive quantities that may poison an organism, they also are considered limiting. Predators are often limiting factors for animal populations. It is usually difficult to determine just what factor or factors are limiting in any given situation, but clearly this information is essential to predicting the effects of environmental changes on ecosystem development and illustrates the critical importance of sound environmental knowledge for informed resource management.

Minimize Disruption by Mimicking Nature

A third principle of ecological management of natural resources is that human resource systems should mimic nature whenever possible. This principle follows from the study of ecosystems and from the experience of unanticipated consequences of human modifications to ecosystems. Although the scientific study of ecosystems advanced tremendously in recent years, we still cannot predict ecosystem states and processes with any degree of accuracy. This is due to the enormous complexity of food webs and interactions between individuals, species, and communities.

The study of ecosystems illustrates that these natural systems and the organisms they include developed over millions of years of evolution and adaptation. Each and every species in a community has adapted characteristics that allow it to exist in a specialized niche. Species interact with other species in a community, sometimes competitively, sometimes symbiotically, but always in ways that continue the exchange of energy and matter within the system and perpetuate its existence.

Experiences in recent decades showed that attempts to control nature, more often than not, ended with problems larger than those we hoped to solve initially. Control of predators results in a mushrooming of prey populations. Control of agricultural pests with insecticides results in the development of strains resistant to pesticides. In the face of this unpredictability and recurring management failures, many ecologists conclude that we are not likely to be able to control natural systems with any reliability, and we are better off to let these systems manage themselves to the extent possible.

One way to promote the systems' self-management is to use management strategies that exploit natural processes by mimicking them, working with nature rather than against it. For example, the history of agricultural pest control is one of many failures and only modest successes. The largest problems are those associated with human health effects, estimated to include 20,000 deaths per year worldwide; in the United States alone about 67,000 people are poisoned annually by pesticides (Pimentel et al. 1992; National Research Council 2000). Additional problems include damage to nontarget organisms, development of pesticide resistance, crop product losses, and water pollution.

Although these problems are balanced against the benefits of pesticide use, many argue that unintended negative effects can be reduced by adopting an alternative strategy called *integrated pest management* (IPM). This approach utilizes more intensive management, combining use of crop rotations and crop combinations that reduce pest infestations, biological control methods such as introduction of pest predators, and limited use of pesticides. The IPM approach recognizes that some pest losses are inevitable, but if the natural pest-resisting capabilities of an agricultural ecosystem are exploited, much less chemical pesticide need be used, minimizing the unintended negative impacts of pest control.

Close the Loops

Biogeochemical cycles are, ultimately, closed loops at the global scale. Carbon that is taken up in photosynthesis passes through the food chain, is released to the atmosphere, and is taken up in photosynthesis again. Nitrogen in the atmosphere is fixed in the soil, taken up in plants, converted to nitrate when plants decay, leached by soil water and transported to a river, converted to a gas through denitrification, and released to the atmosphere.

The gross imbalances in biogeochemical cycles that are resulting in a steady increase in atmospheric CO_2 concentrations and are causing typical nitrogen concentrations in human-impacted rivers to be 100 times their natural levels are a result of human-caused transfers of materials. Many of these transfers go just one way. Carbon is taken out of the Earth in the form of fossil fuels but is not put back again. Atmospheric nitrogen is converted to fertilizer and pumped into the soil, but our crops do not remove all of this nitrogen from the soil and much is left to contribute to water pollution.

If, every time we moved a quantity of matter from one place to another or changed it from one form to another, we made sure that an equivalent quantity was transferred in the other direction, then these problems would be alleviated. Carbon would not accumulate in the atmosphere, water would not be polluted, and resource use would be indefinitely renewable.

Of course, this is impossible in many cases. Fossil fuels cannot be replaced; to do so requires energy, which would defeat the purpose of using them. In fact, the second law of thermodynamics means that we would use more energy putting the fuels back in the ground than we could get out! But failure to close the loop means that we are creating an imbalance in Earth's biogeochemical cycles, which ultimately will damage other parts of the environment. The only solution, then, is to limit our use of resources to those that can be replaced, recycled, or reused. In the case of energy, this means abandoning the use of fossil fuels and relying only on the various forms of solar energy. In the case of nitrogen fertilizers, it means limiting our use to just the amount that can be taken up by the crops we plant and harvest, so that no excess remains to pollute water.

CONCLUSIONS

It will never be possible to manage all resources according to such strict ecological principles, and probably we wouldn't want to do so. But an ecological viewpoint on natural resources holds that environmental problems derive from violating these principles, and the more we are able to change how we use resources and return to production and consumption practices that are more like the natural processes that existed before humans occupied the Earth, the more likely we are to protect ecosystems and resources for future generations.

REFERENCES AND ADDITIONAL READING

Abell, R. A., D. M. Olson, E. Dinerstein, P. T. Hurley. 2000. *Freshwater Ecoregions of North America: A Conservation Assessment*. Washington, D.C.: Island Press.

Asner, G. P., T. R. Seastedt, and A. R. Townsend. 1997. The decoupling of terrestrial carbon and nitrogen cycles. *BioScience* 47:226–234.

Bureau of Land Management. 1999. *Public Land Statistics 1999*. http://www.blm.gov/

Carson, Rachel. 1962. *Silent Spring*. Boston: Houghton Mifflin.

Census Bureau. 2001. *Statistical Abstract of the United States*. Washington, D.C.: U.S. Bureau of the Census. http://blue.census.gov/statab/www/.

Colburn, T., D. Dumanoski, and J. P. Myers. 1996. *Our Stolen Future*. New York: Dutton.

Council on Environmental Quality (CEQ). 1994. *Environmental Quality, 24th Annual Report*. Washington, D.C.: U.S. Government Printing Office.

Cutter, S. L. 1993. *Living with Risk*. London: Edward Arnold.

Detwyler, T. R. and M. G. Marcus, eds. 1962. *Urbanization and Environment*. Belmont, CA: Duxbury Press.

Fish and Wildlife Service. 2002. *National Wildlife Refuge System Lands Database*. http://realty.fws.gov/nwrs.htm.

Food and Agriculture Organization. 2002. *FAOSTAT*, http://apps.fao.org/.

Graham, M. and C. Miller. 2001. Disclosure of toxic releases in the United States. *Environment* 43(8):8–20.

Goudie, A. and H. Viles. 1997. *The Earth Transformed*. New York: Blackwell.

Houghton, R. A. 1994. The worldwide extent of land-use change. *BioScience* 44:305–329.

Hynes, H. 1989. *The Recurring Silent Spring*. New York: Pergamon.

Intergovernmental Panel on Climate Change, 2001. *IPCC Special Report on Land Use, Land-Use Change and Forestry*. http://grida.no/climate/ipcc/land_use/index.htm.

Kormondy, E. J. 1969. *Concepts of Ecology*. Englewood Cliffs, NJ: Prentice-Hall.

Krimsky, S. 1999. *Hormonal Chaos: The Scientific and Social Origins of the Environmental Endocrine Hypothesis*. Baltimore: Johns Hopkins University Press.

Lovelock, J. E. 1979. *Gaia, a New Look at Life on Earth*. Oxford: Oxford University Press.

Marsh, G. P. 1864. *Man and Nature, or Physical Geography as Modified by Human Action*. New York: Scribner.

Myers, N., ed. 1989. *The Gaia Atlas of Planet Management*. London: Pan Books.

National Park Service. 2002. *Public Use Statistics Office*. http://www2.nature.nps.gov/stats/.

National Research Council. 1994. *Science and Judgment in Risk Assessment*. Washington, D.C.: National Academy Press.

National Research Council (NRC). 1996. *Understanding Risk: Informing Decisions in a Democratic Society*. Washington, D.C.: National Academy Press.

———. 2000. *The Future Role of Pesticides in U.S. Agriculture*. Washington, D.C.: National Academy Press.

National Resource Conservation Service, 2002. *State of the Land*. http://www.nrcs.usda.gov/technical/land/.

Pimentel, D., et al. 1991. Environmental and economic effects of reducing pesticide use. *BioScience* 41:402–409.

———. 1992. Environmental and economic costs of pesticide use. *BioScience* 42:750–760.

Postel, S. L., G. C. Daily, and P. R. Ehrlich. 1996. Human appropriation of renewable fresh water. *Science* 271:785–788.

Ricketts, T. H., E. Dinerstein, D. M. Olson, C. J. Loucks. 1999. *Terrestrial Ecoregions of North America: A Conservation Assessment*. Washington, D.C.: Island Press.

Robertson, G. P. 1986: Nitrogen: regional contributions to the global cycle. *Environment* 28(10): 16–20, 29.

Rojstaczer S., S. M. Sterling, and N. J. Moore, 2001. Human appropriation of photosynthesis products. *Science* 294:2549–2552.

Simmons, I. G. 1981. *Ecology of Natural Resources*. 2nd Ed. Philadelphia: W. B. Saunders.

Smil, V. 1997. Global population and the nitrogen cycle. *Scientific American* July:76–81.

_____. 1999. *Cycles of Life: Civilization and the Biosphere*. New York: W. H. Freeman.

Turner, B. L., W. C. Clark, R. W. Kates, J. F. Richards, J. T. Matthews, and W. B. Meyer, eds. 1990. *The Earth as Transformed by Human Action*. Cambridge: Cambridge University Press.

United Nations Environment Programme. 2002. *Global Environment Outlook 3*. London: Earthscan.

U.S. Bureau of the Census. 2001. *Statistical Abstract of the U.S.* Washington, D.C.: U.S. Government Printing Office.

U.S. Environmental Protection Agency. 2000. *TRI Public Data Release Report.* Washington, D.C.: Office of Prevention, Pesticides, and Toxic Substances, U.S. EPA. www.epa.gov/tri/tridata/tri99/index.html.

Vitousek, P., et al. 1986. Human appropriation of the products of photosynthesis. *BioScience* 36: 368–373.

Vitousek, P. M. 1994. Beyond global warming: ecology and global change. *Ecology* 75:1861–1876.

Vitousek, P. M., H. A. Mooney, J. Lubchenco, J. M. Melillo. 1997. Human domination of Earth's ecosystems. *Science* 277:494–499.

Wilderness Information Network. 2002. *Wilderness Information System.* http://www.wilderness.net/.

World Resources Institute. 1996. *World Resources 1996–97*. New York: Oxford University Press.

For more information, consult our web page at *http://www.wiley.com/college/cutter*.

STUDY QUESTIONS

1. Using your local community as an example, list some of the adverse effects of biogeochemical cycle disruptions. Are these disruptions produced locally (e.g., activities that are going on in your community), transferred from someone else's activities either upwind or upriver, or resulting from larger-scale global phenomena? Why is it important to consider the likely source of the disruption?

2. Keep a diary of the chemical products you use in any given day, such as shampoos, laundry detergent, dish soap, and deodorant. Examine each for chemicals used. How many of these products contain synthetic substances? Can you think of nonsynthetic substitutes (hint: vinegar, baking soda, lemon juice, and baby powder) that might accomplish the same thing? How many alternatives did you discover?

3. How does geographic scale help us understand the differences between communities, ecotones, and ecosystems?

4. Why is the concept of bioregions useful in environmental management? What are some of the major characteristics of the world's major biomes?

5. What might land uses look like in the year 2010? In the year 2030? How might human activities affect them?

CHAPTER 5

THE HUMAN POPULATION

The world's population is increasing, but not as rapidly as it was a decade ago. In 1985 global population was estimated at 4.84 billion, and by 2002 it had reached 6.1 billion. In 2002 one-fifth of this total, or 1.3 billion, lived in China, the world's most populous country, with another billion living in India. The United States had 283 million people (Population Reference Bureau 2001).

As the world's population increases, so does its use of resources. Environmentalists and population theorists such as Paul Ehrlich and Lester Brown and economists such as H. E. Daly and N. Georgescu-Roegen warn of increased population pressures on natural resource consumption. People must be fed, housed, and clothed, and the more people there are, the more food resources, housing materials, and fibers must be produced. Other theorists are confident that population growth does not mean a drop in the standard of living.

The population problem, or crisis to some, is not a recent phenomenon. In 1798, the British economist *Thomas Malthus* foresaw some of the world's current population problems. Malthus wrote that populations increase in size geometrically; that is, they double in size in a fixed time period. Geometric growth is shown by the upward-trending curve in Figure 5.1. Malthus also wrote that food supplies increase arithmetically; that is, they increase by addition of a fixed amount in a given time period. Arithmetic growth is shown as a straight line in Figure 5.1. Eventually, he said, population growth would outstrip the food resources (assuming, of course, no new resources were developed), with catastrophic consequences—mass starvation, poverty, and economic and social collapse. The history of population growth and food production since Malthus' time has been very different from what Malthus predicted. Population has indeed grown rapidly, but food production has grown even faster. The crisis he predicted has not materialized.

Debates continue today over the relationship between population growth, resource consumption and scarcity, and environmental degradation. *Neo-Malthusians* take the same perspective as Malthus, yet they argue for strong birth control measures to postpone or delay population growth to a level below the limit of resource availability. Advocacy of birth control to stabilize population growth, instead of expecting nature to do the job through famine and war, differentiates neo-Malthusianism from the original Malthusians.

A nineteenth-century critic of Malthus, Karl Marx, stressed that there was no single theory of population growth and resource use. Increased population growth did not by itself, as Malthus suggested, result in excessive resource use and a lowered standard of living. Marx believed that poverty was caused by the economic system, which exploited labor for the benefit of the elite. The cause of poverty was economic, he stated, not solely an increase in numbers of people.

In the twentieth century, Esther Boserup (1965, 1981) and Julian Simon (1981, 1986) suggested that population growth may be beneficial in providing a stimulus for improving the human condition. They argue that population growth intensifies land use, resulting in increased agricultural production. The end result is that all individuals benefit from the increased production and thus achieve a higher standard of living. They suggest that Malthus incorrectly assumed that increased population led to increased poverty, starvation, and war. A third perspective suggests that population

88 CHAPTER 5 THE HUMAN POPULATION

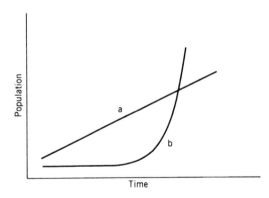

Figure 5.1 Arithmetic versus geometric growth. Arithmetic growth (*a*) follows a straight line, increasing by the same number each year. Geometric growth (*b*) follows a curve, increasing by the same percent each year.

growth in and of itself is not the problem; rather, it is the population growth plus affluence and technology that are the major contributors to natural resource degradation (McKibben 1998). Simply slowing population growth may not reduce natural resource degradation unless we also slow consumption of resources (energy and materials). In other words, it's not only about the numbers of people but about their wants and desires as well (Kates 2000).

Today some population experts anticipate some form of population catastrophe in the near future. Others, however, are confident that human needs can be met no matter how large the world's population becomes. The following historical perspective and some of today's population numbers may help you decide where you stand on this issue.

A BRIEF HISTORY OF POPULATION GROWTH

Although we cannot accurately measure the world's human population in the distant past, demographers and archaeologists, among others, have developed low to high ranges for population size and growth over thousands of years. These estimates change, of course, with the unearthing of new evidence such as ancient communities, new dating methods, and new theories. The world's human population at the end of the most recent ice age, about 10,000 years ago, was somewhere between 2 and 10 million people. It had taken perhaps 1 to 2 million years for the population to grow to this size. When we consider that population has burgeoned to 6.1 billion in the past 10,000 years—and most of that in the last 300 years—it is clear that extraordinary changes have taken place in all aspects of human life in order to adjust to this astonishing growth.

From 8000 B.C. to A.D. 1 the population doubled almost six times, to between 200 and 400 million (Fig. 5.2). The doubling time was more than 1000 years. One should realize that this

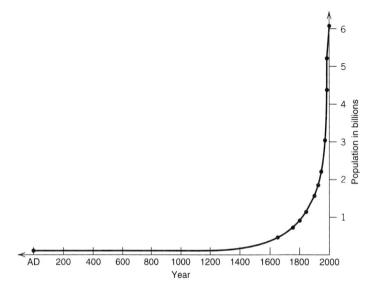

Figure 5.2 World population growth. World population growth was slow and steady until about 1700 to 1750. Since then, the world's population has expanded rapidly.

growth was not steady and smooth. A close look would reveal sudden rises and drops owing to the vagaries of famine, war, and disease over small and large regions. The tendency, however, was toward growth. Between A.D. 1 and 1750, growth continued at about the same rate, ultimately reaching 750 million by 1750. Though scholars differ, the technological developments of agriculture and irrigation no doubt had much to do with this population increase. After 1750 the world's population began its modern climb, starting in Europe. It took only 150 years, from 1750 to 1900, for the world's population to double from 750 million to 1.5 billion. The population doubled once again in 65 years between 1900 and 1965 and nearly doubled again between 1965 and 2000, a mere 35 years!

Where was the growth, and where is the growth likely to occur in the future? After 1750, Europe's population mushroomed, and the resulting crowding and poor conditions had a great impact on the settlement of the Americas, Australia, and New Zealand and on European imperialism worldwide. Europe needed raw materials to support its astonishing human and economic growth, and those resources were available in Africa, Asia, the Americas, and the Indian subcontinent. During the twentieth century, the benefits of the Scientific Revolution that led to Europe's boom—the germ theory of disease, ideas about cleanliness, vaccinations, agricultural improvements—were introduced to the rest of the world, and population growth shifted to Europe's former colonies.

Doubling time is the number of years it takes a population to double in size, assuming a constant rate of natural increase. The lower the number of years to double in size, the faster the growth rate. In the mid-1960s, when the world's population was growing at 2.0 percent per year, the doubling time of world population was about 35 years. By 2001, with a growth rate of 1.3 percent per year, it had increased to about 47 years. This change in doubling time shows the weakness of the assumption of a constant rate of increase. For example, in 2049, a date 47 years from 2002, the world's population will almost certainly be less than double the 2002 level. If population growth trends existing in 2002 are projected through the twenty-first century, the world's population would reach 9 billion by 2050 and 10 billion by the year 2100. At the national and regional level, great differences in growth rates exist (Table 5.1). Some countries, notably in Africa and southwest Asia, have growth rates well over 3 percent per year, with doubling times under 25 years. Other areas, particularly Europe, have virtually zero population growth.

World population growth has slowed significantly since the late 1960s, and in the 1990s the growth rate slowed even more, surprising many demographers. Fertility rates are dropping rapidly in much of the world, and estimates of future population have been revised downward repeatedly. There are now clear signs that the period of rapid population growth centered on the 20th century is coming to an end. Present growth rates cannot be extrapolated very far into the future, however. Many factors, such as war, disease, social pressure, government programs, immigration, and emigration, may affect the doubling time on a year-to-year basis, as population projections illustrate (Issue 5.1). Since we don't have crystal balls to peer into the future, these population forecasts are often more art than science.

BASIC DEMOGRAPHICS

Many factors have contributed to this dramatic increase in world population. One is the development of a broader worldwide food base because of increased trade; another is humanity's rise in overall disease resistance, which was also a result of increased trade and travel. It has been suggested as well that population began to rise in the wake of better medical technology and theory, leading to a drop in infant and child mortality rates and an increased life span for large segments of the world's population. Let us discuss the possibilities and implications of present and future growth rates.

We will examine two dimensions of the world population picture. The first includes the rate and causes of population growth and is called *population dynamics*. The second is the location of growth, that is, its spatial distribution around the world. Both are essential to our understanding of current and future population trends. In examining population dynamics, the first question is: How and why do populations grow? Two factors are important in understanding global population growth: rates of natural growth/decline and the

Table 5.1 Demographic Characteristics of Selected Countries, 2001

Country	Population (millions)	Rate of Natural Increase (%/year)	Crude Birth Rate (births/1000 population)	Crude Death Rate (deaths/1000 population)	Total Fertility Rate
World	6,137	1.3	22	9	2.8
Argentina	37.5	1.1	19	8	2.6
Australia	19.4	0.6	13	7	1.7
Bangladesh	133.5	2.0	28	8	3.3
Brazil	171.8	1.5	22	7	2.4
Canada	31.0	0.3	11	8	1.4
China	1,273.3	0.9	15	6	1.8
Colombia	43.1	1.8	24	6	2.6
Congo (Dem. Rep.)	53.6	3.1	47	16	7.0
Cuba	11.3	0.6	14	7	1.6
Egypt	69.8	2.1	28	7	3.5
Ethiopia	65.4	2.9	44	15	5.9
France	59.2	0.4	13	9	1.9
Germany	82.2	0.1	9	10	1.3
Guatemala	13.0	2.9	36	7	4.8
India	1,033.0	1.7	26	9	3.2
Indonesia	206.1	1.7	23	6	2.7
Iran	66.1	1.2	18	6	2.6
Israel	6.4	1.6	22	6	3.0
Italy	57.8	0.0	9	10	1.3
Japan	127.1	0.2	9	8	1.3

age structure of a population. A third factor, immigration/emigration, is an important consideration at the national or regional level but does not influence global trends.

Birth, Death, and Fertility

Natural growth (sometimes called natural increase or decrease) is a simple measure of population growth that examines the differences between births (fertility) and deaths (mortality) in a given group. *Birth* and *death rates* are normally expressed as rate of occurrence per 1000 people. For example, if 3000 babies were born in a population of 150,000, then the birth rate would be 20 per 1000. *Natural increase* is the difference between birth and death rates and is expressed as a percentage figure. In the United States the rate of natural increase in 2000 was 0.5 percent. In other words, about 3.8 million people were born in the United States during the year and about 2.4 million died, causing a net natural increase of 1.4 million. Net legal immigration added another 0.8 million (not counting undocumented immigrants) to the U.S. population during this time, raising the net growth rate to 0.8 percent.

Annual growth rates must be considered in combination with the actual population figure. A low annual growth rate of a small population is significantly different from a comparable annual figure for a much larger population. India and Colombia, for example, had similar annual growth rates of 1.7 and 1.8 percent, respectively, in 2001. India, with a base population of 1.03 billion, increased by 17.5 million annually. Colombia, on the other

Table 5.1 (*Continued*)

Country	Population (millions)	Rate of Natural Increase (%/year)	Crude Birth Rate (births/1000 population)	Crude Death Rate (deaths/1000 population)	Total Fertility Rate
Mexico	99.6	1.9	24	5	2.8
Myanmar	47.8	1.6	28	12	3.3
Nigeria	126.6	2.8	41	14	5.8
Pakistan	145.0	2.8	39	11	5.6
Peru	26.1	1.8	24	6	2.9
Philippines	77.2	2.2	29	6	3.5
Poland	38.6	0.0	10	10	1.4
Romania	22.4	−0.1	9	15	1.3
Russia	144.4	−0.7	9	15	1.2
South Africa	43.6	1.2	25	14	2.9
South Korea	48.8	0.9	14	5	1.5
Spain	39.8	0	10	9	1.2
Tanzania	36.2	2.8	41	13	5.6
Thailand	62.4	0.8	14	6	1.8
Turkey	66.3	1.5	22	7	2.5
Ukraine	49.1	−0.7	8	15	1.1
United Kingdom	60.0	0.1	12	11	1.7
United States	284.5	0.6	15	9	2.1
Vietnam	78.7	1.4	20	6	2.3
Yemen	18.0	3.3	44	11	7.2

Source: Population Reference Bureau, 2001.

hand, with a population base of 43.1 million, added only 776,000 to its population that year.

One of the most important factors in population growth today is the birth rate. The more babies are born, the more a population grows. Birth rates in the world today are typically between about 10 and 50 per 1000 population. Birth rates are controlled mostly by the *fertility rate*, a measure of the average number of children a woman has in her reproductive years (ages 15 to 49 years). Birth rates are also affected by the age structure of a population. If there is a large number of young women in a population, the birth rate will be higher, while in an older population birth rates are lower because more women are beyond child-bearing age.

On a global scale, the total fertility rate was 2.8 in 2001. The rate is significantly higher in the less industrialized nations (3.6) than in the more industrialized nations (1.6). There are also some very distinct regional differences. For example, Africa had a total fertility rate of 5.2, versus 1.4 in Europe.

Death rates are more closely related to the age structure of a population than are birth rates. Countries with older populations have higher death rates than those with younger populations. Death rates are also affected by factors such as nutrition and the availability of health care, and so high death rates also occur in very poor countries, even though their populations may be relatively young. Death rates in the world tend to be between about 5 and 20 per thousand, much lower than typical birth rates. This difference between birth rates and death rates results in population growth.

ISSUE 5.1: AIDS AND POPULATION GROWTH IN AFRICA

The HIV/AIDS pandemic is now 20 years old, and since it began, more than 60 million people have been infected with the human immunodeficiency virus (HIV). Mortality from the disease is high; it ranks as the fourth largest cause of death worldwide. In sub-Saharan Africa, it is the leading cause of death. More than 2.3 million African people died from AIDS in 2001, with another 28.1 million (9 percent of the population aged 15–49 years) now living with the virus. To put this in perspective, the death toll is equivalent to the entire metropolitan population of Tampa, Florida, or Portland, Oregon.

In the hardest-hit nations (see figure), more than 5 percent of the population is affected, mostly in southern Africa. The highest prevalence of HIV in the world is in Botswana, where 36 percent of the adult population is infected. However, more than one-fifth of Africans living in southern Africa are infected. In many of these nations, life expectancies also have plummeted. A baby born in Botswana, Malawi, Mozambique, or Swaziland now has a life expectancy of less than 40 years. In all of sub-Saharan Africa, life expectancy is normally 62 years, but with the HIV/AIDS pandemic it has dropped to 47 years.

HIV/AIDS affects not only the sexually active population (males and females) but also, increasingly, newborns who are born HIV-positive. Child mortality rates are rising throughout sub-Saharan Africa. In one of the most affected nations, Zimbabwe, nearly 70 percent of the deaths among children under the age of five years are due to AIDS. If children are lucky enough to be born without HIV infection, ultimately they may be orphaned; more than 12.1 million children lost a mother or both parents to the epidemic in 2000. Among all the world's children orphaned by the AIDS-related death of a parent, the overwhelming majority live in sub-Saharan Africa (95 percent). In the Congo (Democratic Republic), Ethiopia, Kenya, Nigeria, Tanzania, Uganda, Zambia, and Zimbabwe, more than 5 percent of children (under age 15 years) are orphaned by AIDS.

The strains on these nations to cope with crisis is apparent, with enormous impacts on income, poverty rates, and economic growth. Per capita growth in the most affected sub-Saharan nations is falling by one percent yearly as a direct result of AIDS. Poverty is on the rise, with per capita income levels declining by 13 percent in Botswana, for example. Not only are incomes falling, but households are expected to take on more dependents as well. The cycle of poverty worsens as households cut food consumption and sell what remaining assets they have to cover health care and funerals, becoming more destitute in the process (UNAIDS/WHO 2001).

The plight of sub-Saharan Africa is the most striking example of the HIV/AIDS pandemic, but it is not the only area of concern. Exploding rates of HIV/AIDS in Asia among drug users and sex workers (Cohen 2001) and in Eastern Europe among young people (via sexual transmission and also high rates of injection drug use) are also of concern. To combat the epidemic, the United Nations General Assembly in June 2001 agreed on a framework for national and international accountabilty in the fight against the HIV/AIDS epidemic. In addition to a reorientation of spending priorities among affected nations, new World Bank loans, and a global fund for AIDS, the financial resources are beginning to appear to combat this deadly disease and its impact on national and international security.

Recent trends in world population growth rates and average fertility rates are shown in Figure 5.3. The global population growth rate was relatively low in 1950, because death rates in poorer countries were still high. But improved medicine and hygiene caused significant reductions in mortality, and population rose rapidly. Average fertility rates at this time were about 5. Between 1957 and 1962, population growth plummeted as a result of famine in China brought on by crop failures during the Great Leap Forward. This was the government's attempt to dramatically increase food pro-

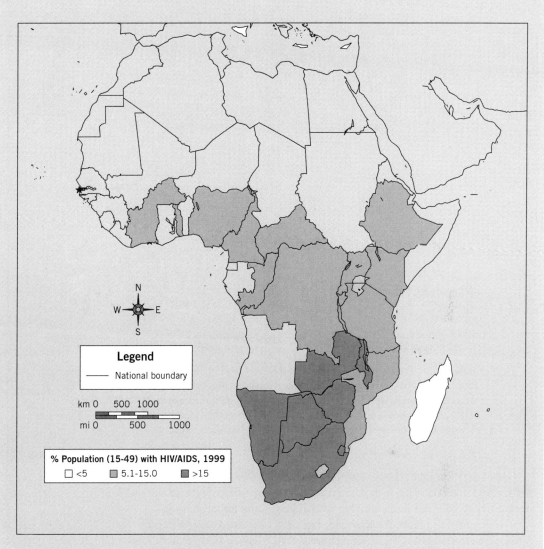

Percentage of African population aged 15 to 49 years with HIV infection or AIDS, 1999. *Source:* Population Reference Bureau, 2001.

duction by planting even where conditions were not favorable. Population growth rebounded quickly, however, and remained over 2 percent per year through the rest of the 1960s. Significant declines in fertility began in the early 1970s. By 1980 the average world fertility rate was below 4, and in 1997 it dropped to 2.8. This decline in fertility directly affected birth rates and thus population growth. In Japan and some European countries fertility has dropped well below replacement level, and if it remains low, populations may decline.

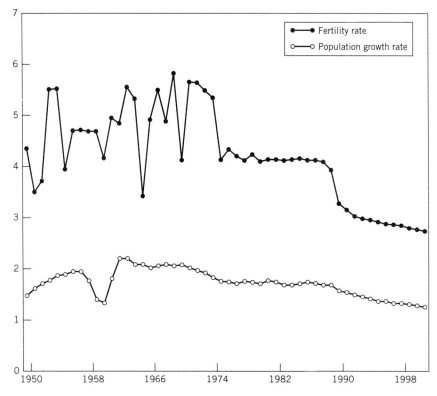

Figure 5.3 Trends in population growth rate and fertility. Fertility has gradually declined since the mid-1960s, and the population growth rate has followed a similar course. The dramatic dip in the world's population growth rate is a result of a dramatic increase in mortality and a decrease in fertility that occurred during the Great Leap Forward in China. *Source:* U.S. Bureau of the Census, International Data Base, 2002.

One of the most important reasons for a decline in the rate of natural growth in the United States has been a steady decrease in the number of children born per family. In 2000, for example, the total fertility rate among American women was 2.1. This means that, on the average, every American woman has at least two live births during her lifetime. The total fertility rate in the United States has steadily declined since the 1950s (Table 5.2), although from 1985 until 1995, a small increase occurred, as members of the *baby boom* generation neared the end of their reproductive years.

For decades, debate has persisted on the question of whether or not high-growth nations should work to decrease their fertility rates. As an example, the People's Republic of China (now with a total fertility rate of 1.8) instituted a one-child limit for all families (Fig. 5.4) during the 1970s. Although the policies varied from province to province, in some areas the birth of a second child resulted in the loss of state medical and school aid, ostracism, and official criticism. A third pregnancy brought strong pressure from the community for an abortion.

Accustomed to large families and desirous of having boys, many Chinese found it very difficult

Table 5.2 U.S. Fertility Rates, 1950–2000

Year	Rate
1950	3.3
1960	3.4
1970	2.1
1975	1.8
1980	1.8
1985	1.8
1990	2.1
1995	2.0
2000	2.1

Source: U.S. Census Bureau, 1999; Population Reference Bureau, 2001.

Figure 5.4 One-child family policy in China. The Chinese government has encouraged late marriage and has imposed economic penalties on families with more than one child. This has resulted in a significant lowering of the country's fertility rate. This wall poster in Chengdu, Sichuan Province, praises the merits of one-child families.

to live with this policy. Reports of female infanticide, though true, remain very controversial, because those who favor population control do not like to admit to such abuses, which often provide ammunition to those who would abolish all limits to growth. In recent years, Chinese policies have eased somewhat as urban-rural and ethnic differences have begun to play a role in family planning.

For the past 25 years India has struggled to reduce birth rates by providing various economic and social incentives, as well as by involuntary measures. Its most recent census revealed, however, that growth rates did decline somewhat between 1980 and 2000. In some countries, fertility has declined primarily as a result of government-sponsored family planning programs, while in others the decline is attributed to the increased educational and economic levels of women.

Zero population growth (ZPG) is a term indicating the number of births that will simply replace a population, without further growth. It takes a total fertility rate of about 2.1 in developed nations or 2.7 in developing nations to maintain a population at a constant size, assuming a stable age structure and no net migration. The difference is explained by higher mortality rates in the developing nations, which require a higher birth rate to offset losses. A total fertility rate of less than 2.1 would eventually lead to population decline, assuming no net immigration. To achieve ZPG globally, a total fertility rate of about 2.5 is needed, but a much lower rate (below *replacement levels*) would be needed in some regions to stabilize population. Even if a ZPG fertility rate were reached, which is highly unlikely in this century, the population would continue to expand simply because of the age structure of the world population. In other words, a disproportionate number of young people have yet to reach their child-bearing years, and even if they have only two children, population growth would be maintained for decades (Bongaarts 1998).

Age Structure

The second factor that contributes to overall population change is the age structure of a population. A short discussion of the profiles of populations by age and sex will give us the key to understanding whether a country has an expanding, declining, or stable population.

The *population pyramid* (or more correctly, the age-structure diagram) is a visually striking representation of the age and sex structure of a population (Fig. 5.5). To make up a pyramid, a population is classified in age groups or *cohorts* by five-year intervals (0–4, 5–9, 10–14, etc.) and by sex. The actual number of people or the percentage of a population that falls into each of these age categories is then graphed. The percentage of males in specific age groups is shown on the left and the percentage of females on the right. The general shape of the pyramid indicates the relative growth of the population.

For example, a rapidly expanding population such as the Congo's (Fig. 5.5) has a very broad base because a large percentage of the population is younger than 14 years of age; that is, many children are being born to each mother. The average fertility rate of Congolese women is 7.0, and the birth rate is 47 per 1000 population. Many of the

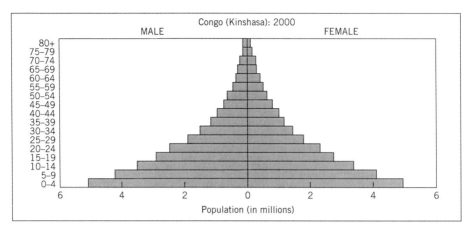
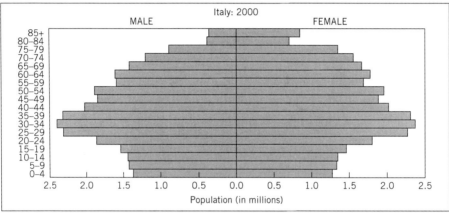

Figure 5.5 Population pyramids for Congo (formerly Zaire) and Italy in 2000. Congo's population pyramid is broad at the base, indicating that it is young and growing rapidly. Italy's pyramid has near-vertical sides, indicating that the number of children is about the same as the number of middle-aged people, and the population is not growing. *Source:* U.S. Bureau of the Census, International Data Base, 2002 (*http://www.census.gov/cgi-bin/ipc/idbagg*).

developing nations have population pyramids of this shape. One hundred years ago, the U.S. population had a similar shape, as did the countries of Europe.

Stable or declining populations such as those of Italy (Fig. 5.5) have population pyramids with narrower bases. As the population ages, the actual number or percentage in each cohort declines because of mortality. Italy's fertility rate is 1.3 (below replacement level), causing the pyramid to be narrow at the base. The low birth rate of 9 is attributable to both low fertility and a relatively small proportion of women in child-bearing years.

The effect of changing fertility on changing age structures is evident in data from Brazil (Fig. 5.6). In 1975 Brazil had high fertility rates characteristic of developing countries at that time and a similarly broad-based population pyramid. But during the next 25 years, both fertility and birth rates declined rapidly, and by 2000 a distinct narrowing of the pyramid's base was evident. The pyramid for 2025, based on projections, shows continued low fertility—a pyramid that is approaching the near-vertical pattern characteristic of stable populations. This is just a projection, and we really don't know exactly what Brazilian fertility rates will be in the next few decades. But we do know how many young girls are alive today, and this number, when multiplied by the fertility rate for each cohort, helps us predict the total number of children who will be born to those women as they enter child-bearing years.

The three pyramids in Figure 5.6 provide an excellent illustration of how the total population

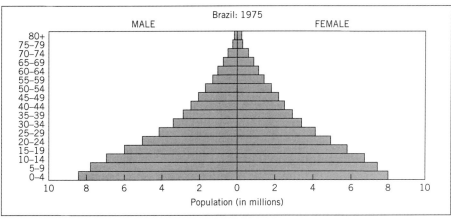

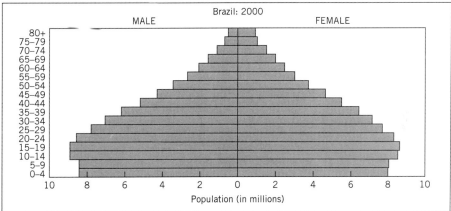

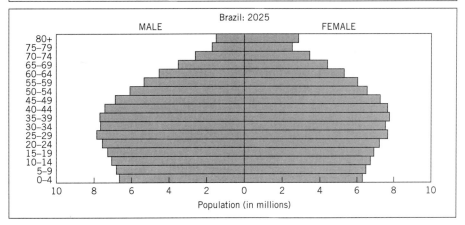

Figure 5.6 Population pyramids for Brazil: 1975, 2000, 2025. Initially characterized by high birth rates in 1975, Brazil's population growth in 2000 reflected a more stable population. Projected growth also indicates a stable population. *Source:* U.S. Bureau of the Census, 2002.

grows as a consequence of a period of high birth rates and increasing longevity of the population resulting from improved nutrition and medical care. With high death rates and low life expectancy, a pyramid shaped like Brazil's in 1975 could represent a stable population (although by this time Brazil's population was already growing rapidly). The lower death rates and greater longevity typical of wealthier nations today mean that the pyramid of a stable population is shaped more like Brazil's in 2025.

The United States has seen significant variations in fertility in the last few decades. Right after World War II and into the 1960s, the U.S. fertility rate was at a modern-day high, as explained earlier. The baby boomers have not been reproducing at the high rate of their parents, and even with their huge numbers, the number of people under age 14 years has decreased since 1970. As the baby boom cohort ages, so does the overall age of the U.S. population. The median age in 1970 was 28 years; in 1980 it was 30, by 1995 it was 34, and at the end of the century it was 35. Declining birth rates and the general aging of the baby-boom cohort of the 1950s, coupled with a long life expectancy, account for this trend.

Migration

Migration, which includes immigration and emigration, is the movement of individuals from one location to another. It has no influence on global population projections, but it does have a significant impact at national, regional, and local population levels. Migration flows are caused by differences in economic opportunities, group conflicts between and within nations, and environmental disruption. These flows can be permanent or temporary.

The populations of many countries are affected by migration. In the short run, refugees may flee or be forced from their homes during war or famine and migrate to nearby nations, with the intention of returning home when conditions improve (Fig. 5.7). Often such short-term migration becomes permanent. In the long run, regions of high population growth or poor economic conditions tend to be areas of net emigration, while those with strong economies and a need for labor tend to be regions of net immigration. In the world today, Europe and North America have substantial net immigration.

Early migration to North America was of two basic types. Free immigrants came in search of a better life, and forced immigrants, or slaves, helped make life easier for others. By the late nineteenth century, mass migration greatly increased the flow as entire ethnic or regional groups decided collectively to move to the United States and Canada, largely in search of better economic opportunities, as conditions became too crowded in "the old country."

Since World War II, the international migration flow has to a large extent been reversed, with the largest percentage of migrants moving from less industrialized nations to European and North American urban areas. The principal destinations of transnational migrants from 1950 onward have

Figure 5.7 Afghan refugees wait for humanitarian aid in northern Afghanistan, November 2001, following the U.S. invasion.

been the United States, Canada, Australia, and New Zealand, with large numbers moving to Europe and Great Britain as well. All of these countries currently have low birth rates, and the native population is stabilizing or shrinking. These new immigrants account for much of the increase in population in these countries. This trend has been accelerated by war and genocide; for example, the Vietnamese population in the United States has expanded since the end of the Vietnam War, and in the Congo, the number of refugees from Rwanda has increased following the 1994 civil war.

As a result of increasing numbers of migrants to the industrialized world, some countries have devised restrictive policies for immigration, maintaining that the immigrants are stealing wages from native-born citizens. Of course there is usually a large component of fear and dislike—racism—among natives toward these new residents. Though often disliked, the peoples of the Middle East, Asia, and Africa are allowed to enter industrialized nations on a temporary basis to work the jobs that declining populations cannot or will not fill. In Europe these temporary residents are called *guest workers*. Fears of an unending flood of illegal immigrants from Mexico into the United States is a perennial source of controversy in the United States.

This situation is only a symptom of a larger problem, that is, population pressure and perceived lack of opportunity in many less industrialized nations. People are both pulled off the land into the cities by the attractive opportunities available there and pushed off the land because economic opportunities in rural areas are limited. A move to the regional or capital city brings little satisfaction because these areas are overcrowded and cannot provide basic services (sanitation, electricity, and housing) for the new immigrants.

Migration will continue as long as rural populations in poorer countries perceive that economic opportunity remains greater elsewhere. When the population boom hit Europe in the eighteenth century, the overcrowded countries had empty lands for the population to spread into; where the land was not empty, as with the native populations of the Americas, the Europeans simply conquered by force. Today the peoples of Asia and Africa have no empty lands into which to expand. The situation is somewhat better in Central and South America, although even there the industrialized, environmentally sensitive countries have chastised the governments for their environmentally destructive methods of settlement. Many peasants still remain landless, thus increasing the likelihood of poor stewardship of the land.

Trends in Population Growth

When we examine current rates of population change due to natural increase and migration, we see significant population growth at rates that, if continued, will cause dramatic changes in all aspects of life in just a few decades. But how are rates of population growth changing? Can present rates be extrapolated into the future? The answer is yes and no. Significant changes in population growth are occurring, and these cannot be predicted beyond a few decades. On the other hand, we know some important parts of the population puzzle, especially the numbers and ages of people alive today, and so we can predict the near future with relative certainty.

The demographic transition is a widely used generalization of past population change that can be used to illustrate these processes (Fig. 5.8). It is a pattern generalized from the history of population in Europe in the last two centuries. Prior to the Industrial Revolution with its attendant social changes and medical advances, most human populations had both high fertility rates and high mortality rates, together with relatively stable populations. In recent centuries in the industrialized nations, mortality rates began to decline as a result of increased standards of living and medical care. This decline preceded the decline in fertility. The result was a period of time in which birth rates were substantially greater than mortality rates, and population increased accordingly.

This period of increase was characterized by a particularly broad-based population pyramid. It

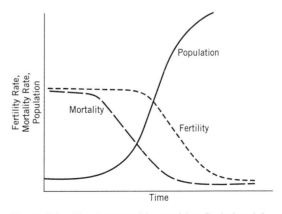

Figure 5.8 The demographic transition. Preindustrial populations have generally had high birth rates and death rates. When death rates fall before birth rates, population rises rapidly.

was also during this era, the eighteenth and nineteenth centuries, that the large populations of Europe found an outlet by settling in North America, Australia, and other frontier areas. Later, social changes brought about in part by the Industrial Revolution led to a decline in fertility in the industrialized nations, so that the base of the population pyramids narrowed and populations stabilized. Thus, Europe and the Americas have passed through their own period of intense population growth.

Since the mid-twentieth century, much of the developing world has been experiencing a similar demographic transition, though much more rapidly than occurred in Europe. Death rates declined dramatically in midcentury, and the decline in birth rates followed a few decades later. The various positions of countries in the demographic transition can be seen in Figure 5.9, which shows birth rates and death rates for countries listed in Table 5.1. Countries like Ethiopia and Nigeria are in the middle of the transition. They have high birth rates because both have high fertility rates and relatively young populations. At the same time, they still have high death rates because of poor nutrition and health. Countries like Brazil, Mexico, and the Philippines are further along in the transition. They have much lower fertility rates than Ethiopia or Nigeria, and so their birth rates are lower. Economic conditions in these countries are also somewhat better, and this, in combination with very young populations, means that death rates are very low. Finally, countries such as Germany and Russia have very low birth rates, as a result of low fertility and older populations. But an older population also means higher death rates, as can be seen in the graph.

Although it appears that the period of great difference between birth and death rates (and hence rapid population growth) in developing nations will be relatively brief, the base population was large to begin with and the growth rate was also large.

THE DISTRIBUTION OF POPULATION AND POPULATION GROWTH

The world's population is far from evenly distributed. It is concentrated in five major regions of the world: east Asia (especially China and Japan), southeast Asia (Indonesia, Vietnam, Myanmar), south Asia (India, Pakistan, Bangladesh), Europe (European Union, Russia), and eastern North America (United States, Canada). Sixty percent of

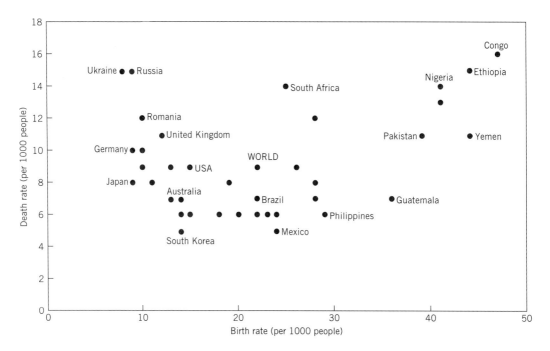

Figure 5.9 Birth rates and death rates for the countries listed in Table 5.1. Countries early in the demographic transition plot on the upper right. Countries in mid-transition show low death rates and declining birth rates and plot in the lower center of the graph. As population growth slows and the population ages, death rates rise; countries in this phase of the transition plot at the left side of the graph. *Source:* Population Reference Bureau, 2001.

THE DISTRIBUTION OF POPULATION AND POPULATION GROWTH   101

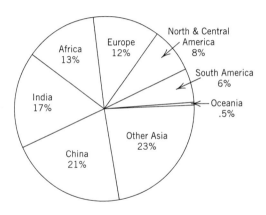

Figure 5.10 World population distribution. Most of the world's population is concentrated in Asia, with the remaining population more evenly distributed among the regions except Oceania, which is the least populated. *Source:* Population Reference Bureau, 2001.

the world's population lives in Asia (Fig. 5.10); China and India together account for about 38 percent of the world's population.

Regional Disparities

Currently, the most rapid population growth is taking place in the less industrialized countries. Based on the medium variant of recent United Nations projections, population worldwide will rise to 8.67 billion by 2035, an increase of 2.96 billion. This growth is overwhelmingly concentrated in the less industrialized nations. Only about 70 million of the 2.96 billion increase in world population expected between 1995 and 2030 will take place in the more industrialized countries; the other 2890 million, nearly 98 percent of the projected growth, will occur in the less industrialized countries (United Nations 1994). Between 1970 and 2000, fertility declined significantly in all parts of the world, but in the industrialized nations this decline brought fertility levels to below replacement levels in many countries. In the less industrialized nations, however, population continues to grow, because of both high fertility (though lower than past levels) and relatively young populations. Thus, while less industrialized countries in 2001 represented about 80 percent of the world's population, in 2025 this portion will be about 84 percent, and it will reach 86 percent in 2050 (Population Reference Bureau 2001). Europe and Africa in 2001 had similar populations, representing 12 and 13 percent of the world population, respectively. In 2025 Europe will represent only 9 percent, while Africa's share of world population will be 16 percent.

Increasing Urbanization

Much of the world's population growth since the 1960s has occurred in urban places in both the industrialized and less industrialized nations (Fig. 5.11), and this trend is expected to continue. Movement away from rural areas, where in most cases

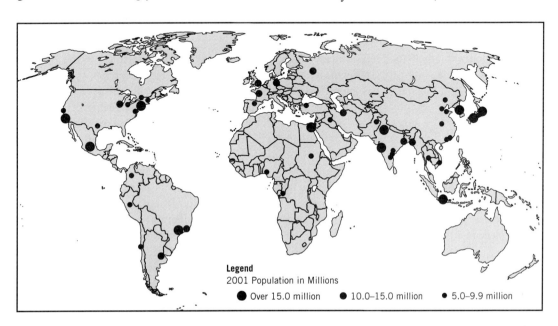

Figure 5.11 Major metropolitan areas of the world, 2001. In the early 1990s, the total urban population of the world surpassed the number living in rural areas. *Source*: World Gazetteer, 2001.

ISSUE 5.2: MEGACITIES: THE NEW URBAN DEMOGRAPHIC TRANSITION

At the close of the twentieth century, nearly half of the world's people lived in urban areas. In the developed countries, three-fourths of the population lived in urban places. It is expected that nearly all of the population growth in the next three decades will occur in urban areas. Why is this important? In order to understand the geographic pattern of the urban demographic transition, we need to first review the brief history of urban growth.

In 1950, there was only one urban area in the world that exceeded a population of 10 million people—New York City. Two decades later, when the United Nations coined the term *megacities* to refer to urban agglomerations with more than 8 million people (raised to 10 million in 1990), the list expanded to five (Tokyo, New York City, Shanghai, Mexico City, and São Paulo). In 2000, there were 19 urban areas that met the megacity criteria, and by 2015, the number is expected to top 23. Geographically, the majority of the world's megacities are located in developing countries. Growth in these cities is expected to be quite rapid (2.3 percent per year for the next 30 years), for a doubling time of 30 years! In 2015, only two of the 10 largest cities in the world (Tokyo and New York) will be in developed nations. It is estimated that the world's largest city in 2015 will be Tokyo (26.4 million), closely followed by Mumbai (Bombay), India (26.1 million), and Lagos, Nigeria (23.2 million).

While the rates of urban growth have slowed somewhat, it is still unclear how many of these megacities will absorb the new residents without significant impacts on already stressed services (sanitation and clean water) and infrastructure (housing and jobs). Overcrowding, poverty, crime, and a host of social ills will most certainly accompany this explosive growth. At the same time, pollution levels and health problems will rise as well, making the urban poor even more vulnerable. An additional twist is the relative location of many of the world's megacities. Many of them are found in coastal areas (Mumbai, New York, Manila, Calcutta) and thus are subject to sea level rise, hurricanes, typhoons, and flooding. Others (Tokyo, Mexico City, Los Angeles, Istanbul) are found in seismically active regions and thus are prone to natural hazards. The location of the megacities and their sheer size increase the vulnerability of the people who live and work there, as residents not only confront the daily hardships of earning a living and raising a family but also must be wary when a storm approaches or anticipate when the next earthquake will occur (United Nations 2000).

---

the population was largely food self-sufficient, has complicated the population pressures in the less industrialized nations. It certainly has intensified the pressures on usable resources, including space, water, and food, and it taxes national abilities to promote social and economic welfare. This trend toward increasing urbanization is a major problem facing the less industrialized nations in the future. At present about half of the world's people live in urban areas. About 90 percent of the world's population increase between 1995 and 2025 is expected to take place in urban areas (Issue 5.2).

POPULATION CONTROL STRATEGIES

Since the 1960s, when the population problem was recognized internationally, various programs have been instituted to control population, primarily through controlling fertility. Some of these programs have been quite coercive, such as forced sterilization or tax penalties for large families. Others have focused on increasing opportunities for women through education and economic development. Some strategies have concentrated on individual behavior, particularly use of contraceptives, whereas others have taken a broader social view, emphasizing economic factors. We now know that many different factors affect fertility rates, and it is often difficult to say which are the most effective in reducing population growth.

Family planning programs in the United States were first institutionalized in the form of birth control clinics in the 1920s. The first birth control clinics were established in New York City by Margaret Sanger, a leader in the women's rights movement. Sanger was concerned with the suffering of women who had too many children, spaced too close together, with no options for fertility control other than illegal abortions. The first birth control clinics were designed to liberate women from the traditional roles of wife and breeder and to allow women to exercise active control over the number and spacing of the births of their children. These early clinics met with strong resistance, in much the same way that the equal rights movement would meet resistance over half a century later. For a time in the United States, diaphragms were illegal and had to be smuggled in from Europe by birth control advocates.

The link between a woman's role in her society and her desire to increase or decrease the size of her family is a strong one, however. An understanding of women's roles in society is a prerequisite for any successful family planning effort.

Socioeconomic Conditions and Fertility

In those nations where standards of living and increased literacy have helped to improve women's status, fertility has declined. Several specific factors have influenced the decline in fertility, including increased educational levels of women, increased female employment outside the home, and marriage at a later age. All of these changes create conditions that motivate women to limit the size of their families. In much of the world social conditions that affect fertility, such as women's employment outside the home, are changing as people move to cities and industrial employment expands. This may be one of the factors contributing to the decline in fertility.

Increased educational levels have opened opportunities other than motherhood to many women. The acquisition of more knowledge not only has made these women more aware of family planning and contraceptive information but also has increased their knowledge of the need for family planning. In addition, increased education influences one's goals and aspirations, and as a woman receives more education she may seek new alternatives or lifestyles and perhaps a career. Last, educated women realize the value of education and want to have their own children educated as well. To do so effectively and in some cases economically, these women feel the need to limit the size of their families.

Increased female employment outside the home is another factor that has contributed to declining fertility rates. The need for two income earners in North American households is a response to increasing economic pressures. With a large percentage of women now in the labor force, large families are not as easily sustained as when women were mainly homemakers. According to the U.S. Census, over half of the nation's married women work outside the home.

Delaying the age of marriage has also contributed to declining fertility. If a woman is marrying later in her life, in her late twenties or early thirties, the number of children she can bear will be smaller. This practice has been used for years in the People's Republic of China as a family planning method and has shown mixed results.

Contraception and Family Planning

Family planning programs are generally voluntary. Worldwide the use of contraceptives averaged around 60 percent for married women (ages 15–49 years) in 2001, but there are large differences between more developed nations (73 percent) and less developed ones (57 percent) in their patterns of use. In the less industrialized nations, for example, many countries have found it necessary to implement compulsory programs or incentives to encourage family planning. Prior to 1965, only a handful of less industrialized nations had officially supported family planning programs. The international spread of family planning began after 1965 and quickly gained momentum in the next 10 years (Fig. 5.12). By 1975, only 3 of the 38 less industrialized nations with populations greater than 10 million had no officially adopted family planning programs. These were Myanmar, Peru, and North Korea. Five years later,

Figure 5.12 Family planning in Jendouba, Tunisia. Education is the key to success for family planning efforts. In Tunisia, the emphasis is on integrating health care (maternal and child), nutrition, and family planning. Notice the absence of fathers in this family planning class.

all but a few developing-world countries had family planning policies, although some had such policies only on paper (Brown et al. 1989). More than half of all couples in the developing world, for example, now use contraceptives. Yet, in many countries, women (and men) still lack access to affordable methods for controlling their fertility (Potts 2000).

Indonesia's example illustrates the use of modernism mixed with tradition. By 1980, after 10 years of work, 40,000 information and distribution centers had been set up in the country's villages. Free contraceptives were accompanied by a relentless public relations campaign. These and related efforts have been successful. Although abortion is illegal in Indonesia, the fertility rate has dropped from 5.6 to 2.7 today. The proportion of married Indonesian women using contraceptives today is 57 percent (United Nations Development Programme 2000).

Experts report that those programs receiving good governmental support and economic assistance will be the most successful in reducing the fertility rate. A well-planned program must include not only the distribution of contraceptives but also increased education about their use. The bitter reaction to India's IUD (intrauterine device) program is a case in point. Many Indian women adopted the IUD as a method of contraception. Family planning personnel would visit rural villages and, after medical examinations, insert the IUDs. The doctors, however, failed to warn the women of potential side effects. When these became apparent, rumors spread and the program virtually collapsed overnight. Similar reactions to the IUD followed in the United States, with class action suits pertaining to one particular type, the Dalkon shield.

In the United States, fertility rates have risen slightly since the mid-1980s, from 1.8 to 2.1 in 2001. The number of abortions has been fairly steady since 1990, although the rate per 1000 women has declined. Women are seeking more educational opportunities and postponing marriage to a later age. They also have more choices for when and if to start a family, choices that were not as socially acceptable two decades ago.

POPULATION GROWTH AND AFFLUENCE

Earlier in the chapter we suggested that it is not only the absolute growth in population that is problematic, but also where that growth is occurring. There is a third factor that is important as well, and this has to do with the impacts of that growth on natural resources and environmental quality. Within the scientific and environmental communities there is a consensus that growths in population, affluence, and technology are jointly responsible for environmental degradation. While there is still some disagreement on the strength of each individual component in contributing to environmental decline, the identity known as IPAT has endured for three decades since it was initially proposed by Paul Ehrlich and John Holdren (1972).

IPAT is a simple formulation that states that resource or environmental impact (I) equals population (P) times affluence (A) times technology (T). In reality, this identity considers consumption patterns a product of affluence (normally measured as per capita income) and technology (the technical ability to exploit resources). For example, smaller, more affluent, and technologically sophisticated nations may use more resources with greater levels of adverse environmental impacts than larger, less affluent, and less technologically oriented ones. This identity helps explain, in part, why the United States (with slightly less than 5 percent of the world's population) consumes about 23 percent of the world's energy resources, while China (21 percent of the world's population) consumes half of that (11.5 percent). More important, it describes the increasing divide between rich and poor nations, where one-fifth of the people in high-income countries account for 86 percent of the private consumption of resources. What Americans and Europeans spend on cosmetics ($8 billion annually) and ice cream ($11 billion) is more than enough to provide basic education and water and sanitation to the more than two billion people worldwide who go without either or both (Crossette 1998).

For our efforts toward a sustainable future, noted geographer Robert Kates suggests a reformulation of the original IPAT identity to one that closely resembles a population/consumption variant (Kates 2000). In this formualtion, I (resource depletion or environmental impacts) equals P, or population (number of people or households), times C, or consumption per person (the transformation of energy, materials, and information), times the impacts of that consumption. In order for us to achieve a modicum of sustainability, population growth needs to be slowed, consumers must be satisfied with what they have (and sublimate additional desires for the greater good), and the individual impacts of consumption must be shifted to less harmful patterns (fossil fuels to renewable energy, for example) or reduced through substitution (see Chapter 13).

CONCLUSIONS

The rate of population growth worldwide is clearly dropping. The reasons vary from one culture to another, and in many cases the causes for declining fertility are not clear. It is impossible to say whether population control programs, industrialization, employment opportunities for women, access to birth control, or other factors are most responsible for the change. One thing is certain, however: fertility rates can and do change very rapidly. At the moment they are declining in most of the world, and most demographers expect this trend to continue.

If current trends persist for the next few decades, we can expect world population to grow at a much slower rate than it did in the 1960s and 1970s. The next doubling of population will probably take much longer than is implied by the current doubling time. In fact, many predict that world population will stabilize at the level of 9 billion, in which case it would never double its current size. Demographic trends often hold surprises, however, and such predictions must be read with caution.

Even though population growth is slowing, it is still a major factor in the world resource picture. A 30 percent increase in world population in the next three to four decades is a near certainty, and if new people consume resources at rates similar to the present ones, the environmental consequences will be dramatic. We will turn our attention to these consequences in the remainder of this book.

REFERENCES AND ADDITIONAL READING

Bongaarts, J. 1998. Demographic consequences of declining fertility. *Science* 282:419–420.

Boserup, E. 1965. *The Conditions of Agricultural Growth: The Economics of Agrarian Change under Population Pressure*. London: Allen & Unwin.

_____. 1981. *Population and Technological Change*. Chicago: University of Chicago Press.

Brown, L. R. and H. Kane. 1994. *Full House: Reassessing the Earth's Population Carrying Capacity*. New York: W. W. Norton.

_____ et al. 2001. *State of the World, 1989*. Washington, D.C.: Worldwatch Institute.

Brown, L. R., G. Gardner, and B. Halweil, 1998. *Beyond Malthus: Nineteen Dimensions of the Population Challenge*. Worldwatch Paper no.143. Washington, D.C.: Worldwatch Institute.

Chen, L. C., W. Fitzgerald, and L. Bates. 1995. Women, politics, and global management. *Environment* 37(1):4–9, 31–33.

Cohen, J. 2001. HIV gains foothold in key Asian groups. *Science* 294:282–283.

Crossette, B. 1998. Most consuming more, and the rich much more. *The New York Times*, September 13: A3.

Daly, H. E. 1977. *Steady-State Economics*. San Francisco: W. H. Freeman.

Durning, A. B. 1989. *Poverty and the Environment: Reversing the Downward Spiral*. Worldwatch Paper no. 92. Washington, D.C.: Worldwatch Institute.

Ehrlich, P. R. and A. H. Ehrlich. 1990. *The Population Explosion*. New York: Simon & Schuster.

Ehrlich, P. 1968. *The Population Bomb*. New York: Ballantine Books.

Ehrlich, P. and J. Holdren. 1972. Review of the Closing Circle. *Environment* April:24.

Georgesu-Roegen, N. 1979. Comments on papers by Daly and Stiglitz. In V. K. Smith, ed. *Scarcity and Growth Reconsidered*. Baltimore, MD: Johns Hopkins University Press.

Greenhalgh, S. and J. Bongaarts. 1987. Fertility policy in China: Future options. *Science* 235:1167–1172.

Hardin, G. 1993. *Living Within Limits: Ecology, Economics, and Population Taboos*. New York: Oxford University Press.

Harkavy, O. 1995. *Curbing Population Growth: An Insider's Perspective on the Population Movement*. New York: Plenum Press.

Harvey, D. 1974. Population, resources, and the ideology of science. *Economic Geography* 50: 256–277.

Hohm, C. F. and Lori Justin Jones. 1995. *Population: Opposing Viewpoints*. San Diego, CA: Greenhaven Press.

Jacobson, J. L. 1988. *Environmental Refugees: A Yardstick of Habitability*. Worldwatch Paper no. 86. Washington, D.C.: Worldwatch Institute.

Kates, R. W. 2000. Population and consumption: what we know, what we need to know. *Environment* 42(3):10–19.

Lutz, W. 1994. *The Future Population of the World: What Can We Assume Today?* London: Earthscan Publications.

Malthus, T. R. 1976. An essay on the principle of population. In P. Appleman, ed. *An Essay on the Principles of Population: Text, Sources and Background, Criticism.* New York: W. W. Norton.

Mazur, L. A. 1994. *Beyond the Numbers: A Reader on Population, Consumption, and the Environment.* Washington, D.C.: Island Press.

McKibben, B. 1998. A special moment in history. *The Atlantic Monthly* 281(5):55–78.

Mink, S. D. 1993. *Poverty, Population, and the Environment.* Washington, D.C.: World Bank.

Moffett, G. D. 1994. *Global Population Growth: 21st Century Challenges.* Ithaca, NY: Foreign Policy Association.

Newland, K. 1977. *Women and Population Growth: Choice Beyond Childbearing.* Worldwatch Paper no. 16. Washington, D.C.: Worldwatch Institute.

_____. 1980. *City Limits: Emerging Constraints on Urban Growth.* Worldwatch Paper no. 38. Washington, D.C.: Worldwatch Institute.

Population Reference Bureau, 2001. *World Population Data Sheets.* http://www.prb.org/.

Potts, M. 2000. The unmet need for family planning. *Scientific American* January:88–93.

Preston, S. H. 1986. Population growth and economic development. *Environment* 28(2):6–9, 32–33.

Sen, G. 1995. The world programme of action: A new paradigm for population policy. *Environment* 37(1):10–15, 34–37.

Simon, J. 1981. *The Ultimate Resource.* Princeton, NJ: Princeton University Press.

_____. 1986. *Theory of Population and Economic Growth.* Oxford/New York: Basil Blackwell.

_____. 1990. *Population Matters: People, Resources, Environment, and Immigration.* New Brunswick, NJ: Transaction Publishers.

Stein, D. 1995. *People Who Count: Population and Politics, Women and Children.* London: Earthscan.

Stokes, B. 1980. *Men and Family Planning.* Worldwatch Paper no. 41. Washington, D.C.: Worldwatch Institute.

United Nations. 1994. *The Sex and Age Distribution of the World Populations: The 1994 Revision.* New York: United Nations.

_____. 1995. *The World's Women 1995: Trends and Statistics.* New York: United Nations.

_____. 2000. *World Urbanization Prospects: The 1999 Revision.* www.un.org/esa/population/wup1999/wup99.htm.

United Nations Development Programme. 2000. *Human Development Report 2000.* New York: Oxford University Press.

United Nations Joint Programme on HIV/AIDS (UNAIDS)/World Health Organization. 2001. *AIDS Epidemic Update, December 2001.* http://www.unaids.org/epidemic_update/report_dec01/index.html#full.

U.S. Bureau of the Census. 1999. *Statistical Abstract of the U.S.* Washington, D.C.: U.S. Government Printing Office.

_____. 2002. *International Data Base.* www.census.gov/ipc/www/idbnew.html

World Gazeteer. 2001. Available online: http://www.world-gazeteer.com/hom.htm.

World Resources Institute. 2000. *World Resources 2000–2001.* Washington, D.C.: World Resources Institute.

For more information consult our web page at ***http://www.wiley.com/college/cutter.***

STUDY QUESTIONS

1. What is the significance of geometric versus arithmetic growth?
2. Why is the spatial distribution of population growth so important? Which nations or regions are growing most rapidly? Why? Which regions are not increasing in population?
3. Why do some people resist birth control measures? What incentives do some nations provide to use them?
4. What is the relationship between population, affluence, technology, and resource depletion? Does affluence always mean increasing consumption?
5. If you gaze into your crystal ball, what will the world population patterns be like in 2015? In 2030?

CHAPTER 6

AGRICULTURE AND FOOD PRODUCTION

Several resource issues are of particular importance in predicting and planning for future world supplies of food. These issues include the variability of rainfall patterns in lands already marginal for agriculture, overgrazing, competing land uses, ancient cultural traditions, and the efficiency of using meat for food. In this chapter we will look at the quantity and quality of agricultural resources and food production globally and within the United States.

Although food production is a global issue, local conditions vary tremendously around the world. Some regions chronically underproduce and either face widespread malnutrition or must import a large portion of their food needs. Others, such as the United States, have ample productive capacity and only limited ability to sell surpluses overseas, so that much land is idle and the industry as a whole continues to shrink relative to the rest of the national economy.

International trade in agricultural commodities will become more significant in the coming decades. Trade and specialization have transformed agriculture in most of the world, and this trend will continue. In the United States, this means the continued decline of the traditional small family farm and a shift toward large farms run as corporate businesses. American farmers today serve markets around the world, and production is thus affected by world market and trade factors. Although American agriculture is immensely productive and the United States is a major food exporter, it has yet to make a significant contribution to alleviating food shortages in the developing world, because most food shortages are the result of failures of distribution systems rather than production.

Food production is probably the most important natural resource issue facing the world today. World population growth, though slower than in the 1960s, is still more rapid than at any other time prior to the twentieth century. This growth necessitates a continuing expansion of agricultural output. The expansion must take place in the context of severe constraints on the availability of agricultural land and on the ability of natural biogeochemical cycling processes to supply agricultural inputs (such as water) and absorb waste products (such as nitrogen). Thus, we must achieve a 50 percent increase of agricultural output over the next half-century or so through more intensive use of existing lands while minimizing the environmental impacts of that intensification.

Plants that are now our major crops were bred carefully for generations, just like our major food animals. For optimal production, herders and ranchers need to provide high-quality food for livestock and ensure that grazing land is not overused. Most of the world's grazing lands are in semiarid climates, for in these low-rainfall areas the land generally cannot support arable agriculture without irrigation. Palatable grasses and low shrubs, however, are available to livestock. Animals are grazed on other lands too, usually mountainous areas and forests, both of which are unsuited for agriculture. In the United States, cattle grazing was the early backbone of the western states' economy and in fact contributes to much of the region's folklore. Today raising and selling cattle is part of the nation's agribusiness industry. Often, conflicting demands are made on the land available for grazing, and a debate ebbs and flows over the comparative value of meat versus grain production for national and global food supplies.

FOOD PRODUCTION RESOURCES

The world's food supply is based on about 30 major crops. The top 7 (maize, rice, wheat, potatoes, cassava, sweet potatoes, and barley) have annual harvests of 100 million metric tons or more, and these 7 account for about three-quarters of the harvest of the top 30 crops. In terms of global meat supply, 6 species provide 95 percent of all meat production. Pigs, chickens, and cattle supply most of our meat, followed by much smaller contributions from lambs, turkeys, and goats. Total world production of poultry has been growing rapidly and recently surpassed global beef production.

Great disparities exist in the amounts and kinds of food consumed in rich and poor nations, with rich nations eating not only more but better (Fig. 6.1). In the wealthy countries of North America and Europe, the average consumption of food (as measured by the amount that reaches the consumer—not all of this is necessarily eaten) is typically 120 to 140 percent of the needed caloric intake. In most of the poorer areas of the world, average food consumption is closer to 100 percent of need. This is the average, of course, and some people in a society consume above it, whereas others are inevitably underfed. The following paragraphs describe these patterns in greater detail.

Crops

The total land area of the world is about 13 billion hectares. Of this total, only about 12 percent is presently *arable*, or suitable for cultivated agriculture. Figure 6.2 shows the distribution of arable land in the world. Arable land is not uniformly distributed in relation to population. Asia has the most arable land, but it also has an immense population to feed. On the other hand, North America has only 12 percent arable land and a relatively small population. When we consider that American agriculture generally produces higher yields per acre than systems where fertilizers and pesticides are less available, these disparities from one area to another are even greater.

Agricultural production has increased steadily since the 1960s. Between 1960 and 2000 world food production doubled and per capita production

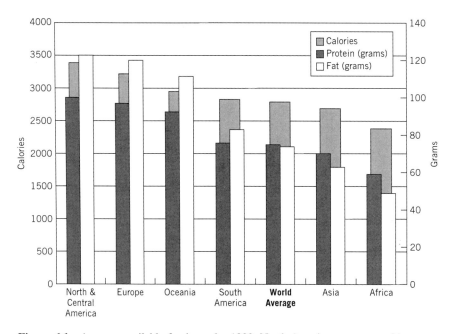

Figure 6.1 Average available food supply, 1999. North Americans consume 21 percent more calories than the world average, while Africans consume 15 percent less than the world average. North Americans and Europeans also have diets that are much richer in protein and fats than most of the rest of the world. These averages mask variations within continental regions. *Source:* FAO, 2001.

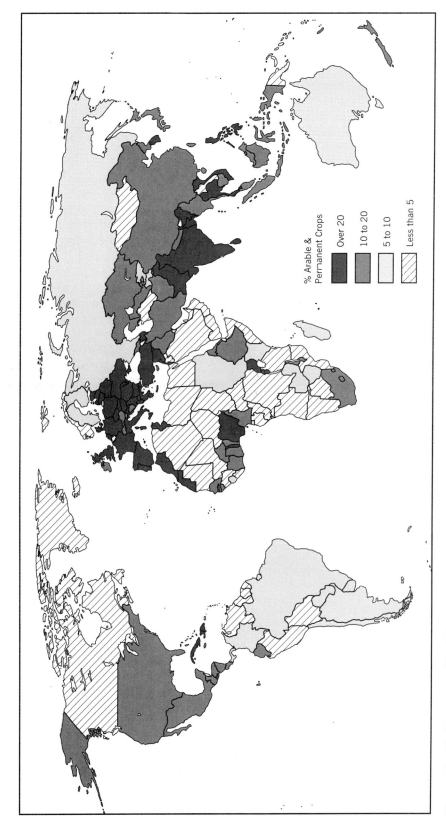

Figure 6.2 World agricultural land. Percentage of land that is arable land and land in permanent crops, 1999. *Source:* FAO, 2002.

Table 6.1 Per Capita Food Production, as a Percentage of that During 1989–1991

Region	1961	1971	1981	1991	2000	2001
Asia (excluding USSR)	70.1	74.9	83.2	101	125.7	124.8
Europe (excluding USSR)	71.9	82.4	94	99.6	94.2	91.8
Africa	107.6	112.3	100.1	102.4	100.2	97.8
Latin America & Caribbean	81.5	85.5	97.6	100.9	116.7	117.9
North & Central America	85.2	95.7	108.6	99.4	110	105.9
Oceania	91.2	103.6	106	99.3	120.6	120.3
South America	78.7	81.7	95	100.7	121	122.6
USSR/former USSR	83.7	96.4	87	90.3	63.2	NA
World	84	89.9	94.7	99.3	108	106.9

NA = not available.
Source: FAO, 2002.

increased about 25 percent (Table 6.1). The rate of increase in both total and per capita production was largest in Europe and Asia, where increases in production dramatically outpaced population growth. Food production also grew faster than population in South America. In Africa production increases have not kept pace with population growth, and as a result, per capita food production there declined about 10 percent between 1960 and 2000. In North America production generally increased, although there were wide variations in output from year to year.

Much of the increase in food production since 1960 is attributable to the *Green Revolution*. The Green Revolution is a name applied to a series of agricultural innovations developed between 1945 and 1965 for the purpose of increasing food production in poor countries. These innovations in-

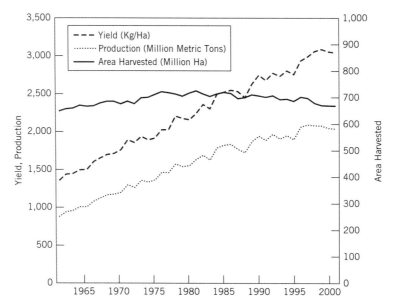

Figure 6.3 Trends in world cereal production, 1961–2001. Production has increased steadily since the 1960s. Area harvested has remained relatively steady, so that the increase in production is primarily a function of yield. *Source:* FAO, 2001.

cluded the development of new varieties of wheat and rice that were capable of much higher yields than traditional varieties, especially when fertilizer is applied (Easterbrook 1997). The effect of the Green Revolution is clearly seen in data for cereal grain production (Fig. 6.3). The amount of land used for grain production has remained fairly constant, while yields per unit land area have more than doubled. The main driving force behind those yield increases was the use of fertilizers. Irrigation also was significant (Fig. 6.4). The Green Revolution is largely responsible for preventing widespread famines that were predicted in the 1960s.

Since 1990 global trends have changed, largely as a result of the collapse of the former Soviet Union. Food production in the former Soviet Union dropped nearly 40 percent between 1990 and 1994, and world grain inventories were reduced substantially. Thus, while food production continued to increase in most parts of the world, events in the former Soviet Union caused world averages to remain fairly constant during the past decade.

Livestock

About 12 percent of the world's population is dependent on livestock production for both food and income. Patterns of life dependent on livestock developed thousands of years ago in Asia and Africa and are still important in many world regions. Most of the world's *pastoralists* are located in Africa, Asia, and parts of Central and South America. The traditional *pastoral nomad* is not oriented toward the production of large quantities of meat and dairy products for market. Instead, products of the sheep and goat herds of Asia and the cattle herds of Africa are used locally or traded on a small scale. The past 100 years witnessed a dramatic decline in the number of these pastoral nomads as a result of settlement in permanent locations. The process of *sedentarization* was enforced by governments that encouraged herding groups to settle in villages and also controlled the crossing of national borders. In addition, many nomads left their herds and families for work in urban areas.

With the nineteenth-century settlement of drier areas in the Americas and Australia, it became environmentally and financially feasible to raise and transport large numbers of livestock for transport to distant markets. This production is primarily oriented toward meat rather than milk production. It is also more highly dependent on a technologically complex set of elements, including truck transport, antibiotics, and other food supplements.

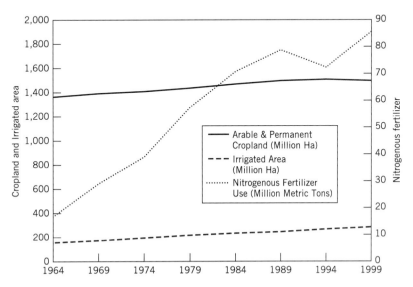

Figure 6.4 Trends in world irrigation and fertilizer use, 1964–1999. The total amount of cropland has increased only slightly, mostly as a result of irrigating land that was previously not arable. The increase in fertilizer use in this time period is the main factor driving increased food production. *Source:* FAO, 2001.

Over time, improvements in the overall standard of living and increased use of animal protein in fulfilling dietary needs have resulted in a subsequent increase in *ruminant* resources (cattle, sheep, and goats) (Fig. 6.5). In the late twentieth century, livestock increasingly were concentrated in fixed locations called feedlots, where food was brought to them, instead of letting the animals graze on live vegetation. While cattle, sheep, and goat populations increased after the 1950s, the rate of growth slowed during the 1990s. Sheep populations declined 12 percent between 1990 and 2000, while cattle populations increased by only 5 percent (Fig. 6.6). However, other livestock populations have increased dramatically in the 1990s, led by chickens (38 percent) and pigs (8 percent). Regionally, substantial increases in livestock resources occurred in Asia (chickens, pigs, sheep, and goats), Latin America (cattle, buffalo, and chickens), and Africa (cattle, buffalo, chickens, sheep, and goats). Decreases in ruminant populations were found in North America, Oceania (sheep and goats), and western Europe (cattle, sheep).

This increase in animal populations means a rise in the use of grazing lands and is thus a major cause of land degradation in places such as Africa. However, about three-fourths of the world's ruminants no longer are raised on range resources, as countries increasingly move toward intensive livestock production. Farming and the use of forage crops such as alfalfa now contribute to the animals' feed. This is the pattern in the developed world but even more so in the developing countries. For example, in Asia, 18 percent of cereal production is fed to animals, compared to 62 percent in North America (Fig. 6.7).

The U.S. Agricultural Land Resource Base

About 70 percent of U.S. land is classified as agricultural land. "Agricultural land" is a broad term, however, and includes cropland, rangeland, forestland, and pastureland. Cropland, or land on which crops are presently grown, comprises only about 25 percent of agricultural land. The cropland area of the United States amounts to about 0.63 hectares per person, as compared with the global average of 0.25 hectares per person.

America always has been a land of agricultural abundance, with a large area of arable land and a relatively small population. Through the first two centuries of European settlement in North America, agriculture was based largely on a fertile soil resource, supplemented in some sectors by cheap slave and indentured labor. In the mid-nineteenth century, the opening of the Great Plains was facil-

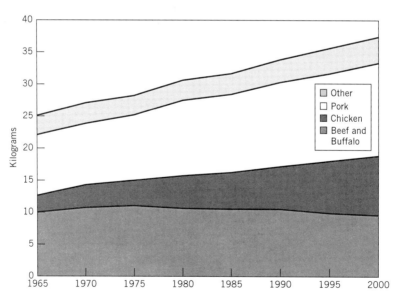

Figure 6.5 World per capita meat production, 1965–2000. Beef and buffalo meat production per person has declined slightly, while production of other meats, especially chicken, has increased dramatically. *Source:* FAO, 2001.

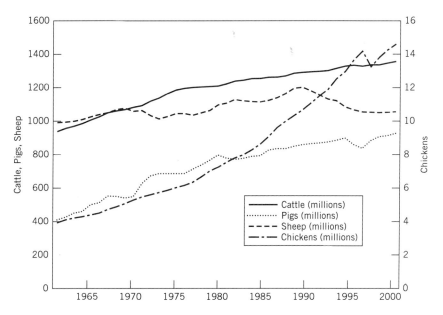

Figure 6.6 World populations of cattle, sheep, pigs and chickens, 1961–2001. Chicken and pig populations grew, while numbers of cattle and sheep had slower increases during this time period. *Source:* FAO, 2001.

itated in part by the development of the steel plow, one of many major innovations contributing to the growth of U.S. agricultural production. Today the U.S. agricultural system is among the most technologically advanced in the world.

American agriculture is based on high and continual inputs of capital and replacement of animal power and human labor by machines, which has resulted in drastic reductions in the number of farms and farm workers. This culminated in an agricultural system that, unlike those in traditional societies, is operated as a complex of small and large businesses whose primary goal is earning money rather than producing food for subsistence. This system has come to be known as *agribusiness*.

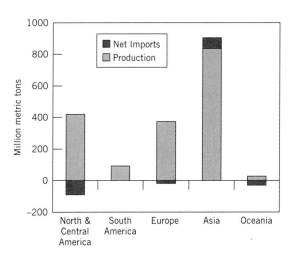
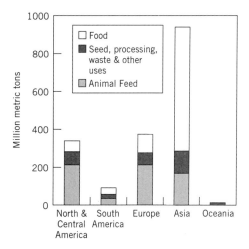

Figure 6.7 Production and disposition of cereal grains, 1999. Worldwide about 37 percent of cereal production is fed to animals; the portion is much higher in wealthy countries, where large amounts of meat are consumed, while in Asia most cereal production is consumed by humans directly. *Source*: FAO, 2001

U.S. agriculture dominates food exports in the world economy, with American grain exports accounting for over half of the world total. American agriculture feeds many more people than live in the United States, and there is potential for even greater food exports. In addition, the average American consumer pays a much smaller portion of disposable income for food than the average consumer in most of the rest of the world.

Farming in the United States has been profitable at times and unprofitable at others; the effects of these variations are seen in recent trends in area harvested, yields, and total production of corn (Fig. 6.8). In the early 1970s, increasing world demand for grain and government policies favoring exports stimulated farmers to increase production. The area planted expanded, and with relatively favorable weather, yields and total production rose substantially. This period was also one of considerable investment in agricultural land and machinery. But bumper harvests in the late 1970s kept prices down, while inflation continued to force costs and interest rates up. In the mid-1980s, a decline in profitability, coupled with government policies aimed at reducing production, led to declines in area harvested, with associated production declines. In the late 1980s, the situation improved as a result of a combination of lower interest rates, lower acreages planted, and stable prices for inputs such as fuel and chemicals. In the 1990s, farming was again becoming profitable, but deregulation of production led to higher production that depressed commodity prices and reduced farm profits.

Because the United States exports so much food, American agricultural prosperity is linked closely to world agricultural markets. These markets exert an important influence on agriculture in both the rich and the poor nations. In recent decades, the general pattern was one of government policies that encouraged production by

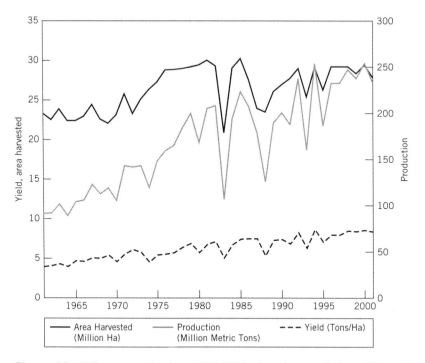

Figure 6.8 U.S. corn production, 1965–1995. Area harvested (in millions of hectares) has remained relatively steady, while yields (expressed as metric tons per hectare) have increased slightly. Year-to-year variations in production (in millions of metric tons) are mostly caused by yield variations, although some fluctuations (such as the dip in area harvested in 1993) are attributable to government policies. *Source:* FAO, 2001.

keeping domestic prices high and also encouraged exports by subsidizing such sales. This kept world market prices low, thus making it difficult for poorer nations to export agricultural products. At the same time, many developing nations followed policies that also keep their domestic prices low in order to keep food within reach of their poorer citizens. Low domestic prices discourage production, and most developing nations lack the foreign currency needed to import enough food to meet their needs. The situation is therefore one of considerable imbalance: the United States and other wealthy nations overproduce and the developing nations underproduce in a world market that makes it increasingly difficult for rich nations to sell to poor ones.

Modern American Agricultural Systems

American agriculture today includes both mixed and monocultural cropping systems. *Mixed cropping* systems are agricultural systems that combine several different crops in a single farm unit. They usually include crops for both human consumption and fodder (animal feed). Mixed cropping is used in areas where several different crops can be grown with roughly the same profit per acre and in dairy farming areas. A typical dairy farm in the northeastern United States includes corn and alfalfa grown for fodder for dairy cattle, with milk marketed for cash income. By producing several different animal and vegetable products simultaneously, mixed farming systems make efficient use of the land resource while minimizing susceptibility to unfavorable weather or market conditions for any one commodity. Mixed systems benefit from crop rotation, in which the crops grown on a given parcel of land are changed from year to year or season to season, thus reducing depletion of particular nutrients. These farmers also make greater use of plants by feeding otherwise unused parts of plants to animals and by returning some organic matter to the fields in the form of manure. *Monoculture* is an agricultural system in which one crop is cultivated repeatedly over a large area. Other, much more distinctive characteristics of modern monocultural systems include a reliance on technology (in the form of machines), specialized plant varieties, fertilizers, and pesticides. This technological agriculture is most fully developed in the United States but is found in most of the other wealthy nations of the world as well. Monoculture is necessary to take advantage of the labor-saving benefits of machines. Additional benefits include economies of scale and more efficient marketing. The plant varieties bred not only produce high yields but also have a large degree of uniformity in plant dimensions and ripening time. If an entire 160-acre field is to be harvested at one time by machines that harvest eight rows at a pass, then clearly all the plants must mature at the same time. A food production system that is primarily monocultural is a highly specialized one. It must have the capacity to store produce over long time periods and transport it efficiently from one area to another.

One consequence of the uniformity of plants in monocultural systems is their susceptibility to disease and pest infestation. If an insect that can successfully attack a particular variety takes hold, large fields can be devastated quickly. For this reason, substantial inputs of pesticides are required normally. Many of these specialized plant varieties also require large inputs of fertilizer or irrigation water to realize maximum yields. In addition, the machines, fertilizers, and chemicals require large amounts of energy, in the form of fossil fuels and electricity, to operate or produce.

Monoculture under these conditions produces large yields of uniformly high quality. As a result, the value of crops on a per acre basis is very great. But to achieve this high-value harvest, much capital is needed to purchase the inputs of production—land, machines, seed, fertilizer, and so on. It is a capital- and energy-intensive system rather than a labor-intensive one. In recent years, capital became an increasingly costly input for American farmers, and at the same time there was growing concern about the negative aspects of agricultural chemicals. There are indications that American agriculture is turning away from intensive use of chemicals, although it is not clear whether other factors of production (such as labor) will be substituted or whether yields per acre will be allowed to decline. These trends are discussed in more detail later in this chapter.

NATURAL RESOURCES FOR AGRICULTURAL PRODUCTION

The connection between natural resource condition and food production is an intimate one, but production is not dependent on natural resources alone. Rather, it is a function of both natural conditions and human management. For example, plants obviously require soil, water, and sunlight to grow. But the soil also must be managed in ways that promote plant growth; seeds that will produce the food we need must be planted at the right time and tended as they grow. Crops must be harvested, processed, stored, and distributed to consumers. All of these things are essential to the food production system. In examining the geographical variations in food production today, as well as the possibilities for increasing that production in the future, we need to consider the role each of these factors plays in food production. In this section, we examine soil, water, fertilizers and pesticides, seed, labor and machines, animals, and the food distribution infrastructure.

Soil

Soil is the uppermost part of the earth's surface, which has been modified by physical, chemical, and biological processes over time. It is the essential medium for plant growth, and it is a complex and dynamic mixture of solid and dissolved mineral matter, living and dead organic matter, water, and air. Soil is formed over long periods of time, usually thousands of years. Soil is not a uniform material and there are literally thousands of different soil types around the world. There are a about a dozen or so major groupings or categories of soils. These soil groups reflect the geographic variability in parent material and in soil-forming processes and roughly correspond to the location of biomes described in Chapter 4.

Many factors influence soil formation, including climate, parent material, topography, erosion, and biological activity. Climate affects soil by determining the amount of water that enters the soil from rainfall and the amount that is drawn from the surface by evaporation. Climate also determines soil temperatures, which are important in regulating chemical reactions in the soil as well as influencing plant growth. *Parent material* is the mineral matter on which soil is formed. It affects the soil by supplying the mineral matter that forms the bulk of the soil. Parent material has a fundamental influence on *soil texture*, which is the mix of different sizes of particles, and on the chemical characteristics of the soil.

Topography influences soil primarily by regulating water movement within and over the soil. On slopes, water moves down and laterally through the soil, providing drainage. In low-lying areas water accumulates, and soils become waterlogged. Along stream courses sediment accumulates, producing fertile alluvial soils. Topography also affects the rate of erosion. *Erosion* is the removal of soil by running water or wind. It is a natural process that is accelerated greatly by human influences such as vegetation removal. The rate of erosion relative to the rate of new soil formation is an important determinant of soil characteristics.

Finally, biological activity makes soil the distinctive, living, dynamic substance it is, rather than just an accumulation of sterile rock particles. Biological activity includes the growth and decay of plants and animals in and above the soil. It contributes organic matter to the soil, which constitutes the basic storage of nutrients for most ecosystems. Biological activity also aids in the physical modification of the mineral soil by contributing organic acids that break down rocks, by exerting physical forces that fracture rocks, and by stirring and aerating the soil so that water and air penetrate below the surface. Vegetation cover regulates water losses by evapotranspiration and protects the soil from excessive erosion. The type of vegetation is also important in determining soil characteristics. For example, the thick grass cover of a prairie leads to the development of an organic-rich and uncompacted topsoil that is not usually found in forest soils. In short, biological activity plays a fundamental role in soil formation and helps maintain the ability of soils to support life.

The way in which individual soil particles group together in aggregates is called *soil structure*. Soil structure is important in determining the water-holding capacity of soil and the speed with which water soaks into and through the soil. Plowing, compaction, oxidation of organic matter, extraction of nutrients by plants, and desiccation can sometimes destroy soil structure to the detriment of its water-holding properties.

Soils develop slowly from unweathered bedrock to a complete soil profile. The rate of new soil formation varies from place to place, but as a rule it is very slow in human terms, requiring hundreds to tens of thousands of years. Erosion is part of the natural soil system, and soil eroded from the surface must be replaced by new soil formed from parent material. If the soil erosion rate is high, then the soil may not develop as thickly or as completely as in areas of low erosion rates.

Soils vary greatly in their ability to support agricultural production, depending on their fertility, water-holding characteristics, temperature, and other factors. *Soil fertility* is defined as the ability of a soil to supply essential nutrients to plants, which is dependent on both the chemical and textural properties of the soil. Generally, soils that have a high proportion of clay and organic matter also have high fertility, for these substances have the ability to store and release nutrients. Sandy soils, on the other hand, generally have lower amounts of nutrients available to plants, although as little as 10 to 15 percent fine particles may be sufficient to supply the needed nutrients. The abundance of various elements in the soil, such as phosphorus and calcium, is also important. This is often controlled by the chemical composition of the bedrock and the degree of leaching of these nutrients by water percolating through the soil.

The *water-holding capacity* of a soil is determined primarily by its texture. As a rule, coarse-textured soils have low capacities to store water, whereas clayey soils hold large volumes of water in the upper parts of the profile. At the same time, however, clay soils often are poorly drained, with the tendency to become waterlogged, which prevents air from reaching plant roots. Waterlogging is particularly common in humid areas, on floodplains, and on other flat land. In addition, irrigated areas often experience waterlogging if irrigation causes a rise in the local groundwater table.

The productive capabilities of soils are thus of fundamental importance in determining what plants are grown, what particular management techniques are used, and what typical yields are expected. Soil inventories and maps are essential in evaluating the soil resource, and the U.S. Natural Resource Conservation Service produces soil maps and conducts inventories of the condition of the nation's soils. In its 1997 National Resource Inventory (the most recent available), it found that of 378 million acres (153 million hectares) of cropland, 86 million, or 56 percent, were classed as "prime farmland"— land that "has the best combination of physical and chemical characteristics for producing food, feed, forage, fiber, and oilseed crops and is also available for these uses" (NRCS 2001). It should not be surprising that most of U.S. cropland is in this "prime" category, because land that is less desirable for crops is in other uses, such as forest and range. This added potential cropland is not being used largely because it is not the best land for agriculture. Nonetheless, the loss of prime land is significant: about 500,000 acres (200,000 hectares) of prime land are lost to development each year in the United States.

Water

Water is essential to plant growth and is supplied to crops naturally by rainfall and artificially by irrigation. Rainfall farming is entirely dependent on the weather to provide a sufficient but not excessive supply of moisture to plants. With not enough moisture, plants wither and die; with too much, the soil cannot be worked, crops rot in the field, or roots suffocate from lack of air. The amount and timing of rainfall are major determinants of what can be grown, where, and when. In areas without irrigation, rainfall variability is a major cause of year-to-year changes in yields. Droughts, early rains, and late rains take their tolls in crop failures or low yields in many regions of the world each year. Some scientists feel that in the future, climates will become more variable; if this is the case, yield fluctuations will become an even greater problem than they are today (Issue 6.1).

In those parts of the world where rainfall is sufficient and reasonably dependable, crops are grown without irrigation. But about 17 percent of the world's arable lands are productive because water was applied artificially to plants. Also, much of the land deemed potentially arable became so only through irrigation. Most of the world's irrigated land is in the seasonally dry regions of Asia. China and India together contain 40 percent of the world's irrigated land; the United States has about 8 percent of all irrigated land in

ISSUE 6.1: AGRICULTURE, CO₂, AND CLIMATE: THE ONLY CERTAINTY IS CHANGE

When extreme weather occurs, one of the most obvious and immediate effects is on agriculture. Farming is vulnerable to a wide variety of bad weather: cool spring weather delays growth, wet spring weather delays planting, autumn rains prevent harvesting, and drought and early frost damage crops. All of these take their toll on crop yields. At the same time, there can be especially good weather: summer rain, dry weather to help crop ripening, and warm spring weather sometimes contribute to bumper crops in the United States. Weather variability has obvious impacts on agricultural production, and although the American consumer rarely notices these effects directly in other parts of the world, these variations mean, literally, feast or famine. In countries vulnerable to food shortages, a drop in production has an immediate impact on diet and health.

Climatologists are predicting changes in global weather, largely as a result of increasing atmospheric CO_2 concentrations. What does this mean for agriculture? Will the changes be good or bad? Will higher atmospheric CO_2 levels mean more plant growth? The effects of CO_2-induced climatic change on agriculture are difficult to predict, but it seems likely that they will be significant.

It is generally recognized that global warming associated with the greenhouse effect will include a wide variety of climatic shifts and not simply a rise in temperature. The temperature rise will be significant but perhaps not the most important aspect of climate change. There may be shifts in storm tracks and in the prevailing wind and pressure patterns. In some locations this may mean warmer or wetter weather, and in other places cooler or drier weather. Many forecasters anticipate hot and dry conditions in the central United States, which would probably be damaging to corn production. But a warming trend would also make many northern areas that now have summers that are too short and cool for crops more favorable for agriculture.

Although climatologists working with global climate models may make predictions of how the climate in a given place may change, a great degree of uncertainty surrounds such specific predictions. Only recently has a consensus emerged that average temperature worldwide will increase, and as yet there is little agreement on conditions in specific agricultural areas. It is reasonable to assume, however, that within the next decade or two we will have a clearer idea of how regional-scale climatic patterns will change (or are changing right now). When this knowledge is available, we will be in a better position to analyze likely shifts in growing regions for various crops. But these changes will probably take place over decades, and our experience of world agricultural patterns

---

the world. The distribution of irrigated lands in the United States is illustrated in Figure 6.9.

Four major types of *irrigation* are used in the world today: flood, furrow, sprinkler, and drip. *Flood irrigation* involves inundating entire fields with water or allowing water to flow across entire fields. The most widespread use of flood irrigation is in growing paddy or sawah rice, primarily in southern and eastern Asia. Rice fields are quite flat and bounded by dikes. During the wet season the fields are flooded, and rice seedlings are planted in standing water. Later the fields are drained, and the rice ripens in a relatively dry field. Flood irrigation is also used to irrigate pastures in some areas. *Furrow irrigation* (Fig. 6.10a) also requires very flat land, but in this case the water flows between rows of plants, which are grown on low ridges. Water is delivered to the furrows by small ditches or in pipes and applied as needed throughout the growing season.

Sprinkler irrigation (Fig. 6.10b) requires substantially more equipment than flood or furrow irrigation. In sprinkler irrigation, water is pumped under pressure to nozzles and sprayed over the land. Nozzles are fixed or moved across a field manually or automatically. Sprinkler irrigation usually results in much higher evaporative losses than other methods, but in areas of very permeable soils, seepage losses are important, so that sprinklers are preferred to furrow irrigation. *Drip irrigation* is a relatively recent development and is used primarily in orchards and vineyards. Each

makes it clear that climate is only one of several factors affecting agricultural production. Unforeseen changes in other variables, such as mechanization, development of new crop varieties, and economic factors, may contribute at least equally with climate in determining the future geography of crop production. In addition, the complex biochemical interactions between increased atmospheric CO_2 and plant growth add another layer of unpredictability.

Carbon dioxide not only plays a central role in the earth's atmospheric circulation but also is an essential input to photosynthesis. The concentration of carbon dioxide in the atmosphere affects the rate at which plants remove that gas from the air and store it as biomass: the more CO_2 in the air, the faster plants photosynthesize. If plants grow faster, then crop yields may also increase.

To understand the importance of this effect, we must recognize that different plants use different biochemical processes in photosynthesis. Two of these are the C_3 and C_4 processes. C_3 plants, including wheat, barley, rice, and potatoes, respond well to increases in atmospheric CO_2, whereas C_4 plants such as corn, sorghum, and sugarcane do not (Parry 1990). Increases in CO_2 concentration also affect the efficiency of water use by plants: the more CO_2 present, the less water is used per unit of plant growth. This means that in areas of water shortage, higher CO_2 concentrations are likely to have beneficial effects on production. In addition, when plants grow in a CO_2-enriched atmosphere, their stomata are open less, causing them to lose less water to evaporation. This increases the efficiency of plant water use, and in dry areas this could mean that less water is needed for irrigation.

Predicting the effects of these changes on world agricultural production is clearly impossible at this time. The effects at any given place depend on what varieties of crops are grown, whether water availability is an important limiting factor, and how other factors of production (fertilizer, seed, energy, machinery, and so on) change—too many unknowns even for the most careful analyses. But we can make three simple generalizations. First, farmers need to be prepared for change. They should avoid growing crops that are easily damaged by unusual weather and should concentrate on varieties that are relatively robust. Second, we should expect the geography of crop production to change: areas that are important for one crop today may be growing something different in the future, areas that have no crops today may become productive, and currently viable regions may cease to produce. Finally, agricultural research needs to be directed toward developing new crops and adapting old ones to a wide array of possible future climatic conditions.

plant has a small pipe that delivers water at a controlled rate directly to the base of the plant. The water drips out very slowly so that little is lost to evaporation or to seepage. Although it is an expensive system to install, it is cost effective in areas where water is scarce.

Irrigated agriculture generally produces high yields as long as water is available. This is because the other environmental characteristics of dry lands, plentiful sunshine, and warm temperatures are conducive to crop growth. This high productivity is not without its costs, however, and in many areas of the world waterlogging, salt accumulation, groundwater depletion, and disease are serious side effects of irrigation. In some areas, salinization is severe enough that it is forcing abandonment of formerly productive land. In parts of the arid western United States, much of the environmental damage associated with irrigation is attributed to government policies that provided water at artificially low prices. These subsidies encourage inefficient use of water, such as for production of hay. This excessive use has contributed greatly to the salinity problems in the lower Colorado River.

In most of the world's farming regions, water already is being used intensively. Worldwide, agriculture accounts for more than 70 percent of the freshwater withdrawals, but use of the water for irrigation accounts for 40 percent of the world's production of food (UNEP 2002). Water resources are already stressed in much of the

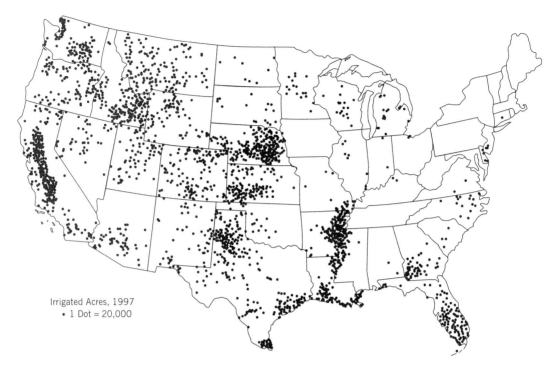

Figure 6.9 Irrigated land in the United States, 1997. *Source:* NRCS, 2001.

world, however, especially in semiarid areas. Increases in irrigated land are therefore limited.

Greater opportunities lie in increasing the efficiency of irrigation, not expanding its current use. The most significant inefficiencies in irrigation water use are seepage and evapotranspiration losses from canals and other conveyance systems and evaporation from irrigated fields. There is great potential for using wastewater, either raw or partially treated, for irrigation. This is already commonplace in some regions—Israel and California have extensive wastewater reuse systems—but poor countries find it difficult to make the needed investments.

Fertilizers and Pesticides

Average yields of grains worldwide more than doubled between 1960 and 2000, owing mostly to increases in the use of *fertilizers*. Fertilizer use increased over four times in that same 40-year period. While fertilizer is used at high rates in the U.S. and Europe, significant production increases from expanded fertilizer use are still possible in poorer countries. One of the main reasons for this prominence is the development of high-yielding plant varieties that require large inputs of fertilizer to realize their potential. In addition, fertilizers make production possible on otherwise marginal land.

The three most important nutrients required by plants are nitrogen, phosphorus, and potassium. Nitrogen is derived ultimately from the atmosphere, but it is made available to plants by nitrogen-fixing bacteria. It is the nutrient that is most often deficient and one that is most widely applied to crops. Additions in amounts of 45 to 90 pounds/acre (50 to 100 kg/ha) increase yields from 1.5 to 3 times, depending on plant variety and inherent soil fertility. Natural gas is an important raw material in the manufacture of most nitrogen fertilizers. The most commonly used forms are ammonia (NH_4) and urea ($CO[NH_2]_2$). Phosphorus is usually present in small quantities in soils, but it is often in relatively unusable forms. It is usually applied as superphosphate or as phosphoric acid, which are manufactured from phosphate rock or from guano (bird dung) deposits found in some coastal areas. In soils, potassium generally is found in larger quantities than phosphorus because it is a more abundant constituent of most rocks. Plants also demand it in large quantities, and in many areas potassium fertilization is important. In some areas, local soil conditions or the particular needs of plants require that

Figure 6.10 Irrigation methods. (a) Furrow irrigation in Texas; (b) center pivot sprinkler irrigation in Nebraska.

other fertilizers be added, with lime (a source of calcium and magnesium as well as a regulator of soil pH) being the most common.

Historically, organic fertilizers (primarily manure) were the most important source of nutrients, especially nitrogen. Organic fertilizers also help maintain good soil structure and water-holding capacity by keeping soil organic matter content high. In the wealthy nations, inorganic fertilizers are now more important, but in the developing nations manure is still a common fertilizer. In most areas, manure supplies are quite limited, and manure is more difficult to apply than other forms of fertilizer. Manure is low in nutrient content relative to that of synthetic fertilizer and is not capable of providing the large inputs of nutrients demanded by high-yielding crop varieties. Increasingly, therefore, inorganic sources of nutrients are replacing organic sources, and this trend is expected to continue.

Pesticide is a general term referring to any of a number of chemical agents used to control organisms harmful to plants, including insects (insecticides), fungi (fungicides), rodents (rodenticides), plants (herbicides), and some types of worms. Thousands of different kinds of pesticides and herbicides are in use, and the vast majority are complex organic compounds manufactured with petroleum as an important raw material. Among insecticides, organochlorines, organophosphates, and carbamates are important types.

The large-scale use of pesticides in agriculture began in the 1950s. Among the first widely used insecticides were organochlorines such as DDT, aldrin, dieldrin, and chlordane. In the 1960s and 1970s, these were largely replaced by organophosphates for most uses, in part because insects began to develop resistance to the effects of organochlorines and in part because organophosphates break down more rapidly and therefore are less likely to accumulate in the environment. Today many different types of chemicals and application methods are used (Fig. 6.11). Among the small-grain crops, pesticides and herbicides are used most intensively on corn, soybeans, and sorghum. Most fruits and vegetables are susceptible to damage by insects, fungi, and other pests, and various pesticides are used, depending on specific circumstances.

The dangers of pesticides began to be discussed in the early 1960s (as noted in Chapter 4), and as a result, their application is regulated in many ways. But in general their use has increased continuously since their introduction, as we discuss in a later section.

Seed

Ever since the development of agriculture, farmers have practiced crop improvement through seed selection. Observant of the variations in a single crop, farmers saved seed from those plants that possessed the characteristics they preferred and planted them the following year. Over a period of thousands of years, this process led to the development of the world's major modern crops. One of the most spectacular examples of this long-term process is maize, or corn. Since 6000 B.C., corn has been altered from a small grain head to its present size and productivity. The success of hybrid corn varieties was a major factor contributing to a large increase in acreages of corn planted in the United States over the past few decades.

Figure 6.11 Aerial application of insecticide to a lettuce field in California.

Since the early twentieth century, the application of Mendel's laws of genetics accelerated the process of selection. Following the successes of hybrid corn, new varieties of wheat and rice were developed as part of the Green Revolution. Many of these new varieties were bred to produce high yields when adequately fertilized. They were bred to be shorter so that the plants would not collapse under the weight of the heavier grains they supported. More recently, advances in bioengineering are making it possible to manipulate the genetic material directly, so that alterations are made in a single generation of a plant species. Not only does this allow the creation of new plant varieties, but also, because hybrid seeds tend to produce more vigorous plants than nonhybrid varieties, substantial increases in yield have been achieved. Those farmers who can afford to pay for these sophisticated seeds choose among a wide array of disease-, insect-, and drought-resistant varieties and for specific fruit or grain size, flavor, ripening time, packing and processing qualities, and so on.

There is some evidence that, at least in the United States and other countries with technologically advanced agriculture, the "miracle seed" productivity is leveling off. This change is in large part due to diminishing returns from the addition of more fertilizer and the continuing use of high-yield varieties. There is still room for greatly increased productivity in the developing world, where the use of chemical fertilizers and improved seed varieties is increasing.

The next wave of improvement in seeds already has begun, in the form of *genetically modified* (GM) crops. Many GM crops are already in widespread production in the United States. Among the most widely used modification is one that causes plants to produce a toxin derived from the bacterium *Bacillus thuringiensis*, or Bt. This bacterium and the toxin it produces occur naturally and help control pests that affect crops. Corn and potatoes are two of the crops widely produced with Bt-modified seeds. Their use is so widespread that new strains of insects that are resistant to the toxin are beginning to appear, so that in a few years the advantage of this genetic modification and the beneficial effects of natural Bt toxin may be significantly reduced.

In recent years controversy in the seed development industry focused on intellectual property rights, protected by patents on new varieties. Such patents allow seed developers to collect royalties for use of seeds and thus provide a greater economic incentive for research and development into new crop varieties. This is particularly important in the case of genetically modified crops, for which the research and development costs are substantial and largely funded by private capital rather than in government research programs. One company, Monsanto, developed a gene that prevents crops from producing fertile seeds, meaning farmers would have to buy new seeds, rather than saving part of their harvest to plant the next year's crop. Under mounting criticism from many directions, Monsanto in 1999 agreed to halt development of that technology. But some see this development as yet another cost for farmers and consumers and another potential barrier to increasing production. In addition, the abandonment of older low-yield varieties is threatening some of these varieties with extinction, though efforts are being made to preserve seeds in gene banks (see Chapter 8).

Perhaps the most significant barrier to the expansion of genetically modified crops is resistance to growing and consuming them in some regions, most notably western Europe and Japan (Gaskell et al. 1999). In those regions there is widespread concern about GM foods, both from environmental and human health perspectives (Brown 2001). It is feared, for example, that GM foods may produce allergic reactions that non-GM foods would not. Because of this resistance, governments of some countries have not approved GM foods for human consumption. This means that if a farmer wants to grow crops that might be sold to those countries, he or she must not only grow non-GM crops but also keep those crops separate from GM crops all through the transportation, storage, and marketing chain from farm to consumer. This can be very difficult, with the result that many food distributors are either unwilling to sell to countries that refuse GM foods or unwilling to buy foods that cannot be certified as GM-free. The eventual success or failure of biotechnology hinges on human attitudes about these technologies rather than on agronomic factors. In 2000, the most common GM crops were corn, potatoes, soybeans, cotton, and canola. Globally, about 44.2 million hectares are in GM crop production, with 68 percent of the acreage in the United States (Brown 2001).

Labor and Machines

To grow crops, soil must be tilled, weeds removed, and plants harvested. Until the nineteenth century in North America, most of this work was done by human and animal labor, with use of simple tools such as plows and hoes. In much of the developing world, the work still is done by human and animal power. But increasingly, machinery driven by fossil fuels predominates (Fig. 6.12). In the nonindustrialized countries, an impetus toward mechanization was introduced as part of the Green Revolution "package."

In the United States, Canada, and other relatively wealthy countries, labor is one of the most expensive inputs to production, and technological change was driven by the need to reduce labor costs. Such reductions were achieved by replacing humans with machines, with the result that the percentage of the labor force employed as farmers has dropped steadily during the past century. In 1850, 64 percent of the U.S. labor force was made up of farmers; this figure had dropped to 2.5 percent by 2000. In the early days of mechanization, these millions of workers were absorbed by the availability of industrial jobs in cities. However, during the past three decades, the demand for underskilled labor in industry also dropped, aggravating problems of unemployment. In addition, much of the remaining nonmechanized farm work is performed by seasonal and permanent legal and illegal immigrants to the United States.

The technological developments in agriculture involving farm machinery had their greatest effects in the industrialized, wealthier nations. For example, grain production was revolutionized through a series of inventions. The plow evolved from a wooden horse- or ox-drawn implement to today's gang plow with up to 16 blades, capable of plowing 10 acres (4 ha) per hour. The reaping and threshing of grain were once done with sickles, scythes, and human muscle power. Today the diesel-powered combine both harvests and threshes grain at a rate of up to 12 acres (5 ha) an hour. These machines need a single human operator, who rides in an air-conditioned cab (Issue 6.2).

The amount of petroleum products required to fuel this mechanical transformation is enormous. In comparing total energy inputs with total energy outputs per hectare of land, the energy efficiency of modern, machine-powered, chemical-intensive agriculture is substantially lower than that of traditional animal-powered methods, but yields per acre are higher under energy-intensive techniques. The largest uses in mechanized agriculture are fuels for tillage, planting, harvesting, transport, and water pumping for irrigation. In addition, heat and light are needed for livestock and poultry production, crop drying, and frost protection. Energy is also utilized in the form of agricultural chemicals, and this amounts to more than a third of the total energy used in U.S. agriculture. The depen-

Figure 6.12 Combine harvesting of wheat in Washington.

ISSUE 6.2: THE DIGITAL FARMER

Precision agriculture is carefully tailored soil and crop management that fits the different conditions found in each field. It allows the farmer to customize chemicals, fertilizers, and seeds to achieve the greatest output per acre while decreasing input costs. Improving the farmer's marginal return requires more precise data about crop yields and their variability within individual farm fields, and this goal has ushered in a new "high-tech" approach to farming. Some people call it global positioning system (GPS) farming, and others simply refer to the new agriculturalists as digital farmers. For geographers, it's an exciting application of spatial knowledge and use of geographic information-processing technology.

In many parts of the Midwest today, you can see GPS antennas on farm equipment and computers with GIS (Geographic Information Systems) packages in the air-conditioned cabs. Accuracies between one and two feet make it possible to develop field-specific inputs to maximize yields and minimize costs. Farm equipment has been fitted with a variety of sensors to monitor everything from rate of fertilizer application to amount of corn cut.

During spring planting, for example, data on crop inputs (fertilizers, pesticides, seeds) are recorded. The sensing devices attached to the planter or the fertilizer and pesticide applicators are electronically connected to GPS units that geocode the information for the real-time computer input maps. The specific longitude and latitude of the application are stored in the computer. At the end of the day, the farmer can download the data and produce an input map of his field. Throughout the season, the farmer can record additional fertilizer and pesticide applications, soil moisture, and so forth and again create maps of inputs for the season.

When harvesting begins, yield monitors (attached to the combine) transmit the data to disk but also display it on a computer screen in the cab. Data are received every 1 to 3 seconds so that the farmer can see (in real time) how much corn is being yielded. Later, these stored data can be used to make a digital map of yields at a 1- to 3-meter scale.

The most exciting development for the farmer is the yield map, which illustrates the variability in yield in the fields. Low-yielding spots can be identified and inputs altered for the next crop or growing season. The ability to see the distribution of yield can result in significant savings in fertilizers, for example. Why add more fertilizers to an area that is already highly productive? In this way, the farmer can begin to think about the agronomic relationships that might produce such variability such as mistakes in fertilizer or pesticide applications, drift from a neighbor's pesticide application, diseases, or micro-zonations in soil fertility or nutrients.

Farming in the twenty-first century will require advanced degrees not only in agronomy but possibly geography. Hopefully, students with interests in agriculture can see how this spatial information helps in the management of our food resources. Perhaps others are wondering what geography can contribute to natural resources management. One answer is digital farming.

dency of U.S. agriculture on fossil fuels leads farm interests to resist energy-conservation measures or carbon emission controls that would involve increases in fuel prices.

Animals in the Food Production System

Meat production plays an important role in many agricultural systems. One important function of meat is to provide a source of concentrated protein in the diet. While adequate protein can be obtained from purely vegetable diets, the availability of animal protein adds some flexibility in meeting human nutritional requirements. A second important function of meat production is to take advantage of biomass that is not digestible by humans. Grass and shrubs, for example, are not acceptable food for humans, but cattle, sheep, and goats have the ability to digest these plants and convert them to meat and dairy products that humans can use. Other animals, especially pigs and chickens, are able to consume human food wastes and convert them into usable food.

In many ways, then, meat production complements agriculture, thus increasing the total amount of food we can harvest from a given amount of land. At the same time, in some animal production systems, domestic animals compete with humans for food. This is particularly the case in North America and Europe, where most meat is produced not by feeding grass and other inedible materials to animals but by feeding them the same grains that human eat, grown on land that produces grain for direct human consumption. The efficiency of converting vegetable food energy to animal food energy in the United States is about 5 percent; that is, 20 calories of food energy as feed (range grasses and grains in the feedlot) are needed to produce one calorie of food energy as meat on the table.

If the food used to produce meat is grown on land that could be used to grow grain for human consumption, then we are in effect substituting a substantial amount of vegetable food for a small amount of meat, which would seem to be unwise in a world short of food. About two-thirds of the grain grown in the United States is fed to livestock; the proportion is similar in Europe. On the other hand, meat produced by grazing animals on land that is not otherwise usable for food production represents an important increase in available food supplies. Thus, range animals are a means of converting otherwise unusable vegetable matter to valuable food, even if the efficiency of conversion is relatively low (Fig. 6.13). It is when livestock are fed high-quality corn and other grains that questions of efficiency and equity arise.

Most of the world's prime grazing lands are natural grasslands found in semiarid and subhumid areas. The ability of grasses to grow rapidly when conditions are favorable, combined with the variability of precipitation in semiarid areas, leads to seasonal and annual variations in the amount of grass that grows in grassland areas. This means that the number of herbivores supported by the land also varies. Under natural conditions, populations such as rodents and deer are kept in check by competition for available food. But the population levels of domestic animals are controlled by humans, not by natural conditions. To maximize animal production in the short run, herders often exceed the carrying capacity of the land, which results in damage to the vegetation.

Grazing affects plants in several ways. Plants are reduced in size, which usually inhibits their

Figure 6.13 These cattle are converting grass into food on semiarid rangeland in Colorado that probably would not be agriculturally productive as cropland.

ability to photosynthesize and grow. Animals damage plants by trampling, which is particularly detrimental to young plants. The reduction in plant cover caused by grazing leads to a deterioration in soil conditions that also inhibits plant growth. On the positive side, the seeds of some plant species are spread by grazing animals, either by being eaten and excreted or by being attached to fur. Other plant species are inhibited from reproducing under grazing pressure, for example, if their seeds are digested. Finally, grazing stimulates plant growth by reducing competition among plants for moisture or nutrients. Thus, the plants are able to replace some of the biomass taken by animals.

Range plant species vary in their ability to survive and reproduce under grazing pressure. Some species quickly regenerate leaves lost to animals, whereas others cannot. Some are relatively unaffected by trampling, and for others trampling is fatal. Most important, some species are more palatable to grazing animals and thus are eaten first, while others are less palatable and are eaten only after the desirable forage is consumed. These differences in susceptibility of plants to grazing impact are the basis of the ecological changes that result from grazing (Fig. 6.14).

Range ecologists classify plant species in a particular area as decreasers, increasers, or invaders. *Decreasers* are plant species that are present in a plant community but decrease in importance (as measured by numbers of plants or percentage of

Figure 6.14 Much of the rangeland of the western United States has been degraded by a century of grazing. In many areas grass cover has been replaced with shrubland like this in Nevada.

ground covered) as a result of grazing. They are generally the most palatable plants, but they may also include species that are negatively affected by animals in other ways, such as by trampling. *Increasers* are species that were present prior to grazing and increase in importance as a result of grazing. They may be less palatable species, or they may increase simply because there is less competition from the decreasers for water or nutrients. *Invaders* are species that were not present prior to grazing but are able to colonize the area as a result of the change in conditions.

No species is classified as decreaser, increaser, or invader without reference to a particular site. A species may be an increaser in one area and an invader in another, or it may be a decreaser in one area and an increaser in another. For example, big sagebrush (*Artemisia tridentata*) is present on much of the rangelands of the United States, and over most of this area it is an increaser (Fig. 6.15). Although it is high in nutrients, it is unpalatable to cattle and sheep except when other forage is unavailable. Big sagebrush also is unaffected by trampling. In some areas of

Figure 6.15 Sagebrush rangelands in New Mexico.

the American West, it was insignificant or not present prior to the onset of grazing in the nineteenth century, but today it is a dominant species. In these areas it is an invader.

ENVIRONMENTAL IMPACTS OF FOOD PRODUCTION

Agricultural activities represent an enormous transformation of natural ecosystems occurring over very large portions of the earth's surface. As such, they inevitably cause vast environmental impacts. Many of these impacts occur beyond the farm field, such as water pollution from agricultural runoff, air pollution from blowing dust, or dispersal of agricultural chemicals. Off-site impacts such as these are considered in Chapters 10 and 11 of this book. Here we consider those impacts that degrade the quality of agricultural resources for their primary function of producing food. We focus on two processes: soil erosion and rangeland degradation.

Soil Erosion

Soil erosion on agricultural lands takes place through three major processes: overland flow (or runoff), wind, and streambank erosion. Of these processes, overland flow erosion is the most visible and widespread, and in most agricultural areas it is quantitatively the most important. Wind erosion occurs on exposed soil if strong winds blow at times when the soil surface is relatively dry. Streambank erosion is limited to fields that border streams, and though locally significant it is not a major factor in soil erosion worldwide.

How Erosion Happens Accelerated *overland flow* erosion is primarily the result of intense rainfall on bare ground. Raindrops striking a bare soil surface break up clumps of soil into individual particles. These are moved by the splash made on raindrop impact and compact the soil surface so that water is less able to soak in, thus reducing the *infiltration capacity*, or the maximum rate at which soil absorbs water. When the precipitation intensity exceeds infiltration capacity, water flows across the surface as overland flow rather than soaking in. When overland flow rates are particularly high, small stream channels called *rills* may be formed to carry the water away.

Soil erosion due to overland flow varies considerably from place to place. Some areas are extremely susceptible to erosion, whereas in other areas it is only a minor problem. Some of the major factors influencing the severity of erosion are topographic factors such as slope steepness, the inherent susceptibility of the soil to erosion, the intensity and frequency of rainfall, and the cropping and management practices of farmers.

Cropping and management practices are very significant to overland flow erosion. Cropping practices include what kinds of crops are grown and when they are planted. Row crops, such as corn and soybeans, tend to allow more erosion than do continuous cover crops such as wheat or hay. If the field is bare during part of the year, then that is the time when there is the greatest susceptibility to erosion. As the plants grow and mature, they cover a greater amount of the ground and erosion susceptibility decreases. In most cases, the time of planting is dictated by plant characteristics and weather, but in some situations particular crops that provide greater cover at times of more erosive rainfall are chosen. In addition, the decision of whether or when to plow stubble under has effects on erosion susceptibility. This is an example of a management practice. Others include the choice of various conservation techniques such as contour plowing, terracing, or minimum tillage, practices that are discussed in the following sections.

Wind erosion occurs when there is a combination of high wind velocities and a soil surface that is easily eroded. High wind velocities are obviously dependent on weather conditions, and some areas are windier than others. Wind erosion is greater at times when the soil surface is dry, and for this reason wind erosion is generally a greater problem in arid than in humid regions. Vegetation cover is important, for it controls wind speed at the soil surface. Thus, plants are very effective in reducing surface wind velocities, and wind erosion is negligible under vegetation cover.

Extent of the Problem Accelerated soil erosion is found on virtually all cultivated lands, but the extent of the problem varies greatly from place to place (Pimentel et al. 1995; Crosson 1995, 1997; Trimble and Crosson 2000). The results of a U.S. Department of Agriculture survey of sheet and rill erosion on U.S. lands (NRCS 2001) are summarized in Figure 6.16*a*. The greatest rates of soil erosion are found in the Southeast and in the Corn Belt states, with average rates ranging from 4 to 6 tons per acre (8 to 12 tons/ha) per year. Wind erosion is mapped in Figure 6.16*b*. It is most severe in the western Great Plains states, at rates similar to those of rill erosion in the East. The sum of wind and water erosion is mapped in Figure 6.16*c*. Rates of 2 to 4 tons/acre (5 to over 10 tonnes/ha) per year correspond to a loss of about 0.2 to 0.4 inch (0.5 to 1 cm) from the soil surface in 10 years. If we recognize that the top 4 to 12 inches (10 to 30 cm) of most soils holds the greatest proportion of the nutrients and organic matter, then these rates may lead to removal of much of this very fertile layer in just a few decades. However, a portion of the eroded soil is redeposited in areas close to where it was eroded and may still support profitable planting. Net erosion (erosion minus local redeposition) is somewhat less than the rates shown in Figure 6.16.

Soil Conservation Techniques Over the years, many techniques have been developed to control erosion while still allowing efficient agricultural production. Some of these techniques, such as crop rotation, have been known for centuries, whereas others have been developed fairly recently. Some of the important ones are discussed in this section.

Crop rotation is a farming method that is aimed primarily at maintaining soil fertility, but by so doing plant growth is enhanced and erosion reduced. Crop rotation simply means that over a period of several years the crops grown on a field change in a systematic pattern. Some crops demand more of some nutrients than others, and by changing crops from year to year excessive depletion of nutrients is prevented. Typical rotation patterns include fallow periods or plantings of crops like alfalfa that restore some nutrients to the soil. Often such crops are plowed under rather than harvested; this technique is called green manuring. Crop rotation also allows the ground to be covered a greater percentage of the time, thus reducing erosion. Some rotation patterns repeat one or more crops, such as a rotation of corn, corn, oats, and hay. In the short run, this technique requires planting less profitable crops in some years, which may reduce the farmer's ability and willingness to use it.

Contour plowing, another important soil conservation method, involves plowing across a slope, or on the contour, rather than up and down a slope. Contour plowing reduces erosion by trapping water in the furrows where it can soak in, rather than running down the furrows and causing erosion. It is also a good water conservation technique. On hilly land, contours are rarely straight, and so contour plowing requires plowing in curvy lines across a field. Similarly, field boundaries rarely follow the contours, and so this usually results in irregular-shaped patches and some unused land, making it more time-consuming and perhaps less profitable for the farmer than plowing in straight rows parallel to the borders of a rectangular field.

Contour plowing is often used in conjunction with *terracing*. Terracing involves constructing ridges or ditches parallel to the contours, which trap overland flow and divert it into drainage channels, thus preventing it from continuing downslope and causing erosion. In some parts of the world, notably Southeast Asia, terraces are constructed with flat surfaces for ponding of water and rice production. In the United States most terraces are subtle ridges on sloping fields, and in many cases the farmer plows and plants on them as if they were not there. Contour plowing, terracing, or crop rotation often is combined with *strip cropping*, in which crops are planted in parallel strips along the contour or perpendicular to prevailing winds. One strip is planted with a less protective crop or left fallow, while the adjacent strip has a protective cover. Soil eroded in one area is deposited nearby, and little of the soil is lost.

With regard to wind erosion, the most useful prevention methods are stubble mulching and windbreaks. Stubble mulching simply means leaving plant residue on the ground between growing seasons rather than plowing it under immediately after harvest. Windbreaks are lines of trees planted perpendicular to the most erosive winds. Both of these techniques act to reduce wind velocity at the soil surface and thus reduce erosion. Some of these methods are illustrated in Figure 6.17.

Another conservation measure that is rapidly gaining acceptance is *minimum tillage*, or *conservation tillage*. It includes a variety of techniques that seek to reduce erosion by maximizing the amount of crop residue that is left on the soil surface and by maintaining soil structure conducive to infiltration rather than runoff. Most do this by reducing

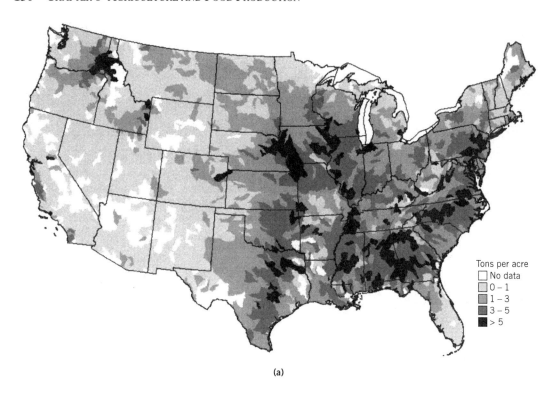

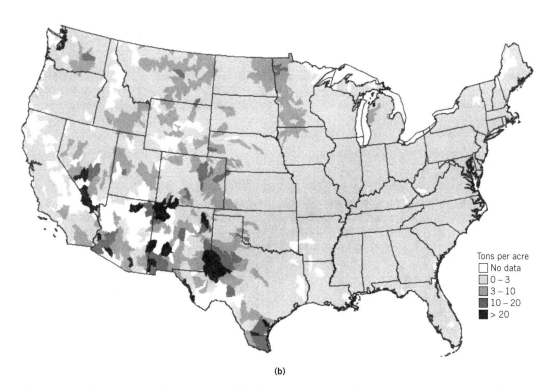

Figure 6.16 Soil erosion on U.S. cropland, 1997. Rates of sheet and rill erosion (*a*) and wind erosion (*b*) combine to demonstrate the extent of total erosion (*c*), geographically dominant in the western states. *Source:* NRCS, 2001.

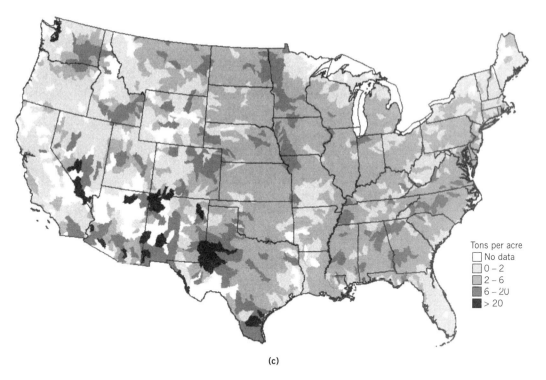

(c)

Figure 6.16 (*Continued*)

the amount of plowing, which also conserves energy and reduces the fuel costs of operating a tractor. One of the major drawbacks of minimum tillage is that instead of physically removing weeds by plowing, herbicides are used for that purpose, which increases the risk of the harmful side effects of these chemicals. Conservation tillage is adopted in much of North America and is helping to reduce erosion without taking land out of production.

Decisions on how to manage agricultural land are made by individual farmers, who, like any individuals operating businesses, are concerned primarily with maximizing their incomes, at least in the short run. Farmers are aware of the threat of erosion to the productivity of their lands (which in the long run destroys their capital base), and most feel that more should be done to prevent it. The problem is that their management decisions must consider the costs of inputs of seed, fertilizer, irrigation water, fuel, pesticides, and so on, relative to the value of the crops that are produced. Inputs of fertilizers and pesticides generally result in substantial increases in crop yields, and so variations in these inputs are much more significant to short-term returns than long-term reductions in inherent soil fertility caused by erosion. Most farmers in the United States today, especially those with smaller operations, are struggling under enormous debts, and farm prices are not high enough to pay off loans and provide a comfortable profit. This profit is necessary if farmers are to restrict planting, invest in erosion control structures, or otherwise constrain their activities.

Government officials concerned with soil conservation recognized this problem and since the 1930s have implemented programs to ease the economic burden of soil conservation. One important method is payments to the farmer in exchange for planting soil-conserving cover crops that keep the soil covered year-round instead of more profitable but less protective crops. Such payments serve a second purpose, that of reducing farm production so as to increase prices. In addition to these payments, subsidies are made available from time to time to pay for capital improvements such as construction of terraces or gully control. Finally, Natural Resource Conservation Service extension agents provide technical assistance to farmers by making available information on different erosion control methods and ways to incorporate them into a profitable farm management plan.

(a)

(b)

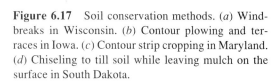
(c)

(d)

Figure 6.17 Soil conservation methods. (*a*) Windbreaks in Wisconsin. (*b*) Contour plowing and terraces in Iowa. (*c*) Contour strip cropping in Maryland. (*d*) Chiseling to till soil while leaving mulch on the surface in South Dakota.

As a result of these programs, erosion has been reduced dramatically since the early twentieth century. The most severe erosion problems that were evident in the 1920s and 1930s are now more controlled. Land that was degraded by excessive erosion is no longer being tilled and instead is in woodland or pasture or in a conservation-type program. In the past few decades continued progress has been made, as illustrated in Figure 6.18, which shows a 30 percent reduction in erosion between 1982 and 1997.

In many areas the erosion problem persists, and more remains to be done. In the late 1970s and early 1980s, economic difficulties led farmers to plant on more and more land, rejecting cover crops in favor of more profitable row crops. The

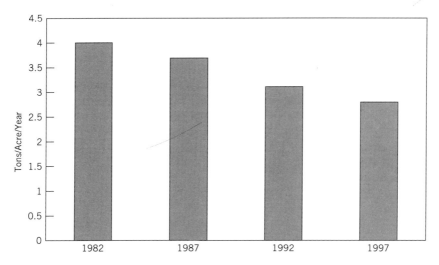

Figure 6.18 Reductions in soil erosion rates on U.S. cropland, 1982–1997. This trend is a result of several factors, including removing highly erodible lands from production through the Conservation Reserve Program and increasing use of conservation tillage methods on cropland. *Source:* NRCS, 2001.

large machinery in use today is not well suited to terraced fields, and in many areas terraces were destroyed. These trends prompted new warnings on the severity of the erosion problem, but no new government programs emerged to deal with it. Although farmers might be willing to accept mandatory programs to control erosion, they would be willing to do so only if society as a whole, through the government, compensated them for reduced incomes. On the other hand, if new agricultural technologies increase yields at the same time as erosion control reduces either yields or profits, few farmers will do much to control erosion. This is the case with some conservation tillage systems, which have significant initial capital costs but significantly lower operating costs because the soil is not plowed as frequently. Much of the recent reduction in erosion rates illustrated in Figure 6.18 is a result of the adoption of conservation tillage practices. In some cases this adoption was subsidized by government erosion-control programs and in others it was purely at the initiative of the farmers.

Rangeland Degradation

Range, or grazing land, provides forage for limited numbers of domestic animals. *Overgrazing* occurs when the number of animals on these lands exceeds carrying capacities. Several areas of the world, notably the dry lands around the Mediterranean Sea, have a long history of overgrazing, with resulting problems of devegetation, erosion, and ultimately the threat of desertification. *Desertification* is land degradation in dryland regions resulting mainly from adverse human impact. It occurs in parts of all the major semiarid regions of the world and affects some of these regions more than others.

The Global Perspective A number of factors contribute to the degradation of the world's rangelands. Overgrazing, the expansion of cultivation, the conversion of rangeland to cropland, and increased human population pressures are all taking their toll on rangeland conditions worldwide. There are four types of soil degradation: water erosion and wind erosion, which strip away nutrients from the soil and the soil itself, and physical and chemical processes, which reduce the productivity of the soil (Table 6.2).

One of the first baseline studies to estimate world land degradation patterns, the Global Assessment of Soil Degradation Desertification (GLASOD), found that in 1990, 1.9 billion hectares (17 percent of all vegetated land) showed signs of human-induced degradation, 11 percent of it labeled as moderate,

Table 6.2 Types of Human-Induced Soil Degradation

Type	Definition	Processes Involved
Water erosion	Loss of topsoil through the action of water	Sheet or surface erosion, rill and gully formation
Wind erosion	Decrease in vegetative cover, thus exposing topsoil to potential removal	Overgrazing, agricultural practices, deforestation, fuel-wood removal
Chemical deterioration	Loss of soil nutrients	Salinization, acidification, pollution
Physical deterioration	Reduction in soil quality for agriculture	Compaction, waterlogging

severe, or extreme (World Resources Institute 1992). By 2002, nearly 23 percent of the world's land resource based was degraded (UNEP 2002). Worldwide, about 35 percent of the human-induced soil degradation is caused by overgrazing, 30 percent by deforestation, 27 percent by agricultural activities, 7 percent by overexploitation for fuel-wood use, and the remaining 1 percent by industrialization (UNEP 2002). Regionally, these causes vary in their importance. In Africa, land conversion and the cultivation of marginal areas are the leading causes of soil degradation; in North and Central America it is agricultural expansion, in Asia it is desertification and overgrazing, and in South America it is deforestation. Figure 6.19 shows areas of moderate, severe, and extreme degradation. As you can see, those areas of serious concern are in Eurasia, the Middle East, China, and North and Central America. The areas of major concern include small areas with severe degradation and larger areas with moderate degradation.

Rangeland Degradation in the United States Rangeland decline is widespread in the United States, where serious problems of degradation exist on much of the nation's range, especially in the Southwest. It has been estimated that 37 percent of North America's arid lands are in a state of "severe" desertification. Within the United States, perhaps 225 million acres (91 million ha; about 25 percent of the rangeland total) are in a state of "severe" or "very severe" desertification (Dregne 1997; Sheridan 1981a). In the early 1990s, the NRCS began a program intended to reduce problems of rangeland degradation, and in 1997 39 percent of the U.S. rangeland no longer had serious resource problems, while 61 percent of rangeland and 46 percent of permanent pasture had serious ecological problems (Natural Resources Conservation Service 1997).

One case study helps illustrate the contribution of poverty, carelessness, and fluctuating climatic conditions on rangeland degradation. In the areas occupied by the Navajo Indians in Arizona and New Mexico, very severe degradation is caused by too many people, with few economic alternatives, being crowded onto too little land. During the past century, the Navajo population has multiplied by a factor of ten, while the land available has only tripled. In the late 1930s, when the U.S. government first tackled the overgrazing problem, the Navajo's sheep population was an estimated 1.3 million, but it increased to 2.17 million by the early 1980s (Sheridan 1981b). When one considers that the U.S. government believes that the 15-million-acre (6-million-ha) Navajo reservation has a carrying capacity of 600,000 sheep, the problem of overgrazing becomes quite clear.

What are the environmental effects of this overgrazing, and why do the Navajo continue this practice? Most of the land surface of the reservation is badly eroded, and the plant cover the animals rely on for food is not being maintained. Until recently the Navajo have had few alternatives, and conflicts between traditional livelihood patterns and high unemployment rates (over 60 percent) forced the Navajo to continue their overgrazing practices. Since sheep and cattle herding remain the major source of income for many Navajo, they are naturally unwilling to reduce animal population levels to reach what they see as a U.S. government–generated ideal (carrying capacity) that could threaten their own security.

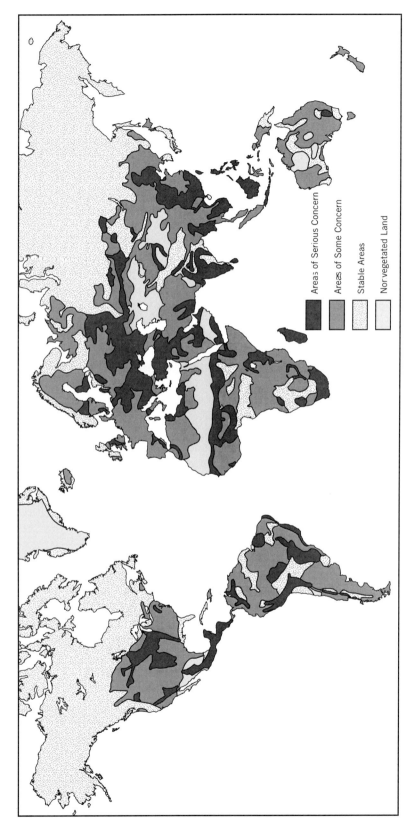

Figure 6.19 Soil degradation. Areas of serious concern for land degradation are located in the world's prime agricultural belts as well as highland areas. *Source*: World Resources Institute, 1992.

AGRICULTURAL POLICY AND MANAGEMENT

Throughout the world, in capitalist and socialist economies alike, agriculture is fundamental to national welfare and security. As a result, virtually every government in the world intervenes in their domestic agricultural system. These interventions take many forms, including production subsidies, price regulation, commodity management, research and education programs, and direct regulation of farm practices. Some programs are more successful than others, but nearly everywhere such management has profound impacts on how food is produced and distributed.

Subsidies

Some of the most successful policies in recent decades were those that stimulated production by providing financial subsidies to farmers. In the United States, for example, the thrust of most U.S. agricultural policies was to subsidize agricultural income and maintain higher prices than occurred in a free market, while restraining production and pursuing conservation goals. Most of these policies were instituted in the 1930s as part of the New Deal programs and continued in place until the mid-1990s. The subsidies took many forms, including loans that were forgiven if a crop failed, direct payments in compensation for taking land out of production, insurance against crop failure resulting from bad weather or disease, and targeted subsidies to help with conservation-related investments.

Subsidies also are responsible for the enormous increase in agricultural production that took place in Europe from the 1950s to the 1990s. The Common Agricultural Policy instituted there established uniform prices for all member nations in the

---

ISSUE 6.3: AGRICULTURAL SUBSIDIES, TRADE, AND POVERTY IN THE DEVELOPING WORLD

During the Great Depression of the 1930s, poverty among the rural population reached proportions never before seen in the United States. A central element of the New Deal legislation was a new system of government controls on agricultural production. The main goals of this legislation were to reduce problems of overproduction and depressed crop prices, encourage soil conservation, and subsidize farm income. These goals were achieved through a variety of programs, but most guaranteed that farmers would receive at least a certain minimum price for their crop, regardless of market fluctuations or low yields. In return, farmers agreed to limit the amount of land that was planted and to farm in ways that would help conserve the soil. The New Deal programs have been in place for nearly 70 years, and they have become an integral part of the American agricultural system.

Although the principal crops of the United States in the 1930s were grown for mostly domestic consumption, there was important international trade in agricultural commodities between the more developed countries of Europe and North America and their colonial dependents in Asia, Africa, and Latin America. That trade had been established under the colonial system and was strictly controlled by the imperial nations. For example, sugar and bananas were grown in the so-called Banana Republics of the Caribbean and Central America and shipped to North American markets, while cotton and tea grown in India and East Africa supplied English industrial and domestic demands. In addition to controlling the terms of trade to their own benefit, the colonial powers often actively restricted or prevented economic development in their colonies, as a means to protect industrial interests at home and enhance their own wealth.

Today the formal and informal colonial relationships that existed in the 1930s have ended, but the fundamental terms of trade remain. Markets are dominated by the wealthy countries, producers have little influence on prices, and poverty in the producing countries persists. Despite considerable industrial development in some regions, the economies of most of the world's poorer countries remain based in agriculture. As such, agricultural commodities are fundamentally important in gaining the foreign exchange income needed to finance economic development.

When a government subsidizes production of crops, it not only increases the income of the producer but also usually raises food prices in the pro-

European Union. Prices were set high enough so that the least efficient farmers (or those farming the poorest land) could still stay in production. At these high prices, the more efficient farmers operating on more fertile land made substantial profits. They cultivated all the land they could, invested in new equipment, and prospered. These policies generated more production than the market could absorb, and thus the government accumulated surpluses of agricultural commodities. Some of these surpluses were sold on the world market, usually at prices below domestic levels, while others were given to famine relief and welfare programs. These policies also created new erosion problems where there were few before; for example, hilly land formerly in pasture was brought into grain production to take advantage of high grain prices.

In the mid-1980s these problems became more acute, and in the context of government budget-cutting and the trend toward reduced governmental intervention in markets, many of these subsidies were reduced. In the United States, the most dramatic change took place with passage of the Food Security Act of 1996, also known as the Freedom to Farm Bill (Issue 6.3). However, the cut in subsidy was short-lived, for in 2002, a new Farm Bill was passed, which reversed the previous efforts to wean American farmers from federal subsidies. In fact, the Farm Security and Rural Investment Act, as it is known, actually increased subsidies for large grain and cotton farmers; the larger the farm, the greater the subsidy (Becker 2002). The smaller family farms receive very little in subsidies (Environmental Working Group 2002). Agricultural subsidies in the United States and other wealthy nations also are seen as a significant barrier to economic development in poor countries (Issue 6.3).

ducing nation. In the United States the price of a hamburger or a loaf of bread comprises mostly processing and marketing costs, so a modest increase in the price of beef or wheat at the farm gate has little impact on the consumer, but it means a lot to the farmers and those trading in the basic (unprocessed) commodities.

Another effect of government subsidy programs is that they tend to create surpluses. The government may set a price for a commodity that is higher than the market price and promise to buy the farmer's produce at that higher price, even if the market didn't demand it. This results in greater amounts of commodities being sold on world markets and thus reducing prices in those markets, and thus reducing income in poor countries that export those commodities.

In recent years the rules governing international trade have been the focus of much debate and controversy. These rules are fundamental to economic relations and development in a world of increasingly interwoven trade relationships. In general there has been a move toward reducing barriers to trade, such as import tarriffs imposed by a nation for the purpose of protecting domestic producers. Such tarriffs are seen as unfair barriers to competition; if they were removed, then producers in other countries would have an easier time exporting their goods. Agricultural subsidies have the same effect. They make it easier for domestic producers to grow a crop at a profit while not raising the price of the good on the international market.

The elimination of agricultural subsidies is seen as key to encouraging development in the agriculture-dependent economies of Africa, Asia, and Latin America. Negotiations in the World Trade Organization have focused on the issue, as did debate at the Johannesburg Summit. In 2002 the European Union pledged to reform its system of subsidies, focusing on cutting the link between production and income subsidies and instead tying subsidies to environmental measures such as reducing erosion and improving food safety. In the United States, farmers remain heavily dependent on subsidies. An attempt to phase out subsidies and production controls in the 1995 farm bill failed when production increased and farm commodity prices fell, reducing farm income. The 2002 farm bill retains or increases most subsidies, and it is unlikely that anything will change in the near future.

Sustainable Agriculture

In every agricultural region of the world, the negative impacts of agriculture are apparent. The effects of soil degradation and agricultural pollution are more severe in some areas than in others, but there is universal recognition that while food production must increase to meet the needs of a growing population, the productive capacity of agricultural lands must be preserved and nurtured. Farm practices that seek to balance production and preservation are known as *sustainable agriculture*.

Although it is difficult to define precisely those farming techniques that constitute sustainable agriculture, they generally involve intensive soil conservation measures and minimal use of pesticides and inorganic fertilizers. The number of farmers using these methods in the developed world is relatively small, mostly in specialized crops such as fruits and vegetables. But the number of farmers producing grain, meat, and dairy products by sustainable methods is increasing, and this trend is likely to continue as long as the costs of chemical inputs remain high and the need to maximize yield per acre is low.

One key feature of the drive toward sustainable agriculture is the reduced use of farm chemicals, especially pesticides. Pesticides were a boon to modern technological agriculture, but they also have harmful side effects. The side effects of most concern are health hazards to agricultural workers using the pesticides, health effects on the general population through contamination of food, water, and air, and adverse ecological effects.

The most severe human health hazards of pesticides are those associated with the occupational exposure of farm workers handling the substances. Workers in the field at the time of application are exposed, as are those handling crops at harvest time. Accidental exposure is also a major concern. Although pesticides are regulated to limit effects beyond the farm as well as on it, major problems remain. For example, the nonpersistent pesticides that are used today generally do not accumulate in high concentrations in the environment. However, they are highly toxic at the time they are applied, hence those in contact with pesticides at that time are most at risk.

The general population is also exposed to agricultural chemicals, primarily through consumption of foods containing these substances but also through transport of pollutants in water and air. Although these exposures are less acute than those faced by agricultural workers, the number of people affected is much greater.

Many scientists and others concerned with the use of agricultural chemicals have long warned of dangers and advocated alternative approaches. But for the most part the governmental and university researchers and advisers who guide U.S. agricultural technology maintain that productive agriculture depends on substantial chemical inputs. Examination of recent trends in pesticide use, for example, suggests that insecticide use is on the decline. Herbicides, which are used intensively in modern minimum-till and no-till methods, are still used intensively, but their use does not appear to be growing, even though more and more farmers are adopting reduced tillage techniques. Agriculture's reduced use of these chemicals is likely to improve many aspects of the food system, from the health of farm laborers to reduction of pesticide residues in processed foods, as well as reductions in groundwater pollution. The passage of the Food Quality Protection Act (1996) is one example. This act requires that pesticide tolerance settings reflect risk standards that are definable by modern scientific methods rather than absolutely zero risk. The legislation also includes both dietary and nondietary exposures (e.g., lawns) and pays particular attention to certain high-risk groups such as children, infants, and farm workers—groups that might need extra precautions if they ingest too much of the pesticide (Wargo 1996).

Rangeland Management

The best way to manage rangelands is to maintain animal populations at or below carrying capacity. Carrying capacity varies from location to location and from year to year. Range ecologists developed techniques for estimating carrying capacity on the basis of vegetation type, average precipitation, soil characteristics, and other data. Because it remains difficult to predict rainfall months in advance, planning herd sizes to suit range conditions is sometimes a problem. Under ideal conditions we can determine fairly closely just how many animals should be allowed to occupy a given piece of land and under what conditions. The problem is that herders sometimes have limited flexibility when they have to decide how many animals they can keep on a given piece of land, and there is usually an incentive to own more rather than fewer animals than the land can sustain. Overgrazing is the result.

Rangeland that is damaged by overgrazing is not lost forever; it will recover if grazing pressure is reduced or if rains improve plant growth. In contrast to this passive method, several active techniques can be used to improve range quality. These include mechanical, chemical, and biological control of undesirable plants, as well as burning, seeding, fertilization, and irrigation. Mechanical brush control by plowing, bulldozing, or dragging heavy chains across the land is used to control sagebrush, mesquite, juniper, and other undesirable species. In some cases, the slash from such operations is burned, and in others it is left as an erosion control measure. In some areas, goats are used to control woody species, for they strip plants of leaves and eat entire seedlings, while cattle do not. Care must be taken that the goats do not overgraze the range, however, or desirable species will also be lost.

Most range management issues are complicated by the fact that populations are sparse and resources are often commonly owned. This is certainly the case in the United States, where much rangeland is under the control of the Bureau of Land Management.

The availability of free or low-cost forage on federally owned lands was essential to the profitability of the developing cattle industry. In addition to leasing the water and forage rights on government property, many ranchers obtained title to vast areas of federal land by evading the acreage limitations imposed by the 1862 Homestead Act (see Chapter 3). By the early twentieth century, the use of barbed wire had created boundaries to the great American pasture, and the days of open-range cattle and sheep drives wound down. These drives were replaced by truck and train transport to regional markets, notably Chicago, Minneapolis, Omaha, Dallas, and Denver.

The federal government imposed several constraints on the traditional independence of the western rancher, beginning with the founding of the National Forest System (1905), passage of the Taylor Grazing Act (1934), and the establishment of the *Bureau of Land Management* (*BLM*) in 1946. Initially, ranchers were not permitted to let animals graze in the National Forest reserves, but a nominal fee system later permitted the land to be opened for use. Battles have raged ever since over the government's right to control numbers of cattle and the seasonality of grazing. The Taylor Grazing Act organized federal lands into a system of 144 grazing districts for joint management by the federal government and local stock raisers. Some cattlemen resisted this scheme and attempted to turn management of these lands over to the state governments. However, the effects of overgrazing were evident, and the Taylor Grazing Act's range rehabilitation plans received congressional support. The act was passed, but it remained underfunded and subject to constant disagreement between federal and local interests (Stegner 1981).

Today, 22,000 ranchers lease grazing rights from the BLM to graze about 3 percent of the nation's beef cattle (BLM 2000). In general the fees charged for grazing on BLM lands are a small fraction of those charged by private landowners, and some environmentalists argue that these low fees encourage overgrazing on public lands.

The BLM took over the administration of these and other public lands from the General Land Office and administered them in a similar manner until 1976. In that year Congress passed the *Federal Land Policy and Management Act (FLPMA)*, which brought together thousands of pieces of legislation related to public land management. It also increased the BLM's power to manage its 270 million acres (109 million ha) for the public good, with *multiple use* and *sustained yield* as guiding principles. From their inception, the BLM and predecessor agencies have struggled to reduce overgrazing. Today's grazing allotments (number of animals permitted per acre) were established according to 1930s estimates of historical rangeland use.

Rangeland improvement measures are used on both private and public lands. The federal government, particularly the BLM, plays a major role in attempting to repair some of the damage done by overgrazing in the past. In 1979, BLM began a long-term program to assess and improve the rangeland it administers. Progress has been slow, and as of 2000, only 27 percent of 153 million rangeland acres had been assessed. Of these, 66 percent were found to be meeting all standards or making significant progress toward meeting the standards. Most of the remaining assessed lands were below standard, but action has been taken to ensure progress toward meeting the standards (BLM 2000).

CONCLUSION

Debate continues regarding the future of world agriculture. There are millions of hungry people in the world today, yet crop production increases are leveling off in the "breadbasket" nations of the world. Self-sufficiency in food production is far from a reality in many nations. Furthermore, cropland losses continue to decline globally as agricultural land gives way to expanding cities or the land is degraded to such an extent that it loses its fertility.

Many food production experts in the wealthier nations encourage further transferral of mechanized agriculture to the less affluent, with the implication that the American system should be adopted worldwide. This would have to be accompanied by a massive industrialization program to employ the billions of farmers and families pushed off their land by mechanization. Such a transformation has immense social, environmental, and economic implications.

Many countries are attempting to stimulate a return to small-scale, subsistence-oriented farming for their rural residents but are fighting a massive flow of people to the cities. It is unlikely that these plans can be effective without a strong guarantee to farmers that they can turn a profit growing food for themselves and others. This is impossible under most present-day land-ownership and cash-crop-oriented systems.

In the United States, technological improvements will continue to produce increases in crop yields, and the country will continue to be a major food exporter. Concerns for the future of agricultural resource development revolve mainly around the problems of energy efficiency, pesticide use, and soil erosion. Energy prices are particularly important to farm income. Although energy prices fluctuate, it seems clear that in the long run they must increase. There is growing concern over the ecological and health effects of pesticide use, as pests become resistant to them and as farmers turn more to chemical than mechanical means for controlling weeds. Nonetheless, soil erosion continues at unacceptable rates in many areas of the United States. Although it will probably not be an acute problem causing abandonment of land, it is a chronic problem that ultimately will force farmers to increase inputs of fertilizer and possibly organic matter to replace lost nutrients and to improve deteriorating soil structure. All of these factors further increase the cost of producing food and fiber. Thus, it seems that the United States will continue its role as an important food provider, but food prices will have to rise substantially in the next few decades to allow this production to take place.

REFERENCES AND ADDITIONAL READING

Becker, E. 2002. House passes the Farm Bill which Bush says he'll sign. *The New York Times*, May 3, 2002:A19.

Brown, K. 2001. Genetically modified foods: are they safe? *Scientific American* April:51–57.

Bureau of Land Management. 2000. *State Assessment Report—Standards for Rangeland Health.* http://www.blm.gov/nhp/what/00standards.htm.

Conway, G. 2000. Food for all in the 21st century. *Environment* 42(1):8–18.

Crosson, P. 1995. Soil erosion estimates and costs. *Science* 269:461–462.

———. 1997. Will erosion threaten agricultural productivity? *Environment* 39(8):4–9, 29–31.

Dale, E. E. 1930. *The Range Cattle Industry.* Norman: University of Oklahoma Press.

Dodd, J. L. 1994. Desertification and degradation in sub-Saharan Africa. *BioScience* 44(1):28–34.

Dregne, H. 1977. Desertification of the world's arid lands. *Economic Geography* 52:332–346.

Easterbrook, G. 1997. Forgotten benefactor of humanity. *The Atlantic Monthly* 279(1):74–82.

Environmental Working Group. 2002. *Farm subsidy database.* http://www.ewg.org/.

Food and Agriculture Organization. 2001. *FAOSTAT.* http://apps.fao.org/.

Gardner, G. 1996. *Shrinking Fields: Cropland Loss in a World of Eight Billion.* Worldwatch Paper no. 131. Washington, D.C.: Worldwatch Institute.

Gaskell, G., M. W. Bauer, J. Durant, and N. C. Allum. 1999. Worlds apart: the reception of genetically modified foods in Europe and the U.S. *Science* 285: 384–387.

Haslam, G. 2002. Growers and greens unite. *Sierra* May/June:56–61.

Hilyard, N. and S. Sexton. 1996. Too many for what? The social generation of food "scarcity" and "over population." *The Ecologist* 26(6):282–289.

Holdrege, C. and S. Talbott. 2001. Sowing techonology: the ecological argument against genetic engineering down on the farm. *Sierra* July/Aug:34–39.

National Academy of Sciences. 1972. *The Genetic Variability of Major Crops*. Washington, D.C.: National Academy of Sciences.

National Research Council. 1994. *Rangeland Health: New Methods to Classify, Inventory, and Monitor Rangelands*. Washington, D.C.: National Academy Press.

Nothdurft, W. E. 1981. The lands nobody wanted. *Living Wilderness* 45(153):18–21.

NRCS. 2001. *The 1997 National Resource Inventory*. U.S. Natural Resources Conservation Service. www.nhq.nrcs.usda.gov/NRI/1997/.

Paarlberg, R. 2000. Genetically modified crops in developing countries: promise or peril. *Environment* 42(1):19–29.

Parry, M. 1990. The potential impact on agriculture of the greenhouse effect. *Land Use Policy* 7:109–123.

Pimentel, D., et al. 1995. Environmental and economic costs of soil erosion and conservation benefits. *Science* 267:1117–1123.

Schneider, K. 1989. Science Academy says chemicals do not necessarily increase crops. *The New York Times*, September 8, p. A1.

Sheridan, D. 1981a. Can the public lands survive the pressures? *Living Wilderness* 45(153):36–39.

_____. 1981b. Western rangelands: Overgrazed and undermanaged. *Environment* 23(4):14–20, 37–39.

Soil Conservation Service. 1980. *America's Soil and Water: Condition and Trends*. Washington, D.C.: U.S. Government Printing Office.

Stegner, W. 1981. Land: America's history teacher. *Living Wilderness* 45(153):5–7.

Steinbrecher, R. A. 1996. From Green Revolution to gene revolution: The environmental risks of genetically engineered crops. *The Ecologist* 26(6):273–281.

Trimble, S. W. and P. Crosson. 2000. U.S. soil erosion rates—myth and reality. *Science* 289:248–250.

United Nations Environment Programme. 1992. *World Atlas of Desertification*. London: Edward Arnold.

_____. 2002. *Global Outlook 3*. London: Earthscan.

U.S. Department of Agriculture. 1981. *RCA Appraisal Parts I and II*. Washington, D.C.: U.S. Government Printing Office.

_____. 1989. *The Second RCA Appraisal, Soil, Water, and Related Resources on Nonfederal Land in the United States, An Analysis of Condition and Trends*. Washington, D.C.: U.S. Government Printing Office.

_____. 1996. *1995 Farm Bill: Guidance of the Administration*. Washington, D.C.: U.S. Government Printing Office.

U.S. Forest Service. 1981. *An Assessment of the Forest and Rangeland Situation in the United States*. Forest Research Report No. 22. Washington, D.C.: U.S. Government Printing Office.

U.S. Natural Resource Conservation Service. 1997. *A Geography of Hope*. Washington, D.C.: U.S. Government Printing Office.

_____. 1997b. *National Resource Inventory*. http://www.nhq.nrcs.usda.gov/NRI/intro.html.

Vallentine, J. F. 1989. *Range Development and Improvements*. 3rd Ed. San Diego: Academic Press.

Wargo, J. 1996. *Our Children's Legacy: How Science and Law Fail to Protect Us from Pesticides*. New Haven: Yale University Press.

World Resources Institute, 1992. *World Resources 1992–93*. New York: Oxford University Press.

_____. 1996. *World Resources 1996–97*. New York: Oxford University Press.

For more information, consult our web page at ***http://www.wiley.com/college/cutter***.

STUDY QUESTIONS

1. Americans pay much less for food (as a percentage of their income) than do most other people in the world. Is it a good thing that our food prices are low? Should we pay more so that farmers can operate in ways that conserve resources more effectively?
2. Most of the world's food is produced in the same country in which it is consumed, although international trade in food is increasing. What are the impacts of increasing international food trade on food availability in poor countries? In rich countries?
3. Does the meat production system in the United States and other wealthy countries, in which meat is produced by feeding grain to animals, affect the availability of food in other parts of the world? If we ate less meat, would this make more food available?
4. How do you feel about pesticides used in the production of the food you eat? How much more would you be willing to pay for food produced without the use of pesticides?
5. Identify the major categories of environmental impacts of agriculture (both cropland and rangeland) where you live.

CHAPTER 7

FORESTS

Forests are the natural vegetation in biomes that have ample soil moisture and a sufficient growing season. They occupy areas of poor soil and steep slopes as well as high-quality lands. Because trees can survive on marginal lands, many forests remain intact even when there is great demand for agricultural land. Forests are found in virtually all humid and subhumid regions of the world, from the tropics to the margins of the tundra.

About 9.6 billion acres (3.9 billion hectares), or about 30 percent of the Earth's land surface (excluding Antarctica), are covered with forest and woodland (UNEP 2002) (Fig. 7.1). This portion has changed considerably over the centuries, generally decreasing as cultivated land has expanded. The original forest cover of the Earth approached 50 percent of land area (again, excluding Antarctica).

The remaining forests of the world are generally inhabited at much lower densities than farmland, but the lack of dense population does not mean a lack of economic importance. Forests supply a wide range of commodities for humans and other organisms, including fiber for fuel, lumber and paper, wildlife habitat for both hunting and biodiversity purposes, carbon uptake and storage, water purification and recycling, and recreation.

In addition, forests are under threat from many sources, and as a result Earth's forestland area is decreasing. Trees are being harvested for fiber, mostly for fuel. In many areas, forests are being cleared for other purposes, primarily agriculture (both crops and pasture). In industrial regions forests are being damaged by air pollution, especially acid deposition. Consequently, forests are the subject of considerable competition and controversy worldwide. In this chapter we explore some of the contrasting uses of forests and the ways these uses impinge on each other.

FORESTS AS MULTIPLE-USE RESOURCES

The fact that forests occupy lands that are not always suited to agriculture, grazing, or other uses does not mean that they are low-value lands. Rather, forests are among the most widespread, versatile, and easily exploited of the world's natural resources. They are used for fuel, construction materials, and paper. The water that is shed by forests in humid areas is usually clean and plentiful, inviting the construction of dams and other facilities to capture and use the water. Forests are the home to diverse wildlife, both plant and animal. Forests serve as food sources for people who occupy them and those who live nearby, providing sustenance for birds, monkeys, deer, squirrel, and many other animals.

In the twentieth century, new uses of the forest emerged, with values sometimes even greater than fuel and fiber. These new uses focus on the qualities of the relatively undisturbed forest, especially its biodiversity. Trees are valued not just for the quality of the wood they contain but also for their roles in forest ecosystems and food chains. An old dead tree, once seen simply as lifeless wood best used for lumber, is now the home to myriad species that depend on such dead trees for their existence. Trees that were once only curiosities or botanists' academic interests are now potential sources of medicines to fight diseases in distant cities. Other users of the forest include tourists from distant cities, who travel to the forest to view its diversity and uniqueness, to experience its solitude, or to respect

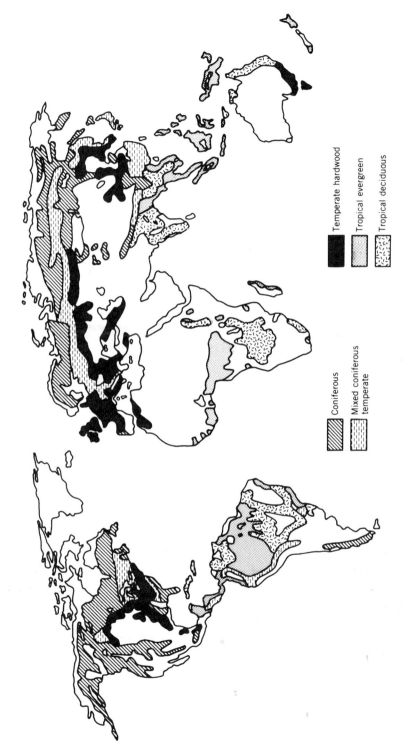

Figure 7.1 Current distribution of forests in the world. *Source*: UNEP 2000.

its role in absorbing the excess carbon dioxide emitted by industrial society.

Forests thus present a paradox: it is a resource that is at once vastly abundant and scarce. Forests and their lands are used for an extraordinarily wide range of activities, and those who use and manage the forest resource disagree sharply over how much forest there is and how it should be used. For the decision-maker in a paper products company, there may be more forest than could ever be needed, because forests are renewable when properly managed. In contrast, the wilderness preservationist believes that a second-growth forest is profoundly different from a forest that was never cut and maintains that forested wildlands should be protected from the ax forever. Prospectors and miners, yet another interest group, do not care about the trees so much as what lies underneath them. These groups often have incompatible management and use goals for forestlands.

The existence of multiple uses is particularly significant because most forests are publicly owned (Brooks 1993). In the United States 40 percent of forestland is publicly owned; in Canada 94 percent is public. The average for industrial countries is around 56 percent. Ownership of forests in less industrialized countries in the tropics is less clear. Nominally, the vast forests of the Amazon, central Africa, and southeast Asia are publicly owned, but public management and authority are difficult to establish at the local level. In much of the world's forests, however, management decisions are made by government agencies that are subject to political pressure from various interest groups (Issue 7.1). Forest management is often as much a political process as a scientific one.

FORESTS AS FIBER RESOURCES

In 2000, an estimated 3.3 billion cubic meters of wood are harvested annually from the world's forests. About 50 percent of this total is consumed for fuel and charcoal; the remainder is processed into sawn lumber, plywood and other panel products, and paper. Sawn lumber represents more than half the industrial (nonfuel) use of wood, whereas paper production accounts for about 30 percent of industrial wood production. Total harvesting of wood increased steadily in the past several decades, although the rate of timber harvesting appears to be leveling off.

In many parts of the world the total area of forest is declining. Nevertheless, the ultimate goal of the timber industry should be to harvest forests on a sustained-yield basis: to take only that amount that can be replaced by new forest growth each year. Unfortunately, actual practice doesn't always conform to this objective.

Principles of Sustainable Forestry

Forests are renewable resources. Trees are harvested, and new trees grow to replace them. Sustainable forest management means managing a forest in such a way that it will produce a given amount of timber each year indefinitely. If we harvest more than that amount, then we are exceeding the sustainable yield of the forest.

The key question for a forest manager is: How much timber can be cut each year without exceeding the sustainable yield? The answer depends on many factors, principally relating to how fast a forest replaces itself. These factors include tree species characteristics, climate, soil characteristics, and management practices. Growth rate is not only important to the productivity of a given forest but also critical to the question of sustainability.

If a forest is managed sustainably, then trees must have sufficient time to replace themselves before they are harvested. As an illustration, let us assume that we have a tract of original forestland, say, 2000 acres. Moreover, let us assume that this forest takes 100 years to replace itself. Suppose we cut 20 acres of land each year (Fig. 7.2a). After 10 years we have 200 acres, or 10 percent of the land, that was cut. If it begins to regrow immediately, then these 200 acres have forest that is up to 10 years old, and the remaining 90 percent of the land still has mature forest more than 100 years old. After 20 years, if we cut at the same rate, we have 200 acres that are 11 to 20 years old and 200 acres that are less than 10 years old. If we continue to harvest at this rate for 100 years, then eventually we have the same portion of land in each age class. Furthermore, every 10 years 200 acres of forest reach maturity and are available for harvest.

Now consider the example in Figure 7.2b. In this case we cut 400 acres of forest each year. After 10 years, we have 400 acres of forest that is less than 10 years old and 1600 acres of mature forest. If we continue to cut at this rate, after 50 years we have no more mature forest, with our oldest trees only 50 years old. If we wait until the trees are 100

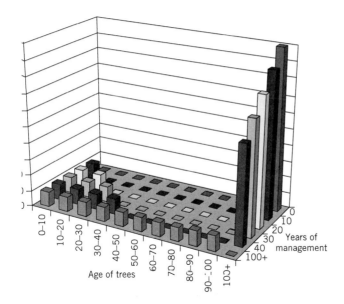

(a)

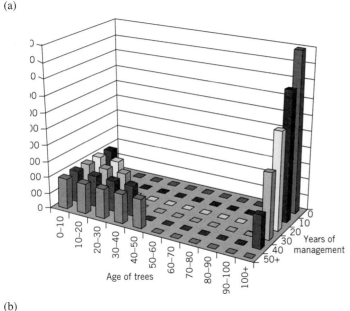

(b)

Figure 7.2 Effect of forest harvesting on the age distribution of forests. As forests are harvested, new trees may grow in their place. (a) If the rate of harvest in area per year is less than 1/n times the total forest acreage, where n is the number of years required for a forest to reach maturity, then the forest can be harvested at that rate indefinitely. (b) If forests are harvested at a high rate, then the area of harvestable forest declines and eventually is depleted, requiring an interval without harvesting before trees reach maturity and harvesting can begin again.

years old to harvest them, we then have another 50 years before we can harvest here again.

Thus, the area of forestland that can be harvested each year on a sustainable yield basis is:

$$\frac{\text{Area of forestland available}}{\text{Number of years to maturity}}$$

The numerator in this equation obviously is limited by the total amount of forestland in any given area. In general this amount decreases as land is allocated to other uses such as agriculture, grazing, recreation, and habitat preservation. Increasing demand for forest products therefore requires that the denominator of the equation decrease to make more forestland available for harvesting. This is achieved in two ways: by increasing growth rates through various forest management techniques and by decreasing the size of trees that we harvest. In the following paragraphs we explore the variables in this equation in more detail.

Forest Management

Forests are managed in many different ways, in terms of both their uses and the specific techniques employed in these uses. If we consider the use of

ISSUE 7.1: CHIPKO: GRASS-ROOTS ENVIRONMENTALISM OR A STRUGGLE FOR ECONOMIC DEVELOPMENT?

Deforestation is a major issue on the southern flanks of the Himalayas. The region is an area of rapid population growth and large demands for wood (mostly for fuel but also for construction). It is also a region of rugged mountains, where slopes are steep and removal of the forest is contributing to greatly accelerated erosion, depleting soil fertility, and causing problems of sediment accumulation and flood management downstream. One area where this deforestation is especially problematic is a region in the Indian state of Uttar Pradesh, known as the Garhwal Himalayas. There a political movement known as Chipko was at the forefront of a struggle over management of these forests. Chipko became an international symbol of grass-roots opposition to unsustainable commercial resource exploitation. But geographer Haripriya Rangan (1996) suggests another interpretation of the Chipko movement and provides some interesting insights into the linkages between environmentalism and other political movements.

The Garhwal Himalayas is a region of northern India with rugged terrain and limited economic opportunities. It is marginal, both within India and within the state of Uttar Pradesh, one of India's most populous states. As such, it is a region of relative poverty and limited political influence. When India won its independence from Britain in 1947, the people of the region supported themselves through subsistence agriculture, forestry, service of pilgrims' needs, and trade across the Himalayas. In the early 1960s, India fought a series of border wars with Pakistan and China, and the Garhwal gained strategic significance. Trade across the Himalayas was eliminated, and the Indian government took control of significant areas of forestlands for defense purposes, reducing the already limited economic opportunities. When Green Revolution agricultural technologies were adopted in the early 1970s, Garhwal was further marginalized. These technologies require significant capital investment for irrigation and purchase of chemicals. In the level land of the Indo-Gangetic Plain, the core of Uttar Pradesh, such investment was feasible. But in the hilly terrain of Garhwal it was more expensive, and capital was even more limited. As a result, cash cropping was less feasible, and subsistence farming remained dominant.

To obtain fuel and raw materials for crafts and to supplement their incomes through selling wood, the people of Garhwal harvested wood from the forests of the area. But these forest resources became more scarce, largely because of commercial harvesting on lands controlled by the state government of Uttar Pradesh. Much of this harvesting was carried out by contractors from outside Garhwal, employing migrant laborers who were also drawn from outside the region. Thus, this harvest did little to contribute to the local economy, and the Uttar Pradesh government was not sympathetic to requests that timber be made available to local harvesters. A protest movement grew, and from 1973 to 1975 villagers staged protests in which they went into the forests and clung to the trees, preventing them from being cut. A local forest contractor, Sunderlal Bahuguna, became a leader of the movement. Environmentalism was a potent force at that time, and Bahuguna cited the need to preserve forests for use by local harvesters against the forces of wanton exploitation in order to maintain ecological integrity for the benefit of all. Chipko means, literally, to stick to something or to hug it. The image of people hugging the trees to prevent oth-

forests as sources of fiber, the most obvious aspect of management is how, when, and at what rate timber is harvested. But the ways in which the forest is managed between harvests are equally important.

Harvest Techniques A variety of harvesting techniques are used in forestry management. Not all of them are used interchangeably because each has specific goals and impacts. The three most important are *shelterwood cutting*, *selective cutting*, and *clear-cutting*. Another method for harvesting also discussed in this section is chipping or biomass harvesting, a technique generally used for less valuable wood on small lots.

ers from harvesting them became a powerful symbol for environmentalists.

At this time, the government of Indira Gandhi was embarking on a program of nationalizing important industries. The Chipko leaders, demanding greater local access to the forests, met little success in their negotiations with Uttar Pradesh officials. So when they appealed for help from the national government, it used the opportunity to take control of the forests and to institute strong controls on forest management. These controls included requiring consent from the national government prior to any large-scale harvests, prohibition of harvesting certain tree species, and a ban on harvesting above elevations of 1000 meters in the Himalayas.

One result of these measures was that both immediate economic opportunities and longer-term development projects in Garhwal were significantly curtailed. Declining administrative budgets in forestry agencies made it difficult for them to respond to local needs. Road construction, for example, cannot be carried out without a permit, and delays in issuing such permits slowed not only road construction but also irrigation development and electrification. Compounding the problem was the fact that while much of the impetus for forest protection came from the national government, the Garhwal forests were still administered by Uttar Pradesh. Despite the victory of the Chipko movement, the people of Garhwal still did not have either access to their forests or a significant political role in their management. The resulting tensions fueled new demands for the separation of Garhwal from Uttar Pradesh, creating a new state of Uttaranchal. Like Chipko, the central issues were economic: control over forest resources and infrastructure for economic development. Unlike Chipko, the movement for a state of Uttaranchal used the threat of violence to make its voice heard.

What was the Chipko movement? What did it mean? From the perspective of environmentalists, it was a heroic struggle of people whose livelihood was directly dependent on natural resources and who sought to prevent excessive exploitation by outsiders interested simply in short-term economic gain. The people hugged the trees, clearly demonstrating their concern for nature and environmental preservation. Chipko is seen as similar to the Rubber Tappers movement in Brazil, a confrontation between those wishing to exploit the forest sustainably (without clearing the trees) and those who would exploit it through deforestation. The message is essentially the same: local control of natural resources is a route to sustainability and that, left to themselves, people with a long tradition of local resource use protect those resources.

On the other hand, the struggle in Garhwal can be seen as one for economic development. The Chipko movement emerged because the people of Garhwal were denied access to a local resource, and instead outsiders were being allowed to exploit that resource without regard for the needs of local inhabitants. Hugging the trees was a means to prevent this exploitation and regain control of the forests for themselves. When the Chipko movement took the direction of increasing external control of the forest resource, it fell apart as a grass-roots movement, although it lived on as an international environmental legend. In its place we have a new movement, this time explicitly focused on local control. The message is that such control of natural resources is necessary for economic development to truly benefit local populations and that economic considerations, not environmental ones, are central (Rangan 1996, 2001).

Shelterwood cutting is a several-stage process requiring thinning and cutting. First, trees of poor quality are removed from both the forest floor and the stand itself. This opens up the forest floor to more light, enhancing seedling growth and reducing competition. The remaining trees provide some shelter for the seedlings. When the seedlings take root and become established, some of the remaining mature trees are harvested. Shelterwood cutting is an efficient technique in small plots with relatively homogeneous tree species. It is costly in terms of labor inputs for larger acreages and so is not practiced widely on large tracts of commercial forestlands.

Selective cutting is normally used only in forests of mixed age or in forests with trees of widely varying economic value. The mature trees of the most desirable species are harvested, while the others are left intact. In an oak–hickory forest, for example, the mature oaks are selectively cut, leaving immature oaks, mature hickories, and other species standing. Selective cutting is used primarily in hardwood forests. When used in mixed-species forests, selective cutting leads to a loss of diversity, which has negative impacts on wildlife and other sectors of the forest environment. Selective cutting is costly and appropriate only when the value of the harvested trees is high relative to those left uncut.

Clear-cutting is the most widely used method of harvesting and also the most controversial. About two-thirds of U.S. timber production is harvested this way. The technique involves cutting all the trees regardless of size or species and is appropriate when the trees are relatively uniform in species and age or when it provides the most desirable form of regeneration. Clear-cutting, however, does remove the entire forest canopy and leads to soil erosion and wildlife habitat destruction. It also leaves a more disrupted and scarred landscape than other harvesting techniques. Clear-cutting produces much more timber per unit of acre harvested than selective cutting or shelterwood cutting.

Loggers recently developed a method for consuming whole trees of any size and shape, from any size tract. This *biomass harvesting* turns trees into wood chips that are used to make pulp or fuel wood-fired power plants. Chips are easier to handle in large quantities than logs and are loaded on trucks, railroad cars, or ships for quick transport to pulp mills or power plants. Loggers cut selectively or consume all standing timber, depending on the requirements of the job and stipulations of the landowner. This method has great economic appeal for harvesting the vast majority of U.S. forestlands—the small parcels in private hands. It also is used in the developing world, where forests are being cut and exported to industrialized countries for processing into pulp and paper.

Silviculture Once a forest is harvested, the land is available for regeneration and growth of new trees. The period of regeneration and subsequent growth is a time for management to maximize growth rates and thus minimize the time needed to produce a harvestable stand of trees. Intensive *silviculture* on productive lands—tree farming—produces much larger yields of timber than occur in natural forests, just as cultivated crops produce much more food than is collected from unmanaged ecosystems.

The incentive to establish intensive silviculture as a forest management strategy results from a shortage of timber available for harvest. In areas where timber resources are abundant, such as North America, it often is cheaper simply to harvest the standing timber and move on to new land rather than invest in regrowing timber for harvest decades in the future. But if the price of standing timber is high enough and if growth rates are fast enough that newly planted land is harvestable in the foreseeable future, then the investment in regeneration and management is justified. In some countries, governments require replanting or invest in replanting themselves, so that timber is available for future generations, even though such investments are not always justified on the basis of short-term returns.

One region where intense silviculture is practiced today is the southeastern United States, especially the Atlantic Coastal Plain. This is an area of warm temperatures and abundant rainfall but relatively poor soils that support only limited agriculture. The soils are quite capable, however, of supporting pine plantations that grow very rapidly in the subtropical climate. The dominant tree species planted in this region is loblolly pine. Trees are planted almost immediately after harvest of the preceding crop. They are fertilized and treated with pesticides to limit insect damage. Stands are thinned to encourage growth, and fire is used to reduce understory growth that competes for moisture and nutrients. By using these techniques, harvestable stands of trees with diameters up to 30 cm (1 foot) are produced in as little 20 to 35 years. The expansion of timber production using these methods was rapid in the twentieth century, encouraged by the development of plywood, fiberboard, and paper manufacturers, all of whom make use of small trees.

Forest Products Technology
Uses of Wood   To better understand the changing supply and demand for timber and their relation to sustainable forestry, it is necessary to un-

derstand how wood is used. In most of the world, harvested wood generally is used for fuel. Wood supplies the domestic heating and cooking needs of half the world's population, and fuel is the dominant use of timber in most developing nations (Brooks 1993; United Nations 1992). In the industrialized world today, wood is no longer a major fuel, and most wood is used for industrial purposes. In the United States, for example, fuel was the major use until the mid-nineteenth century. Gradually, fuelwood was replaced by other energy sources, and this use plummeted after the 1930s (Clawson 1979). In the 1970s and 1980s, fuelwood use expanded again, so that today fuel is among the major uses of wood in the United States.

Beginning around 1900, wood first was used to make plywood and pulp for paper, and these uses have increased steadily ever since. Demand for lumber follows cycles in new housing construction. Today, for example, pulp production demands slightly less wood than lumber in the United States because we are in a home construction boom. In addition, wood chips and sawdust are now used instead of whole pieces of wood to make boards. Production of these fiber-based structural panels continues to grow. Today about 28 percent of all wood harvested in the United States is used for sawn lumber, while about 40 percent is used to make pulp and paper products. About 10 percent of U.S. wood is used for plywood and similar products, while 18 percent is used for fuel (Fig. 7.3) (FAO 2001). The United States also imports a substantial amount of wood and wood products, mostly from Canada.

These different uses require varying kinds of wood. For example, just about any kind of wood is burnable, although hardwoods are better than softwoods for this purpose. Lumber, on the other hand, requires trees that are straight and as great in diameter as possible—in other words, old trees. Most lumbering requires softwoods, but specialized industries such as furniture construction use hardwoods. Plywood is made from large sheets of veneer only millimeters thick that are glued together. These sheets are not sawn from logs; they are peeled from a turning log with a large blade that cuts sheets that are hundreds of feet long. This means that plywood manufacture does not need logs of large diameter. Much younger trees are used, provided they are softwoods, relatively straight, and free from knots. Similarly, pulp production is not dependent on any particular size of log, and the tree species is more important than age. Particle board and similar products are made with waste from other processes.

Impacts on Timber Requirements The transition from lumber as the main nonfuel use of wood to use of manufactured wood products such as plywood, particle board, and paper had profound impacts on the requirements of the timber industry. The most significant of these was the reduced need for large (and therefore old) trees, which allowed the harvesting of small, young

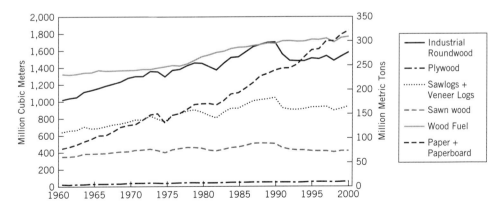

Figure 7.3 Trends in world forest product production. Use of wood particles for panels and papermaking has grown much more rapidly than use of sawn wood and plywood. This allows the use of smaller trees that can be grown in a few decades rather than requiring a century or more to reach commercial maturity. *Source:* FAO, 2001.

trees in their place. For example, in the 1800s most of the timber industry in northern Maine was centered on the white pine. This tree grew straight and tall in the old-growth forest and produced very high-quality lumber: long boards with straight grain and relatively few knots. Loggers operating in Maine at that time took the white pine first and left other species standing. Today, the forest products industry in Maine is dominated by paper production. Much smaller trees are harvested, most of them from second-growth forests and a greater portion of the trees is used. Smaller logs are made into pulp because they are not useful for lumber. Clear-cutting is now the preferred harvesting technique, rather than the selective cutting practiced in the 1800s.

The ability to manufacture large pieces of wood such as plywood from small trees reduces the dependence of the lumber industry on old-growth forest and makes possible a shift toward using "fiber farms" rather than forests to supply our needs for building materials and paper. This shift has taken place already in the southeastern United States, where fast-growing pines are planted, intensively managed, and harvested in cycles as short as 20 to 25 years. The trees grow to diameters of perhaps 30 cm in this time, but this is adequate for pulp, pulpwood, and particle board.

In the United States, lumber prices are low enough that sawn boards remain the material of choice for most housing construction—the main use of such lumber. But if prices for lumber rise significantly, we have the ability to replace it with other materials. Such changes are taking place already. A new home in the 1950s was built with plywood sheathing (the inner layer of the outside walls) and sawn wood siding. Today the sheathing on new homes is either a type of particle board called oriented strand board, made from small flakes of wood, or a foil/paper-covered plastic foam board, and siding usually is vinyl rather than wood. If prices for the 2-by-4 and 2-by-6 boards that form most of the framing in new houses rise, they can be replaced with metal studs, like those already used in most commercial construction. New plastic- and paper-based materials are available to substitute for these boards if needed.

The significance of these changes is that if we decide that nontimber uses of forests, especially old-growth forests, are of significant value, then the technology exists to replace this sawn lumber with other products relatively easily. It is just a matter of having the price of lumber rise sufficiently so that substitute materials become more attractive. Manufactured wood products made from younger trees and nonwood substitutes, if cheaper than lumber, could meet our needs for years to come.

Nonfiber Uses of Forest Resources

Habitat

Trees are more long-lived than other plant types, and they take a long time to mature. They contain large amounts of nutrients stored in living biomass. Forests are ecosystems in which most of the available nutrients accumulate in live trees over a long period of time. Relatively small amounts of nutrients are contained in herbs, shrubs, or the soil. In an experimental forest in New Hampshire, researchers found that annual uptake rates of nutrients were a small fraction of the total amounts in storage, leading to the notion that forests are warehouses or sinks for *biomass*. Restoration of ecosystem nutrients after forests are clear-cut takes several decades (Likens et al. 1978). For most tree species, the rate of growth (as measured by biomass) is relatively slow when the tree is young, because it is small and does not have a large photosynthetic capacity. Growth rate increases as the tree gains a larger total leaf area and declines as it reaches maturity.

In many forests, the amount of stored biomass reaches a steady state in which old trees die and their nutrients are taken up by younger ones. For a large portion of the world's forests, cyclic disturbances kill all or nearly all the trees at the same time, releasing nutrients and beginning a new cycle of forest growth. These disturbances include fire, disease or insect infestation, and windstorms.

The significance of old-growth forest habitat is at the center of debates over the impacts of timber harvesting (Issue 7.2). In many of the world's forest regions today, most forestland was harvested at least once and therefore contains young, second-growth timber. In these areas, relatively little land that was never cut remains. The old-growth forest, undisturbed by humans for at least hundreds of years, contains species that are rare or nonexistent in second-growth forests. Each of these species is associated with a particular microhabitat to which it is best suited. For example, some species nest in or feed in standing dead trees. As old trees decay, they become home to insects and other invertebrates that live in and feed on the rotting wood. These in turn are food for birds, small mammals, and even foraging bears. Others, like the marbled murrelet and the spotted owl, two endangered bird species that inhabit the old-growth forest of the Pacific Northwest, nest only in very tall trees. Without these old tall trees, these and many other species are threatened with extinction (Maser 1994).

Water Resources

Forests are found in the more humid parts of the Earth's land surface, where mean annual rainfall exceeds evapotranspiration. The excess water becomes runoff that is available for human use. Forest vegetation has two contrasting impacts on that runoff. It protects the quality of the water but decreases its quantity. In most cases, maintaining good water quality is a prime concern and thus takes precedence.

Water quality is usually very good in forest areas because forest soils are very permeable. The continuous vegetation cover, high productivity, annual leaf fall, and large amount of standing biomass support a diverse community of organisms that live in the soil and maintain its ability to absorb water. Because most water arrives at streams by subsurface paths rather than by running across the surface, natural soil erosion rates are low.

When a forest is harvested, the soil is disturbed, increasing the amount of overland flow and erosion, with negative impacts on water quality. The extent of impact depends on the kind of disturbance and the terrain. If the forest is clear-cut with heavy machines, with logs being dragged across the ground, the soil is damaged considerably. On the other hand, if the use of machines and dragging is minimized, the damage is minor or insignificant. Steep slopes are much more susceptible to increased erosion caused by logging than are flat areas. Road-building is often one of the most damaging aspects of logging operations, especially in steep mountainous areas where road cuts and fills increase the likelihood of landsliding, which adds much more soil and rock to streams. Landsliding is especially severe in the Himalayas, the Andes, and the Pacific Northwest of North America. Accelerated erosion associated with deforestation is a major water-quality problem in the Pacific Northwest, and this poor water quality caused by logging is one of the causes of the decline in salmon populations there.

In addition to regulating overland flow and erosion, forests play a critical role in evapotranspiration. Like all green plants, trees use water. When water is plentiful, trees use about the same amount as any other vegetation type, but when water is limited, as it is for at least part of the year in most climates, trees use more water than grasses, for example. This is because trees typically have deeper and more extensive root systems, which allow them to use water stored in the soil below the reach of shallow-rooted plants. For this reason trees stay green in a dry season, while grasses turn brown.

If maximizing the amount of runoff is a goal of watershed managers, then forest cover is undesirable. Accordingly, some area managers actually discourage the growth of trees, or replace trees with grasses, as a means of decreasing evapotranspiration and thus increasing runoff. Care must be taken in such operations to ensure that the conversion of vegetation does not degrade water quality.

When a large forest region is deforested, the decrease in evapotranspiration actually reduces the amount of water vapor in the air and thus may reduce precipitation. For decades trees were planted in China in the belief that this practice leads to an increase in precipitation, although there is little evidence to support this.

The interior of the Amazon basin ultimately receives its water vapor from evaporation in the Atlantic Ocean. But before Atlantic water reaches the interior it falls as precipitation and is evapotranspired back to the atmosphere. Most of the water vapor reaching the interior was cycled through vegetation at least once on its way west across South America. Some climatologists have raised concerns that if large areas of the Amazon are deforested and evapotranspiration is reduced, the moisture supply to the western Amazon might be restricted, reducing rainfall there. As yet this

ISSUE 7.2: THE PACIFIC LUMBER SAGA

Probably the most intense environmental debate of the 1980s and early 1990s in the United States was the controversy over management of the old-growth forests of the Pacific Northwest. These forests are the last fragment of old growth that once covered much of the nation and were systematically cleared beginning when the Europeans arrived four centuries ago. As loggers headed westward across the country, they harvested timber faster than it could regrow, knowing that more was always available a little further west. Now that harvest has reached the west coast, and with the end of the old growth in sight, the battle to save what remains has intensified.

The battle between logging and forest preservation was waged throughout the Pacific Northwest, but nowhere more intensely than around a 3000-acre tract known as the Headwaters Forest, owned by Pacific Lumber Company. This tract is both the focal point of a controversy and an excellent illustration of how a capitalist society makes choices about natural resource use.

The story begins in the early 1980s. Pacific Lumber (PL) was a small company whose stock was publicly traded, but it was run as a family operation. Its operations were based in Scotia, a company town on the northern California coast, about 200 miles north of San Francisco, where PL owned 189,000 acres of forestland, including several thousand acres of old-growth redwood. PL was a paternalistic company that took care of its employees, many of whom came from families that worked for the company for generations. PL was the center of the community and accepted certain responsibilities associated with that role. It managed its forests on a sustained-yield basis so that the children and grandchildren of its employees would be able to live and work there, as had their parents and grandparents. Wages were good, work was steady, and the company intended to keep things that way.

But in the early 1980s a new phenomenon emerged on Wall Street, in which financiers sought out companies whose stock was undervalued; that is, they could be acquired for less than the market value of the companies' assets. An undervalued company was vulnerable to becoming the target of a hostile takeover, in which an offer is made to purchase a controlling share of a company's stock, and if sufficient shares are accumulated, then the new owners take control of the company. Corporate raiders, as the purchasers were known, would buy a company, financing the deal with high-yielding "junk" bonds, sell the more liquid assets as quickly as possible to pay off the bonds, and be left with a smaller, yet still profitable, company. A wave of such hostile takeovers swept the nation's business community in the 1980s, facilitated in part by deregulation of the securities and banking industries. Pacific Lumber's vast holdings of forests and a significantly overfunded pension fund were attractive assets, while its stock price was relatively low because of the company's modest profits and stock dividends. It was ripe for the picking.

The raider was Maxxam Corporation, controlled by a Houston financier named Charles Hurwitz. With financing arranged by Michael Milken (who would later be convicted of several federal offenses related to his junk-bond operations), Hurwitz bought the company in 1985 for $840 million, perhaps half of its true value. The acquisition plan called for selling off some peripheral parts of the business, taking about $50 million from the pension fund, and dramatically increasing the rate of harvest to generate sufficient cash flow to pay off the expensive junk bonds quickly. This is exactly what happened, and by the late 1980s the company was running at maximum capacity, with most workers on mandatory 58-hour work weeks, harvesting timber as fast as possible.

The environmental community was outraged, particularly at the potential loss of old-growth forest. The old growth of the Pacific Northwest is home to several endangered species that are unable to survive successfully in young second-growth. The heart of PL's holding, a 3000-acre tract that is known as the Headwaters Forest, was the largest remaining piece of privately held old-growth forest in the region. More than half of the remaining old growth is preserved already in state and National Parks. Even worse, the redwoods were harvested at an accelerated rate to pay off junk bonds rather than to provide a livelihood to the people of Humboldt County. To the environmentalists, it was clearly a case of greed against good, of destroying the forest to create short-term gains for the rich.

Leading the battle for the environmentalists, or at least at the forefront of it, was EarthFirst!, a loose-knit organization of radical environmentalists who believed in protecting the environment at nearly any cost. They used a variety of nonviolent tactics, including demonstrations, street theater, and obstruction, but occasionally aggressively attacked the machinery of destruction, a tactic known as *monkey-wrenching*. Of all the forms of monkey-wrenching used in the Pacific Northwest, the most controversial is tree-spiking. This involves driving a long metal spike deep into a tree, where it isn't visible from the outside but where it will destroy the blade of a saw when the tree is cut or milled, often with great danger to the operator of the saw and others in the vicinity. Once a tree is spiked, it is valueless to a lumber company, and if a sawyer knows it is spiked then the tree won't be harvested. But if a spiked tree is cut, then serious injury or death can be the result. Needless to say, such tactics enraged both timber harvesters and moderate environmentalists, as a result of which EarthFirst! publicly renounced tree-spiking.

Through the late 1980s and climaxing in 1989, demonstrations and counter-demonstrations were held in the forest, at Pacific Lumber's mill in Scotia, and at Maxxam's corporate headquarters in Texas. Some of these demonstrations turned violent, and two of EarthFirst!'s leaders were injured when a pipe bomb exploded in their car. Countless lawsuits were filed to stop the takeover or the logging, some continuing into the late 1990s. Eventually, the Federal government was drawn into the controversy, and a deal was negotiated whereby the Federal government acquired the Headwaters Forest from Pacific Lumber. Administered by the Bureau of Land Management, a draft management plan for the forest has been prepared (BLM 2002).

What are the lessons of the Pacific Lumber saga? There are many, and they vary according to the point of view. One view is that PL was mismanaged before the takeover, with the result that its shareholders were getting a poor return on their investment and the stock was undervalued. The takeover was simply a process whereby a more effective (i.e., more profitable) management strategy was introduced that significantly increased the income that was generated from PL's lands, and the shareholders were rewarded with a significant increase in dividends and the value of their shares. With allowances for some potentially illegal aspects of the way the takeover occurred, this view is essentially correct.

But was PL really mismanaged? If we look beyond just the shareholders and consider a broader community of interested parties, the stakeholders, perhaps PL was very well managed. The stakeholders in this case include those who benefited from the high quality of the ecosystems in the area—its high biodiversity and good water quality, for example. They also include future generations who benefit from the forests, either through employment at PL and businesses that served it and its employees, or through appreciation of the incommensurable values of the forest. In accelerating the rate of timber harvest, Hurwitz and the Maxxam shareholders took profits for themselves, at the expense of the wider community of stakeholders, whose interests were not recognized by the marketplace. Those stakeholders were forced to use other channels—demonstrations, lawsuits, and ultimately government investment through acquiring the forest in order to voice their interests in management of the resource (Harris 1996; Newton 1990).

argument is supported mostly by climate models, with little observational evidence of changing precipitation associated with deforestation.

Recreation

In many countries, especially wealthier ones, forests are important recreational resources. They are relatively undeveloped and have open space available for hiking, camping, and other outdoor activities. They are often in mountainous areas that offer other amenities such as skiing, climbing, and the cooler weather of high elevations. Because they serve as a habitat for game animals, they are preferred areas for hunting and fishing.

Recreational use of forests traditionally has had little impact on the forest and thus is compatible with most other uses. The one use with which recreation is incompatible is logging. During logging operations recreational use is not possible, and after logging the damage to the forest lingers for some years until significant regrowth has occurred. In some cases, recreation and logging coexist in the same forest region. The maintenance of roads for logging is also advantageous in increasing access for vacationers. Lumber companies in such areas leave buffer strips of forest along roads, rivers, and lakes to preserve the impression of forest for visitors. The greatest problem involving recreation and logging is that recreational users and the businesses that cater to them form a significant interest group that usually is opposed to increased logging. They thus increase the likelihood of political conflict over forest management issues.

In northern Maine, for example, much of the forest is owned by paper companies that actively manage the land for fiber production and harvesting. These companies also own many of the important access roads to the area and open them to the general public, greatly increasing recreational opportunities. One of the major recreational resources of the area is the Allagash Wilderness Waterway, a popular wilderness canoe route. The canoeists who travel this route do so out of a desire to see the undeveloped and unaltered landscape, and many of these canoeists also want to preserve that landscape. When controversies develop regarding the use of these lands (which is regulated by state and federal law), these environmentalists are likely on the side of preservation rather than logging, creating some tension between paper companies and recreational users. In most cases, however, the mutual interests of both groups are recognized, and they coexist in the same landscape.

Carbon Storage

Much of the concern about tropical deforestation focused on its impacts on the global carbon cycle. This concern is well-founded, for the tropical forests play a dominant role in this important cycle. At the same time, we should not forget that fossil fuel combustion discharges vastly more carbon dioxide into the atmosphere than does deforestation. When we consider the role of forests in carbon cycling, we need to examine two dimensions of the problem: carbon processing and carbon storage. Both are important, but each is affected differently by forest management practices.

Although they occupy only about 30 percent of the world's land area, forests play a dominant role in biogeochemical cycling between the biosphere and the atmosphere. In the carbon cycle (see Chapter 4), carbon is removed from the atmosphere in photosynthesis, stored in living and dead biomass, and released through respiration. Forests occupy the well-watered parts of the planet and thus have high rates of primary productivity compared to arid and semiarid areas. As a result, forests are responsible for an estimated 70 percent of the total carbon exchange between the terrestrial biota and the atmosphere. Annually, the world's forests absorb and release about 2 metric tons of carbon through net photosynthesis and respiration (Watson et al. 2001). For comparison, total annual emissions of carbon from industrial sources is about 6 metric tons. Although the gross absorption of carbon by forests is significant, the net absorption is much smaller. In fact, for the world's forests as a whole, more carbon is lost by burning than is stored by net photosynthesis.

The total amount of carbon stored in the living biomass in the world's forests and woodlands is about 1150 metric tons—about 200 times the annual amount emitted to the atmosphere by fossil fuel combustion. Of this, about 45 percent is in tropical forests (Watson et al. 2001). When forests are cleared and burned or allowed to decay, this carbon is released into the atmosphere. The total amount of carbon released from the biota each year is estimated at 1 gigaton, most of which comes from tropical forests. When forests are cut and the wood is used for lumber or paper, the carbon contained in those products is not returned to the atmosphere. Instead, at least some of it is stored in the products from which the wood and paper are made. The amount of net storage depends on how the products are used and disposed of. Lumber in buildings and paper in landfills do not return to the atmosphere immediately but may do so over a long period of time.

If a forest is cleared and allowed to regrow, then carbon is taken out of the atmosphere through forest growth. During the early years of forest regrowth, the net storage of carbon is significant: perhaps as much as 1000 tons of carbon per square kilometer of forest per year in a rapidly growing tropical forest. In many parts of the world, particularly in previously cleared areas that are now regrowing such as eastern North America, forests are storing carbon (Fig. 7.4).

Forests are therefore very significant to the world's carbon cycle in both the processing and storage of carbon. When forests are cleared and replaced with other land uses (such as pasture), there is a net loss of stored carbon and an increase in atmospheric CO_2. On the other hand, when forests are allowed to regrow and are managed for sustainable forestry, there is no net release of carbon to the atmosphere in the long run.

The Role of Fire

Trees are susceptible to fire. Annually, tens of thousands of forest fires burn millions of hectares of land in North America, Asia, and other forest regions. Many species adapted to this situation by developing mechanisms for rapid regeneration after fire, including sprouting from the root crown and seeds that are released or germinate only after being heated. Other species have characteristics that protect against fire damage, such as low flammability or particularly thick bark.

In the past, fire was believed to be harmful to forests, but today forest fires are recognized as a natural and important part of most forest ecosystems (Cochrane 2001; Romme and Despain 1989). Fires cause major, if temporary, disruption of the forest ecosystem. They consume dead and living biomass and, if severe enough, kill most or all the trees. Water use and evapotranspiration are decreased by the loss of live trees and shrubs, which results in increased runoff. This runoff causes accelerated erosion and huge losses of nutrients from the forest soil. Downstream, the eroded soil and nutrients contribute to sediment and dissolved solid loads in streams and contribute to *eutrophication* in lakes.

Fires also have beneficial effects. They allow the release of nutrients stored in dead biomass, which stimulates growth after the fire. They also remove old stands of timber that are particularly susceptible to insect or disease infestation, thus inhibiting the spread of pests. Removing the forest canopy allows sunlight to reach ground level and promotes rapid growth of early successional species, thereby beginning the forest reestablishment process. Most important, however, frequent fires allow accumulated fuel to burn off relatively harmlessly, preventing the severe fires that occur in areas of high-fuel buildup. In many commercial forests, such as the loblolly pine forests of the southeastern United States, ground fires are deliberately set from time to time to kill off plants that compete with the pines, so as to maintain an even-aged stand.

Forest ecosystems vary in their susceptibility to fire and thus in how frequently fires occur. *Fire frequency* is the average number of years between successive forest fires at a given site. Some forests, like the chaparral woodlands of Southern California, are susceptible to fire and have natural fire frequencies of 20 to 60 years. The pine forests that grow on areas of very sandy soils along the east coast of the United States also experience frequent fires under natural conditions. Most forests, such as those in the mountainous western United States and Canadian north, have natural fire frequencies of 100 to 400 years. Fire frequency

Figure 7.4 Second-growth forest in Wayne County, Pennsylvania. The trees are relatively young and growing, but not particularly valuable as timber.

depends on many factors, including the rate of fuel accumulation, fuel moisture levels, and ignition sources.

There are three basic kinds of forest fires. *Ground fires* are fires that burn within the organic matter and litter of the soil. They smolder slowly and have little effect on trees. *Surface fires* burn on the ground surface, consuming litter as well as the herbaceous and shrubby vegetation of the forest floor. They burn faster than ground fires and clear all the low vegetation of the forest, but they have little effect on large trees. Finally, *crown fires* burn treetops as well as low vegetation, usually killing all or almost all above-ground vegetation. These fires are the most destructive to timber, wildlife, and the soil. Fires vary greatly in the temperatures that develop within the canopy and at ground level. Crown fires are much hotter than surface fires, but wind, fuel availability, and moisture levels are important influences on fire temperatures. Hotter fires are more destructive than cooler ones, because they consume greater amounts of organic matter. The result is greater post-fire erosion and nutrient loss, which retards the process of forest regeneration.

Prior to 1972, the long-standing policy of the U.S. Forest Service, U.S. National Park Service, and other forest management agencies was to fight all naturally started fires. Paradoxically, this strategy exacerbated damage from fires. The easiest fires to extinguish are those that start during relatively wet or low-wind conditions. These low-temperature fires cause only minor damage to the forest, while performing the valuable function of consuming available fuel.

The fires that are hardest to put out are those that occur during particularly dry, windy conditions or that burn in areas where lots of fuel become available. These fires tend to have relatively high temperatures and are more likely to be crown fires rather than surface or ground fires. By extinguishing the less harmful low-temperature fires, huge amounts of fuel become built up on the forest floor, which eventually leads to the more severe crown fire. As a result, forest fires are less frequent in many areas today than they were prior to human interference. Natural fires occur in some ponderosa pine forests every one to two decades, and now fires do not recur on a given site for centuries.

Figure 7.5 Homes in Los Alamos, New Mexico, burned in the Cerro Grande fire in May 2000. Ironically, this fire was started in a prescribed burn intended to reduce the risk of a large-scale conflagration.

When a fire does strike after a century of fuel buildup, it is likely to be a very damaging one.

Even without the public's general awareness, most U.S. agencies involved in forest management had adopted a "let-burn" policy by the middle of the 1970s. National Parks were conducting scientific studies to understand the fire history of their particular regions, and many used deliberately set fires as a positive part of their management program (Romme and Despain 1989). The National Park Service and the Forest Service formulated a common set of guidelines on the use of fire: (1) it must be tailored to the environmental specifications and history of each site; (2) naturally set fires are allowed to burn if the timing and location are right; (3) unplanned human-caused fires are to be put out; and (4) prescribed burns are used only when conditions are right—cool and wet—so that the fire does not spread beyond the desired area.

The Forest Service policy of using prescribed burns to reduce the risk of intense and uncontrollable wildfires resulted in an unfortunate disaster in 1999. A prescribed burn near Los Alamos, New Mexico, got out of control when the weather changed unexpectedly, and wildfire consumed 43,000 acres of forest plus over 220 structures in the Los Alamos area (Fig. 7.5). Despite this unfortunate set of circumstances, the let-it-burn policy is still in effect. A more recent issue concerns the vast areas in many western states that contain thousands of acres of blackened trees from an especially severe summer fire season in 2001–2002. The Forest Service is considering letting loggers harvest the trees on these public lands, where up to this point logging operations ceased—adding another dimension to the multiple use management goal.

DEFORESTATION AND REFORESTATION: THREE EXAMPLES

Forests are not inexhaustible and are depleted in many areas of the world. In some cases, as in most of the wealthy nations, forests recovered in recent decades continue to be important renewable resources (Rice et al. 1997). But in many of the poorer countries, rapid population growth and rising fossil fuel prices caused increased demand for wood, primarily for fuel. In these areas, wood is in critically short supply, and deforestation has caused accelerated erosion and soil degradation.

The world's forests are altered often dramatically by human activity, to the extent that the amount of forest today is only about two-thirds of the original forest prior to widespread human disturbance. This reduction of forest cover occurred over several thousand years, especially in Europe and east Asia, but deforestation along with population growth in the last 300 years accelerated the process, especially in the New World. The most important reason for forest clearance was the creation of cropland, but forests were cleared for grazing, fuel, and lumber as well.

Much of the forestland that remains today is altered by human activity. Large portions of the tropical forest region are subject to shifting cultivation, and forestlands that were harvested now have regenerated either partially or completely. The following examples help illustrate these points.

The Amazon Forest

More than 45 percent of the world's tropical rainforests have been cleared, and the rate continues at about 1.0 percent annually (UNEP 2002). Ecologists are concerned about the long-term effects of this clearing on soil resources, species diversity, and global biogeochemical cycling as well as the social impacts on native human populations. Rising demand for fuel and building materials, as well as a desire to open new lands for farming, suggests that the worldwide forest resource is in dire jeopardy. Associated with the depletion of the rainforests is a concern that the byproducts of wood burning—for warmth, for cooking, for processing into other products, and for land clearance—are contributing to global warming and loss of biodiversity.

In tropical areas, most notably the Amazon River basin in South America, deforestation is proceeding rapidly to make way for other land uses (Fig. 7.6). The Amazon rainforest covers about 1.7 billion acres (700 million hectares), which is roughly equivalent to 90 percent of the area of the contiguous United States, and extends into nine countries. In Brazil alone, 1.2 billion acres (490 million hectares)—the size of the United States west of the Mississippi River—is classified as Amazonian forest.

Until the 1960s, very little development of any kind took place in the Amazon, except for a few

Figure 7.6 Clearing a tropical rainforest in the western Amazon, Brazil.

Deforestation proceeded rapidly. In the early 1970s, less than 1 percent of the Brazilian Amazon was cleared; by 1989, more than 10 percent was cleared (Moran 1993). The rate of clearance was especially great in the 1980s, reaching rates in excess of 5 million acres (2 million hectares), or about 0.5 percent of the original forest area of the Brazilian Amazon, per year. Some estimates placed the rate of clearance at almost 1 percent per year. In the late 1980s and early 1990s, international concern about the Amazon reached a peak, and under pressure from environmental organizations the Brazilian government began to restrict forest clearance. The rate of forest loss has varied considerably from year to year but remains high (Fig. 7.7).

settlements along the major rivers. But development expanded more or less continuously in the region in the post-World War II years as its resources were seen as the basis for both individual and national economic gain. Construction of the capital city, Brasilia, 900 miles (1500 kilometers) from the Atlantic coast, began in 1957, and a highway connecting it with Belém on the Amazon River was built in 1958. This and other roads opened up the region to development, primarily cattle ranching and small farming.

Many environmentalists argue that Amazonian deforestation is a matter of grave concern and should be prevented if possible. The primary concerns expressed are as follows.

- **Loss of biodiversity** The tropical rainforests constitute only about 7 percent of the world's area yet contain more than half the world's species (Wilson and Peters 1988). Areas as small as a few hectares contain hundreds of tree species and many thousands of animal species. Key to the problem is the fact that most of these species are not even identified, let alone

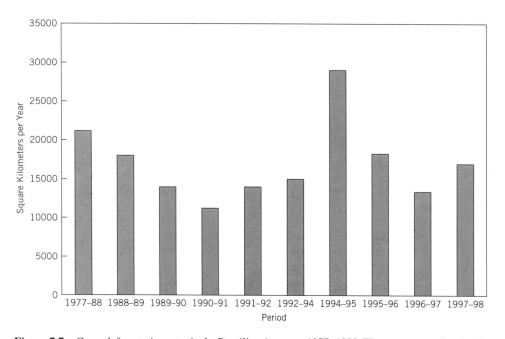

Figure 7.7 Gross deforestation rates in the Brazilian Amazon, 1977–1998. The average rate for the 21-year period is 18,860 km^2 per year. *Source:* CNN, 1999.

having understood life cycles. We don't know the true extent of their range or the conditions that are necessary for their survival. Nonetheless, on the basis of a few limited studies and many extrapolations, we believe that large numbers of species will become extinct as deforestation proceeds.

- **Emissions of carbon dioxide** Tropical forests are important to global carbon cycling for two reasons. First, they store large volumes of carbon in living biomass, and when this biomass is destroyed through clearance and burning the carbon is released into the atmosphere. It is estimated that perhaps 15 percent of the total global emissions of CO_2 are derived from land-use change, with deforestation the largest part of this figure. If forests are replaced with pasture or cropland and are not allowed to regrow, then this release of carbon is permanent. Second, the tropical forests have very high rates of productivity and thus have an enormous capacity to absorb carbon from the atmosphere. Perhaps 20 percent of total global photosynthesis takes place in the tropical forests. Recent measurements suggest that gross uptake of carbon in the Amazon forests themselves is about 0.6 gigaton of carbon annually, or about 10 percent of total global emissions to the atmosphere (Keller et al. 1996). If these are replaced with less productive ecosystems, then removal of carbon from the atmosphere by photosynthesis is reduced.
- **Disruption of the regional hydrologic cycle** The Amazon basin receives over 2 meters of rainfall annually. This water is derived from evaporation in the Atlantic Ocean to the east and is carried inland on the tropical easterly circulation. As water is carried westward, it falls from the atmosphere as precipitation and is returned through evapotranspiration. The interior western portions of the Amazon basin are 2000 to 3000 kilometers from the Atlantic and so derive a substantial amount of their moisture from water evapotranspired from forests to the east. On average about 30 percent of the rain that falls in the Amazon is derived from local evapotranspiration, but in the western portions of the basin the proportion is much higher. If forests were removed, the amount of evapotranspiration would decrease. This would increase the amount of runoff, perhaps causing more erosion, but more important it would decrease atmospheric humidity and thus precipitation in the interior.
- **Destruction of indigenous cultures** The Amazon is home to numerous groups of people who have lived with the forest as their home for millennia. These people developed an intimate knowledge of the forest resources and learned to exploit these resources on a sustainable basis through a combination of hunting, gathering, and low-intensity shifting cultivation. Expansion of Western culture, agriculture, and industrialization not only removes the forest, which is their home, but also exposes these people to the ravages as well as the benefits of global trade. Some find this a positive effect, while others view it negatively. In any case, it represents both a significant disruption of established life-styles and a potential loss of the cultural knowledge of indigenous peoples.

While these concerns are serious ones, proponents of logging and development in the Amazon point to the fact that much of North America, including virtually all the deciduous forest of eastern Canada and the United States, was cleared between the seventeenth and nineteenth centuries. The extent and rate of deforestation were similar to those in the Amazon. Much of what was cleared, especially in the eastern United States, has since regrown, but in the process of deforestation some species went extinct, soil was eroded, and indigenous cultures were destroyed. Some argue that this damage should not be repeated; others suggest that such a decision should be left to the Brazilians.

The Siberian Forest

The Siberian forest is vast, encompassing about 750 million hectares (Lakehead University 2002). This amounts to about 22 percent of the total forest area of the world and about 72 percent of the boreal or northern coniferous biome. It extends from the Ural Mountains in the west to the Pacific Ocean in the east, an expanse of about 8000 kilometers. The extent of forestlands within Siberia is equivalent to about two-thirds the total land area of the United States.

As discussed in Chapter 4, the boreal forest is characterized by relatively low species diversity. Most of the Siberian forest is coniferous forest (predominantly pine, cedar, spruce, larch, and fir), though portions are dominated by birch and aspen, particularly in the east. The climate of the area is one of extremes, and about 65 percent of the forest is underlain by permafrost (permanently frozen ground). Forest growth rates are relatively slow because of the cold climate and low amounts of solar radiation, but the cool climate also serves to reduce rates of organic matter decomposition. As a result, large amounts of carbon are stored in living and dead biomass. The total carbon stored in the Siberian forest is estimated at 30 gigatons, or about six times the total amount of carbon emitted to the atmosphere by fossil fuel combustion worldwide. The forest acts as a net sink of carbon, absorbing about 0.4 gigaton annually (Shvidenko and Nilsson 1994).

Industrial-scale exploitation of the Siberian forest began after World War II, at a time when large amounts of timber were needed to reconstruct Eastern Europe and to support the rapidly expanding economy of the former Soviet Union. During the 1980s, about 800,000 hectares were clear-cut annually—on the order of 0.1 percent of the total Siberian forest (Shvidenko and Nilsson 1994). This rate declined by about half between 1988 and 1993 as a result of the general economic decline in the former Soviet Union/Russia during that period. Harvest fell to only one-third of the amount allowable under forest management practices in place at the time (Korovin 1995).

Throughout this period, forest harvesting was primarily by means of clear-cutting, with heavy machinery and minimal concern for protection of the soil, efficient collection of wood, or regeneration. Soil erosion problems are widespread. Of particular concern is damage to permafrost, in places where it underlies the forest. Removal of trees results in warming of the ground and melting water in the soil. This both encourages erosion and makes revegetation more difficult. Replanting of harvested areas was nonexistent, with the forest left to regenerate naturally. Fires were also widespread, consuming about 1 to 1.5 million hectares annually. As a consequence, the forests became susceptible to insect infestations, which damaged more than 46 percent of the Russian boreal forest during the mid- to late 1990s (UNEP 2002).

Despite these problems, vast resources are available for use. In 1995, of approximately 650 million hectares classed as forest and managed by government forest authorities, about 75 percent was classed as exploitable forest, 20 percent was weakly protected, and 5 percent was protected (Lakehead University 2002). The Siberian forest thus represents an enormous natural resource available for commercial exploitation, and a significant increase in harvest rates is possible. The recent history of the region and its place in the world helped encourage such an increase.

Three factors are at work. First, the transition to a market economy creates opportunities for entrepreneurs to invest in the forest products industry, establishing new businesses and markets. In some cases, the government agencies responsible for forest management themselves encouraged exploitation as a means of securing their own personal funding. In other cases, large tracts of land were leased to foreign forest products companies, which are beginning to harvest timber and export the logs. Second, Russia today has an enormous need for foreign exchange, and timber trade is an obvious potential export. It therefore has welcomed offers from Japanese, Korean, and European companies to develop timber resources. Most of these are joint ventures between Russian and foreign entities. Third, political instability and the increased autonomy of regional governments reduces the authority of central resource management agencies to control exploitation. Some claim that regional or local authorities sold timber that is officially protected by the national government. In some instances, there is no single central authority with clear responsibility for management decisions. In others, the same agency responsible for conservation is the same one that collects the income from timber exploitation (Acharya 1995; Korovin 1995).

The U.S. Forestland

The United States includes about 300 million hectares of forestland. Most of the U.S. forest resources are concentrated in the Pacific Northwest, Alaska, and the East and Southeast regions (Fig. 7.8).

At the time of the European settlement of North America, forests covered about two-thirds of the United States. This forest was both a resource and an obstacle to the early settlers. It provided fuel and building materials but at the same time stood in the way of land clearance for agriculture. Timber was plentiful in most areas in the seventeenth and eighteenth centuries, and no one had to go far for wood. By the mid-nineteenth century, however, population and economic growth began to exert a great demand for wood, and supplies were diminishing. Local or regional shortages developed in the northeastern United States as the focus of timber harvesting moved west to the Great Lakes states. Writer and explorer Henry David Thoreau, traveling by canoe in the wilderness of central Maine, found the presence of towering white pines notable, as these were being harvested from remote areas even in the first half of the nineteenth century. Thoreau also complained of a chronic shortage of firewood in central Massachusetts.

Land area in forest cover continued to decline into the early twentieth century, largely because of clearing of land for agriculture. By 1920, however, forestland clearance slowed. The increasingly productive agricultural lands of the Midwest were competing with eastern agriculture for some

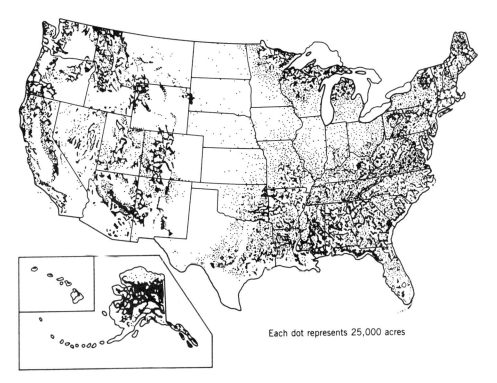

Figure 7.8 Forestlands in the United States. The nation's forests are concentrated in the East, Southeast, Pacific Northwest, and Alaska.

time, and beginning in the nineteenth and continuing into the early twentieth century, farmland was being abandoned rapidly, especially in areas within and east of the Appalachians. This farmland gradually reverted to forest, and as a result forest acreage in the lower 48 states actually increased about 20 percent in the last 60 years.

The history of U.S. forest exploitation is one in which, region by region, forests were exploited to the point of exhaustion and then abandoned as new resources were opened up farther west. In many cases, the forest has regrown after abandonment, although the quality of *second-growth* forest seldom matches what was destroyed in the first place. By the end of the twentieth century, we had cut all but a small amount of the original old growth; the forest today is almost exclusively second-growth, and much of that is relatively young.

Despite this abuse, many of our forests are in good shape in comparison to those in many other nations (McKibben 1995). In fact, after 300 years of continual and often extravagant logging, the U.S. forest resource today is considered abundant and resilient, according to many measures. About two-thirds of the U.S. forestlands are rated as commercial forests. This means that the land is capable of growing at least 20 cubic feet of wood per acre ($1.4 m^3$ of wood per hectare) in a fully stocked stand on an annual basis (U.S. Forest Service 2000). The timber produced is therefore commercially profitable. The remaining one-third of forestland is classified as noncommercial and is categorized as parks, wildlife habitat, recreation, and wilderness.

Commercial forestland in North America is differentiated by type of vegetation or tree species and is divided into *hardwoods* and softwoods. Hardwood forests generally consist of broadleaf and deciduous trees such as oak, maple, and hickory. In the United States, hardwood forests are located primarily in the northern and southern regions of the eastern half of the country (Fig. 7.9). Commercial hardwood stands are used largely for furniture making and flooring, although they are also important for heating. The total acreage of hardwood stands in North America is greater than that of softwoods, but the timber is commercially less valuable because of the greater difficulty in harvesting and lower demand.

Softwoods are conifers, usually evergreens. The primary North American species are spruce, pine,

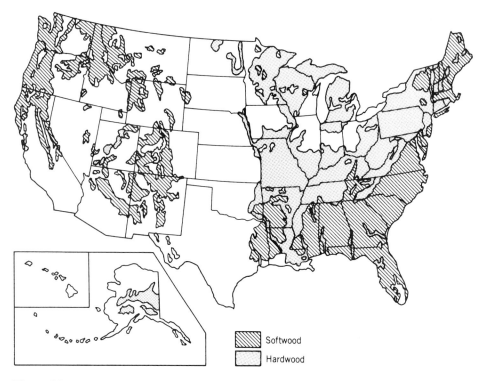

Figure 7.9 U.S. commercial forests, by type and region. Hardwoods dominate in the South and East and softwoods in the Southeast, Northeast, West, and Alaska. *Source:* Haden-Guest et al. 1956.

and cedar, whose wood is softer and grain is farther apart than in hardwoods. Softwoods are used primarily for paper products, lumber, and plywood. Softwood forests are located throughout North America but dominate in the Pacific Coast, Rocky Mountain, and Southern National Forest Service Regions. Softwoods supply about 80 percent of total lumber consumption in the United States.

Forest ownership is critical to forest management and timber supply (Table 7.1). The federal government (primarily the U.S. Forest Service and the Bureau of Land Management) administers 27 percent of the nation's forestlands. Nine percent are in state and county ownership, including state forestlands. The forest products industry owns about 9 percent of the commercial forest, mostly in the Southeast and with smaller holdings in northern New England and the Pacific Northwest. Finally, 54 percent of the commercial forestland in the United States is privately owned in small holdings and on farms. Thus, only a small portion of private, commercial forestland is owned by the forest industry itself. The federal government controls about one-fifth of commercial forestland. When we consider that most of the forests in "other private" ownership are in small holdings that are not easily accessible to the forest products industry, it becomes clear that less than half of the commercial forest is really available for harvest, and most of this is in government ownership.

When ownership patterns are examined on a regional basis, a number of interesting trends emerge (Fig. 7.10). Federal ownership dominates the western forests, which are managed mostly by

Table 7.1 Ownership of Forestlands in the United States

Ownership	Area (Million Acres)	% of Commercial Forestland
Federal	204	27
State and county	69	9
Forest industry	68	9
Other private	405	54
Total	747	100

Source: U.S. Forest Service, 2000.

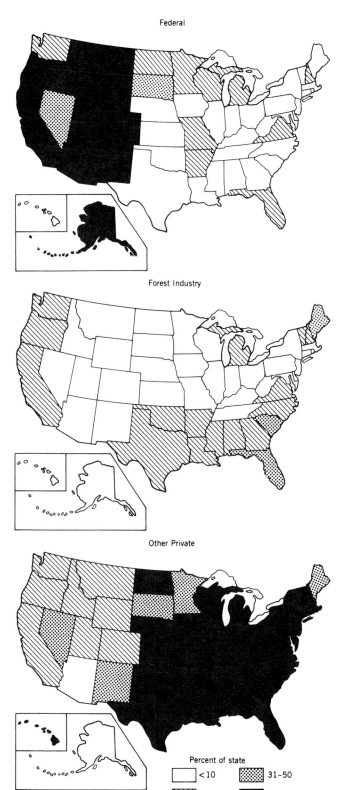

Figure 7.10 Ownership of commercial forestland. Federal ownership is dominant in the West and Alaska. Forest-industry ownership is most evident in the West, South, and Northeast. Other private holdings dominate in the East. "Percent of state" refers to the percentage of commercial forestland in each state by ownership category.

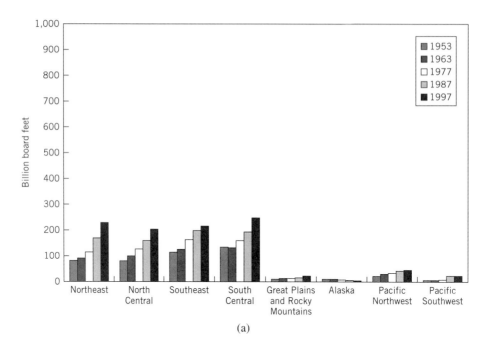

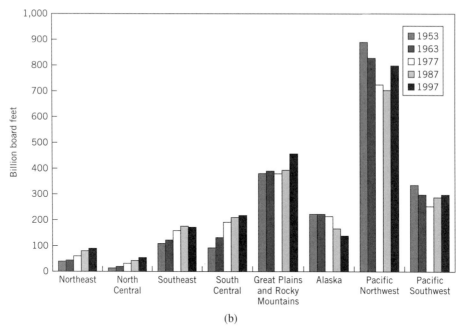

Figure 7.11 Standing sawtimber on timberland in the United States, 1953–1997: (a) hardwoods; (b) softwoods. The volume of standing timber has been increasing in most of the United States for several decades, mostly as a result of the regrowth of forests following initial harvest and/or abandonment of agricultural land. Hardwood volume, mostly in the eastern states, has doubled since 1953. Softwoods, which account for a much greater portion of standing sawtimber, have been intensively harvested in the Pacific region and Alaska, where standing volumes have declined until recently. *Source:* Smith et al. 2002.

the Forest Service and the Bureau of Land Management. The timber industry dominates the ownership of forestland in Maine, South Carolina, and Florida. Private holdings by farmers and other individuals are the dominant type of ownership in the East. As a result of these patterns, public policies on forest issues have a strong regional focus, with most of the controversy between government and the timber industry concentrated in the West.

Because most of the forests in the United States are secondary forests (they have already been cut once and are regrowing), much of the timber is relatively young. Trees are fairly small and thus less valuable for lumber than the larger trees found in old-growth forests. At the same time, younger forests grow faster than old forests. Indeed, an old-growth forest doesn't grow at all—old trees die and decay at roughly the same rate that new trees are produced. Thus, in volume terms, the U.S. forests are growing slightly more timber than is being cut, and the standing volume of timber is steadily increasing (Fig. 7.11). But on an age basis, the amount of timber in mature forests with large trees is decreasing (Fig. 7.12).

Similarly, at the aggregate level, we are harvesting below the long-term sustainable level. But aggregate statistics are based on the assumption that all commercial forestland is available for harvest at some time or another, and this is not the case. Most of the commercial forest is in small private holdings. If a forest products company is to harvest this timber, it must make agreements with individual landowners, one at a time. Because not every landowner wants to sell timber at the same time, this means that if this timber is to be harvested, it must be taken from small, scattered parcels. This greatly increases management and transportation costs and makes harvesting such timber prohibitively expensive.

The forest products industry prefers to concentrate in a few regions of the country where sufficient forestlands exist to take advantage of economies of scale. In the United States, these are principally the Southeast, northern New England, and the Pacific Northwest, with smaller operations in the Rocky Mountains and the northern Great Lakes. In the Southeast and in Maine, the forest products industry controls most of the commercial forestland and manages this land intensively. Trees are replanted after harvest and are cut on a relatively short cycle (as short as 20 to 30 years in the Southeast). These smaller trees are used mainly for plywood and paper manufacture. Sawn lumber operations are concentrated in the Pacific

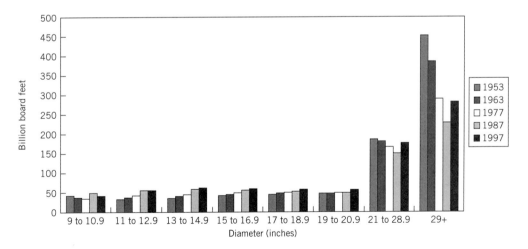

Figure 7.12 Size distribution of standing sawtimber in the Pacific Northwest, 1953–1997. The volume of timber in the largest size class (29 inches or more in diameter) decreased by almost 50 percent between 1953 and 1987 as the forests of the region were intensively harvested. Most of the smaller size classes increased in volume as second-growth forests replaced old-growth ones. The increase in volume of large-diameter timber between 1987 and 1997, the result of a slowdown in harvest rates, is testimony to the high productivity of these forests and the ability of the resource to renew itself given sufficient time. *Source:* Smith et al. 2002.

Northwest, where substantial areas of old-growth forest still exist. The old-growth forest there has been reduced significantly since the 1940s, and environmentalists want to stop logging of the old growth and preserve what remains. In the short run, this forces curtailment of some logging operations and substitution of foreign (mostly Canadian) for domestically produced lumber. But in the long run the forest products industry has to learn to rely more heavily on second-growth forests and the smaller trees they contain.

Conclusions

Forests are areas of low-population densities, because they are found in places that are not as suitable for farming as other areas. But this does not diminish their importance as resources. Expanding populations and resource use are increasing pressures on the forest resource, so conflicts over its management and use will intensify. Because a large portion of the world's forestlands are either government-owned or subject to strong governmental controls of their use, the process of decision-making regarding use of the forest is largely a political one, as is clear in the examples of Amazonian, Siberian, and U.S. forests.

References and Additional Reading

Acharya, A. 1995. Plundering the boreal forests. *Worldwatch* 8(3):20–29.

Bierregaard, R. O. Jr., C. Gascon, T. E. Lovejoy, and R. Mesquita, eds. 2001. *Lessons from Amazonia: The Ecology and Conservation of a Fragmented Forest.* New Haven: Yale University Press.

Bonnie, R., S. Schwartzman, M. Oppenheimer, and J. Bloomfield. 2000. Counting the cost of deforestation. *Science* 288:1763–1764.

Brooks, D. J. 1993. *U.S. Forests in a Global Context.* General Technical Report RM-228 U.S. Forest Service, Rocky Mountain Forest & Range Experiment Station.

Bureau of Land Management. 2000. *Headwaters Forest Reserve Draft Management Plan EIS/EIR.* http://www.ca.blm.gov/arcata/draft_plan.html.

Christensen, N. L., et al. 1989. Interpreting the Yellowstone fires of 1988. *Bioscience* 39:678–685.

Clawson, M. 1979. Forests in the long sweep of U.S. history. *Science* 204:1168–1174.

CNN. 1999. *Aumenta o desmatamento da Amazônia.* http://cnnemportugues.com/latin/BRA/1999/02/11/amazonia/index.html.

Cochrane, M. A. 2001. In the line of fire: Understanding the impacts of tropical forest fires. *Environment* 43(8):28–39.

Dunning, J. and D. Thron. 1999. *From the Redwood Forest: Ancient Trees and the Bottom Line: A Headwaters Journey.* Chelsea Green Publications.

Food and Agricultural Organization. 2001. FAOSTAT, http://apps.fao.org/.

Fearnside, P. M. 1990. The rate and extent of deforestation in Brazilian Amazonia. *Environmental Conservation* 17:213–226.

Gusewelle, C. W. 1992. Siberia on the Brink. *American Forests* 98(5/6):17–20.

Haden-Guest, S., et al. 1956. *A World Geography of Forest Resources.* New York: Ronald Press.

Harris, David. 1996. *The Last Stand.* San Francisco: Sierra Club Books.

Jepson, P., J. K. Jarvie, K. MacKinnon, and K. Monk. 2001. The end for Indonesia's lowland forests? *Science* 292:859–861.

Keller, M., D. A. Clark, C. B. Clark, A. M. Weitz, and E. Veldkamp. 1996. If a tree falls in the forest . . . *Science* 273:201.

Korovin, G. 1995. Problems of forest management in Russia. *Water, Air and Soil Pollution* 82:13–23.

Kremen, C., J. O. Niles, M. G. Dalton, G. C. Daily, P. R. Ehrlich, J. P. Fay, D. Grewal, and R. P. Guillery. 2000. Economic incentives for rain forest conservation across scales. *Science* 288:1828–1832.

Lakehead University. 2002. *World Boreal Forests, an Introduction.* Faculty of Forestry and the Forest Environment. http://www.borealforest.org/world.htm.

Likens, G. E., et al. 1978. Recovery of a deforested ecosystem. *Science* 199:492–496.

Maser, C. 1994. *Sustainable Forestry: Philosophy, Science and Economics.* Delray Beach, FL: St. Lucie Press.

McKibben, B. 1995. An explosion of green. *The Atlantic Monthly* April:61–83.

Moran, E. F. 1993. Deforestation and land use in the Brazilian Amazon. *Human Ecology* 21:1–21.

Mott, W. P., Jr. 1989. Federal fire policy in national parks. *Renewable Resources Journal* 7:5–7.

Newton, L. H. 1990. The chainsaws of greed: The case of Pacific lumber. In W. M. Hoffman, R. Frederick, and E. S. Perry, eds. *The Corporation, Ethics and the Environment.* New York: Quorum Books.

Noss, R. F., ed. 1999. *The Redwood Forest: History, Ecology, and Conservation of the Coast Redwoods.* Washington, D.C.: Island Press.

Rangan, H. 1996. From Chipko to Uttarancahal: Development, environment and social protest in the Garhwal Himalayas, India. In R. Peet and M. Watts, eds. *Liberation Ecologies: Environment, Development, Social Movements.* New York: Routledge.

___. 2001. *Of Myths and Movements: Rewriting Chipko into Himalayan History.* London: Verso.

Rice, R. E., R. E. Gullison, and J. W. Reid. 1997. Can sustainable management save tropical forests? *Scientific American* 276(4):44–49.

Romme, W. H. and D. G. Despain. 1989. The Yellowstone fires. *Scientific American* 261(5):37–46.

Shvidenko, A. and S. Nilsson. 1994. What do we know about the Siberian forests? *Ambio* 23:396–404.

Skole, D. and C. Tucker. 1993. Tropical deforestation and habitat fragmentation in the Amazon: Satellite data from 1978 to 1988. *Science* 260:1905–1910.

Smith, W. B., J. S. Vissage, D. R. Darr, and R. M. Sheffield. 2002. *Forest Resources of the United States, 1997.* Washington, D.C.: USDA Forest Service.

United Nations. 1992. *Conservation and Development of Forests. Rio de Janeiro, Brazil: UN Conference on Environment and Development.* New York: United Nations.

UNEP. 2000. *Original Forest Cover.* http://www.unep-wcmc.org/forest/original.htm.

United Nations Environment Programme. 2002. *Global Outlook 3.* London: Earthscan.

U.S. Forest Service. 2000. *RPA Assessment of Forest and Range Lands.* http://www.fs.fed.us/pl/rpa/rpaasses.pdf.

Watson, R. T., I. R. Noble, B. Bolin, N.J. Ravindranath, D. J. Verardo, and D. J. Dokken. 2001. *IPCC Special Report on Land Use, Land-Use Change and Forestry.* http://grida.no/climate/ipcc/land_use/index.htm.

Wilson, E. O. and F. M. Peters, eds. 1988. *Biodiversity.* Washington, D.C.: National Academy Press, pp. 3–20.

For more information, consult our web page at *http://www.wiley.com/college/cutter*.

STUDY QUESTIONS

1. List the major timber and nontimber uses of forestland. For each use, identify the ways in which the use is compatible or incompatible with other uses of the forest.
2. When the forests of the upper midwestern United States were cleared, the wealth that was generated (and some of the lumber) was used to build cities such as Chicago, which themselves contain the capital goods that their populations use today to generate wealth. To some extent, then, one form of capital (the forest) converted into another (the city). Was this a good thing?
3. List the uses of forest products that you encounter in a day. For each, identify a substitute material that could be used instead of a tree. How would the environmental impacts of using the substitute be different from the impacts of using the tree?
4. How is an old-growth forest different from a second-growth forest? How much is lost (or gained) in the conversion?
5. Forestlands in the United States owned by the forest products industry generally yield more wood per acre than publicly owned commercial forests. This is because many of the privately owned forests are more intensively managed. Should privatization of forestlands be encouraged?

CHAPTER 8

BIODIVERSITY AND HABITAT

Biological diversity refers to both the genetic variability among individuals of a species and the abundance of individuals within a species. Wide variations in genetic traits increase the likelihood that at least some individuals of a given species will survive environmental change. The number of different species, the abundance of individuals in that species, and the number of species present at a particular time within a specific geographic area are also indicators of biological diversity. The most ecologically diverse environments are the tropical forests, where there is a much greater abundance of plant and animal species than in any other single biome.

Species extinction is a fundamental threat to biological diversity. The death of an individual represents the loss of an organism capable of reproducing the same form as other individuals in the species. The death of individuals is a natural process. The death of an entire species is an irreversible process in which both the basic form and the reproductive potential are lost. The contribution of the species to the vitality of the planet is also lost. Species extinctions occur naturally, but humans have dramatically accelerated the pace and process in the last few centuries.

Human activities often result in the reduction of biological diversity through the destruction and simplification of natural habitats. Urban sprawl leads to an increase in the amount of asphalt and concrete at the expense of fields, forests, marshlands, and other valuable habitats. Modern farming and forest cultivation result in single-crop patterns over broad areas, maintained by chemicals that destroy unwanted species. These practices endanger the genetic and ecological diversity of plant and animal communities.

Since the 1960s, the cry has been heard regarding the destruction and alteration of habitats and species. In 1962, Rachel Carson raised the concern in *Silent Spring* that pesticides would cause widespread extinctions (see Chapter 3). Researchers such as Eckholm (1978) and Ehrlich and Ehrlich (1970, 1981) argued that when forests are clear-cut, when meadows are paved, when rivers become sewers, we are destroying species with potential value—both economic and amenity value. These writers maintained that species and habitat destruction have reached epidemic proportions worldwide. By the late 1980s, concern over species destruction—and the implications for life on Earth—were no longer confined to gloom and doom debates on college campuses and in government offices. Newspapers, magazines, and the electronic media dramatically increased coverage of issues such as the destruction of the world's rainforests, the killing of rhinoceroses, and the plight of wolves.

In this chapter, we explore the problems of biodiversity and habitat conservation. We begin by examining why we are concerned with the destruction of a resource that, to some, has little obvious economic value, but that has become a prime focus of the environmental movement. We then review the available information on the extent of species extinctions and the processes causing them. Finally, we focus on efforts to reduce threats to biodiversity.

THE VALUE OF BIODIVERSITY

Loss of biodiversity has several consequences. Ecosystems are undermined when plant and animal species are destroyed or when they move into new areas. The possibility of using as yet untried species for food, fuel, fiber, or medicine disappears when they are eradicated. Human appreciation and understanding of nature also are diminished by species and habitat loss (Cairns 1995; Daily 1997). Many people question the ethics of human beings to deny other species the right to exist (Chapter 1). Most profoundly, even from a completely selfish, anthropocentric point of view, it is feared that removal of even a few species from the web of life could lead to ecological imbalances that result in further extinctions and perhaps even alterations of the biogeochemical cycles on which we depend for food production and waste removal.

Ecological Interactions

The science of ecology has taught us that everything is connected to everything, and a change in one part of an ecosystem inevitably has implications elsewhere. At the level of individual links in a food web, when a plant species is eliminated, either locally or globally, the species that directly or indirectly rely on it, including insects, higher animals, and other plants, can be adversely affected.

The stability of ecosystems, in terms of their ability to maintain populations of organisms, is often enhanced by the diversity of organisms they contain. Not all ecosystems are made more stable by increased numbers of species, but in some, diversity helps contribute to stability by providing a supply of different species that are all capable of carrying on food processing. If one species declines in number as a result of some disturbance such as disease, then other species are available to occupy that niche. The metaphor of not putting all one's eggs in a single basket is an appropriate one.

For example, grasslands are subject to large environmental swings, particularly in available soil moisture. Indeed, grasses dominate midlatitude semiarid environments largely because they are able to lie dormant during dry spells but grow rapidly when moisture is available, as well as returning rapidly after fire. The original grasslands of North America contained a wide variety of grass and other herbaceous species. This diversity helped them withstand wide environmental swings because drought-tolerant species could take over in dry spells, or species resistant to a particular disease or insect could expand when other species were suffering. The North American grasslands of today are much less diverse than were the original ecosystems, and this loss of diversity is believed to have contributed to a loss of stability as well (Tilman 1996).

Diversity does not always lead to stability, however. For example, although the diversity of an ecosystem may contribute to its overall stability in terms of maintaining biomass and energy transfers, greater diversity may not make it easier for individual species to avoid local extinction. The complexity of food webs in diverse ecosystems may cause individual species to suffer wider population swings than would occur in a simpler system (Moffat 1992).

Potential Resources

In addition to maintaining the resource functions of ecosystems currently in use, nature contains many things that we might use at some time in the future. Food and medicine are the most often-cited potential uses of wild plants and animals, but many other uses are imaginable, including chemicals, fiber resources, and erosion control.

Of all naturally occurring species, plant and animal, it is estimated that humans have found uses for less than one-tenth of 1 percent of the total. The enormous majority are untested and their potential beneficial uses are unknown. It *is* known that at least 75,000 species of plants have edible parts, yet the world today relies almost entirely on about 30 plant species for its food supply, mostly wheat, rice, maize, millet, and rye. Given the stress on the world food production system that is expected as population grows in the coming decades, it seems a good idea to reexamine some of the 7000 plant species that humans have used for food during our occupancy of the earth and to conduct research on newly discovered plant species with promising value.

In recent years, researchers have looked at the possibility that previously unused or even despised plant species could be used for food, fiber, and medicine. For example, mesquite, a weedy nuisance on western cattle ranges, is now being promoted as a potential world food source. It produces abundant annual crops of a highly nutritious bean, once a staple for the region's Native Americans. Mesquite wood has become popular as a fuel in gourmet cooking, commanding a high price in some urban markets. Another recent discovery—recent to modern science, at least—is the buffalo gourd, used for at least 9000 years by the Native Americans. This widespread wild plant provides vegetable oil, protein, and starch of high quality and thrives on very little water. Since humans depend on a narrow range of crop species for food, the discovery of new food resources is very important. The Central American amaranthus produces seeds that contain a high-quality protein that could be of use to protein-deficient human societies. Eelgrasses, grown in salt water, offer a potential substitute for grains in some heavily populated coastal areas.

The potential for significant environmental changes in the world caused by climate change or other factors may generate the need for new plant resources. For example, new crop varieties will have to be developed to take over for those currently in use if climate in important agricultural regions becomes unsuitable for varieties produced now (Crosson and Rosenberg 1989; Wilson 1992). To that end, a 16-organization network of agricultural research establishments, the Consultative Group on International Agricultural Research (CGIAR), was formed to promote the development of new crops and agricultural techniques worldwide (Table 8.1).

Many of our most valuable medicines are derived from plants (Cox and Balick 1994). In the United States, for example, of the top 150 most prescribed drugs, 40 percent of them have active ingredients derived from plants. Sales of these totaled about $15.5 billion in 1990 (Reid 1995). Globally, these active ingredients produced $75–$150 billion in revenues (World Resources Institute 2000). Incristine, discovered in the mid-1950s, is an alkaloid found in a Madagascar periwinkle. The chemical causes a decrease in white blood cell counts and has been used to fight cancer and cancerlike diseases. Quinine, an alkaloid in Cinchona bark, was used to treat malaria until synthetic quinine was developed in the 1930s. Digitalis, from foxglove, is widely used to treat chronic heart failure by stimulating the heart to pump more blood and use less energy. A number of well-known painkillers, including morphine and codeine, are derivatives of the opium poppy. Taxol, an anti-tumor agent used in cancer treatment, is derived from the Pacific Yew tree. Bee venom has been used to relieve arthritis, and the venom of the Malayan pit viper is used as an anticoagulant to prevent blood clots and to lessen the danger of heart attack. Animals also provide important models for studying human diseases. The Mexican salamander, for example, is an endangered species that is being used in the study of injured heart muscles. Globally, there are between 4000 and 6000 medicinal plants that are traded (the majority are exported from China). Botanicals (especially ginseng and echinacea) generate $20 to $40 billion annually in the United States alone (World Resources Institute 2000).

The medicinal use of plants and animals is a particularly compelling argument for conservation of not only ecosystems but also the indigenous cultures of people who occupy them. Indigenous peoples have intimate knowledge of the plants and animals of the regions they inhabit, and they use many of them for medicinal purposes. The people of the Amazon not only depend on the forest for their own medicines but also have the knowledge that may help others use the forest beneficially.

In the southwestern United States and adjacent Mexico, native peoples are known to have made use of some 450 wild plants. Anthropologists contend that many of these desert-adapted species could be of value to modern society. Guayule is a shrub grown in northern Mexico and Texas. Before 1910 it supplied 10 percent of the world's rubber (Ehrlich and Ehrlich 1981); the latex in the guayule shrub is very similar to that in the rubber tree. Jojoba, a shrub related to boxwood, has seeds that contain a liquid wax. This wax, which makes up as much as 60 percent of the jojoba bean's weight, can be used for lubricating metal parts and other purposes once served by sperm whale oil, the use of which is now outlawed. Another seemingly unlikely possibility for development is the all-American goldenrod, whose leaves contain up to 12 percent natural rubber. It is easy to grow, can be mowed and baled, and resprouts without annual sowing. These are just a few of the thousands of potentially useful plant species that make a strong economic argument for preserving not only rare but also abundant species.

The Inherent Value of Species

Perhaps the most compelling reason many people are deeply concerned about loss of biodiversity is unrelated to the material benefits of ecological stability, scientific and educational values, or potential resources. Rather, it is the belief that we, as humans, have an obligation to respect the rights of other species to exist (see Chapter 1). While a belief in animal rights is fundamental to many of the world's cultures, it is not a significant part of

Table 8.1 The Centers and Purposes of the Consultative Group on International Agricultural Research (CGIAR) System

Center (Date Established), Location	Purpose
CIAT: *Centro Internacional de Agricultura Tropical (1967), Calí, Colombia*	Improve production of beans, cassava, rice, and tropical forages/grasses in the tropics of the Western Hemisphere
CIP: *Centro Internacional de la Papa (1970), Lima, Peru*	Improve the potato, sweet potato, and other root and tuber plants in the Andes and other mountain areas and develop new varieties for the lower tropics
CIMMYT: *Centro Internacional del Mejoramiento de Maiz y Trigo (1966), Mexico City, Mexico*	Improve maize, wheat, barley, and triticale
IPGRI: *International Plant Genetic Resources Institute (1974), Rome, Italy*	Promote an international network of genetic resources (germ-plasm) centers
ICARDA: *International Center for Agricultural Research in the Dry Areas (1975), Aleppo, Syria*	Focus on drylands agriculture in arid and semiarid regions in North Africa and West Asia
ICRISAT: *International Crops Research Institute for the Semi-Arid Tropics (1972), Hyderabad, India*	Improve the quantity and reliability of production of food, especially sorghum, millet, groundnut, chickpea, and pigeonpea, in the semiarid tropics
IFPRI: *International Food Policy Research Institute (1974), Washington, D.C., U.S.A.*	Address issues arising from governmental and international agency intervention in national, regional, and global food problems
IITA: *International Institute of Tropical Agriculture (1967), Ibadan, Nigeria*	Improve worldwide production of cowpea, yam, cocoyam, sweet potato, cassava, rice, maize, and beans, among others, as well as food security in humid tropical regions
ILRI: *International Livestock Research Institute (1995), Nairoba, Kenya, formed by the merger of the International Laboratory for Research on Animal Disease (1973), Nairobi, Kenya, and the International Livestock Centre for Africa (1974), Addis Ababa, Ethiopia*	Help develop controls for trypanosomiasis (transmitted by the tsetse fly) and theileirosis (transmitted by ticks), conduct research and development on improved livestock production and marketing systems, train livestock specialists, and gather documentation for livestock industry
IRRI: *International Rice Research Institute (1960), Los Banos, Philippines*	Select and breed improved rice varieties and maintain a germ-plasm collection bank
ISNAR: *International Service for National Agricultural Research (1980), The Hague, The Netherlands*	Strengthen national agricultural research systems through institutional innovations
WARDA: *West Africa Development Association (1970), Bouake, Cote d'Ivoire*	Promote self-sufficiency in rice production in West Africa and improve varieties suitable for the area's agroclimate and socioeconomic conditions
ICLARM: *International Center for Living Aquatic Resources Management (1977), Penang, Malaysia*	Responsible for fisheries and other living aquatic resources
ICRAF: *International Center for Research in Agroforestry (1977), Nairobi, Kenya*	Domestication of agroforestry resources, productive land use, and soil fertility replenishment
IIMI: *International Water Management Institute (1984), Colombo, Sri Lanka*	Water use in agriculture and integrated management of water and land resources
CIFOR: *Center for International Forestry Resources (1993), Bogor, Indonesia*	Social, environmental, and economic consequences of tropical forest degradation

Source: Crosson and Rosenberg, 1989:134; CGIAR, 2001.

the Christian tradition that dominates much of the Western world. But today many environmentalists believe in some fundamental rights for nonhuman species, and this is a motivating factor in their concern for loss of biodiversity. This ethical argument centers on the rights of nonhuman entities merely to exist, regardless of any usefulness to humans. Ehrenfeld (1981), in a classic book, argues that all living things have a right to coexist on the planet. Humans, possessing the power to destroy and alter plant and animal species, should exercise stewardship in preserving plant and animal species. This represents only one of the many arguments for preserving biological diversity.

On the other hand, some might take a neo-Darwinian view of extinctions that regards humans as just another species; in this perspective, the extinctions caused by humans are no different from mass extinctions of the past such as the disappearance of the dinosaurs. This view maintains that species should be allowed to die because they have been unable to compete successfully with humans and other species. Furthermore, we should feel no guilt about species extinction because it is a natural process, and we do not need to keep rare species alive at great cost to human society.

The Pace and Processes of Extinction

The number of species on Earth is unknown, but some guesses can be made. At present, about 1.7 million species have been identified and described. We know that a much greater number have not been identified, but actual numbers of unidentified species can only be estimated by extrapolation. One example of such an estimate is based on studies of canopy beetles in tropical trees. One species of tropical tree was found to have 163 species of canopy beetles that lived only in that species of tree. If each of the about 50,000 tree species had 163 specific canopy beetles, then there would be 8 million species of canopy beetles on Earth. About 40 percent of known insects are beetles, so we can guess that there might be about 20 million species of insects on Earth. Obviously, this is only a guess, but it gives us an idea of the amount of diversity that may exist (Pimm et al. 1995).

Some types of organisms are better known than others. Vertebrates are fairly well known, for example. We have identified more than 45,000 species of vertebrates, and it is estimated that the total number of species is 50,000 to 55,000. Insects, on the other hand, are less well known. About 950,000 have been described, and the estimates of the total number of insect species range from 2 million to 100 million! Most estimates of the total number of all species range between about 3 million and 100 million. A recent United Nations Environment Program report suggests a reasonable estimate of about 13 to 14 million (United Nations Environmental Program 1995; World Resources Institute 1996, 2000).

Because we are uncertain about the total number of species present on the planet, we are similarly uncertain about the number of these that have become extinct recently or are threatened with extinction. One recent assessment suggests that about 24 percent of roughly 4600 mammal species and 11 percent of 9700 bird species are currently threatened. This is in addition to 58 mammals and 115 birds known to have become extinct in the past 400 years (World Conservation Monitoring Centre 1992). This same assessment estimates that 3 to 4 percent of fish and reptile species and only a fraction of a percent of insects are known to be threatened. It is obvious that we know much more about some groups of organisms, such as mammals, than we do about others, such as insects.

Extinction is a natural process, and so data about numbers of species estimated to be threatened or to have become extinct recently should be considered in relation to natural extinction rates. Again, much uncertainty exists, but most estimates suggest rates of perhaps one extinction per species per million years over geologic time. If we have 10 million species on the planet today, this would suggest an average of perhaps 10 extinctions per year. Over the past century, rates of extinction may have been 10 to 100 times higher than this estimated geologic rate (Pimm et al. 1995). Between 1600 and 1900, for example, the rate of known species extinction was one species every four years, and this escalated to one species per year between 1900 and 1980. By 1993, the estimates for mammal and bird extinctions were one to three species per year (Council on Environmental Quality 1994). These extinction rates among better-known groups are sometimes used as indicators of overall extinction rates. The actual number of extinctions must be much more than the number we know about, because we have identified only a small fraction of the number of species in existence. It may be that the numbers of insect species known to be threatened are low sim-

Table 8.2 Species at Risk in the United States

	Percentage of Species in Group		
Group	Vulnerable[a]	Imperiled[b]	Extinct or Possibly Extinct[c]
Amphibians	14.3	12.1	0.9
Birds	4.7	2.6	3.3
Butterflies/skippers	12.6	5.0	0.0
Crayfish	17.1	6.4	0.9
Dragonflies/damselflies	9.9	5.3	0.4
Ferns	11.7	4.3	0.7
Flowering plants	17.1	8.5	0.9
Freshwater fish	13.1	10.9	2.1
Freshwater mussels	16.8	14.1	12.5
Mammals	8.6	5.0	0.2
Reptiles	10.6	4.9	0.0
Tiger beetles	12.3	4.4	0.0

[a]Vulnerable populations are rare, typically 3000–10,000 individuals.
[b]Imperiled populations have fewer than 3000 individuals.
[c]Possibly extinct or presumed extinct = known historical occurrences but no current occurrences.
Source: Stein, 2001.

ply because we have much less information about insects than we have about vertebrates.

In the United States, fewer than one percent of all known species are possibly extinct; almost half of these are freshwater mussels, amphibians, freshwater fish, and birds. However, 30 percent of all species found here are extremely vulnerable to extinction, especially flowering plants and freshwater fishes (Table 8.2). Regionally, the greatest number of species at risk are in California and Hawaii (Fig. 8.1),

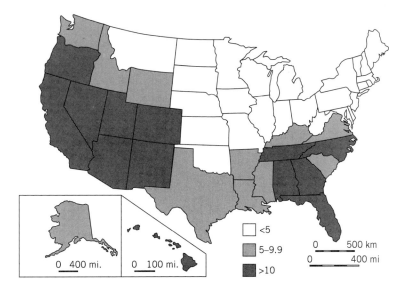

Figure 8.1 Percentage of species at risk, 2000. The highest rates are found in the West, Hawaii, and in portions of the Southeast, while the lowest percentages are found in the Great Plains and Northeast regions. *Source:* Stein, 2001.

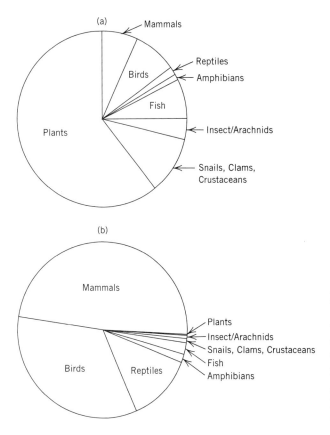

Figure 8.2 Threatened and endangered species, 2000. Plants and mammals make up the largest percentages of threatened and endangered species. However, when comparing the United States (a) to the rest of the world (b), mammals, birds, and reptiles are more at risk globally, while fishes, crustaceans, insects, and plants are more vulnerable in the United States. *Source:* U.S. Bureau of Census, 2000.

states that also have a large number of species and species endemic to the state. The lowest rates are in the upper Great Plains.

In the United States, we make a distinction between *endangered species*, which are defined as those in danger of becoming extinct throughout all or a significant part of their natural ranges, and *threatened species*, which are those species likely to become endangered in the near future. As of 2000, the U.S. Fish and Wildlife Service listed a total of 932 species as endangered in the United States only, with an additional 519 so listed for foreign areas (Fig. 8.2). Another 267 species were listed as threatened in the United States only, and 37 for foreign areas (U.S. Bureau of the Census 2000). Freshwater invertebrates and plants are among the most hard-hit groups.

Globally, the United States has the greatest number of threatened species (997), mostly invertebrates and mollusks (Table 8.3). Malaysia has the next highest, the majority of which are plants (85 percent), followed by Indonesia, where nearly 20 percent of that nation's threatened species are mammals. Brazil leads the world in the number of threatened bird species, while Australia has the most reptile and amphibian threatened species.

Even though a species may not be extinct, zoologists worry about the consequences of inbreeding among the relatively few members of a small population. A small group of animals may not be enough for a breeding population. Within a few generations of inbreeding, negative recessive traits may become prevalent, and the species can die out, a victim of its own genetic weaknesses. For example, some of California's rare Tule elk have short lower jaws, which makes eating difficult. This may be the result of breeding within the small group of animals that biologists used to establish the herd. Today researchers advocate a more sophisticated use of genetics when attempting to reestablish species. For example, "embryo banks" preserve the frozen genes of some vanishing species. Another solution is to use more animals for the ini-

Table 8.3 Threatened Species, by Taxonomic Group and Location

Taxonomic Group	Top Three Countries
Mammals	Indonesia India China
Birds	Brazil Indonesia China
Reptiles	Australia China Indonesia
Amphibians	Australia United States Philippines
Fishes	United States Mexico Indonesia
Mollusks	United States Australia Portugal
Other invertebrates	United States Australia South Africa
Plants	Malaysia Indonesia Brazil
Total	United States Malaysia Indonesia

Source: IUCN, 2000.

tial breeding population; however, there is just not enough room in wildlife refuges to maintain larger populations. Fewer than 5 percent of the world's preserves have the space for a genetically diverse breeding population of large wild mammals. It is probable that, in a crowded world, species survival will depend on human genetic technology.

Causes of Biodiversity Loss

Many different factors play a role in decreasing numbers of individuals and species, but nearly all of them are related to human impacts. Some of these impacts are direct, as through hunting, while others are indirect, caused by habitat modification or introduction of foreign species (Fig. 8.3). More specific causes operate at local to regional scales and include pollution (oil spills and nitrogen deposition), diseases and parasites, human consumption patterns and trade, and global warming (especially increased ocean temperatures, which contribute to declining coral stocks) (UNEP 2002).

One study of the known causes of animal extinctions in the past 400 years showed that 23 percent were caused by hunting, 36 percent by habitat destruction, and 39 percent by species introductions (World Conservation Monitoring Centre 1992). Although these figures are only rough estimates, they do indicate that each of these three processes is significant. In the following paragraphs we examine them in more detail.

Habitat Modification For thousands of years, humans have been altering animal and plant species and the places they inhabit, or *habitats*, and life forms have been affected correspondingly. Concern about human impacts on the biosphere has been mounting in the Western world since at least the nineteenth century. More recently, scientists have used the term *criticality* to characterize environmental zones where human activity has so severely degraded the

Figure 8.3 Harvesting old-growth timber, such as this in the Hoh forest of Washington, threatens species such as the spotted owl and the marbled murrelet. Intense controversy continues in both the United States and Canada over the ecological impacts of logging old-growth forests in the Pacific Northwest.

Figure 8.4 The Aral Sea, a vast inland water body that has shrunken because runoff has been diverted to irrigation use, is one of the more dramatic examples of human impacts on the hydrologic cycle.

natural environment that economic activity and human habitation are virtually impossible. Furthermore, the likelihood of environmental restoration of these regions to their former condition is almost hopeless (Kasperson et al. 1996). A good example of a critical environmental zone is the Aral Sea. Located in semiarid central Asia, the Aral Sea was the world's fifth largest freshwater lake (Fig. 8.4). Large-scale irrigation to produce cotton diverted the sea's main feeding streams. The Aral Sea has lost one-half of its surface and 70 percent of its volume since the 1950s. Fishing, once a thriving industry along the seashore, is no longer viable as salinization has killed most of the fish species. Evaporative salts blow into the agricultural areas, and pesticide and fertilizer use has reduced soil fertility to such an extent that even agriculture is now threatened.

While not formally designated as critical environmental zones, large parts of the world have been altered by human activity. For the world land areas as a whole, 26 percent is estimated to be subject to severe levels of human degradation, 20 percent to moderate levels, and 18 percent to light levels (Table 8.4) (FAO 2001). The causes include intensive agriculture, overgrazing, deforestation that led to increased water and wind erosion, and chemical and physical deterioration of the soil. At the regional level, the greatest amount of degraded land is found in areas of high population density, especially Europe and Asia (Fig. 8.5). In Europe about two-thirds of the land area is subject to moderate to severe degradation. Extensive areas of little-disturbed land are found in Africa (principally the Sahara), Russia (Siberia), and North America. Most of these relatively undisturbed habitats are areas of low biological activity because they are either deserts or very cold regions.

The biological consequences of past land-use changes are dramatic. The European lion was extinct by A.D. 80; wolves vanished along with Europe's forest cover. Similarly, wolves and bears were driven from the eastern United States in the eighteenth and nineteenth centuries by a combination of habitat loss and hunting. As the extent of forests grew in the past century, so too did the range of the bear and the wolf. These species are

Table 8.4 Continental Averages of Degraded Land

	Percentage of Land per Degradation Category				
Continental Area	None	Light	Moderate	Severe	Extremely Severe
Sub-Saharan Africa	34	24	18	15	10
Asia/Pacific	28	12	32	22	7
Europe	9	21	22	36	12
North Africa/Near East	30	17	19	27	7
North America	51	16	16	16	0
North Asia (east of Urals)	53	14	12	17	4
South/Central America	23	27	23	22	5
World	34	18	20	20	6

Source: FAO, 2001.

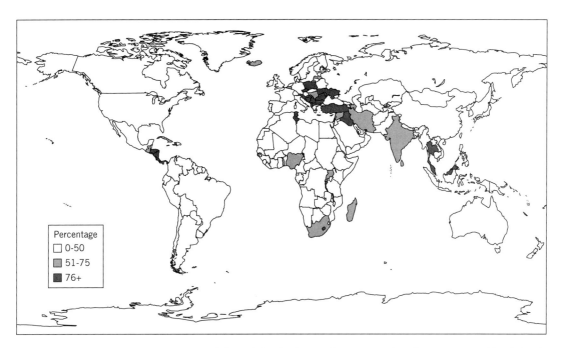

Figure 8.5 Extent of human-induced land degradation in the world, represented as the percentage of land classified as severe and/or very severe. *Source:* FAO, 2001.

growing in numbers today, sometimes generating new conflicts with humans.

The American bison, or buffalo, whose vast herds impeded agricultural settlement and held up trains for hours, was almost wiped out in the second half of the nineteenth century, through a combination of hunting and habitat loss. Today, the American bison lives in protected refuge areas where populations are stable or growing. Often, the populations become so large that they spill over into private lands such as those surrounding Yellowstone National Park. The number of bison being herded commercially for meat production is also increasing, as ranchers in the northern Great Plains learn that the buffalo's natural adaptation to that environment means that they can be produced with lower costs than domestic cattle.

The greatest concern about habitat loss today is focused on the tropical forests. As discussed in the previous chapter, the world's rainforests are being cleared at a rate approaching 1 percent per year, and perhaps 55 percent of the total has already been cleared.

Species Introductions The human role is not always one of destruction; often we are responsible for the swift spread of species to new areas. It is estimated, for example, that one-eighth of California's plant species are exotic species—that is, imported from other places. The dominance of introduced species is particularly severe on islands, which typically have low natural species diversity because of their remoteness from other land. A recent study of Hawaii, for example, found a total of 21,368 species of organisms of all types present on the islands. Of these, about 41 percent are endemic to Hawaii. Twenty percent are known to have been introduced, and the remaining 49 percent are of unknown status (Mlot 1995). The total above excludes about 10,000 nonnative commercial species used in agriculture or as ornamental plants. If these were included, then a minimum of 45 percent of the islands' species would be nonnative.

Such a high density of foreign species can make it difficult for native species, which are not adapted to such competition, to withstand extinction. For example, many bird species are estimated to have gone extinct as a result of the introduction of rats on these islands by Europeans. Hawaii has lost at least 18 species of birds since European settlement (Pimm et al. 1995).

In addition to introducing species that were not present before, humans have created habitats that favor some species over others. The suburbs offer a very comfortable habitat for deer, coyote, squirrel, and raccoon. The white-tailed deer population in the United States had dropped to 500,000 in 1900; with control over hunting, their numbers have risen to at least 13 million, at least as many as when Europeans first arrived. Today many people regard them as a nuisance species, eating crops and suburban shrubbery and creating a hazard on the roadways. Other introduced species that have transformed the landscape include kudzu (the South), eucalyptus trees (California), hydrilla (freshwater ecosystems), and zebra mussels (freshwater lakes) (Issue 8.1). These biological invaders are especially troublesome in Florida, where the state spends nearly $25 million a year to combat exotic flora in order to protect agriculture and tourism (Burdick 1995).

Hunting Hunting is a third major cause of species extinctions. Many large mammals became extinct in North America at the end of the Pleistocene glaciations, including the woolly mammoth and the giant sloth. Some of these extinctions are believed to have been a result of excessive hunting. Settlement of Pacific islands by Polynesian peoples is estimated to have resulted in the extinction of over 2000 bird species, or about 15 percent of the world total, within the past few thousand years.

In spite of the fact that much of the Amazon's forest cover remains relatively undisturbed, many animal species are already significantly affected by humans. Hunting is the principal culprit, being

---

ISSUE 8.1: THE MASS EXTINCTION OF FRESHWATER MUSSELS

Probably the most dramatic example of human-caused extinctions in North America is one that has been going on for decades, is well known to biologists and water-quality specialists, and yet receives very little public attention. It is occurring among freshwater mussels, a group of organisms that is particularly diverse in North America. Perhaps one of the reasons that the fate of mussels receives so little attention is that the problem is both widespread and long-standing. No critical environmental battles will be won or lost over mussels—their fate is already sealed by the land-use patterns that have become established in eastern North America in the past 300 years.

Freshwater mussels were once widespread in the rivers of eastern North America. They are filter feeders, drawing water through their bodies and filtering out fine particles of food—both dead organic matter such as leaves falling from streamside trees and living organisms such as microscopic plants and animals. Individual mussels may live several decades. They have developed a unique reproductive style, in which the larvae attach themselves to the gills or fins of fish and grow there for a few weeks before dropping to the stream bottom to grow and mature to adults. In the process, the mussels are spread up and downstream with the host fish. Many mussel species have become very specialized and require certain species of host fish for their reproduction.

North America has a wide variety of mussels. Roughly 300, or about one-third of all mussel species in the world, are native to North America. Most are found in the perennial streams and lakes of the eastern part of the continent, and half of them are in trouble. Of the 300 originally occurring here, about 20 are considered extinct and another 130 are either listed as threatened or proposed for listing (Cushman 1995). As a group, mussels are in worse shape than any other major type of animal.

Many factors have contributed to the decline of the mussels, all related to human activity and none easy to manage. When Europeans arrived in North America, they found many rivers teeming with mussels and quickly exploited the resource. Some were harvested for food, while others were collected for the shells. A major industry developed in the nineteenth century in the upper Mississippi and other areas, involving the use of mussel shells to make buttons and other mother-of-pearl objects; today most buttons are made

carried out both by indigenous peoples and colonizers. Commercial exploitation of Amazon fauna, including manatees, turtles, caiman, deer, peccary, otters, and various cats, primates, and birds, has increased substantially since European occupation. Several species have become locally extinct as a result (Redford 1992).

Another example is that of the black rhinoceros of Africa, whose population is dropping rapidly because of poaching—from 65,000 in 1970 to fewer than 4000 today. The animals are killed simply to obtain their horns, which are sold for thousands of dollars each and used as daggers and aphrodisiacs in Asia and the Middle East. In Namibia, wildlife authorities are actually dehorning the animals to save their lives. In Zimbabwe, poachers are shot and sometimes killed by wildlife authorities (Rees 1988). In these international examples, the situation is exacerbated by human population pressures on insufficient and marginal lands. When a family is pushed off their land to make way for a wildlife preserve, it is no wonder that they may be tempted to turn to poaching.

Sport hunting is also a problem in some areas, though less so now than in the past. The classic example in North America is the passenger pigeon. The total passenger pigeon population in 1810 was estimated at about 5 billion. During most of the nineteenth century, killing these birds for food or sport was easy, for they could be shot down in the hundreds by aiming into their roosting places at night or by firing at random as they flew overhead. The passenger pigeon was extinct in the wild by 1899, and the last one died in the Cincinnati, Ohio, zoo in 1914.

from plastic. Water quality is a much more pervasive problem. As filter feeders, mussels are exposed to large volumes of water and thus are very vulnerable to pollution. Both chemical pollutants and increased sediment caused by soil erosion have contributed to the mussel decline. Dams are another factor. They are barriers to migration of fish and mussels, fragmenting populations and limiting reproductive potential. If the host fish cannot pass a dam, then the portion of a river isolated by the dam can no longer receive introductions of mussels from beyond the dam. If a disturbance such as a flood or an episode of poor water quality should wipe out the species locally, then replenishment is not possible. Dams also create areas of slow-moving water and sediment deposition, which is detrimental to those species requiring gravel-bed streams and swift currents. Declines in certain host fish species may also contribute to the problem.

Finally, in recent decades the introduction of Zebra mussels has caused a new problem. Zebra mussels, a European freshwater species, were introduced to the Great Lakes in the 1970s and have since spread throughout the Great Lakes and the Mississippi and Ohio River systems. They are relatively small filter feeders that attach themselves to hard substrate such as rocks, piers, boat bottoms, and other mussels. Zebra mussels grow and reproduce rapidly and cause myriad problems wherever they occur. In Lake Erie, Zebra mussels have eliminated 90 percent of the native mussels, and similar problems have been identified in the Ohio River system (Nature Conservancy 1997; Neves 1997).

Because many of the factors contributing to mussels' decline are pervasive and fundamental to the way we use resources today—agricultural runoff and dams in particular—they will probably continue to suffer. They are also not a particularly glamorous type of animal and thus won't receive the attention that wolves, bears, or whales receive. Mussels have been called a "canary species." Like a canary in a mine, warning miners of bad air, they provide an early indicator of an unhealthy ecosystem. They are especially sensitive to a wide range of environmental impacts, and their high sensitivity is the reason they have been so hard-hit in recent decades. They are a sign of the widespread impact humans are having on the environment and, to some, a warning of future problems that should be avoided.

CONSERVATION OF BIODIVERSITY

What can be done to prevent this loss of biodiversity? Unfortunately, so much of the problem is caused by habitat loss and species introductions, our options are very limited. The main causes of habitat loss are the expansion of human settlement (especially agriculture) and deforestation. These processes cannot be reversed or significantly slowed in an era of expanding human population growth and migration. Similarly, while some species introductions are intentional, many more are inadvertent consequences of human travel and trade, which are not likely to be curtailed significantly in the near future. Several important initiatives have been undertaken, with both limited success and, in some cases, much controversy.

It is difficult to estimate the biological, social, and economic impact of biodiversity loss. Because of this uncertainty, the wisest course is to apply the "precautionary principle." This simply means we need to err on the side of reversing the trends in species extinctions and biodiversity reduction. If we don't, then the species are gone forever.

Species Protection

One approach to biodiversity conservation is to protect threatened and endangered species at the species level (Table 8.5). This includes actions such as prohibiting killing or trading in endangered species, efforts to preserve small amounts of habitat required to maintain a local population, and captive breeding programs.

The most significant program aimed at restricting hunting and trade in endangered species is the Convention on International Trade in Endangered Species of Wild Fauna and Flora (CITES). This international treaty regulates and controls commerce in endangered species and other species threatened by overharvest. The treaty negotiations began in 1973, with representatives from 80 countries. By 2001, it had been ratified by 157 countries, including the United States. CITES prohibits international trade in the most endangered species and their products and requires export licenses for some species and their products. Initially, only animal species were included, but beginning in 1992 commercially important timber species were added to the list of banned species. Enforcement is left up to the individual treaty nations and varies according to national motivation, economics, and ability. As a result, international trade in endangered species has increased, despite the efforts of CITES.

Most of the traded species originate in developing countries and are imported to markets in developed nations. The illegal wildlife trade is often as lucrative as illegal drug trafficking but without the risks. Products made of ivory and rhino horn and furs from South American ocelots and jaguars and from North American lynxes, bobcats, otters, and wolves are all protected under CITES, but the trade continues. Collectors of rare birds and animals, such as the South American macaw or the Asian cockatoo, pay up to $8000 for one of these endangered species, thus providing a market for the illegal trade.

The most notorious example of poaching is the insatiable demand for elephant-tusk ivory and the impact this demand has had on the African elephant, *Loxodonta africana*. In 1979, there were 1.5 million of these majestic beasts; today that number stands at around 500,000. It is estimated that a hunter can kill 200 to 300 of these animals daily, although in 1989 the collapsing market for ivory reduced elephant poaching by nearly 80 percent in most of Africa (World Resources Institute 1993). All the countries that ratified the CITES treaty have agreed to a certain legal quota of elephant ivory; however, it is estimated that 80 percent or more of ivory harvesting and trade takes place illegally. Nearly a decade ago, the United States announced a total ban on the importation of any African ivory, even via indirect routes such as Hong Kong. A number of African countries have also called for a total ban on the international ivory trade, and in Kenya, the message was made clear in a dramatic pyre in which tons of confiscated ivory were burned. In 1997, the ban on elephant ivory was partially lifted so that Zimbabwe, Botswana, and Namibia could sell their stockpiled ivory to Japan. That sale, completed in 1999, sent the equivalent of more than 5,000 ivory tusks to Japan for $5 million (Traffic 2001).

In addition to these efforts to protect species in the wild, about 200 species are being conserved in captivity (World Resource Institute 1994). Many of these have only a few dozen of individuals alive in the world. Some, like the California condor program, are attempting to reintroduce these species to the wild, with limited success (Pattee and Mesta 1995). Others are simply preserving the

Table 8.5 Protecting Wildlife by Law

Instrument	Purpose
United States	
The Lacey Act, 1900	Outlaws interstate trade of wildlife harvested or possessed against the laws of that state
Migratory Bird Treaty Act, 1918	Prohibits hunting or injury to migratory wild birds moving between the United States, Mexico, and Great Britain (now Canada)
Migratory Bird Conservation Act, 1929	Authorized purchase of new lands for waterfowl refuges.
Migratory Bird Hunting Stamp Act, 1934	Requires hunters (aged 16 years and over) to buy a federal waterfowl stamp prior to hunting migratory waterfowl
Pittman-Robertson Act, 1937	Raises money for state wildlife conservation programs by means of excise taxes on rifles, shotguns, ammunition, and archery equipment
Marine Mammal Protection Act, 1972	Bans the killing and importing of whales and most marine mammals; the moratorium can be waived for indigenous hunting and scientific takings if the current status of the species warrants it
Endangered Species Act, 1972	Provides federal protection to species designated as threatened or endangered
Fisheries Conservation and Management Act, 1976	Restricts foreign fishing in U.S. territorial waters; established regional fisheries managment councils to determine fisheries conservation and management policies
Major land-based international treaties	
Convention on International Trade in Endangered Species of Wild Fauna and Flora (CITES)	Regulates international trade and transit of certain animals, plants, their parts, and resulting products
Convention on Wetlands of International Importance, Especially as Waterfowl Habitat, 1971 (Ramsar)	Provides a framework for the conservation of wetlands and the designation of wetlands of international importance
Convention on the Conservation of Migratory Species of Wild Animals, 1979	Protects migratory wild animal species
North American Waterfowl Management Plan, 1986	Concluded cooperative agreement between the United States and Canada to restore sufficient wetland habitat to reestablish waterfowl populations to 1970 levels
Convention on Biological Diversity, 1993	Provides framework for international cooperation in conserving biological diversity

Source: World Resources Institute, 1993, 1994.

species in captivity with little hope of establishing a wild population in the near future. The bald eagle presents a different story. Habitat loss, hunting, and poisoning by pollutants (especially DDT) led to such a decline in the population that Congress passed protection plans in 1940 to save our national bird. Twenty years later, the number of these birds was still declining (with only 417 breeding pairs in the lower 48 states). By 1978, the bald eagle was officially listed as endangered. However, recovery was imminent, and by 1993 4000 nesting pairs were recorded (Council on Environmental Quality 1994), and a year later its status was changed from endangered to threatened. Current estimates of bald eagles hover around 50,000.

Such efforts can be extremely expensive. One extreme example is Florida's dusky sparrow. During the 1970s, the U.S. government spent over $2.5 million to buy 6.25 acres (2.5 ha) on Florida's east coast to create the St. John's River Refuge for the dusky sparrows. By 1981, there were five male sparrows living in a large cage (for their own safety), and a sixth male was believed to be alive in the wild. There were no known females in existence. How did this highly artificial situation develop?

The sparrows' original island habitat had been flooded to control mosquitoes around Cape Canaveral. Fires and drainage of marshes had further destroyed the birds' nesting and living area. Scientists proposed that the males be allowed to mate with a close relative, the Scott's seaside sparrow; after five generations, the offspring would be nearly full-blooded dusky sparrows. The suggestion was turned down in 1980, when the U.S. Fish and Wildlife Service decided that this hybrid sparrow would not meet the requirements of the Endangered Species Act. The agency instead gave a "pension" of $9200 per sparrow to care for them until their death.

The last surviving member, Orange Band, an aging male with gout, died in June 1988 in luxurious captivity. Using new techniques, his keepers studied his genetic makeup and found that he and the other duskies were not really a separate subspecies after all—at the genetic level, they were identical to the common seaside sparrow that lives in abundance along the Atlantic. The scientists suggest that nineteenth-century taxonomic classifications should be updated with twenty-first-century biotechnology when species are being kept alive at great cost, to make absolutely sure that these organisms are definitely entitled to their own separate grouping. Before he died, Orange Band was the father of two near-pure "duskies," who were released into the wild to prosper with their genetically identical relatives along the beach.

In contrast to superficially insignificant species such as a sparrow, the need to conserve potential agricultural species is much clearer. Humans depend on only a few dozen plant species for food. A substantial loss of any one of these crops in a given year would almost certainly lead to widespread human starvation. How might such a loss occur? Modern agricultural technology has led to greater uniformity in the world's crops. The seed sown in a field is genetically uniform, minimizing irregularities in the mature crop and making the plants easy to harvest by machine. Unfortunately, a genetically uniform crop also means that the individual plants are all equally vulnerable to attack from pests and diseases. If such a crop is exclusively planted over a wide geographic area, the food supply of an entire region could be drastically reduced in a very short period of time.

Crop species are also endangered by a decrease in their genetic diversity and by the disappearance of wild relatives (Cohen et al. 1991). For example, Mexico is the historical center for the domestication of corn and reknown for its diversity in varieties of corn. Currently, only 20 percent of the varieties of corn found in Mexico in the 1930s are available (Tuxill 1999). It is therefore necessary to maintain germ-plasm banks of the wild relatives of our principal crops. Agricultural experts can interbreed the positive characteristics of these plants, such as resistance to particular diseases, with the high productivity of the crop plants. In the event of an ecological disaster that eradicated the entire crop, we would have a well-preserved, genetically less vulnerable replacement to fall back on. The likelihood of such a large-scale disaster is exceedingly small, however, because experts have learned to provide seed with greater built-in diversity, following a near disastrous failure of the America corn crop in the early 1970s (National Academy of Sciences 1982). Modern crop plants are very sophisticated genetic packages.

Germ-plasm banks are also a refuge of last resort for threatened plant species. Even if the plant dies out, its genetic makeup is preserved in case of a need to revive the species. Unfortunately, this is impossible for the great numbers of species eradicated by the clearing of tropical rainforests, which are not catalogued or noted by the scientific community, much less collected and preserved.

A clearinghouse for germ-plasm collection and research is the National Plant Germplasm System (NPGS), managed by the U.S. Department of Agriculture. It has over 400,000 accessions, and new ones are added at a rate of 7000 to 15,000 a year. The NPGS collects, preserves, evaluates, and distributes U.S. and international plant germ-plasm resources. In addition, 33 U.S. botanical gardens have formed a network called the Center for Plant Conservation. When a rare plant's habitat is destroyed, a number of the plants are moved to these protected gardens. A self-perpetuating 50-plant collection of each species is then grown in the belief that this number of plants will hold most of a species' possible genetic variations. In addition, seeds of each species are stored in freezers, and a few are thawed every five years and tested for continuing germination viability.

Habitat Conservation

Another method of protecting species is to protect their habitats, those areas best suited to species' needs. The amount of land under some form of protection from development has grown dramatically in recent years and today accounts for about 6 percent of the world's land area (excluding Antarctica). A key part of this protection is the Biosphere Reserve Program of the United Nations Educational, Scientific, and Cultural Organization (UNESCO). The number of biosphere reserves worldwide has risen rapidly since the early 1970s, with the main objectives of conserving diverse and complete biotic communities, safeguarding genetic diversity for evolutionary and economic purposes, educating the public and training people in conservation, and providing areas for ecological and environmental research. In 2001, there were 411 biosphere reserves in 94 countries covering more than 264 million hectares, or less than 1 percent of the world's land area. Roughly two-thirds of these reserves are found in developing nations. To qualify as a biosphere reserve, an area must have outstanding, unusual, and complete ecosystems, with accompanying harmonious traditional human land uses. A reserve consists of a largely undisturbed core area surrounded by one or more buffer zones of human occupancy. Scientific research and training are carried on between the core and buffer zones, and local communities are encouraged to involve themselves in this approach to the preservation of older human and natural ecosystems (Fig. 8.6). At present, certain biomes such as mountains are well represented in the reserve system, whereas others, including tropical, subtropical, warm-arid, and intermediate areas, have very little protection (U.S. Man and the Biosphere Program 2001).

The United States has 47 officially designated biosphere reserves. One biosphere reserve is the Pinelands National Reserve (PNR) of New Jersey, which was designated a U.S. National Reserve in 1978 and made part of the international network in 1988. The PNR is also part of the U.S. Experimental Ecological Reserve network. What makes the Pinelands so ecologically valuable?

Often known as the Pine Barrens, this distinctive area covers about 990,000 acres (400,000 ha)

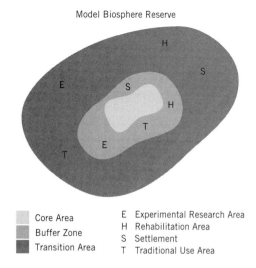

Figure 8.6 Biosphere reserve conceptual map. Biosphere reserves typically consist of a core area with strict protection to achieve conservation objectives; a buffer zone, where limited activities are permitted as long as they are compatible with conservation; and transitional areas, where sustainable resource management practices are developed.

of sandy soils on the coast and inland in south-central New Jersey (Fig. 8.7). Threatened on all sides and from within by accelerated development, the Pine Barrens supports a wide variety of plant and animal life in upland, aquatic, and wetland environments, including 39 species of mammals, 59 species of reptiles and amphibians, 91 species of fish, 299 species of birds, and over 800 different kinds of vascular plants. Of the 580 native plant species, 71 are in jeopardy. Over a hundred of these are at the northern or southern limits of their geographic range, creating a unique and irreplaceable mix of species (Good and Good 1984). The area was long used by Native American tribes, but human population numbers remained low until recent decades. However, pressure has grown to develop the Pine Barrens for residential, retirement, recreational, military, and commercial purposes. The layers of national and international protection will do a lot to ensure the continued integrity of this largely intact natural area.

Habitat protection has a long history in the United States. Theodore Roosevelt was the first President to propose the establishment of national wildlife refuges. During his presidency, the first national wildlife refuge was established in 1903 at

Figure 8.7 The Pinelands National Reserve of New Jersey is an island of protected, relatively undeveloped land in the intensely urbanized eastern United States.

Pelican Island, Florida, for herons and egrets. This was the beginning of the National Wildlife Refuge System, which is currently managed by the Fish and Wildlife Service of the U.S. Department of the Interior. The system provides sanctuaries for endangered and threatened species of plants and animals (Fig. 8.8).

The National Wildlife Refuge (NWR) system currently has a total of 494 units covering 92 million acres. Unfortunately, many of these refuges are severely polluted, to the extent that birds nesting in them are producing deformed or stillborn offspring. The most notorious of these is the Kesterson NWR in California, which has become a collecting basin for the chemical-laden runoff from surrounding farmlands. Other federal protection programs include the National Wilderness Preservation system (104 million acres) and the Wild and Scenic Rivers system (10,700 river miles). Estuarine Research Reserves contribute another 402,000 acres, and if we include our National Parks (80.3 million acres), then the nation's total habitat conservation programs protect more than 275 million acres. As shown in Table 8.6, the expansion of habitat protection has increased significantly since 1960. The federal government is not the only agency in the United States involved in the protection of wildlife habitat. There are many wildlife refuges in the form of state game preserves, as well as many private and public interest organizations, such as the Nature Conservancy, the Trust for Public Lands, and the Izaak Walton League, that purchase critical habitat lands and preserve them from encroachment.

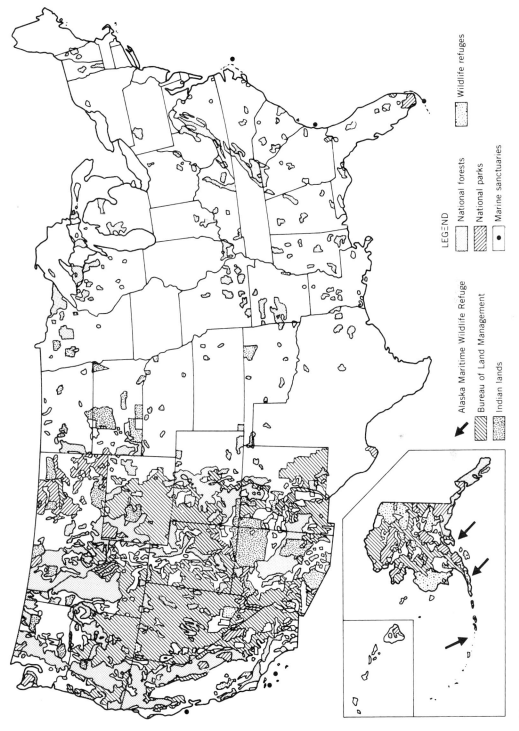

Figure 8.8 Public lands in the United States. Not all of these lands are completely protected from overexploitation. Some are managed for preservation goals and others for conservation goals. *Source*: Council on Environmental Quality, 1989.

Table 8.6 Habitat Protection in the United States, 1960–2002 (in Million Acres, Unless Otherwise Noted)

System	1960	1970	1980	1990	1995	2002
National Parks	26.2	29.6	77.0	80.1	83.2	83.6
National Wildlife Refuges	17.3	29.2	69.9	89.1	92.3	95.2
National Wilderness Areas	...[a]	10.4	79.7	95.0	103.7	105.8
National Wild and Scenic Rivers (miles)		...[b]	868	5,662	9,318	10,734–11,294
National Estuarine Research Reserves		...[c]	...[c]	0.22	0.26	0.4–1.02

[a]Did not exist until 1964.
[b]Did not exist until 1968.
[c]Did not exist until 1975.
Source: Council on Environmental Quality, 1994; updates from administrative agency web sites.

Despite the successes, degradation of species habitat sometimes continues inadvertently (Issue 8.2). The maintenance of biological diversity requires an assessment of existing conditions and the monitoring of trends. One such program is the *Gap Analysis Program (GAP)*, run by the National Biological Survey. With use of remote sensing and Geographic Information System (GIS) techniques, the GAP maps landcover, distribution of wildlife, and landownership patterns to identify potential "gaps" in protection, hence the name (LaRoe et al. 1995). A good example is the Biomap project in Massachusetts (Natural Heritage Program 2002), which provides an interactive web site. The GAP has been quite successful in identifying those areas most at risk by means of these geographical techniques and influencing land acquisition and conservation management programs at the local, state, and federal levels. It also demonstrates the value of a map and of a geographical perspective.

The Endangered Species Act

The United States has taken an active role in the protection of species. The most comprehensive piece of legislation regulating protection of all species of flora and fauna is the Endangered Species Act (ESA), passed in 1973. The ESA essentially bans acts such as hunting that directly affect threatened endangered species, and it also forbids government actions that would result in the loss of critical habitat for threatened or endangered species. The ESA is strong legislation that can significantly limit actions that might affect endangered species. Because of this strength, the law has been controversial ever since its passage.

The ESA requires that the Department of the Interior (in the case of nonmarine species) or the National Marine Fisheries Service (in the case of marine species) identify threatened or endangered species. It makes it illegal to capture, kill, sell, transport, buy, possess, import, or export any species on the endangered or threatened list. The act also requires the Departments of the Interior and Commerce to delineate the habitats of endangered and threatened species and to map these critical habitats, which are a prerequisite for species survival. It then forbids destruction of critical habitats as a result of dam-building, highway construction, housing developments, or other projects supported in whole or in part by federal monies. This final provision is the one that has generated the most controversy.

The controversy started immediately after passage of the act, when a small fish was identified that was believed to be threatened by completion of the Tellico Dam in Tennessee (Fig. 8.9). The Tellico Dam controversy ultimately resulted in amendments to the act in 1978, requiring closer consultation between government offices enforcing the ESA and sponsors of capital improvement projects to avoid long and costly disputes over the fates of both the endangered species and the construction project. The amendments also allowed some exceptions to comprehensive species protection. Specifically, a major project can go ahead if it can be shown clearly that the benefits of the project outweigh and overshadow the species preservation issue. Exceptions to the ESA can be granted by six high-ranking (cabinet and subcabinet) officials and one representative of the state af-

ISSUE 8.2: ECOTOURISM: LOVING WILD PLACES TO DEATH

Many people are eager to experience nature firsthand and often do so in their own backyards, public parks in their neighborhoods, or recreational areas in their communities. The more adventurous travel great distances to commune with nature, visiting remote and relatively unspoiled places in Central and South America or Africa. Some of the most popular destinations are the Galapagos National Park (Ecuador), Amboseli National Park (Kenya), Nepal, and Australia's Great Barrier Reef Marine Park.

Nature travel, or *ecotourism*, has become a big business, with many countries depending on revenues as a source of foreign currency. In Kenya, for example, ecotourism generates more than 30 percent of its foreign currency—more than the exports of coffee or tea, its primary agricultural crops.

The demand for tourism has nearly doubled since the 1970s. Now, annually there are more than 600 million arrivals of international tourists, many seeking ecotourist destinations. The irony is that once "discovered" as tourist attractions, these environments become threatened by too many visitors. Kenya's Amboseli Park, for example, receives more than 200,000 visitors a year, many of them on one-day excursions from Nairobi to look for the safari animals (elephant, lion, leopard, cape buffalo, and rhinoceros). In the United States' Yosemite Park, more than 3 million people visit annually, most of them concentrated in the Yosemite Valley. Traffic is so congested that the National Park Service restricts entrance to the Park during many summer weekends. In the 1990s, more than 275 million people visited our National Parks annually. The most popular were the Great Smoky Mountain, Grand Canyon, Yosemite, Yellowstone, Olympic, and Rocky Mountain parks, all with more than 2.5 million visitors each.

Nature-based tourism is big business, generating as much as $222 billion annually and growing. What has all this attention wrought? Wildlife in many of these protected areas are now threatened. Cheetahs in Amboseli Park, for example, have changed their hunting patterns from dawn to dusk in order to avoid the tourists. Development pressures abound in these areas as entrepreneurs seek to provide tourist-related facilities such as lodging, food establishments, and the like. In foreign areas, many of the cultural resources of the areas are lost, and local people are forced to either relocate or adapt to the new development.

The year 2002 was labeled the International Year of Ecotourism by the United Nations, a designation promoting the use of ecotourism as a tool for sustainable economic development. While the goals of ecotourism are laudable—to satisfy tourists while maintaining environmental protection and helping the local economy—it may be too much of a good thing. Although many tour operators and visitors are sensitive to environmental concerns, others are not. Many question whether ecotourism is, in fact, a sustainable form of development as the increasing numbers of visitors create tensions between conservation and economic development goals of these areas. Are we loving our National Parks and other wild places to death? (Boo 1990; World Resources Institute 2000).

---

fected. Environmentalists have named this group the God Committee.

When it came into office in 1981, the Reagan administration was not interested in enforcing the endangered species legislation. The Secretary of the Interior resisted the inclusion of new species on the endangered and threatened lists, even though the Fish and Wildlife Service had identified 2000 species that were eligible for listing. Amendments to the ESA in October 1982 stated that the Department of the Interior would consider only biological factors, not habitat destruction, in evaluating a species for listing. This resulted in a precipitous drop in the number of species considered for protection (Bean 1983). Nonetheless, in the summer of 1988 the U.S. Senate renewed the Endangered Species Act by the largest margin ever.

Renewal of the act did not end the controversy, however. New debates arose, this time with terrestrial habitats as the focus. Old-growth forest was dwindling in the Pacific Northwest (see Chapter 7), and two endangered species of birds that depend

Figure 8.9 The Tellico Dam, Tennessee.

on old-growth forest habitat were identified: the spotted owl and the marbled murrelet. Environmentalists made the case that continued logging of old-growth timber in the National Forests constituted a violation of the ESA, and the courts agreed, forcing a curtailment of logging activities. Elsewhere in the nation, similar conflicts emerged over such issues as suburban development in the habitat of the California red-legged frog; logging old-growth timber in the Southwest, home of the Mexican spotted owl; expanding populations of wolves and grizzly bears in the northern Rockies, and preservation of habitat for the red-cockaded woodpecker in the Southeast. When the Republicans gained control of Congress in 1994, the ESA was targeted for substantial reform, particularly with the aim of restoring the rights of property owners to manage (develop, clear-cut) their land without government interference. In April 1995, Congress imposed a moratorium on new species listings while it debated reforms. The moratorium was finally lifted by a court order in May 1996. Meanwhile, recovery programs continue with species reintroductions and partnerships with local landowners to develop recovery plans (Scott 2000). Despite some notable successes such as the proposed delisting of the bald eagle in 1999 and the proposed reclassification of the gray wolf to a threatened species in 2001, the ESA remains controversial to all stakeholders. As a result, the Endangered Species Act is one of the most polarizing pieces of legislation ever passed by Congress, and it continues to create an impasse on what to do next (Bean 1999).

The Convention on Biological Diversity

The Convention on Biological Diversity is an international treaty aimed at promoting biodiversity conservation worldwide. It was developed in a series of negotiations leading up to the 1992 United Nations Conference on Environment and Development in Rio de Janeiro (see Chapter 3) and was signed by most of the nations of the world at that conference. The United States initially refused to sign the treaty, but after Clinton's election the United States signed. By 2001, 168 nations had ratified the treaty, which went into effect in 1993. The United States has not ratified the treaty.

The Convention is more an agreement to work toward biodiversity conservation than it is a commitment to binding action; nonetheless, it represents a major step forward in this area. It includes provisions obligating member nations to do the following:

- cooperate with other nations in conserving biodiversity,
- develop plans and programs to conserve biodiversity,
- establish systems of protected areas,
- provide financial support for in situ conservation,
- include biodiversity conservation in national decision-making, and
- establish programs for scientific and technical education and training for biodiversity conservation.

One contentious issue surrounding the treaty was the question of how much money wealthy countries would provide for implementation of the treaty in poorer nations and how these funds would be controlled. But the issue that generated the most controversy concerned control of genetic resources and the profits to be earned from their use. The treaty establishes national sovereignty over genetic resources contained within a nation's borders, prohibiting collection and export of specimens without a government's permission. In addition to establishing this national ownership of genetic material, the treaty calls for unrestricted international transfer of technologies utilizing genetic resources. This would mean, for example, that if a company developed new seed varieties or medicines, these technologies should be made freely available worldwide rather than having the intellectual property rights surrounding the technology held by the developing country. The United States and other wealthy nations with companies involved in biotechnology are concerned about protecting intellectual property rights and argue that genetic materials and the technologies that make use of them should be privately owned and distributed through market mechanisms. This would ensure that those developing these technologies would have the incentive to do so and would also provide for payment of royalties to the countries where genetic resources originate.

Another point of controversy surrounding the Biodiversity Convention is the mechanism for funding biodiversity conservation measures. The convention states clearly that the wealthy nations of the world should be responsible for the lion's share of the funding responsibility. Most wealthy nations have accepted this statement in principle, but debate continues over how much control the donor nations should have over how the money is spent, as opposed to the recipient countries responsible for evaluating biodiversity and developing and enforcing measures to conserve it (Raustiala and Victor 1996).

CONCLUSIONS

One question that frequently appears in the biodiversity debate is whether the central concern is preserving species or preserving habitat. The simple answer is to do both! With the exception of a few animals large and interesting enough to be preserved in zoos, most plant and animal species will not survive unless their native habitats are also preserved. If the entire land area of the world is converted to significantly altered ecosystems such as cities, farms, pastures, and plantation forests, many species will indeed become extinct. We cannot preserve species without also preserving habitat. What remains to be resolved, however, is just how much habitat will be preserved. Substantial areas were set aside during the past decade for biodiversity and habitat conservation, and additional areas came under management that would allow limited development but still preserve some habitat. But in much of the world, the United States included, pressures are mounting to limit the extent of such set-aside land and to allow continued or expanded development and resource extraction on the remaining wild lands. For the most part, this debate will take place at the national level, as it is in the United States through consideration of the Endangered Species Act.

A second key issue is the balance between the desire to maintain biodiversity and the need to maintain or improve standards of living, especially among the world's poor. Without question, biodiversity conservation is primarily a concern of the rich, who have the luxury of being able to worry about preserving species in which they have no material interests. In the United States and other wealthy nations, where public sentiment for biodiversity conservation is strong, mandating preservation of not only individual species but also the ecosystems that support them and other unknown species depends on the political will to legislate such protection. To be sure, such measures do have some economic disadvantages—loggers may become unemployed or landowners may be unable to reap all the potential value from their property. But these people usually have other opportunities. For example, in Washington and Oregon, where the ESA's impacts on the logging industry are of greatest concern, unemployment has generally been below the national average, largely because of growth in the high-technology industries. But in Brazil, or Congo, or India, preservation of the remaining undisturbed or little-disturbed habitats has relatively little weight in comparison to meeting the basic needs of millions of people. Thus, when a large hydroelectric project or forest harvesting scheme is proposed for such areas, the environmental impacts are usually discussed at the international level, while in the areas directly affected the debates usually center around who will gain and who will lose economically or politically. The extreme contrasts in wealth between rich and poor nations will continue to make international cooperation on biodiversity difficult.

REFERENCES AND ADDITIONAL READING

Bean, M. J. 1983. Endangered species: The illusion of stewardship. *National Parks*, July/August, pp. 20–21.

———. 1999. Endangered species, endangered act? *Environment* 41(1):12–18, 34–38.

Bisby, F. A. 2000. The quiet revolution: biodiversity informatics and the internet. *Science* 289:2309–2312.

Boo, E. 1990. *Ecotourism: The Potentials and Pitfalls.* New York: World Wildlife Fund.

Burdick, A. 1995. Attack of the aliens: Florida tangles with invasive species. *New York Times*, June 6, p. B8.

Cairns, J., Jr. 1995. Ecosocietal restoration: reestablishing humanity's relationship with natural systems. *Environment* 37(5):4–9, 30–33.

Cohen, J. I., J. T. Williams, D. L. Plucknett, and H. Shands. 1991. Ex situ conservation of plant genetic resources: Global development and environmental concerns. *Science* 253:866–872.

Consultative Ground on International Agricultural Resources (CGIAR). 2001. http://www.cgiar.org.

Council on Environmental Quality. 1989. *Environmental Trends.* Washington, D.C.: U.S. Government Printing Office.

———. 1994. *Environmental Quality.* Twenty-Fourth Annual Report. Washington, D.C.: U.S. Government Printing Office.

Cox, P. A. and M. J. Balick. 1994. The ethnobotanical approach to drug discovery. *Scientific American* 270(6):82–87.

Crosson, P. D. and N. J. Rosenberg. 1989. Strategies for agriculture. *Scientific American* 261(3):128–135.

Culotta, E. 1995. Bringing back the Everglades. *Science* 268:1688–1690.

Cushman, J. H. 1995. Freshwater mussels facing mass extinction. *The New York Times*, October 3, p. B5.

Daily, G. C. 1995. Restoring value to the world's degraded lands. *Science* 269:350–353.

———, ed. 1997. *Nature's Services: Societal Dependence on Natural Ecosystems.* Washington, D.C.: Island Press.

Eckholm, E. 1978. *Disappearing Species: The Social Challenge.* Worldwatch Paper no. 22. Washington, D.C.: Worldwatch Institute.

Ehrenfeld, D. W. 1981. *The Arrogance of Humanism.* Oxford: Oxford University Press.

Ehrlich, P. and A. Ehrlich. 1970. *Population, Resources, and Environment: Issues in Human Ecology.* San Francisco: W. H. Freeman.

———. 1981. *Extinction: The Causes and Consequences of the Disappearance of Species.* New York: Ballantine Books.

Enserink, M. 1999. Biological invaders sweep in. *Science* 285:1834–1836.

Food and Agricultural Organization (FAO) 2001. Terrastat. http://www.fao.org/waicent/faoinfo/agricult/agl/agll/terrastat.

Good, R. E. and N. F. Good. 1984. The Pinelands National Reserve: An ecosystem approach to management. *Bioscience* 34:169–173.

Gupta, A. 2000. Governing trade in genetically modified organism. *Environment* 42(4):22–33.

International Union for the Conservation of Nature and Natural Resources (IUCN). 2001. http://www.iucn.org/redlist/2000/index.html.

Kaiser, J. 1999. Stemming the tide of invading species. *Science* 285:1836–1841.

Kasperson, J. X., R. E. Kasperson, and B. L. Turner II. 1996. Regions at risk: Exploring environmental criticality. *Environment* 38(10):4–15, 26–29.

LaRoe, E. T., G. S. Farris, C. E. Puckett, P. D. Doran, and M. J. Mac, eds. 1995. *Our Living Resources: A Report to the Nation on the Distribution, Abundance, and Health of U.S. Plants, Animals, and Ecosystems.* Washington, D.C.: U.S. Department of the Interior, National Biological Survey.

Lemonick, M. D. 1994. Winged victory. *Time*, July 11, p. 53.

Luoma, J. R. 1989. Prophet of the prairie. *Audubon* 91(6):54–60.

MacDonald, G. 2003. *Biogeography: Introduction to Space, Time, and Life.* New York: John Wiley and Sons.

Mattoon, A. 2001. Deciphering Amphibia declines. In L. R. Brown et al., eds. *State of the World 2001.* Washington, D.C.: Worldwatch Institute, pp. 63–82.

Mlot, C. 1995. In Hawaii, taking inventory of a biological hot spot. *Science* 269:322–323.

Moffat, A. S. 1996. Biodiversity is a boon to ecosystems, not species. *Science* 271:1497.

Mooney, H. A. and R. J. Hobbs. 2000. *Invasive Species in a Changing World*. Washington, D.C.: Island Press.

Myers, N. 1993. Tropical forest: The main deforestation fronts. *Environmental Conservation* 20:9–16.

National Academy of Sciences. 1982. *Genetic Vulnerability of Major Crops*. Washington, D.C.: National Academy of Sciences.

Nature Conservancy. 1997. Vermont's freshwater mussels: Uniqueness and diversity, now under siege. http://www.tnc.org/infield/State/Vermont/science/mussels.htm.

Natural Heritage Program, State of Massachusetts. 2002. www.state.ma.us/dfwele/dfw/nhesp/nhbiomap.htm.

Neves, R. 1997. *Partnerships for Ohio River Mussels*. U.S. Fisheries and Wildlife Service. http://www.fws.gov/r3pao/eco_serv/endangrd/news/ohio_rvr.html.

Pattee, O. H. and R. Mesta. 1995. California condors. In LaRoe, E. T., G. S. Farris, C. E. Puckett, P. D. Doran, and M. J. Mac, eds. *Our Living Resources: A Report to the Nation on the Distribution, Abundance, and Health of U.S. Plants, Animals, and Ecosystems*. Washington, D.C.: U.S. Department of the Interior, National Biological Survey, pp. 80–81.

Pimentel, D., L. Westra, and R. F. Noss, eds. 2001. *Ecological Integrity: Integrating Environment, Conservation, and Health*. Washington, D.C.: Island Press.

Pimm, S. L., G. J. Russell, J. L. Gittleman, and T. M. Brooks. 1995. The future of biodiversity. *Science* 269:347–350.

Pimm, S. L. et al. 2001. Can we defy nature's end? *Science* 293:2207–2208.

Raustiala, K. and D. G. Victor. 1996. Biodiversity since Rio: The future of the Convention on Biological Diversity. *Environment* 38(4):16–20, 37–45.

Redford, K. H. 1992. The empty forest. *Bioscience* 42:412–422.

Reaka-Kudla, M. L., D. E. Wilson, and E. O. Wilson, eds. 1996. *Biodiversity II: Understanding and Protecting Our Biological Resources*. Washington, D.C.: Joseph Henry Press.

Reid, W. V. 1995. Biodiversity and health: prescription for progress. *Environment* 37(6):12–15, 35–39.

_____. 2001. Biodiversity, ecosystem change, and international development. *Environment* 43(3):20–26.

Sale, K. 1985. *Dwellers in the Land, the Bioregional Vision*. San Francisco: Sierra Club Books.

Scott, C. 2000. Restoring our wildlife legacy. *Endangered Species Bulletin* 25(3):4–7. http://endangered.fws.gov.

Stein, B. A. and S. R. Flack. 1997. Conservation priorities: The state of U.S. plants and animals. *Environment* 39(4):6–11, 34–39.

Stein, B. A., L. S. Kutner, and J. S. Adams, eds. 2000. *Precious Heritage: the Status of Biodiversity in the United States*. New York: Oxford University Press.

Stein, B. A. 2001. A fragile cornucopia: assessing the status of U.S. biodiversity. *Environment* 43(7):10–22.

Tilman, D. 1996. Biodiversity: Population versus ecosystem stability. *Ecology* 77:350–363.

Tolba, M. K., O. A. El-Kholy, E. El-Hinnawi, M. W. Holdgate, D. F. McMichael, and R. E. Munn. 1992. *The World Environment 1972–1992: Two Decades of Challenge*. New York: Chapman and Hall.

Traffic in Illegal Wildlife. 2001. http://www.traffic.org/briefings/elephants-11thmeeting.html.

Tuxill, J. 1999. *Nature's Cornucopia: Our Stake in Plant Diversity. Worldwatch Paper #148*. Washington, D.C.: Worldwatch Institute.

United Nations Environmental Program. 1995. *Global Biodiversity Assessment*. Cambridge: Cambridge University Press.

_____. 2002. *Global Environmental Outlook 3*. New York: Oxford University Press.

U.S. Bureau of the Census. 2000. *Statistical Abstract of the United States*. http://www.census.gov.

U.S. Man and the Biosphere Program. 2001. http://www.unesco.org/mab/.

Vitousek, P. M., H. A. Mooney, J. Lubchenco, and J. M. Melillo. 1997. Human domination of Earth's ecosystems. *Science* 277:494–499.

Wedin, D. A. and D. Tilman. 1996. Influence of nitrogen loading and species composition on the carbon balance of grasslands. *Science* 274:1720–1723.

Wilson, E. O. ed. 1988. *Biodiversity*. Washington, D.C.: National Academy Press.

_____. 1989. Threats to biodiversity. *Scientific American* 261(3): 108–117.

_____. 1992. Biodiversity: Challenge, science, opportunity. *American Zoologist* 34:5–11.

_____. 1992. *The Diversity of Life*. Cambridge, MA: Bellknap Press of Harvard University Press.

World Conservation Monitoring Centre. 1992. *Global Diversity: Status of the Earth's Living Resources*. London: Chapman and Hall.

World Resources Institute. 1993. *The 1993 Information Please Environmental Almanac*. Boston: Houghton Mifflin.

_____. 1994. *World Resources 1994–95*. New York: Oxford University Press.

_____. 1996. *World Resources 1996–97*. New York: Oxford University Press.

_____. 2000. *World Resources 2000–2001*. Washington, D.C.: World Resources Institute.

For more information, consult our web page at *http://www.wiley.com/college/cutter*.

STUDY QUESTIONS

1. Is a native species more important than a nonnative species? Why?
2. Find out what species are listed as threatened or endangered where you live. What kind of habitat do they need to survive? What areas of this habitat remain? What are the major threats to this habitat?
3. For each of the three main causes of extinction—habitat loss, hunting, and species invasions—identify a species in your area that has been affected by each.
4. The eastern United States was deforested between 1600 and 1900, with unknown loss of biodiversity. Does this history mean that we have no right to criticize other countries causing deforestation today? Why or why not?
5. Should it be illegal for a private property owner to alter habitat in such a way as to endanger a species? Should a property owner be compensated by the government for lost income caused by such a restriction?

CHAPTER 9

MARINE RESOURCES: COMMON PROPERTY DILEMMAS

INTRODUCTION

No one nation owns the world's oceans or controls the resources found in them. The oceans, then, are a *common property resource*. Common property resources cannot be managed by a single individual, nation, or corporation because without some form of governmental or international regulation to allocate resources among users, individuals have little incentive to preserve or protect resources for future generations (Chapter 2). Historically, those nations that could exploit the world's marine resources, such as oil, fish, whales, and minerals, simply did so.

Although we usually think of the earth and its resources in terms of land area, about 71 percent of the earth's surface is covered by water, most of it in the oceans. Earth is a water planet. Virtually all living and nonliving resources are somehow influenced by the oceans.

The living and nonliving resources of the sea have slightly different characteristics than those found on land. First, they are often unseen and thus unmeasurable and uncountable. It is impossible, for example, for a fisheries biologist to know exactly how many fish there are in a given ocean area. It is also difficult to know the size of an oil field in deep water offshore, for exploration technology used on land will not work in the marine environment. Second, the oceans are the ultimate diffuser and therefore the ultimate pollution sink. Oceanic pollutants, for example, can travel immense distances, confounding attempts to identify and regulate the polluter.

Finally, despite a number of international treaties, the question of who owns the majority of the oceans and the resources found within them is still unanswered. On land, governments and individuals claim, occupy, and defend areas based on legally binding boundaries, with use of easily recognized geographic features. Ownership of the oceans is less clear and depends on the current use of the ocean area or the political, technological, or military power of a country or private corporation. For example, U.S. companies seeking to mine deep ocean minerals cannot obtain commercial financing until legal ownership of sections of the deep ocean bottom is established, either by international treaty or by unilateral action by the U.S. government. Such disputes are common, and resource managers frequently focus on who should have access to ocean resources rather than on how those resources should be allocated and used.

THE MARINE ENVIRONMENT

Physical Properties
The physical properties of seawater, the rotation of the earth, and the hydrologic cycle shape the distribution of marine resources and control the ocean's impact on terrestrial ecosystems. We will discuss three important properties of seawater—salinity, temperature, and dissolved oxygen. Salinity and temperature are especially important in determining the circulation of the oceans, while dissolved oxygen is necessary for animals to survive in the ocean.

Salinity Seawater is a solution of minerals and salts of nearly constant composition throughout the world. Sea salts, a product of billions of years of terrestrial erosion, contain at least traces of most elements found in the earth's crust. Six elements, however, comprise more than 98 percent of all sea salts (Table 9.1). On the average, a kilogram of seawater contains 35 grams of salt, or 35 parts per thousand (ppt). These salts are dissolved in variable amounts of water, and slight differences in the *salinity* of seawater can influence the speed and direction of ocean currents and the vertical mixing of surface and bottom waters.

Salinity change may also have a major impact on the ocean's living resources. For example, it governs the spawning time of oysters and other shellfish on the east coast of the United States and the shrimp migrations in the Gulf of Mexico. Juvenile shrimp can tolerate the wide-ranging salinities (0–25 ppt) found in coastal areas; adult shrimp can survive only in ocean waters of 35 ppt salinity. Thus, the success of the shrimp fishing season is largely dictated by rainfall and freshwater river discharge, which dilute seawater and thus affect salinity.

As salinity changes, the density of seawater also changes. Fresh water, or low-salinity water, will float on top of heavier, saltier water to create *stratified estuaries* in coastal areas and *haloclines* in the open ocean. Such stratification can complicate efforts to protect shellfish beds and to monitor pollutants and can even threaten public drinking-water supplies. For example, at the mouths of rivers, freshwater flows over denser seawater, and a wedge of salt water usually underlies freshwater at the surface. At the mouth of the Delaware River, near Philadelphia, the movement of the salt wedge is dictated by the volume of freshwater flow in the Delaware River. If river flow is low, salty ocean water creeps up Delaware Bay, threatening Philadelphia's drinking-water intake. Conversely, seasonal high flows of fresh water lower the salinity in the oyster beds downstream of Wilmington, Delaware, discouraging the spread of oyster parasites and predatory oyster drills. The size and movement of the salt wedge in this partially stratified estuary affects everything from commercial fishing to the drinking-water supplies for over 3 million people.

Temperature Water temperature and water temperature gradients (*thermoclines*) are also physical aspects of the ocean environment that influence the conservation and management of marine resources. The worldwide distribution of the ocean's surface-water temperature depends on the general supply of heat available from the sun. Surface temperature is highest at the equator and declines northward and southward, toward the poles. Total heat loss from the ocean waters (as opposed to the temperature of the water itself) also declines as one moves away from the equator, but not at the same rate. The difference between a surplus of heat at the equator and relatively little elsewhere results in the global heat-transfer mechanisms (air and water currents) that shape our weather. The oceans, then, can be viewed as a giant weather machine. The major ocean currents in the world are illustrated in Figure 9.1.

A change in ocean water temperatures can have a worldwide impact. A phenomenon called the Southern Oscillation describes the interannual fluctuation between warm *El Niño* conditions and cold *La Niña* ones. In the late fall of each year, a warm current, which local fishermen call El Niño, develops along the coasts of Ecuador and Peru. At irregular intervals, a much larger ocean warming occurs at the same time of year along the same coast but stretching westward along the equator, two-thirds of the way across the Pacific Ocean. This large-scale warming completely reverses the wind and current systems of the Pacific Ocean, influencing worldwide weather patterns and causing rare winter/spring hurricanes, floods, and droughts. La Niña has the opposite effect in that warm surface waters are driven westward, thus drawing cold water to the surface in the east.

Finally, the oceans act as a heat sink, absorbing much of the global warming during the past century. The average sea surface tempertaure has risen by 1°F over the past 100 years, and during the past

Table 9.1 Composition of Dissolved Sea Salts in Seawater

Element	Percentage
Chlorine	55.0
Sulfur	7.7
Sodium	30.6
Magnesium	3.7
Potassium	0.7
Calcium	0.7
Minor elements (bromine, carbon, strontium)	1.6
Total	100.0

Source: Gross, 1971:57.

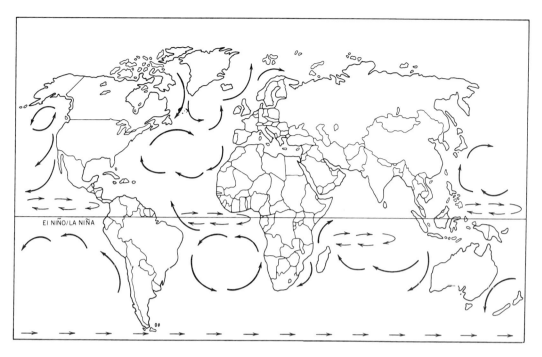

Figure 9.1 Major ocean currents of the world. The large ocean bodies have circular flow patterns, called *gyres*, which are clockwise in the Northern Hemisphere and counterclockwise in the Southern Hemisphere. Superimposed on this pattern are smaller currents, such as the equatorial countercurrents. Periodic disruptions of circulation, labeled El Niño/La Niña, occur in the eastern Pacific.

25 years, the speed of that surface warming has accelerated (Stevens 2000). This increase in temperature causes the seawater to expand slightly, causing sea level to rise. What is unclear is how this increase in ocean-stored heat will affect climate.

Variations in temperature and salinity also drive global-scale ocean circulation patterns. For example, in the North Atlantic, water cools and sinks, then flows southward, around Africa, across the Indian Ocean, and rises in the tropical western Pacific. This deep-water flow has a surface counterpart that returns water to the North Atlantic. Such flows are important in redistributing heat from one part of the planet to another. Changes in such currents over time are believed to play a role in climate variability.

Dissolved Oxygen The last important physical feature of the oceans that affects marine conservation and management is *dissolved oxygen*, the total amount of oxygen present within a body of liquid, in this case water. Dissolved oxygen is absolutely essential for aquatic life.

The distribution of dissolved oxygen is controlled by exchanges with the atmosphere, photosynthesis of phytoplankton, and respiration of oxygen-consuming biota. The solubility of a gas such as oxygen is a function of water temperature: the lower the temperature, the more dissolved oxygen. The vertical distribution of dissolved oxygen in the oceans is also a function of currents and of photosynthetic activity of phytoplankton in the *euphotic zone*. Dissolved oxygen levels generally decline with depth. The deep oceans, however, are rarely devoid of oxygen (or *anoxic*) because cold, deep water generally contains more oxygen than is consumed by the limited populations of animals in deep water. Dissolved oxygen is a key variable in determining the distribution of living resources in the sea and the sensitivity of the oceans to pollutants. For example, many of the fisheries die-offs in the ocean in recent years have been linked to a combination of factors, including weather, currents, and pollution. The pollution stimulates algal growth and creates conditions that are very similar to the *eutrophication* processes in shallow lakes. The results are localized anoxic conditions, resulting in massive fish and shellfish kills.

The so-called Dead Zone in the Gulf of Mexico is an example of the effects of low-oxygen (*hypoxic*) conditions. In 1999, billions of marine creatures suffocated in the Gulf waters along the Louisiana coast because of the lack of oxygen. An area the size of New Jersey (approximately 7,000 square miles) became completely devoid of marine life (Yoon 1998). But this was not the first time. Hypoxia has been documented every summer for the past 20 years in the northern Gulf of Mexico, but the size of the Dead Zone has been increasing every year. The cause of the hypoxia is the use of fertilizers in the Mississippi Basin and the excess nutrients (primarily nitrogen from agricultural runoff) that make their way into the Mississippi and Atachafalaya rivers, both of which empty into the northern Gulf. The excess nutrients set off an ecological chain reaction (Chapter 4). Unfortunately, the Gulf of Mexico's dead zone is not the only documented instance of hypoxia. The Black Sea has an even larger dead zone that has been documented since the 1960s.

ISSUE 9.1: SALMON IN THE PACIFIC NORTHWEST

The Seattle region is home to Starbucks, Microsoft, and many species of salmon. Salmon are anadromous fish. They spend most of their adult lives in the open ocean—the North Pacific in this case—but they breed in freshwater. The salmon of the Pacific Northwest swim upstream in rivers draining to the Pacific, and they spawn in gravel-bed headwaters of streams draining the Rockies, Cascades, and coast ranges. Several species of salmon exist in the area, including chinook, chum, coho, pink, and sockeye. The salmon spawn in the same streams in which they were born and raised, making them particularly susceptible to local extinction.

The Pacific Northwest of North America is a region with a long history of reliance on natural resources, especially forests, fisheries, and hydroelectric power. These three resource bases are highly interdependent. Forests protect the soil from excessive erosion, helping to maintain high water quality. Hydroelectric power is dependent not only on ample supplies of water but also on keeping reservoir sedimentation rates low. Salmon need clean water and unobstructed, free-flowing streams and are suffering under the double threat of excessive sediment loads and impounded rivers. The natural resource base of the region is in trouble, and salmon are the most acute symptom of the problem.

In California, Idaho, Oregon, and Washington the salmon problem is severe. Salmon are gone from about 40 percent of their historical breeding ranges, and the remaining population levels are much reduced (CPMPNAS 1996). Five species of salmon are now listed as threatened or endangered in their ranges (NMFS 2001). Several factors are involved in the decline of salmon, including overfishing, dams, freshwater habitat loss, hatcheries, and natural variations in ocean conditions.

Fishing, in both the ocean and rivers, is one factor in the decline of the salmon. The peculiar life cycle of the salmon makes fisheries management especially difficult. The ocean catch of salmon draws from populations that spawn in Alaska and Canada, as well as the Pacific Northwest. When a salmon is caught in the open ocean, the fisher has no way of knowing where that fish breeds. Similarly, the number of fish in a given region of open ocean may be high, even though those fish include individuals from an endangered population. Fishers from northern ports oppose restrictions on their catch because populations that breed in more northern streams are still plentiful. While salmon catches in the Pacific Northwest are governed by a bilateral treaty between the United States and Canada, indigenous fishers are exempted.

The Pacific Northwest is a region of abundant water, and many dams are used for generating hydroelectricity and storing irrigation water. These include a few very large dams, such as the Grand Coulee, Hells Gate, and Dworshak dams in the Columbia River system, which are so large that they prevent all passage of fish upstream or downstream. Salmon are extinct from areas upstream of these dams. In addition, there are many smaller dams, some of which prevent salmon migration and some of which allow fish to pass. Fish ladders are helpful but do not completely replace the natural channel that was obliterated by the dam. In most of the larger dams some mortality is associated with migration, so that the cumulative effect of a series of dams may be quite severe.

Water-quality and habitat degradation is another serious problem. Salmon require clear, cool,

Also caused by heavy fertilizer runoff, the Black Sea rebounded once chemical fertilizer use was halved, and by the middle of the 1990s, the Black Sea's dead zone was no longer there (Ferber 2001).

Habitat and Biological Productivity
Major Productive Regions One of the keys to managing marine resources is determining where the resources are found and in what quantities. In terrestrial environments, you can see, count, and often map the precise location or habitat of a particular wildlife species or plant. In the ocean, you cannot. Measurement is always indirect, and you must rely on limited data and educated guesses.

Several key physical features tend to influence the distribution of resources, including topography, currents, upwelling areas, salinity

---

gravel-bed streams for spawning, while the young fish (smolts) require both clean water and a diverse habitat, with large organic debris (logs) and vegetation near the channel. Human activities, especially agriculture and forestry, have severely altered habitats in many salmon streams. Timber harvesting, for example, results in greatly accelerated erosion, especially in steep terrain. This adds sediment to the rivers, sometimes covering the stream bed with sand and silt instead of the gravel that salmon must have for spawning. Clearing vegetation along streambanks also contributes to the problem. Vegetation shades the water, keeping it cool and oxygenated, and logs that fall in streams create habitat diversity, which the smolts require for both finding food and hiding from larger predators.

Hatcheries have been used for decades to increase the salmon population, but now it appears, ironically, that hatchery fish have had a negative impact on natural salmon populations. The introduction of large numbers of hatchery fish has reduced genetic diversity in natural populations, created competition with the natural population for food resources, and caused displacement of natural populations from spawning streams.

Finally, conditions in the open ocean determine the growth and survival rates of salmon and thus the number that return to spawn. These ocean conditions are not well understood, but it is known that they vary over time. It appears that the recent decline in salmon populations is at least partly caused by natural variations in the ocean environment. Thus, the population may recover partly, but it also may decline again in the future as those ocean conditions change again.

Several alternatives are available for protecting the salmon from further decline and potentially restoring populations, but nearly all involve constraints on other resource uses. Because dams are such an important contributor to the problem, significant changes in dam operations would be necessary to help solve it. Among the changes that have been proposed are structural modifications to the dams, drawing down reservoirs when the salmon are migrating seaward to speed water flow and thus shorten the smolts' journeys, diverting water around hydroelectric turbines, and even transporting smolts around dams in trucks or barges and (most controversial of all) actual removal of the dams (Verhovek 1999). All of these alternatives involve either very large costs or significant losses in hydroelectric generation, or both. As regards land use, reduction of sediment and other pollutant inflows to streams would mean restrictions on agricultural activities to conserve soil and reduce nutrients and chemicals in runoff. In fact, the listing of salmon under the Endangered Species Act necessitates habitat protection in one of the nation's fastest-growing metropolitan regions, Seattle-Tacoma. Growth restrictions, increased taxes, and more restrictive land use policies will now guide Seattle's future.

The salmon problem provides a powerful argument for an ecosystem approach to resource management. It is clear that the natural resources of this region are closely interdependent, and effective management of the salmon problem will inevitably require consideration of a broad range of issues affecting many different environments, regions, and governments as well as ethical choices about urban development, ecosystem protection, and the desire for a Starbucks on every corner.

gradients, water depth, thermoclines, and prevailing weather conditions. For living resources, mapping these physical features provides data for delineating key habitats of individual species. The fact that many living marine resources tend to be strongly influenced by one of these physical features is not coincidental.

The biological productivity of the oceans is highly variable and dictated by a combination of bottom topography, salinity, water temperature, sunlight, and currents. We can spatially delimit three major productive regions of the oceans—estuaries, near-shore and continental shelf waters, and the deep ocean. Each of these has a different level of importance to marine fisheries and aquatic food resources.

An *estuary* is an enclosed coastal water body that has a direct connection to the sea and a measurable dilution of seawater by fresh water from the land. Estuaries are transition zones where fresh water and salt water mix in a shallow environment that is strongly influenced by tidal currents. Estuaries can be classified as ecotones (see Chapter 4) or transitional areas between two distinct natural systems, terrestrial and marine. Ecotones generally have a greater diversity of species and higher biological productivity than the natural systems on either side. This "edge effect" is especially true in estuaries. The primary productivity of estuaries is 20 times higher than that of a typical forest (Odum 1971).

Since phytoplankton and other primary producers serve as the basis for most marine food chains, the majority of fish and shellfish caught for human consumption are dependent on estuaries during at least a portion of their life cycle. A few species are permanent estuarine residents, but the majority of fish species migrate between estuaries and near-shore and continental shelf waters to spawn or feed. Estuaries are also important nursery areas for immature fish and shellfish. Many fish move between marine and freshwater environments. *Anadromous* fish breed in freshwater and live their adult lives in the sea, whereas *catadromous* fish breed in the sea and live in rivers as adults. Estuaries play a key role in the migration of anadromous and catadromous fish such as Pacific salmon, American eel, and striped bass (see Issue 9.1). Fisheries biologists estimate that 75 to 90 percent of all fish and shellfish caught by commercial and recreational fishermen in the United States are in one way or another dependent on estuaries.

Near-shore and *continental shelf* waters are the second geographic division. They encompass a much larger portion of the world's total ocean area (7 to 8 percent) than estuaries (which occupy 2 to 3 percent) yet are less biologically productive. They slope from the shoreline out to a depth of approximately 656 feet (200 m) and are affected by geologically recent changes in sea level. The continental shelves are the submerged coastal plains that were above water as late as the last Ice Age, 15,000 years ago. Often low and marshy in prehistoric times and subject to repeated burial and changes in pressure as sea level fluctuated, today the continental shelves are a major source of petroleum reserves.

The waters above the continental shelves are also the site of the majority of the world's fisheries. Close to highly productive estuaries, continental shelves are subject to wind-driven and tidal currents and are shallow enough to permit constant mixing of warm surface waters and cool, nutrient-rich bottom waters. The primary productivity of the shelves is approximately double that of the open ocean. One key feature of the continental shelf region is *upwelling*, in which wind-driven surface currents move away from the shoreline and deep, nutrient-rich water is drawn to the surface, creating exceptionally productive areas (Fig. 9.2). Upwelling occurs on a large scale along the western edge of continents and, to a lesser extent, in specific portions of the shelf such as Georges Bank off the coast of Massachusetts. Nearly 99 percent of all fish production occurs in estuaries and continental shelves, and the majority of continental shelf fisheries are concentrated in upwelling regions.

Compared to the continental shelves, the *deep oceans* are a biological desert, even though they encompass 90 percent of the earth's ocean area. They are only half as productive as continental shelves, with most biological activity concentrated in the euphotic zone, where sunlight penetrates. At present, only a few important fisheries are found in the deep ocean, and the most valuable of these is tuna. As coastal nations have claimed the fishery resources of their continental shelves, deep ocean fisheries are becoming increasingly important to nations such as Japan and Russia, which have relatively little continental shelf area that is under their direct control.

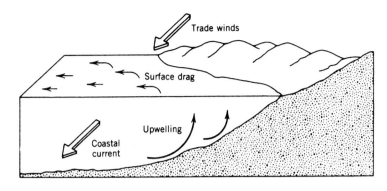

Figure 9.2 Upwelling. This natural event is caused by wind-driven currents that move away from a coastline. Nutrient-laden water is drawn upward to replace the water moved at the surface. Regions of upwelling are generally areas of very productive fisheries.

FISHERIES

The importance of the ocean as a potential food resource is increasing. In the 1970s, about 9 percent of the protein consumed by the world's population came from marine fish and shellfish (Tolba et al. 1992). In 1997, the figure was 16 percent overall but 20 percent in the developing world (World Resources Institute 2000). This rising demand for protein and the leveling off of agricultural production in many parts of the world resulted in a rapid climb in the world's catch of fish (Fig. 9.3); however, the overall importance of marine fish catch to the total fish catch declined somewhat because of the significant increase in aquaculture. In the late 1990s, for example, one-quarter of all fish for food was produced by aquaculture. Fish farming has become an increasing source of protein. Aquacultural production averaged about 10 percent of beef production in 1970, but by 1998 it was 55 percent of it (Worldwatch Institute 2001). Freshwater fish (carp and catfish) and crustaceans and molluscs (scallops and oysters) are among the most dominant types of fish in aquacultural production.

Fisheries Production

Nearly 80 percent of all marine fish caught are eaten by people as fresh, frozen, or canned fish. The remaining 20 percent of the catch is reduced to fish meals, which are made into fertilizers, animal feed (primarily chicken feed), or oils that are used in paints and other industrial products.

Fishery resources are unevenly distributed around the globe in both fresh water and salt water, and nearly 90 percent of the total fish catch is in marine waters (Food and Agriculture Organization 2001). The leading fishing nations in the world are China, Peru, Japan, Chile, and the United States, while China, India, and Japan are the leading aquaculture nations. Ten nations account for nearly 65 percent of the world's total marine fish production (Table 9.2). Regionally, the northwestern Pacific Ocean is the primary fishing region, followed by the southeastern Pacific Ocean and the northeastern Atlantic Ocean. Many of the leading fishing nations—Japan and China, in particular—do not fish solely in their own territorial waters. For example, Russia and Japan utilize all the world's marine fishing areas except the Arctic Ocean. Other major fishing

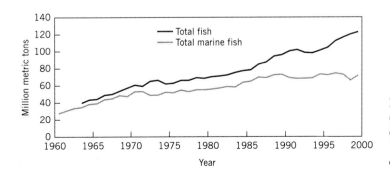

Figure 9.3 World fish catch. (a) Total fish catch 1961–1997 and (b) total marine fish catch, 1960–1999. *Source:* UN Food and Agriculture Organization, 2001.

Table 9.2 Leading Marine Fishing Nations, 1999

Country	Metric Tons Caught
China	11,506,000
Peru	8,257,000
Chile	4,887,000
Japan	3,962,000
Russia	3,467,000
Indonesia	3,423,000
United States	3,330,000
Norway	2,551,000
Thailand	2,340,000
India	2,249,000

Source: U.N. Food and Agriculture Organization (FAO-STAT), 2001.

countries tend to concentrate their fishing activities in regional waters, though not exclusively so. Chile does utilize the southern Atlantic Ocean off the coast of Antarctica, and China also uses the western Africa fishing region. Access to fish is therefore a key issue in ocean management, with countries such as the United States using fisheries as a "food weapon" when we disagree with the internal politics or political controversies in other nations. Disagreements over marine resources have even resulted in armed conflicts between nations, such as the 1970s Cod Wars between Iceland and Great Britain and the continued disputes over territory such as Antarctica.

Commercial landings in the United States totaled 4.1 million metric tons in 2000 (U.S. National Marine Fisheries Service 2000). The five major species by volume caught were Alaskan pollock, menhaden (the oil is used primarily in industry), Pacific cod, Pacific whiting, and crab. In terms of dollar value, the most important species were shrimp, crab, salmon, lobsters, and pollock. In 2000, U.S. commercial landings were valued at $3.6 billion (USNMFS 2000).

Fisheries in Distress

Worldwide, most of the important fisheries either have been severely depleted or are being heavily exploited. In the 1990s, the Canadian cod fishery closed, placing 40,000 people out of work. In addition, the groundfish (cod, haddock, and yellowtail flounder) fishery in Georges Bank, one of the United States' most productive fishing areas, collapsed because of overfishing. In the early part of the decade, the Food and Agriculture Organization (FAO) estimated that two-thirds of the marine fishing stocks were fished beyond their maximum productivity (World Resources Institute 1996). Overfishing (Issue 9.2) and degradation of coastal ecosystems by pollutants and toxic microbes, like *Pfiesteria piscicida*, were the primary causes of fisheries depletion in the past decade.

Although there was an increase in global fish catch after the 1980s, marine harvests declined from 80 percent of the global catch in 1990 to 67 percent in 1999. Regionally, the northern Atlantic fishing grounds have seen dramatic declines in cod catches since the mid 1980s. In the Northwest Pacific region, all fish stocks are exploited beyond sustainable levels. In response to fisheries depletion, the type of species caught has shifted from high-value fish such as cod and haddock to lower-value fish such as sardines and anchovies. Unfortunately, these lower-value species are not used for human consumption but are processed into fish oil and fish meal.

The gobal trends don't show this decrease because aquaculture is increasing at a faster rate. Fish farming is the dominant production method for salmon and acounts for nearly one-quarter of the world shrimp production. Aquaculture now accounts for over 25 percent of all fish consumed by humans (Naylor et al. 1998). While it may appear that fish farming would relieve some of the pressure on wild fisheries, it turns out that shrimp and salmon are fed nutrient-rich diets (mainly from fishmeal and fish oil extracted from wild fisheries). Largely because of their dependence on wild fisheries (inputs are 2–4 times the volume of outputs), shrimp and salmon aquaculture in reality deplete wild fisheries rather than conserve them.

MINERALS FROM THE SEABED

Energy Resources

The United States and other nations are becoming increasingly dependent on the oceans for energy production. The share of traditional energy sources such as oil and gas from offshore areas is

ISSUE 9.2: STRIP MINING THE OCEANS

The number of people involved in commercial and subsistence fishing is declining, yet the fish catch is increasing. Fishing technology has improved our ability to catch more fish with greater efficiency than ever before. The vast majority of people involved in fishing (either for individual consumption or for profit) live in Asia. Similarly, the number of commercial fishing vessels is concentrated in Asia, especially in China and Japan, two of the world's leading fishing nations (World Resources Institute 2000).

Fishery vessels are defined by the United Nations Food and Agricultural Organization as floating fish factories that are used to find, harvest, preserve or process, transport, and land fish and shellfish. They include trawlers, purse seiners, gill netters, and long liners—devices that permit the harvesting of all marine resources within the net's or line's reach. This practice is similar to the clearcutting of forests, where all the trees in the area are removed regardless of size or species. Unfortunately, the efficiency of catching fish with these methods comes at the expense of the resource itself. In fact, it is estimated that the current fleet has 50 percent more capacity than it needs to meet current and anticipated demand (McGinn 1998). Moreover, much of the harvesting occurs in fishing regions far distant from the fishing nations themselves.

Because the nets cannot discriminate between species, much of the catch is thrown overboard—the fish are undersized, the wrong species, of inferior quality, or exceed quotas. This unintended catch (or *bycatch*) is estimated to be around 25 percent of the global marine fish catch (World Resources Institute 2000). Tuna characteristically swim with dolphins, so when the seine nets are launched for tuna, many dolphins get caught in the process, resulting in an outcry from American consumers who want "dolphin-free" tuna. Tuna long-line fisheries entangle not only dolphins but also birds such as the albatross. Improvements in the nets and greater awareness of quick releasing of nontargeted species (such as sea turtles and dolphins) have improved the situation somewhat.

Another practice that is quite destructive is trawling. The trawling equipment is dragged along the sea floor to catch bottom-dwelling species such as cod, flounder, and shrimp. Trawling occurs in the continental shelf waters, and one estimate places the area swept by trawlers at close to 15 million km^2 (Walting and Norse 1998), an area slightly smaller than the size of the United States and Canada combined (around 18 million km^2). Bycatch in shrimp trawls, for example, is the leading cause of sea turtle mortality in the United States (Mulvaney 1998). It is also an incredibly wasteful practice. Because the shrimp are so small, a fine-grained netting must be used to capture them (and everything else). It is estimated that for every pound of shrimp caught, 5 pounds of bycatch is discarded (World Resources Institute 2000). Deep ocean coral reefs are also at risk from trawling equipment, which break and damage the reefs and scar the ocean bed (Revkin 2000).

The destructive fishing practices are not limited to large commercial fleet operations. In Southeast Asia, cyanide fishing is used to harvest reef fish for food or as ornamental aquarium fish sold in pet shops in North America and Europe. The live reef fish trade with exports to Hong Kong (food) and North America and Europe is a $1.2 billion business (Barber and Pratt 1998). It has been estimated that since the 1960s more than a million kilograms of cyanide have been dumped into the coral reefs in the region. The cyanide (hand squirted into the water by local fishers) stuns the fish, making them easier to catch. The primary environmental effect of cyanide fishing is on the coral reefs and the ecosystem that supports them. The Philippines, where cyanide fishing originated, has banned the practice in favor of more sustainable approaches that protect not only the ecosystem but the livelihoods of the local fishers as well.

growing, and several nontraditional sources of energy from the oceans are under development (Chapter 14).

Approximately 30 percent of the earth's exploitable hydrocarbons are believed to exist beneath marine waters, with 90 percent of these unexplored. The majority of the explored oil and gas deposits are found on the continental shelf, near land-based oil and gas reserves. The U.S. outer continental shelf, for example, is roughly 1.8 million square miles (4.7 million km^2) in size. This is comparable to the 1.7 million square miles (4.4 million km^2) of geologically favorable land in the United States that currently supports most domestic oil and gas production. There is a good chance, then, that the offshore regions under U.S. control could produce at least as much oil and gas as is currently produced on land. A similar pattern can be found worldwide.

The contribution of offshore oil to total production has been growing steadily for many years in the North Sea and Persian Gulf regions. In the United States (particularly the Gulf Coast), offshore oil production as a percentage of total crude oil production rose from 4 percent in 1960 to 12 percent in 1980 to 30 percent in 2000 (U.S. Energy Information Agency 2000). In 1998, U.S. offshore wells produced $26.5 billion in sales of crude oil and natural gas (U.S. Bureau of the Census 1999).

The growth of offshore oil and gas production depends on the ability to drill for hydrocarbons in ever-deeper water. Early efforts at offshore drilling were simple extensions of land-based techniques in water less than 20 feet (6 m) deep. Advances in drilling technology, semisubmersible drilling-rig designs, offshore pipelines, and other equipment now permit construction of conventional drilling platforms in up to 1000 feet (305 m) of water (Fig. 9.4). In fact, the technology for drilling temporary exploratory wells has outstripped the technology for building permanent platforms needed to bring an area into regular production. Exploratory wells have been drilled in up to 5000 ft (1600 m) of water, thus opening up large areas of the continental slope and even deep ocean areas to possible development.

Deep-Seabed Minerals

For hundreds of years, sand, gravel, coal, tin, gold, and diamonds have been mined from the sediments beneath shallow-water areas around the world. By the early 1970s, mineral deposits found in the deep ocean became technically and economically exploitable. Many of these mineral deposits

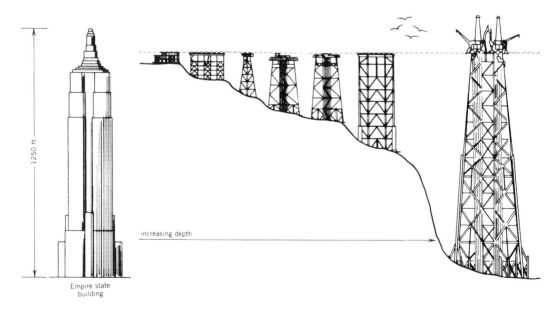

Figure 9.4 Changes in offshore oil drilling technology since 1940. Production oil drilling platforms were installed in 20 feet of water in the 1940s. By 1983, permanent platforms exceeded 1250 feet, approximately the height of the Empire State Building.

contain strategic minerals such as cobalt, which is currently imported by industrial nations, including the United States (see Chapter 13). Of particular interest to the developed nations are manganese nodules. These potato-sized lumps are common features of the sea floor in water from 13,000 to 18,000 ft (4000 to 6000 m) in depth. The nodules are composed of hydrated oxides of iron and manganese, which often form around a nucleus of shell, rock, or other material, just as the pearls that oysters create form around a grain of sand. Manganese nodules are found in all the world's oceans, although the grade (percentage of various metals) and coverage (weight and number per area) vary. The eastern Pacific Ocean several hundred miles south of Hawaii appears to feature exceptionally dense nodule deposits, containing minerals in sufficient quantity to permit commercial exploitation (Albarede and Goldstein 1992; Broad 1994; Knecht 1982; Glasby 2000) (Fig. 9.5).

Exploitation of seabed minerals requires an investment of many millions of dollars. For example, U.S., Japanese, and European mining companies are interested in gaining access to the billions of tons of deep-seabed minerals that are expected to become available as technology improves. But in order to protect their investment, these mining companies would demand guaranteed and probably exclusive access to this resource. The exploitable fields of seabed nodules, however, are in deep ocean waters, beyond any single nation's jurisdiction. Problems such as the lack of scientific knowledge, huge capital investments required for mining, and the vagaries of the minerals marketplace have all contributed to the lack of development of these seabed resources.

Recent developments have thwarted the economic viability of deep sea module mining. The depressed worldwide prices for metals and recent discoveries of nickel deposits in Canada have combined to reduce interest. However, new commercial interest may develop in the future because of perceived shortages in countries such as China, Japan, Korea, and India (Glasby 2000).

MANAGEMENT OF MARINE RESOURCES

The Problem of Ownership

As discussed earlier in this chapter, most of the world's oceans are a common property resource. Part of the ocean is globally common property, while some is nationally common property, although the boundaries between what is controlled internationally and nationally can be difficult to identify. The ownership problem may well be the largest single contributor to overexploitation of marine resources.

First, it is important to understand where control of marine resources by a single nation stops and where international control begins. Diplomats and international law experts make a distinction between control over ocean space and control over the use of ocean space. For example, international treaties recognize a 12-nautical-mile (22.2-km) territorial sea along a nation's coastlines as the

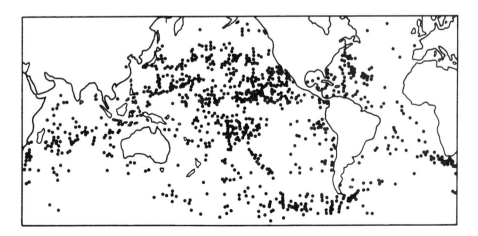

Figure 9.5 Manganese nodules in the oceans. Scientific research vessels have found accumulations of manganese nodules at these locations over the past 100 years. *Source:* Heath, 1982.

exclusive territory of that nation. Both ocean space (including bottom sediments) and the use of that space by fishing vessels, navy ships, mineral companies, or anyone else are controlled by the individual nation. Other types of jurisdiction or "ownership" are less clear. Many coastal nations claim control over all fishing resources within 200 nautical miles (370 km) of shore. Yet national control over other activities in this 200-mile area, such as the transit of military vessels of hostile nations, is not recognized by international law. In some cases, even particular types of fish, such as tuna, do not fall under the jurisdiction of individual coastal nations. In short, the definitions of jurisdiction and ownership of marine resources can vary with the distance from shore and the type of marine activity (Fig. 9.6).

Most of the world's nations accept four general types of jurisdiction over marine resources. *Internal waters* include bays, estuaries, and rivers and are under the exclusive control of the coastal nation. In the United States, jurisdiction over the resources found in these internal waters is shared between federal and state governments. Most states and some municipalities control fishing and shellfishing, while the states and the federal government share jurisdiction over water pollution, dredging, and other activities.

The *territorial sea*, a band of open ocean adjacent to the coast, is measured from a baseline on the shore out to a set distance. Most countries, including the United States, claim a zone that is 12 nautical miles (22.2 km) wide. A few nations, including Somalia, Peru, Ecuador, El Salvador, and Benin, even claim a territorial sea width of 200 nautical miles (370 km). A coastal nation controls all activities, such as fishing, within its territorial sea—except for the right of "innocent passage" by foreign vessels. A Japanese fishing boat can, for example, pass between the Aleutian Islands off Alaska without permission from the state or federal government, but it cannot drop nets and proceed to fish. A foreign military vessel also may pass through territorial waters unhindered, as long as it remains outside the internal waters of the coastal nation. A curious exception to the right of innocent passage through the territorial sea is aircraft. Foreign aircraft must seek permission to enter U.S. air space before moving within 12 nautical miles (22.2 km) of shore. In the United States, the coastal states manage fisheries and oil drilling in the territorial sea, while the federal government patrols and protects it.

The third type of jurisdiction is the 188-nautical-mile (348 km) *exclusive economic zone* that was created by the *Law of the Sea Treaty* (UNCLOS). For a discussion of UNCLOS, see the following section. The exclusive economic zone (EEZ) is a special-use area where activities such as fishing and oil drilling are controlled by the coastal nation, while other activities are not. In 1946, the United States claimed exclusive jurisdiction over its outer continental shelf, which extends out to a depth of about 660 feet (200 m) of water, to control oil and gas development. In 1976, the United States created a 200-nautical-mile Fishery Conservation Zone, claiming control over all fish and shellfish except the highly migratory tuna. Many other nations subsequently adopted this idea and made similar claims. The Law of the Sea Treaty creates a single EEZ that includes fishing and all forms of mineral extraction, no matter what the water depth. Control over other activities in the EEZ is less clear. Some nations (including the United States) claim jurisdiction over ocean dumping and water pollution, and others (excluding the United States) claim jurisdiction over the movements of vessels and oil spills in this zone. When taken together, the territorial sea plus the EEZ gives a nation control over 200 nautical miles of the oceans bordering them.

Finally, the *high seas* are those ocean areas that are beyond the jurisdiction of any individual nation. Traditionally, the limits to activities on the high seas are set by international treaties, such as the Law of the Sea.

The Law of the Sea Treaty

There is a long history of confusion over who owns or controls the oceans. Hundreds of conflicting territorial claims have been made by coastal nations, many of which either clash with or ignore international treaties signed in 1958 on territorial seas and the outer continental shelf. There are over 90 independent nations in the world today that did not exist when the 1958 treaties were signed. Many of these nations are landlocked, underdeveloped, or both, and most are ex-colonies with boundaries and economies that were originally designed to benefit only the colonial power.

Maritime boundary problems, coupled with the desire of many newly independent countries to allocate or reallocate marine resources, led to

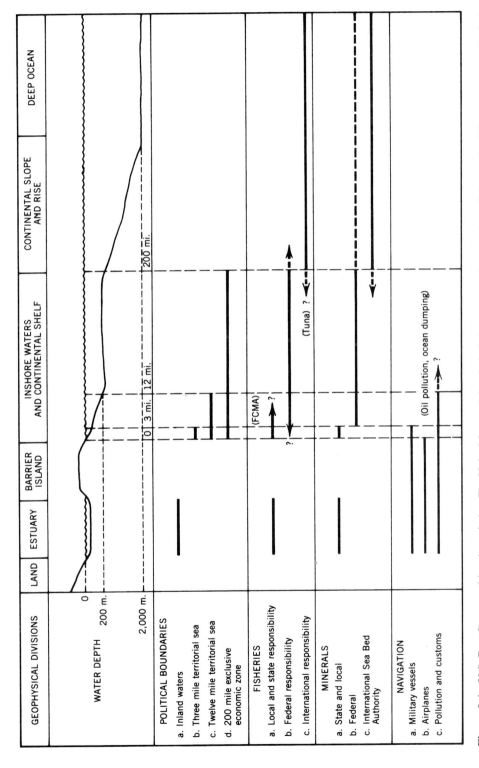

Figure 9.6 U.S. East Coast maritime boundaries. The United States claims jurisdiction; over various portions of the sea for different purposes. Dashed lines indicate shared jurisdictions, question marks indicate disputed or ambiguous boundaries.

negotiations over a new Law of the Sea Treaty. Negotiations started in 1974, with over 160 nations participating, and lasted until 1982, when a final version was approved. A total of 135 nations initially signed the treaty, and 148 have ratified it. The treaty did not take full effect until 1994 and is now in force for 132 nations that have ratified it. The U.S. is not party to this treaty.

The Law of the Sea Treaty clarifies boundary claims by establishing internationally agreed-upon limits on the territorial sea, continental shelf, and exclusive economic zone. It establishes a universal 200-nautical-mile (370-km) economic zone, giving the coastal nation exclusive control over exploitable resources, such as fish, oil, and gas. It also preserves the right of free navigation through this zone by foreign ships, airplanes, and submarines.

Without right of innocent passage, a universal 12-nautical-mile (22.2-km) territorial sea would effectively close off 175 straits or narrow passages through which the majority of shipping travels. The most important of these are the straits of Gibraltar, Dover, Hormuz, and Malacca. The Strait of Gibraltar is 8 nautical miles (15 km) wide and is the major point of access to the Mediterranean Sea. The Dover Strait is the easternmost part of the English Channel and is 17.5 nautical miles (32 km) wide. Virtually all oil imported into northern Europe by ship passes through this narrow portion of the English Channel. The Strait of Hormuz, while only 20.7 nautical miles (38 km) wide, is perhaps the most important in the world. It is located between Iran and Oman in the Persian Gulf. All the oil transported by ship from the Middle East must pass through this vital and strategic strait. The Straits of Malacca (8.4 nautical miles or 16 km wide) lies between the Malay Peninsula and the island of Sumatra. Ship traffic between the Indian and Pacific oceans must use either this strait or others in Indonesian waters. All of Japan's imported oil from the Middle East traverses this strait.

The rights of neighboring nations to regulate traffic, including aircraft, through straits have never been clear. At the insistence of the United States and other nations, the Law of the Sea Treaty creates an internationally recognized "right-of-transit passage" through straits, permitting unimpeded access as long as ships and aircraft comply with minimal navigational rules. This right-of-transit passage is one of the most significant provisions of the treaty. The industrial nations with large navies insisted on this provision, at the expense of less developed countries, which border most of the straits in question.

The treaty also provides for the regulation of pollution and the conservation of living resources, including increased protection of marine mammals. Finally, the treaty allows for the exploitation of deep-seabed minerals but states that these resources constitute the common heritage of humankind. Although the developed nations possess

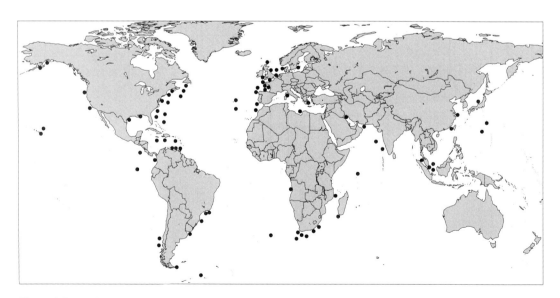

Figure 9.7 Major tanker oil spills, 1962–2001. Each dot represents a spill of over 5000 tons.

the technology and knowledge of deep-seabed mining, the fruits of this expertise would have to be shared by all nations. In other words, those developed nations would not have exclusive right to the resources, even if they were the only ones with the technology and knowledge to get to them.

Marine Pollution Problems

The oceans are so large that we often think of them as a place where we can discard the unwanted byproducts of civilization. This attitude runs the gamut from passively allowing pollutants in streams and rivers to make their way to the ocean to actively and deliberately burying wastes in the ocean. There are three major sources of marine pollution: oil spills, hazardous materials spills, and ocean dumping.

Oil Spills Many major transportation accidents involving oil tankers have occurred (Fig. 9.7). The most famous of these include the wrecks of the *Torrey Canyon*, which spilled 700,000 barrels (95,200 metric tons) of oil off the southern coast of England in 1967; the *Amoco Cadiz*, losing 1,628,000 barrels (221,410 metric tons) off the French coast in the English Channel in 1978; and the wreck of the *Exxon Valdez* in 1989, spilling 260,000 barrels (35,374 metric tons) of oil into Prince William Sound in Alaska (Fig. 9.8). The resulting pollution of the ocean from these spills, including land-based effects, was quite extensive and well documented (Audubon 1989; Fairhall and Jordan 1980; Winslow 1978). The recent grounding of an Ecuadorean tanker released bunker oil into one of the world's most pristine marine preserves, the Galápagos Islands.

Tanker accidents are important pollution episodes (Table 9.3), yet only 21 percent of accidental tanker oil spills are the result of accidents; the majority of oil spilled into the marine environment from tankers is from routine operations. Oil spills also occur in ports and harbors, from stationary offshore drilling platforms, and through runoff from land-based facilities (Freudenburg and Gramling 1994). In the United States alone, there were roughly 8300 reported accidents in 2000, with more than one million gallons (34,000 barrels or 4,600 metric tons) of oil spilled (USCG 2001). More than half of these were in marine waters, but less than half (42 percent) of the volume of oil spilled (2000 tonnes) affected marine environments, a significant decrease from

Figure 9.8 Oil spilling from the *Exxon Valdez* (at left); unspilled oil is being transferred to the ship at right.

earlier decades. Offshore oil exploration and drilling accounts for less than 5 percent of the total world's oil pollution of the oceans, although at times the local impacts are quite severe. For example, in 1979–1980, over 2.7 million barrels of crude oil were spilled in the Gulf of Mexico as a result of the blowout of IXTOC-1, a Mexican-owned oil well in the Bay of Campeche. The drifting oil spill caused an international dispute when it washed ashore on beaches in Texas, destroying wildlife and habitat and severely affecting recreational and fishing industries in the region. During the 1991 Persian Gulf War, Iraq intentionally spilled 3 million barrels in the northern Gulf.

Hazardous Materials Spills Transportation of hazardous materials on the high seas is increasing as developed nations seek new disposal options for the byproducts of their industrialization. In the United States the number of reported hazardous materials spills in marine waters has been

Table 9.3 Top Ten Major Oil Spills, 1970–2000

Ship Name	Year	Location	Oil Spilled (Metric Tons)
Atlantic Empress	1979	Tobago, West Indies	287,000
ABT Summer	1991	Angola	260,000
Castillo de Bellver	1983	Saldanha Bay, South Africa	252,000
Amoco Cadiz	1978	Brittany, France	223,000
Haven	1991	Genoa, Italy	144,000
Odyssey	1988	Nova Scotia, Canada	132,000
Torrey Canyon	1967	Scilly Isles, UK	119,000
Urquiola	1976	La Coruna, Spain	100,000
Hawaiian Patriot	1977	300 nautical miles off Honolulu	95,000
Independenta	1979	Bosphorus, Turkey	95,000

Source: International Tanker Owners Pollution Federation, Ltd., 2001.

relatively constant, although the quantity spilled has been highly variable. The majority of these incidents occurred in river channels, but the amount spilled in the territorial sea is more significant. Most spills take place in coastal waters. Internationally, marine hazardous materials spills are on the rise as well. In 2001, for example, 700 metric tons of styrene were released in the coastal waters at the mouth of the Yangtze River in China when two vessels collided in the fog, and less than two weeks later, another ship capsized in the Straits of Singapore, releasing 600 metric tons of phenol into the environment (Mariner Group 2001).

Ocean Dumping Ocean dumping, one of the major contributors to ocean pollution, includes the disposal of sewage sludge, industrial and solid waste, explosives, demolition debris, radioactive materials, and dredge spoils (Fig. 9.9). Dredged material, by weight the most significant material currently being dumped in the oceans, accounts for over 90 percent of the waste disposed of in the marine environment. This material comes from the removal of bottom sediments from rivers, harbors, and intercoastal waterways to allow navigation in these water courses. Uncontaminated sediment (or spoil) poses little environmental risk. In the United States, dredged material was traditionally dumped in estuaries and on tidal wetlands. Concern over protecting these productive areas has led to a shift to near-shore and continental shelf dumping of dredged materials, although this varies by region.

Contaminated sediment containing toxic materials such as cadmium, lead, copper, and polychlo-

Figure 9.9 A garbage barge in New York harbor, target of a Greenpeace protest.

rinated biphenyls (PCBs) poses another problem. Bottom sediments in many ports and rivers contain high levels of toxic pollutants left over from decades of uncontrolled dumping and pollution. These sediments are usually immobile if undisturbed, but during dredging activities they become suspended in the water. The cost of disposing of such polluted sediments on land is significantly greater than ocean disposal, yet scientific evidence on the safety of various ocean disposal techniques is lacking.

In some parts of the world, sewage sludge and industrial wastes are a major source of contaminants in the marine environment. Although great efforts have been made to upgrade sewage treatment plants onshore to reduce water pollution, the result has been a higher volume of sewage sludge that needs disposal. In 1972, the U.S. Congress passed the Marine Protection, Research, and Sanctuaries Act, which, among other things, requires a permit for ocean dumping. In 1977, the act was amended to encourage land disposal of sewage sludge after 1981. Ocean disposal of sewage sludge therefore declined rapidly throughout the United States, with the exception of the New York Bight, the area of the Atlantic just off New York City, because of a temporary exemption granted to them (Squires 1983; Swanson and Devine 1982). After years of wrangling, New York finally ended ocean dumping of its sludge in 1991.

Nonpoint pollution also contributes to marine pollution. In coastal areas, the problem results from septic tanks and combined sewer/stormwater systems. Although *nonpoint sources* (pollution that comes from diffuse sources such as atmospheric dustfall or moving ships) are difficult to measure, officials estimate that they are an even greater threat to the marine environment than are *point sources* (sources that are easily identified on a map, such as pipelines). Nutrient pollution (especially nitrates and phosphates) is now a major threat to many of the world's coastal zones, especially in Europe (vanDeveer 2000).

Finally, there is one other major contributor to the pollution of marine waters. The accumulation of nonbiodegradable plastics in the oceans has increased during the past two decades. Originating from land-based activities as well as trash thrown overboard from ships, plastics float on the ocean's surface for decades (Wilber 1987). The primary effect of plastic is on marine animals. It has been estimated that 100,000 marine mammals die each year either by ingesting the plastic or by becoming entangled in it (U.S. Congress 1987). Many more seabirds meet their fate in this way. The plastic pollution problem of the world's oceans has now become a global concern.

Protecting Marine Ecosystems

Control of Marine Pollution Because of their common property nature, international efforts to control pollution at sea are also important. The London Dumping Convention of 1972 is the primary international treaty on marine waste disposal and, more important, the only one to which the United States is signatory. The treaty prohibits the dumping of "blacklisted" substances (organohalogens, mercury, cadmium, plastics, oils, radioactive materials, and biological and chemical warfare agents) in waters seaward of the inner boundary of the territorial sea. The Oslo Convention of 1974 regulates the dumping and incineration of wastes by most European countries, and this treaty applies to Arctic, northern Atlantic, and North Sea waters. At-sea incineration is no longer practiced in the North Sea.

One of the most effective international treaties on marine pollution is the International Convention for the Prevention of Pollution from Ships, 1973, 1978 (referred to as MARPOL 73/78). This treaty attempts to reduce pollution from ships, including oil, chemicals, and plastics. Also as a response to the *Exxon Valdez* spill in 1989, the International Convention on Oil Pollution Preparedness, Response, and Cooperation was signed. This treaty sets requirements for oil spill contingency plans and mechanisms for cooperation between transboundary spills, and an annex on hazardous and noxious substances also includes protocols for handling these types of materials. Finally, the United Nations Environment Program (UNEP) Regional Seas effort has been instrumental in developing regional action plans for marine pollution from ocean dumping, oil spills, and land-based sources (Table 9.4).

In U.S. territorial waters, four different federal laws govern pollution and wastes in the marine environment. The Clean Water Act (see Chapter 10) was instrumental in banning sludge dumping in the New York Bight. The act also provided construction grants for sewage treatment plants and combined sewer/stormwater

Table 9.4 Pollution Reduction Elements in Regional Seas Program

Regional Seas	Emergency Response	Control Ocean Dumping	Control Land-Based Pollution	Protected Areas
Antarctic[a]	●			
Baltic Sea[a]	●		●	●
Arctic[a]			●	●
North Atlantic[a]		●	●	
Black Sea (1993)	●		●	
Mediterranean (1975)	●	●	●	●
Kuwait Region (1978)	●		●	
West Central Africa (1981)		●		
Southeast Pacific	●		●	●
Red Sea & Gulf of Aden (1976)	●			
Caribbean (1981)	●		●	●
Eastern Africa (1985)	●			●
South Pacific (1982)	●	●		
East Asia Seas[b] (1981)				
South Asia[b] (1995)				
Northeast Pacific[b] (2001)				
Northwest Pacific[b] (1994)				
Upper-Southwest Atlantic[b]				

[a]Antarctic through North Atlantic Sea programs were in effect prior to the establishment of UNEP's Regional Seas program.
[b]Treaty not yet in force or elements not ratified.
Source: United Nations, 2002.

overflow systems. In addition, the act provided funds for the development of management programs for nonpoint source pollution and the control of toxic pollutants. The 1990 amendments to the Coastal Zone Management Act also require local and state governments to implement controls for coastal pollution control, especially nonpoint sources. The Shore Protection Act of 1988 regulates the deposit of trash and medical debris in U.S. coastal waters.

The Marine Protection, Research, and Sanctuaries Act (MPRSA) was originally passed in 1972 and was designed to (1) regulate the disposal (dumping and pipeline discharge) of wastes into the marine environment and (2) control the level of pollution in these waters. Marine waters are defined as those waters seaward of the territorial sea (3 nautical miles or 5.6 km). The act restricts the transportation and dumping of wastes in the open ocean and regulates the dumping and discharge of solid waste, sludge, industrial waste, dredged materials, radioactive waste, and biological and chemical warfare agents in marine waters. The act was amended and reauthorized in 1985 and became known as the Ocean Dumping Act.

Finally, the 1990 Oil Pollution Act provides coordinated federal assistance for oil spills in U.S. waters and was enacted shortly after the *Exxon Valdez* spill. The act has a number of major provisions, including the requirement for double-hull tanker ships, and the development of oil spill emergency response plans by vessel and facility owners. The act also establishes liability for oil spills shared between shipowners, onshore facilities, deep-water port owners, and offshore facility owners. The most important provision, however, is the creation of the Oil Spill Liability Trust Fund. This fund, based on a five cent per gallon tax on oil, is designed to pay damage claims to injured third parties.

Living Marine Resources In the United States there is a history of protection of marine fisheries. In 1976, the Fishery Conservation and Management Act established the requirement for fisheries management plans, extended governmental control of living resources out to 200 miles, and severely limited foreign fishing in this economic zone. After numerous reauthorizations, this act (also known as the Magnuson-Stevens Act) today is the primary legislation governing fishery resources in the United States. One important amendment was the 1996 Sustainable Fisheries Act. This act is designed to improve fisheries management by moving more toward ecosystem and habitat protection approaches. It addresses the need to eliminate overharvesting and the use of damaging fishing practices as well as the protection of fish habitat. In the waning days of the Clinton Administration, another law was signed, the Oceans Act of 2000. The purpose of this legislation was to establish a commission to make recommendations for a coordinated and comprehensive national ocean policy, one that promotes stewardship of ocean and coastal resources, protection of the marine environment, enhancement of maritime commerce, and expansion of human knowledge about the marine environment.

At the international level, the Ramsar Convention protects wetlands of international importance. In 2001, there were 130 nations that were parties to this international treaty (Ramsar 2002). Slightly more than 1000 sites have been protected worldwide, covering more than 73 million hectares. The majority of these are in Europe (especially the United Kingdom), but a significant number are in the developing nations. For example, Botswana has 6.8 million hectares of wetlands protected under Ramsar (World Resources Institute 2000).

Marine Sanctuaries The Marine Protection, Research, and Sanctuaries Act (1972) also had some preservation elements. It designated marine sanctuaries in the oceans and Great Lakes with the intent to preserve or restore these areas for their conservation, recreation, ecological, or aesthetic value. The National Marine Sanctuary Program (administered by the National Oceanic and Atmospheric Administration) was established to designate and manage nationally significant marine areas. Such areas would be classified on the basis of specific criteria, including the representativeness of the marine ecosystems, research potential, recreational or aesthetic values, or uniqueness (historical, geological, ecological, or oceanographic). Sanctuaries range in size from less than 1 square mile (2.6 square km) to over 5327 square miles (1400 square km). Although they are managed under multiple-use guidelines, human use is balanced with the maintenance of the health and viability of the ecosystem. The first two sanctuaries included in the program in 1975 were the *U.S.S. Monitor* (off the coast of North Carolina) and Key Largo (off the Florida Keys). The act was amended in 1984, and by 2001, 15 units were included in the sanctuary program, covering roughly 200 square miles (Ogden 2001).

Within the United States, there is also a National Estuarine Research Reserves program that is designed to protect and restore the ecosystem health of important estuaries (Council on Environmental Quality 1995). Municipal discharges, urban runoff, and agricultural and industrial sources of pollution all contribute contaminant loads in estuaries affecting shellfish production but recreational value as well. The rate of decline in some of these most productive estuaries—Chesapeake Bay, Puget Sound, and the Mississippi Delta—is cause for concern.

At the international level, marine protection programs take many forms. There are about 3600 coastal and marine protected areas in over 110 countries (World Resources Institute 2000). The International Union for the Conservation of Nature (IUCN) drafted policy guidelines in 1988 for setting up marine protection programs. As is the case with many resources, protection is determined by the individual country. Some areas are protected in parklike settings where wildlife protection and tourism are stated management objectives, and other marine parks are more preserve-oriented, with limited or restricted human use. Australia's Great Barrier Reef Marine Park is the best example of a protected marine resource. It was established in 1975, and the governing authority has the right to regulate and prohibit activities inside the park as well as those activities outside the park that may pollute or otherwise harm the reef ecosystem. The primary purpose of the park is to promote the human use of it in a manner consistent with the preservation of the ecosystem.

Example: Exploitation and Protection of Marine Mammals

Exploitation of whales and other marine mammals has been taking place for thousands of years but became a problem when it reached industrial scales in the eighteenth and nineteenth centuries. Commercial whale-hunting brought several species to the brink of extinction, and in response international agreements have resulted in significant reduction, but not elimination, of whale harvesting. Whale conservation problems provide an excellent illustration of the scientific, social, and political dimensions of marine resource management.

Whale Populations The exact population sizes of marine mammals occupying the world's oceans are unknown, although we do have estimates for specific species. Marine mammals include great whales, small whales, dolphins and porpoises, sirenians and otters, and seals and sea lions.

Many factors influence the number of marine mammals in any given region. Often, the best knowledge we have regarding their population is derived from historical data on whaling and harvesting and current information about their reproductive biology, natural mortality, and habitat. In the case of whales, defining habitat is problematic. For example, the distribution of the California gray whale is relatively easy to determine because their breeding areas are limited to several shallow lagoons on the west coast of Mexico, and portions of their north-south migration routes follow the edge of the continental shelf along the western coast of North America. Both the lagoons and continental shelf are physical features that can be easily mapped. On the other hand, the distribution of other great whales such as the fin, right, and blue whales is greatly dependent on the concentration of food supplies such as krill and plankton. In turn, the distribution of krill and plankton is dependent on currents and weather conditions, thus making it highly variable. Therefore, the key feeding areas for these species could change from year to year.

Estimates of whale populations are extremely difficult to produce, although we can make some educated guesses based on historical and catch data. For example, Table 9.5 provides the best estimates of the virgin population of selected species before harvesting began, as well as current population estimates.

Table 9.5 Whale Abundance

Whale	Population	
	Virgin[a]	Current
Blue	175,000–228,000	<5000
Humpback	115,000	20,000
Bowhead	30,000–54,700	<8500
Right	100,000–200,000	1000–4000
Fin	448,000–548,000	50,000
Sei	256,000	65,000
Sperm	2,400,000–2,700,000	NA[b]
Gray	15,000–20,000	21,000
Minke	140,000	904,270
Pilot	NA[b]	780,000

[a]Estimates of population before harvesting began.
[b]NA = not available.
Sources: Brownell, Ralls, and Perrin, 1989; Council on Environmental Quality, 1982; U.S. Department of Commerce, 1985; World Resources Institute, 1988; UNEP, 1993; Perry et al., 1999; World Wildlife Fund, 2001.

The international regulation of whaling is a classic example of the problems with managing a common property resource. The introduction of the harpoon gun and steam-powered whaling vessels in the late 1800s, coupled with the advent of the seagoing factory ship in the early twentieth century, revitalized the whaling industry of Moby Dick fame. This new technology permitted the exploitation of larger, faster species of whales, such as the blue whale, and the processing of oil, bone, and meat at sea. Several European nations developed fleets of small vessels, called whale catchers, centered around a large factory ship. These fleets caught several different species in both Arctic and Antarctic waters, raising concerns about overfishing as early as 1920. The first international whaling treaty, signed in 1931, proved ineffective. The International Whaling Commission (IWC) was therefore established in 1946, ostensibly to protect and ensure species survival. Currently, 39 countries are members of the IWC.

Regulation of Whaling The wording of the preamble to the 1946 International Convention for the Regulation of Whaling, which created the

IWC, is a good example of the difficulty associated with using a scientifically rational management approach on a common property resource such as whales. The treaty directs that the IWC safeguard for future generations the great natural resources represented by the whale, while also increasing the size of whale stocks to bring the population to an "optimal level" to make possible the orderly development of the whaling industry. Is the objective of the IWC to protect whales, encourage industry development, or both? More than half a century's effort to manage whales by the IWC indicates that nations generally will seek to protect whales only when it is in their national interest to do so. The 1946 treaty creating the IWC permits member nations to object to and then legally ignore the quotas established by the Commission and its scientific committees. Thus, in 1986 when the IWC formally declared a complete moratorium on all commercial whaling at the urging of the United States and other nonwhaling member nations, Japan and the Soviet Union formally objected and were initially allowed to set their own quotas. In 1987, Norway and the Soviet Union ceased commercial whaling, as did Japan in 1988. Under the IWC moratorium, however, countries can still obtain special permits to harvest whales for subsistence purposes for native populations (Inuit) and for scientific research.

International efforts to protect whales are complicated by the slow reproductive rates of these marine mammals. For example, one of the reasons cited in support of the international whaling conventions was the Pacific bowhead whale. The world catch for this species peaked in the mid-1800s, and the IWC's first act was to ban all further commercial harvest. However, right and bowhead whale populations have not recovered despite 40 years of complete protection. It appears that the slow reproductive rate of these species is responsible for the inability of the small, dispersed populations to grow rapidly.

Recent whaling follows a pattern of heavy exploitation of one species, leading to a dramatic population decline and a shift to another species. For example, 11,559 blue whales were caught in 1940, declining to fewer than 2000 in 1960 and 613 in 1965. The catch of the slightly smaller fin whale peaked at 32,185 in the mid-1950s and dropped to 5057 by 1970. The harvest of the still smaller sei whale was minimal until 1960, when it then rose and subsequently declined (Council on Environmental Quality 1981). Similarly, the minke whale catch totaled about 60,000 during the early 1980s. The recovery of the minke whale population since the moratorium has been rapid. Current estimates place the population at around 103,000–204,000, a population large enough to support small-scale whaling, which was in fact recommended by the IWC's scientific committee but not formally adopted by the entire IWC because of strong opposition by environmentalists. The trends in whale harvesting (Fig. 9.10) during the past decade illustrate the effectiveness of the IWC moratorium.

The failure of international efforts to regulate whales has caused an interesting reaction. In 1976, the United States adopted legislation linking the permitted harvest of fish within U.S. waters by foreign nations to their adherence to IWC quotas. Thus, when Japan refused to abide by the IWC's moratorium in 1983, the United States withheld 100,000 metric tons of Japan's 1984 allocation of fish to be caught off Alaska and threatened to limit fishing further if Japan's objections to the IWC were not withdrawn. Several other nations adopted a similar strategy.

In the United States, marine mammals are protected under the Marine Mammal Protection Act, which was passed in 1972. This legislation is similar to the Endangered Species Act (see Chapter 8) but applies only to marine species. The act places a moratorium on harvesting these animals in U.S. territorial waters and prohibits the import of animals or animal products except for public display or scientific purposes. Currently, 20 marine mammals are listed as endangered and 2 as threatened.

Multinational agreements on marine mammal protection are limited, however. Over 92 countries have some type of law regarding marine mammals, but only three (New Zealand, the United States, and the Republic of the Seychelles) have comprehensive protection of marine mammals. The Republic of the Seychelles has gone so far as to declare itself a marine mammal sanctuary and can impose a sentence of up to five years in prison on anyone who kills or harasses a marine mammal. Unfortunately, an agency such as the IWC has no real enforcement powers and must rely on economic sanctions and on the diplomacy of its member nations. Nonwhaling nations such as the United States often cite the need to protect whales as an aesthetic resource. But this perception does

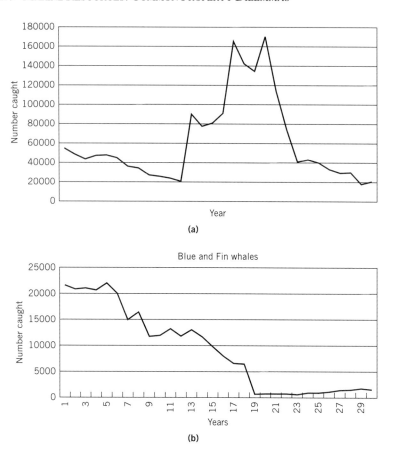

Figure 9.10 Whale (blue, fin, sperm, pilot) catches, 1970–1999. Since the 1990s whale catches have steadily declined, and they have remained flat as a consequence of the 1986 IWC moratorium on whaling. *Source:* United Nations FAO, 2001.

not carry much weight in whaling nations, such as Japan, Norway, and Russia, where whales have economic value. Thus, international management of common property resources such as whales often depends on resolving differences over the perceived value of the resource in question.

Conclusions

As a common property resource, the ocean is accessible to all nations. Responsibility for allocation and management of the ocean's vast resources is clouded by lack of certainty surrounding ownership of the resource. International efforts, such as the IWC and the Law of the Sea Treaty, are important steps in recognizing this dilemma and in appreciating the value and wisdom of conserving this resource.

One of the most critical issues facing our marine environment is coastal development, which destroys valuable wetlands and coral reefs. In addition, municipal sewage discharge and dredging sediments add to the destruction of estuarine environments, the most biologically productive of all. Oil and hazardous material releases in coastal, territorial, and international waters continue to pose problems. Last, and perhaps most important, the overexploitation of marine fisheries is a major worry. Because of this overexploitation, many nations are turning to fish farming (or aquaculture) to meet increasing food demands. Aquaculture normally takes place in near-shore estuarine environments, precisely those environments that are subject to intense land-based development and pollution.

Historically, the oceans were thought of in superlative terms, capable of providing food, transport, recreation, solitude, and dumping grounds in an apparently unlimited supply. As with other once-"infinite" natural resources, however, the oceans, like the air, are now an increasingly regulated part of the global village. Marine science research indicates that, like the atmosphere above, the oceans are integrative and interactive. Pollutants travel worldwide, and the effects of overfishing are global as well. However, marine scientists are also quick to admit that we still have very little understanding of how the oceans operate and what their role is in climate change. Hence, there is a two-sided battle between, on the one hand, the conservation and political leaders of the world's wealthy nations, who call for research and ocean-use restrictions, and, on the other hand, the emerging countries of the world that are trying to stake a claim to their share of the last great common property resource.

REFERENCES AND ADDITIONAL READING

Albarede, F. and S. L. Goldstein, 1992. World map of Nd isotopes in sea-floor ferromanganese deposits. *Geology* 20:761–763.

Audubon. 1989. Wreck of the *Exxon Valdez* (special issue). *Audubon* 91(5):74–111.

Barber, C. V. and V. R. Pratt. 1998. Poison and profits: cyanide fishing in the Indo-Pacific. *Environment* 40(8):4–9, 28–34.

Bohnsack, J. A. 1996. Marine reserves, zoning, and the future of fishery management. *Fisheries* 21(9):14–16.

Broad, W. J. 1994. Plan to carve up ocean floor riches nears fruition. *The New York Times*, March 29, 1994, p. C1.

Brownell, R. L., Jr., K. Ralls, and W. F. Perrin, 1995. Marine mammal biodiversity: Three diverse orders encompass 119 species. *Oceanus* 38(2):30–33.

Burger, J. and M. Gochfeld. 1998. The tragedy of the commons 30 years later. *Environment* 40(10):4–13, 26–27.

Cicin-Sain, B. and R. W. Knecht. 2000. *The Future of U.S. Ocean Policy*. Washington, D.C.: Island Press.

Clark, J. R. 1996. *Coastal Zone Management Handbook*. Boca Raton, FL: Lewis Publishers.

Committee on Protection and Management of Pacific Northwest Anadromous Salmonids (CPMPNAS). 1996. *Upstream: Salmon and Society in the Pacific Northwest*. Washington, D.C.: National Academy Press.

Council on Environmental Quality. 1995. *25th Annual Report*. Washington, D.C.: U.S. Government Printing Office.

_____. 1981. *Environmental Trends*. Washington, D.C.: U.S. Government Printing Office.

_____. 1982. *Environmental Quality 1982. 13th Annual Report*. Washington, D.C.: U.S. Government Printing Office.

CIESIN. http://sedac.ciesin.org/entri.

Fairhall, D. and P. Jordan. 1980. *The Wreck of the Amoco Cadiz*. New York: Stein & Day.

Ferber, D. 2001. Keeping the Stygian waters at bay. *Science* 291:968–973.

Food and Agriculture Organization. 2001. *Fisheries*. http://www.fao.org/waicent/faoinfo/fishery/fishery.htm.

Foster, N. M. and J. H. Archer. 1988. The National Marine Sanctuary Program: Policy, education, and research. *Oceanus* 31(1):5–17.

Freudenburg, W. R. and R. Gramling, 1994. *Oil in Troubled Waters*. Albany: SUNY Press.

Glasby, G. P. 2000. Lessons learned from deep-sea mining. *Science* 289:551–552.

Gross, M. G. 1971. *Oceanography*. 2nd ed. Columbus, OH: Merrill.

Heath, G. R. 1982. Manganese nodules: Unanswered questions. *Oceanus* 25(3):37–41.

International Tanker Owners Pollution Federation Ltd. 2001. *Accidental Tanker Oil Spill Statistics*. http://www.itopf.com/datapack2001.pdf.

Knecht, R. W. 1982. Deep ocean mining. *Oceanus* 25(3):3–11.

Mariner Group, 2001. *Oil Spill History*. http://www.marinergroup.com/oil-spill-history.htm.

McGinn, A. P. 1999. *Safeguarding the Health of Oceans. Worldwatch Paper #145*. Washington, D.C.: Worldwatch Institute.

_____. 1998. *Rocking the Boat: Conserving Fisheries and Protecting Jobs. Worldwatch Paper #142*. Washington, D.C.: Worldwatch Institute.

Mulvaney, K. 1998. A sea of troubles. *E Magazine* 9(1):28–35.

National Marine Fisheries Service. 2001. *Endangered Species Act: Status Reviews and Listing Information.* http://www.nwr.noaa.gov/1salmon/salmesa.

National Research Council. 1999. *Sustaining Marine Fisheries.* Washington, D.C.: National Academy Press.

———. 2001. *Marine Protected Areas: Tools for Sustaining Ocean Ecosystem.* Washington, D.C.: National Academy Press.

Naylor, R. L., R. J. Goldburg, H. Mooney, M. Beveridge, J. Clay, C. Folke, N. Kautsky, J. Lubchenco, J. Primavera, and M. Williams. 1998. Nature's subsidies to shrimp and salmon farming. *Science* 282:883–884.

Odum, E. P. 1971. *Fundamentals of Ecology.* Philadelphia: W. B. Saunders.

Ogden, J. C. 2001. Maintaining diversity in the oceans: issues for the new U.S. administration. *Environment* 43 (3):28–37.

Pauly, D., V. Christensen, R. Froese, and M. Lourdes Palomares. 2000. Fishing down aquatic food webs. *American Scientist* 88:46–51.

Perry, S. L., D. P. DeMaster, and G. K. Silber. 1999. The great whales: history and status of six species listed as endangered under the U.S. Endangered Species Act of 1973. *Marine Fisheries Review* 6(1):1–6. http://spo.nwr.noaa.gov/mfr611.htm.

Ramsar. 2001. http://www.ramsar.org.

Revkin, A. C. 2000. Deep peril for deep-sea corals. *The New York Times*, September 19, 2000:D5.

Russell, D. 1996. The world's fisheries: State of emergency; The crisis comes home. *E Magazine* September/October:38–41.

Sissenwine, M. P. and A. A. Rosenberg. 1993. U.S. fisheries. *Oceanus* Summer:48–55.

Squires, D. F. 1983. *The Ocean Dumping Quandary: Waste Disposal in the New York Bight.* Albany, NY: SUNY Press.

Stevens, W. K. 2000. The oceans absorb much of global warming, study confirms. *The New York Times*, March 24, 2000:A14.

Swanson, R. L. and M. Devine. 1982. Ocean dumping policy. *Environment* 24(5):14–20.

Tolba, M. K. et al., eds. 1992. *The World Environment 1972–1992.* London: Chapman & Hall.

United Nations Environment Programme. 2002. *Regional Seas.* http://www.unep.ch/seas/rshome.html.

United Nations Environment Programme. 1993. *Environmental Data Report 1993–1994.* Cambridge, MA: Blackwell.

U.S. Bureau of the Census. 1999. *Statistical Abstract of the United States.* Washington, D.C.: U.S. Government Printing Office.

U.S. Coast Guard. 2001. *Oil Spill Compendium Data Tables.* http://www.uscg.mil/hq/nmc/response/stats/summary.htm.

U.S. Congress. Office of Technology Assessment. 1987. *Wastes in Marine Environments.* OTA-0334. Washington, D.C.: U.S. Government Printing Office.

U.S. Department of Commerce, National Oceanic and Atmospheric Administration, National Marine Fisheries Service. 1985. *Annual Report 1984/85 Marine Mammal Protection Act of 1987.* Washington, D.C.: U.S. Government Printing Office.

U.S. Energy Information Agency. 2000. *Energy Data Reports.* http://www.eia.doe.gov/emeu/aer/txt/tab0502.html.

U.S. National Marine Fisheries Service. 2002. *Fisheries of the U.S. 2000.* http://www.st.nmfs.gov/st1/fus/fus00/index.html.

VanDeveer, S. D. 2000. Protecting Europe's seas: lessons from the last 25 years. *Environment* 42(6):10–26.

Verhovek, S. H. 1999. Returning river to salmon, and man to the drawing board. *The New York Times*, September 26, 1999:A1.

Walting, L. and E. A. Norse. 1998. Disturbance of the seabed by mobile fishing gear: a comparison to forest clearcutting. *Conservation Biology* 12(6):1180–1197.

Weber, M. L. 2001. *From Abundance to Scarcity: A History of U.S. Marine Fisheries Policy.* Washington, D.C.: Island Press.

Wilber, R. J. 1987. Plastic in the North Atlantic. *Oceanus* 30(3):61–68.

Winslow, R. 1978. *Hard Aground. The Story of the Argo Merchant Oil Spill.* New York: W. W. Norton.

World Resources Institute. 1988. *World Resources 1988–89.* New York: Basic Books.

———. 1996. *World Resources 1996–97.* New York: Oxford University Press.

———. 2000. *World Resources 2000–2001.* Washington, D.C.: World Resources Institute.

World Wildlife Fund. 2001. *Current Status of the Great Whales—July 2001.* http://www.panda.org/species/whales/status.cfm.

Worldwatch Institute. 2001. Fish farming may soon overtake cattle ranching as a food source. *Worldwatch*

Issue Alert Data and Graphs (IA9). http://www.worldwatch.org/chairman/issue/001003d.html.

Yoon, C. K. 1998. A "dead zone" grows in the Gulf of Mexico. *The New York Times*, January 20, 1998: B11.

STUDY QUESTIONS

1. What importance do estuaries and coastal areas have for marine fisheries?
2. What are the major factors contributing to the decline of marine fisheries?
3. Describe some of the major achievements and failures of the International Whaling Commission. What are the major factors that have contributed to these successes/failures?
4. What waste materials are commonly dumped in the oceans? Is this dumping any worse than land disposal of these wastes?
5. How effective is the protection provided, both internationally and in the United States, for marine mammals?

CHAPTER 10

WATER QUANTITY AND QUALITY

The necessity of water as a resource for all life, humans included, is obvious. All life forms require it to some degree, and they require water within certain ranges of quality. Water is essential to life in part because it is an excellent transporter of other substances. It is sometimes called the universal solvent in that a wide range of chemicals (though certainly not all chemicals) are readily dissolved in it. As a result, water serves to deliver nutrients to organisms, especially plants, as well as remove waste products. The movement of water between the atmosphere, the oceans, and the land is a fundamental part of earth's biogeochemical cycling systems. Water carries heat to the atmosphere when it evaporates from the ocean surface and condenses in the atmosphere to form precipitation. It also carries many other substances with it as it flows from the land to the sea or when it falls as precipitation.

As described in Chapter 4, humans have dramatically altered the hydrologic cycle, both locally and globally. These alterations come from our use of water for irrigation, our modification of the earth's vegetation cover, and our withdrawal of water from rivers, lakes, and subsurface aquifers for domestic and industrial use. The quantity of water we are withdrawing is enormous. In some regions, no more water is available for withdrawal, and once-large reservoirs, both underground and on the surface, are significantly depleted. In other regions, what water is available is so polluted it is not usable.

Water-quality and water-quantity issues are closely linked. The more water we use, the more wastewater we generate. If water is withdrawn from a river for human use, then less water remains in the river for waste dilution, and stream quality suffers. Poor water quality limits the human use of water supplies, and demands for improved water quality are placing significant constraints on expanding water use.

The combination of significant water-supply depletion and water-quality degradation means that water resources are increasingly stressed, especially in the more populated areas of the world. In the Middle East and south and central Asia, for example, water-quality and -quantity problems are already acute, straining international relations. In many wealthier countries, water issues are major constraints on future resource development. In this chapter we examine these problems by first reviewing issues associated with water quantity and quality and then discussing the interactions between the two.

WATER SUPPLY AND ITS VARIABILITY

Water storages on the land are of two basic types: surface water and groundwater. Both of these function as important storages in the hydrologic cycle. *Surface water* is liquid water and floating ice above the ground surface, in rivers, swamps, lakes, or ponds. It is derived from direct precipitation or from subsurface sources.

Groundwater is water below the ground surface, in a saturated zone below the *water table* (Fig. 10.1). The water table is simply the top of the saturated zone in which water fills pore spaces and cracks in rocks or sediments. Soil moisture above the water table is not considered part of groundwater. Groundwater is derived from downward percolation of rainfall through the soil and in some areas from seepage of surface water. In addition, many areas of the world have substantial "fossil" groundwater storages that were derived from past

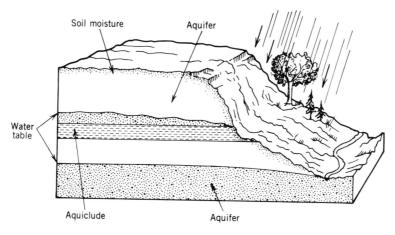

Figure 10.1 Groundwater and surface water. Water that is below the surface, filling spaces in rocks or sediments, is called groundwater. Surface water is water in lakes, streams, and reservoirs.

humid conditions but are not being significantly replenished today. A porous body of material containing groundwater is called an *aquifer*. If the water table is free to rise with additional water, the aquifer is said to be unconfined; if there is an impermeable layer overlying the aquifer, it is described as confined. Such impermeable layers are called *aquicludes*, and they are particularly important in segregating relatively clean groundwater from brackish or contaminated groundwater.

Surface water and groundwater flow from high to low elevations. Surface water flows according to the shape of the land, following channels to the sea. But groundwater flows according to the slope of the water table and the permeability of the materials through which it moves. *Permeability* refers to the speed with which water will flow through a porous medium such as rock or sediments. The steeper the slope of the water table, or the greater the permeability of the ground, the faster water will flow. Usually, the shape of the water table approximately parallels the shape of the land, so that groundwater flows from upland areas toward lowlands, but this is not always the case. For example, variations in the permeability of subsurface materials may affect flow rates and directions, sometimes causing drainage divides for groundwater to be different from those for surface water. The locations of permeable and impermeable areas of rock or sediments not only affect groundwater flow patterns but also determine where water is stored and available.

Spatial Variation in Surface Supply

The renewable supply of fresh water is directly determined by precipitation and evapotranspiration rates (Chapter 4), with runoff being the difference between the two (Table 10.1). The greatest water availability may not be in areas of high precipitation,

Table 10.1 Surface Water Balances

Continent	Precipitation (km^3/yr)	Evapotranspiration (km^3/yr)	Runoff (km^3/yr)
Europe	8,290	5,320	2,970
Asia	32,200	18,100	14,100
North America	22,300	17,700	4,600
South America	18,300	10,100	8,180
Australia/Oceania	7,080	4,570	2,510
Antarctica	2,310	0	2,310

Source: Gleick, 2000.

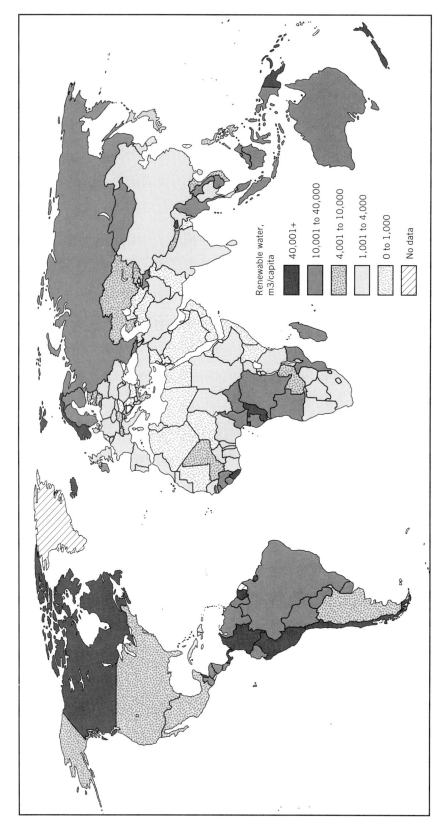

Figure 10.2 Renewable supply of water per capita, cubic meters per year, 2000. Data from World Resources Institute, 2000.

if these areas also have high evapotranspiration. Figure 10.2 is a world map of renewable water supply. Rainfall in some areas, such as the Amazon basin, is so great that very high amounts of runoff occur despite high evapotranspiration rates. Tropical areas with seasonal precipitation patterns such as the monsoon areas of south Asia have high amounts of water available in the rainy season but limited water supply in the dry season. Relatively high amounts of runoff occur in humid portions of the midlatitudes, where evapotranspiration rates are lower.

Unfortunately, the world's population is not evenly distributed with respect to water availability; thus, much of the world's water is not accessible to large population concentrations. About one-sixth of the world's renewable fresh water flows in the Amazon, a relatively unpopulated region. In contrast, India and China have about 40 percent of the world's population, yet their climates yield only about 12 percent of the world's fresh water. On a per capita basis, Asia is the driest continent, with an average of fewer than 4000 cubic meters of water per person per year (Table 10.2). In reality, most Asians have much less water available than this, because much of that continent's runoff is found in relatively unpopulated portions of Siberia. North and Central America have, on average, about eight times as much water per person as Asia. But this average also is deceiving, for most of the runoff from North America is found in the northwestern and eastern parts of the continent, with relatively little in the southwestern United States and Mexico. In the mid-1990s, 40 percent of the world's population lived in water-stressed (<1700 m^3 per capita per year) and water-scarce (<1000 m^3 per capita per year) nations (Gleick 2001a). It is estimated that slightly more than 2.3 billion people live in water-stressed regions, and this is projected to increase to 3.5 billion in 2025 (World Resources Institute 2000). Other estimates suggest a more serious picture—one-third of the world's people already live in water-stressed

Table 10.2 Global Water Availability

Region/Country	Renewable Water Supply (km^3/yr)	Renewable Water per Capita (m^3/yr)
Africa	4,040	5,152
Kenya	20	672
Congo, Dem Rep	935	18,101
Asia	13,508	3,949
China	2,812	2,201
Japan	430	3,393
Europe	2,900	3,981
Italy	160	2,804
Norway	382	85,560
Oceania	2,400	7,886
Australia	352	18,638
Papua New Guinea	801	166,644
North/Central America	7,770	25,105
Costa Rica	112	27,936
United States	2,460	8,838
South America	12,030	34,791
Brazil	5,418	31,849
Chile	928	61,007
World	42,650	7,044

Source: World Resources Institute, 2000.

regions, and this proportion will increase to two-thirds by 2025 (UNEP 2002)!

Temporal Variability

Runoff, the renewable supply of fresh water, is extremely variable in time, and usually the water is least available when it is most needed. Mean monthly discharges for rivers typically vary by one or even two orders of magnitude, depending on seasonal amounts of precipitation and evapotranspiration. This means that if the average flow in the driest month is 5 m^3/sec, the flow in the wettest month may be 50 or 500 m^3/sec. Flows are more variable in small rivers and less variable in large rivers. In the tropics, seasonal variations in river flow usually correspond to seasonal patterns of rainfall. In midlatitude climates, low-flow periods usually occur in the summer because plants are using more water at this time. Summer is also the time when demand for water is higher, because crops and lawns must be irrigated.

Because of this temporal variability, the amount of water we can count on withdrawing from a river is much less than the total amount that flows in it over the year. In addition, precipitation variations from one year to the next further reduce the amount of water we can depend on from rivers. An example of short- and medium-term variation in peak stream flow on the Missouri River is shown in Figure 10.3. The drought years of the early 1930s stand out and serve as the baseline record of low flow, because there has not been such a prolonged period of drought since that time. Similarly, the high-flow periods of 1993–1995 with the associated floods are also apparent.

Water Supplies and Storage

As you have seen, the natural supply of water in the world is highly variable, with some countries (like Brazil) having very large renewable supplies of water per capita and others (like China) having to divide a modest amount of water among a large number of people. But natural supply alone does not ensure water availability. As Figure 10.4 shows, any water-supply system must have the following four components: collection system, storage facility, transportation system, and distribution system. Water-supply engineers design and construct water systems in a variety of ways and, where possible, incorporate natural features in one or more of these components. In virtually all cases, the collection system is natural: it is the *drainage basin* of a river, a groundwater aquifer, or some combination of the two. Rivers are particularly efficient concentrators of surface runoff. As a result, usually little is done to modify collection systems, although vegetation conversion to increase water yield or improve water quality has been used in some areas. Aquifers are much more dispersed conveyors of water. Water flows toward low points in the water table, and when a well is drilled to pump water out, the local water table is depressed. This causes water to flow toward the well, which is exactly what is desired. By drilling

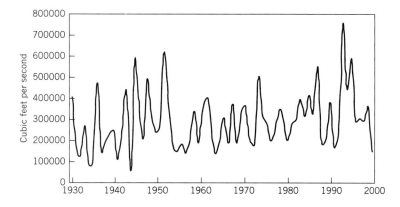

Figure 10.3 Average annual peak discharge of the Missouri River at Hermann, Missouri, 1929–2000. Seasonal variations in discharge have been smoothed out of the data. Flows in wet years are more than four times greater than those in dry years. *Source:* U.S. Geological Survey, 2002.

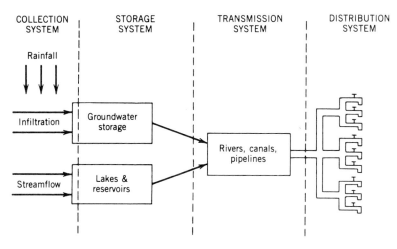

Figure 10.4 Components of a water-supply system. The component with the lowest capacity limits the capacity of the entire system.

wells in particularly porous, permeable underground materials, we can tap into aquifers that have a ready supply of available water.

Storage is necessary to smooth out the natural variations in water availability and to save surplus water from high-rainfall seasons for dry seasons or periods of high demand. Under ideal conditions, a storage facility can allow average withdrawals to equal long-term average flow, and short-term withdrawals can far exceed average flows. In practice, however, average withdrawals are rarely this large. Storages cannot trap all the water during times of flood, and water must be left for in-stream uses. Nonetheless, short-term withdrawals—for periods of weeks or less—frequently exceed average inflows in large reservoirs.

Surface-water storage is accomplished by constructing dams on rivers and impounding water in artificial lakes behind the dams. The amount of water that can be stored is a function of the shape of the valley and the height of the dam. The ideal dam site is a relatively narrow, deep valley (where the dam is built), with a broad, deep valley just upstream. In addition, the valley that is to be inundated should be underlain by impermeable rocks and be relatively unpopulated, and the land should have lower long-term value than the reservoir replacing it.

Transportation and distribution systems can be of many types, depending mostly on the distance between collection site and use area and the nature of the final use. In many cases, transportation distances are so short that the entire system is essentially just a distribution system. These facilities include canals, pipelines, and natural river channels, or any combination of these (Fig. 10.5). The choice of which type of conduit to use depends primarily on terrain, the volumes of water to be

Figure 10.5 The California Aqueduct. This aqueduct carries water from the northern Sierra Nevada to agricultural lands in California's Central Valley.

carried, the distances involved, and the need to protect against seepage or quality deterioration along the way.

It should be clear by now that, although water supply is constrained by natural factors, water development in the form of engineering works also affects water availability. The extent of water development can be evaluated only relative to what is naturally available, and that in turn is subject to debate because there are different definitions of "available water." One indication of the extent of water use can be gained by comparing withdrawals to natural runoff (Table 10.3).

Before proceeding, let us define a few terms that describe these aspects of water use. *Withdrawal* is the removal of water from a surface or groundwater source for a variety of purposes such as municipal, industrial, or irrigation use. *Consumptive* use is the use of that water in such a way that it is not returned to the stream or aquifer; instead, it is returned to the atmosphere by evapotranspiration. *In-stream* uses do not require removal of the water from a river or lake; these include navigation, wildlife habitat, waste disposal, and hydroelectric power generation.

Withdrawals can exceed stream flow because not all of the water withdrawn is consumed; some is returned to the stream. Nonetheless, these withdrawals place a heavy demand on water resources, particularly because they compete with in-stream uses. In the Colorado River basin, for example, a series of power plants at major dams generates about 4 percent of the nation's hydroelectric power, and plants in the Pacific Northwest (mostly in the Columbia River basin) generate almost 50 percent of U.S. hydroelectric power. If water is withdrawn and consumed rather than returned for this in-stream use, energy production will be drastically reduced.

In more densely populated areas of the country, the most important in-stream use is maintenance of water quality. Sufficient flow must be available to dilute and transport sewage effluents and other pollutants, as well as to provide habitat for aquatic life. The U.S. Fish and Wildlife Service estimates the flows necessary to support aquatic habitat and recreation. It has found these flows to be generally 80 to 90 percent of total stream flow in the eastern United States and 40 to 60 percent of total flow in most of the western states (Water Resources Council 1978). Navigation is another important in-stream use that competes with other in- and off-stream uses for the water in our rivers. Depletion of stream flows caused by consumptive off-stream use, particularly irrigation, is a major problem in semiarid and arid portions of the United States (Fig. 10.6).

One way to overcome this variability is to build reservoirs to store excess water from the high-flow season for use in the low-flow season. Around the world, over 45,000 large dams have been built for a variety of purposes, including hydroelectric power generation, flood control, and water storage for irrigation use (World Commission on Dams 2000). About half of these dams are in China, and about 5500 are in the United States. The total storage capacity of the reservoirs they impound is over 6300 km^3, or 15 percent of the world's total annual runoff, and this is growing rapidly. In some regions, the amount stored in reservoirs is much greater than the annual flow. In the Colorado River basin, for example, reservoirs store about four years' average runoff.

In addition to these artificial storages, many natural lakes provide freshwater supplies. The Great Lakes of North America are the principal water supplies of Chicago, Cleveland, Toronto, and other cities on their shores. Unlike reservoirs, however, these lakes cannot be easily drawn down to provide water during dry spells without severely affecting ecosystems and coastal communities.

Groundwater is a more important storage of water for human use. The total volume of water stored in relatively accessible groundwater aquifers is estimated at about 9000 km^3, or roughly one-fourth of global annual runoff. Much more—perhaps as much as 4 million km^3—exists in deeper aquifers, though most of this amount is not economically accessible. Most small-scale and domestic

Table 10.3 Global Water Withdrawals

Continent	Available Water[a] (%)
Europe	16.4
North and Central America	7.8
Africa	3.7
Asia	14.9
South America	1.2
Oceania	1.0
Total	8.0

[a]Withdrawal of surface water/natural availability, expressed as a percentage.
Source: Gleick, 2000; World Resources Institute, 2000.

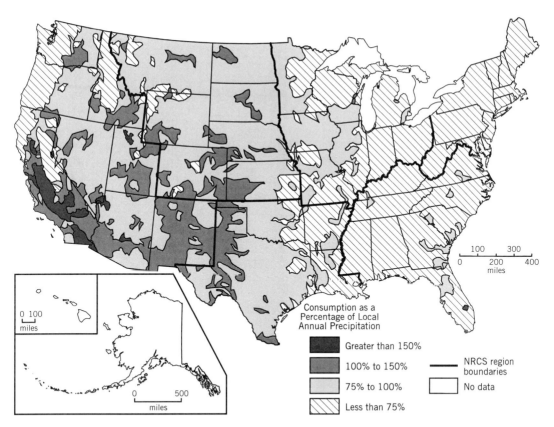

Figure 10.6 Water-depletion areas of the United States. The east generally has a surplus of water, while the arid west is rapidly depleting its water supply. *Source:* U.S. Department of Agriculture, 1989:71.

water-supply systems use groundwater, whereas large industrial and commercial users depend mainly on surface water.

Typically, groundwater storages are replenished relatively slowly, taking years to centuries or more to replace the total volume of a given aquifer. As a result, it is possible to withdraw water much faster than it is replaced, a practice known as *groundwater mining*. This practice is all too common, especially in the semi-arid and arid parts of the world. In a few countries in the Middle East, total withdrawals of water exceed the renewable supply, indicating significant *overdraft* of groundwater at the national level. In many other countries, including the United States, groundwater overdrafts are common at the local or regional level.

One impact of groundwater overdraft is declining well levels, often requiring that wells be deepened for withdrawals to continue. In coastal areas, usually a boundary exists between fresh water and salt water in the ground. Salt water is denser and thus is found underneath the fresh water. A decline in the elevation of the freshwater table causes *saltwater intrusion*, an inland movement of the salt/fresh boundary, which contaminates wells and makes them unusable for drinking water. When this happens, generally the only recourse is to close the wells and find alternative sources of water, most often wells farther inland. This problem is particularly acute on the coastal plain of the eastern United States and in some areas of coastal California (Fig. 10.7). There are also examples of saltwater intrusion into inland aquifers in areas where saline groundwater underlies fresh water. In some areas, notably coastal Texas, southern Arizona, and central California, groundwater overdrafts are causing *subsidence*, or sinking of the land. In Texas, this is contributing to coastal flooding, particularly in suburban Houston, while in Arizona large fissures have opened in the ground.

In the western Great Plains, the most important overdraft problem is in the area underlain by the

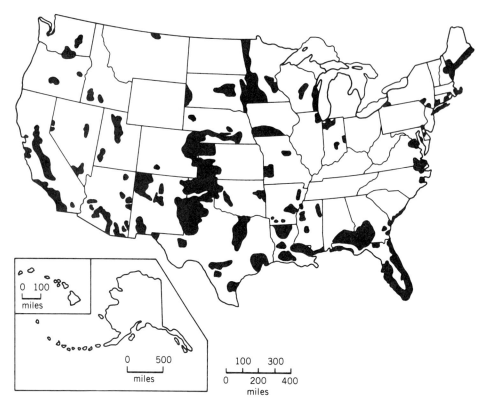

Figure 10.7 Areas of the United States where groundwater decline is of state or local concern. *Source:* U.S. Department of Agriculture, 1989:71.

Ogallala aquifer. This aquifer is a thick, porous layer of sand and gravel that underlies an extensive area from Nebraska to Texas. It contains a large amount of water but has an extremely low recharge rate. Because most of this area is too dry for rainfall farming, groundwater-based irrigation has been rapidly expanding since the 1950s. The rate of withdrawal is enormous, exceeding the recharge rate by 100 times in some areas. The Ogallala aquifer initially allowed the rapid development of irrigated agriculture in the High Plains, but its depletion will lead to an end of irrigated agriculture in the region and hasten the decline of economic growth. In the Arkansas-White-Red rivers region, which includes much of the Ogallala aquifer, groundwater overdrafts represent over 60 percent of all groundwater withdrawals. Another area of extreme overdraft is the Texas-Gulf region, where overdrafts account for 77 percent of all groundwater withdrawals. For the nation as a whole, about 37 percent of all groundwater withdrawals are overdrafts.

THE DEMAND FOR WATER

Water demands fluctuate from year to year, depending on weather patterns. In wet or cool years, demand is usually lower, whereas in dry years demand is greater. To evaluate long-term trends, it is useful to average these short-term fluctuations. For example, in the United States, both water withdrawals and water consumption have risen steadily since the 1960s (Fig. 10.8). Since the mid-1980s, freshwater withdrawals and consumption have declined somewhat. Surface water remains the primary source of water withdrawals (81 percent), a pattern evident from the 1950s. Groundwater withdrawals have increased since the 1950s.

Regional U.S. demand is greatest in the western states, especially Idaho, Montana, and Wyoming (Fig. 10.9). These states have the largest per capita withdrawals, with water used for irrigating sugar beets, potatoes, corn, and beans. The smallest per capita withdrawals are in the Northeast,

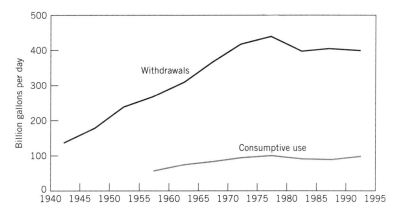

Figure 10.8 Trends in U.S. freshwater withdrawals and consumption, 1940–1995. *Source:* U.S. Bureau of the Census, 2000.

where most of this water is used for industrial purposes and steam electrical generation. However, with the growth of metropolitan areas in the arid West, more demands will be placed on this scarce resource (Issue 10.1).

Off-Stream Uses

Withdrawal and consumptive uses of water are often defined by specific types of use. These include public supply, rural supply (domestic and livestock), industrial supply, irrigation, and hydroelectric power generation (an in-stream use).

Public and rural supplies include both domestic and commercial uses of water, including those familiar to us in our everyday lives at home or at work—washing, cooking, drinking, lawn watering, sanitation, and the like. Nearly 90 percent of the U.S. population is served by municipal water-supply systems; the remainder have individual domestic systems (usually wells).

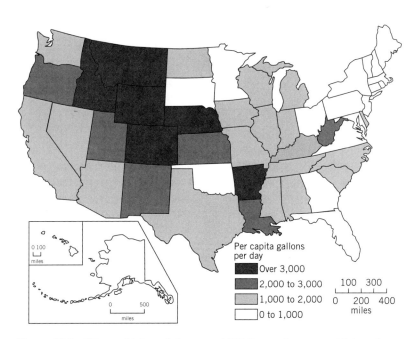

Figure 10.9 Water withdrawals by state, 1995. Per capita water withdrawals are highest in the upper Great Plains states, primarily because of irrigation. *Source:* U.S. Geological Survey, 2001.

ISSUE 10.1: WATER POLITICS IN THE WESTERN UNITED STATES

When questions on the availability and cost of water arise, elected officials in the Southwest often transform themselves into reptile-like creatures with the fangs of a rattlesnake and the changing colors of a chameleon. Water is perhaps the most parochial issue in Western politics and can force urbane, socially activist liberals to demand that government get off the backs of the people, while free market conservatives demand more government regulation over natural resources.

Access to and use of both surface water and groundwater has historically been governed by "use it or lose it" water rights laws in states like Arizona, California, Colorado, and Utah. On the basis of water rights granted to riparian property owners at the time of statehood or early land grants by state and federal governments, access to water supplies is generally governed by historic water uses. These water rights can be bought and sold independently of actual land-ownership. In addition, the water itself is a commodity subject to sale.

Two basic systems of law govern the allocation of water. *Riparian doctrine* is applied to all states east of the 100th meridian and is derived from English common law. The water is owned and controlled by those who own the riparian land, which is defined as the land adjacent to the stream or upon which the stream flows. Riparian land-owners are allowed to use the water as long as their use does not substantially reduce the quantity or quality of water available to other riparian users. Furthermore, the rights cannot be lost because of nonuse.

In the West, a different system prevails. There water is allocated with a *prior appropriation system* based on a "use it or lose it" principle. Prior appropriation recognizes that the water is a finite resource and allocates the amount of available water to users on a predetermined basis. This protects agriculture and, with the heavy governmental subsidies up until the 1980s, has allowed water prices in the West to remain low.

Since the 1930s, the federal government has been the primary developer of new water supplies in the West. Federal agencies such as the Bureau of Reclamation and the Bureau of Land Management have had significant impact on the Western waterscape, ranging from major dam construction to more localized irrigation and canal systems. These large-scale water development projects helped populate the West and brought agriculture to the region. However, the days of large-scale, government-sponsored water projects are over, and the West is now entering a new phase in water conflicts: city dwellers versus farmers and ranchers.

Demand for water in the West has shifted from the agricultural to the urban sector. Because of a virtual halt in new federal water projects, the politics of water periodically heat up, pitting farmers and ranchers against city dwellers. Since urbanites are willing and can pay more for water than farmers and ranchers, they are winning the battle over water at the moment. In Arizona, for example, urban water use exceeds natural runoff, creating a deficiency. In 1980, the state forbade further

---

Agricultural uses, principally irrigation, consume more fresh water than any other use. Worldwide, agriculture uses about 71 percent of total freshwater withdrawals (Table 10.4). This portion tends to be higher in developing than in industrialized countries, even those with relatively high amounts of rainfall. In addition to constituting the largest withdrawal of water, irrigation is by far the most consumptive of all major uses. Typically 80 percent or more of irrigation water is used consumptively rather than being returned to surface water.

In the United States, about 42 percent of water withdrawals are for irrigation. To conserve more of the water, irrigation systems must become more efficient. *Irrigation efficiency* is defined as the volume of applied water in the root zone that is used by the crop. It is expressed as a percentage of the volume of water diverted from surface sources or pumped from groundwater supplies (U.S. Department of Agriculture 1989:75). Drip irrigation is one of the most efficient application methods (90 percent), while flood, furrow, and sprinklers average between 60 and 80 percent efficiencies.

Industry takes the second-largest share of the world's water withdrawals, about 20 percent. The portion withdrawn for industry is highest in Europe, where it averages 45 percent. Industrial uses include a wide range of activities, including water used for washing products in the manufacturing

groundwater overdrafting in the Phoenix and Tucson regions. Urban development continued unabated, however, as the urban centers merely found other sources of drinking water for their residents. Phoenix, for example, bought 50,000 acres (20,235 ha) of farms (many in adjacent counties) including their water rights, and pumped the groundwater into the Central Arizona Project canals for use as municipal drinking water. Salt Lake City and St. George, Utah, have bought more than 100,000 acre-feet of water by purchasing shares of canal and ditch companies. To quench their thirst, metropolitan regions in the arid West look to more rural counties and purchase agricultural land to get the water. The water is now worth more than the land itself. A precedent for such water grabs was fictionalized in the movie *Chinatown*, which was based on a real event. Owens Valley, California, is located about 300 miles (483 km) northeast of Los Angeles. Between 1920 and 1950 the city of Los Angeles bought 75 percent of the valley's land and thus its water. The water was then sent by canal to Los Angeles. Some argue that this event helped shape the destiny of Los Angeles, making it one of the country's largest metropolitan regions.

What can be done about the thirsty West? Although some Westerners argue for halting development and sending the new immigrant easterners packing, this is unlikely. Alternative sources of water, conservation, or new strategies for water allocation must be developed to meet current and projected demands. Marketing water that has already been allocated is now viewed as a substitute for government-subsidized water projects. Water has long been traded among Western irrigation districts, but now urban centers are vying for some of the action. Speculators and private brokers are actively involved in water marketing. Western water rights are being purchased at increasing rates by investment firms (which anticipate that their initial investment can only increase in value over time). Oil companies that originally purchased the water rights for oil shale development during the 1970s soon realized the value of this water for urban uses and to private developers. Farmers were encouraged to sell water as a commodity, not just the rights to the water. In this way, water can be bought and sold like other natural resources such as grain or minerals. However, many states have reacted strongly. Texas, for example, completely outlaws interbasin transfers of water, and New Mexico and Wyoming prohibit out-of-state water transfers, largely to prevent thirsty Californians from taking their water. So while the metropolitan regions of the West increase in population and thirst, their search for water continues. Only time will tell how effective water marketing is in quenching the insatiable demand for water in the West (Reisner 1986, Steinhart 1990).

process, removing waste materials, and cooling. The greatest withdrawals of water in the industrial sector are for cooling thermal electric power plants, typically in once-through cooling systems. Very little of this water is consumed, however.

The manufacturing of paper and steel, the production of petroleum, coal, and chemicals, and food processing are also important industrial uses of water. Most of this water is used for cooling or washing purposes and thus is not heavily consumptive. Although industrial withdrawals are generally increasing, in some cases they are actually decreasing as water-quality controls force factories to remove wastes before discharging water and to release less. Industrial users are turning away from once-through systems toward water-recycling systems in which they use water over and over, reducing discharges of waste to the environment.

Domestic uses take the least water, generally less than 10 percent, except in urbanized regions with relatively less industry and irrigation, such as South America and Oceania. Among the important domestic uses are cooking, laundry, bathing, toilet flushing, and, in North America, lawn irrigation. Total water use in public water systems in the United States averages about 200 gallons (750 liters) per person per day, including commercial uses such as restaurants and other service businesses. Domestic water use is not

Table 10.4 Water Withdrawn for Various Purposes[a]

Region/Country	Per Capita (m³/year)	Purpose of Water Withdrawal (%)		
		Domestic	Industry	Agriculture
Africa	242	9	6	85
Kenya	87	20	4	76
Congo, Dem Rep	8	61	16	23
Asia	675	7	9	84
China	439	5	18	77
Japan	735	19	17	64
Europe	660	14	45	41
Italy	1,005	17	37	45
Norway	488	27	68	38
Oceania	919	57	14	30
Australia	839	12	6	70
Papua New Guinea	28	29	22	49
North/Central America	2,189	11	42	47
Costa Rica	1,540	13	7	80
United States	1,677	8	65	27
South America	477	20	11	69
Brazil	359	21	18	61
Chile	1,634	5	11	84
World	648	9	20	71

[a]The agriculture category includes plant irrigation and livestock, industry includes water for cooling thermoelectric plants, and domestic includes residential and commercial users, as well as municipal providers.
Source: World Resources Institute, 2000.

heavily consumptive—only about 8 percent in the United States, and much of this is in irrigating lawns.

Within the United States, the states with the highest consumptive use are Arizona, California, Kansas, Nebraska, Nevada, and New Mexico (Table 10.5). More than half of the fresh water withdrawals in these states are for plant and animal agriculture. California has the highest percentage consumptive use (70.2 percent), while Tennessee has the lowest (2.3 percent). It should be noted that the high consumptive states derive a substantial portion of their freshwater supplies from groundwater sources.

In-Stream Uses

In addition to these off-stream uses, many important water uses take place in rivers or lakes, without withdrawing water from them. While these uses do not result in any removal of water from the environment (except reservoir evaporation), they do require considerable amounts of water, and thus they compete with off-stream uses. Within a river basin, water taken in one area may not be available in another. For example, if water is held in storage in a reservoir, then it is not part of downstream flow. If water must be released from a reservoir to generate hydroelectric power, then it cannot be held as storage for dry-season use.

Waste Dilution The most important in-stream use of water is for waste dilution. Virtually all rivers in populated areas are used to remove wastes. The more water present and flowing in a river, the lower the concentration of pollutants will be, and thus the better water quality will be. So just removing water from a river reduces water quality, regardless of whether any pollutants are

Table 10.5 Water Source and Consumptive Use, by State, 1995

State	Source (%) Surface	Source (%) Ground	Consumptive Use (%)[a]	State	Source (%) Surface	Source (%) Ground	Consumptive Use (%)[a]
Alabama	94	6	7.5	Nebraska	41	59	66.9
Alaska	60	40	11.8	Nevada	61	39	59.3
Arizona	58	42	56.2	New Hampshire	94	6	7.8
Arkansas	38	62	54.3	New Jersey	91	9	9.8
California	68	32	70.2	New Mexico	51	49	56.4
Colorado	84	16	37.9	New York	94	6	4.6
Connecticut	96	4	7.6	North Carolina	94	6	9.2
Delaware	93	7	9.4	North Dakota	89	11	16.2
Florida	76	24	38.6	Ohio	92	8	7.5
Georgia	80	20	20.3	Oklahoma	40	60	40.2
Hawaii	72	28	53.7	Oregon	87	13	40.6
Idaho	81	19	28.7	Pennsylvania	91	9	5.8
Illinois	95	5	4.3	Rhode Island	93	7	14.0
Indiana	92	8	5.5	South Carolina	95	5	5.2
Iowa	83	17	9.6	South Dakota	59	41	54.1
Kansas	33	67	69.1	Tennessee	95	5	2.3
Kentucky	93	7	7.2	Texas	70	30	43.2
Louisiana	86	14	19.6	Utah	82	18	51.2
Maine	75	25	21.7	Vermont	91	9	4.2
Maryland	97	3	10.3	Virginia	96	4	4.0
Massachusetts	94	6	15.6	Washington	80	20	34.9
Michigan	93	7	5.5	West Virginia	97	3	7.6
Minnesota	79	21	12.3	Wisconsin	89	11	6.1
Mississippi	19	81	50.8	Wyoming	95	5	39.8
Missouri	87	13	9.8	Total	81	19	29.3
Montana	98	2	22.1				

[a] The ratio of consumptive use as a percentage of total freshwater withdrawals.
Source: U.S. Geological Survey, 2001.

added in the process. This use will be discussed in more detail in the context of water quality.

Navigation The major rivers of the world, especially in industrialized countries, carry large amounts of freight. In the United States, for example, inland waterways carry about the same amount of freight (on a weight basis) as is delivered to or from ocean ports. Barge operations on rivers are constrained by river depth, for rivers are typically shallow and large barges may require 10 ft (3 m) or more of water depth to operate. Withdrawals of water from a river reduce depth and thus the usefulness of the river for navigation.

Hydroelectric Power Hydroelectric power is generated by storing water behind a dam and releasing it through turbines when electricity is needed. Hydroelectricity supplies about 11 percent of U.S. electric production, or 5 percent of total energy production. Because electricity cannot be stored in large quantities, timing of hydroelectric power production is relatively inflexible. If the peak demand for electricity occurs at the same time as peak flow (as in the Colorado River basin, for example), all the better. But if peak demand occurs when flow is low, then water availability limits electric production. In addition, the large dams best suited to generating electricity inundate large areas and alter river habitats, causing additional economic and ecological dislocations.

Wildlife Habitat and Fisheries Although many rivers are severely degraded by pollution, these systems contain habitats necessary for the maintenance of important ecological communities and sport and commercial fisheries. These habitat values depend on maintaining good water quality, which in turn depends on water quantity. If the flow in a river is depleted to the point that additions of waste cause high pollutant concentrations, then habitat suffers. One common consequence is lowered dissolved oxygen levels and subsequently impoverished fish communities. In extreme circumstances fish kills occur, caused by low dissolved oxygen or by introduction of toxic substances.

Recreation In many rivers, recreational uses—mostly fishing and boating—are significant. These uses normally require good water quality for maintaining reasonably natural conditions, good fish habitat, and safe swimming and minimizing odors. They also require adequate flow, both for maintaining water quality and for floating boats. Recreational uses also make demands on streambank areas, in addition to the water in the channel.

Much competition prevails among in-stream and off-stream uses in the populated parts of the world. Water quality has degraded, and public concern about this degradation has risen to the point that further increases in pollutant concentrations are unacceptable. The combination of population growth and rising standards of living causes increased demand for off-stream uses of water as well as increased waste generation, yet rivers cannot absorb more waste. Nor can we afford to reduce rivers' waste-assimilating capacity by withdrawing more water. Finally, concern about the ecological impacts of large dams has made such projects increasingly controversial, if not impossible. The United States virtually ceased constructing large dams around 1980; such projects continue in developing countries but with much opposition from international environmental organizations. These factors together mean that further growth in water supplies will be very limited and that per capita water use must decrease significantly in the coming decades.

WATER QUALITY

Impurities in water come from many different sources, both natural and human, and it is often difficult to separate the two. When we speak of *pollution* or pollutants, we are usually referring to substantial human additions to a stream or lake's load of an impurity or impurities. A polluted stream must be defined relative to its condition unaffected by human activity rather than in absolute terms. Similarly, acceptability of given levels of contamination depends on what use we make of the water. For drinking water, absolute levels are important, and standards for drinking water are established by governmental and other agencies.

Pollutants come from diverse human-made and natural sources. One way to classify pollutant discharges is by point versus nonpoint sources. A *point source* is a specific location such as a factory or municipal sewage outfall. A *nonpoint source* is a source that, as far as we know, originates from a large, poorly defined area. Runoff, subsurface flow, and atmospheric sources of water pollution are the primary nonpoint sources.

Some pollutants, such as iron or suspended particulates, may have very large natural sources, so that human activities only marginally increase concentrations. Other pollutants, such as synthetic pesticides, are produced only by humans. Most common impurities, however, are contributed by both human and natural processes. Therefore, except in extreme cases, human pollution is difficult to define quantitatively. Furthermore, in a complex system such as a drainage basin, a given pollutant may have many different sources, including urban runoff, industrial effluents, municipal sewage, and even atmospheric precipitation. Once in a stream system, pollutants are removed by deposition or broken down or combined with other impurities to make new substances, or their concentrations are increased by chemical or biological processes. If a known quantity of a substance is put in a waterway, the amount that leaves may be greater or less, depending on the nature of the substances and the processes acting on it. Under these circumstances, it is virtually impossible to determine accurately the relative contributions of pollutant sources or to predict future contamination levels with confidence.

Major Water Pollutants and Their Sources

The list of substances that are of concern in water-quality assessments is getting longer every year. In part, this expansion is the result of advances in the analytic capabilities of laboratories and the increasing availability of water-quality data. But still there are so many substances that could be measured in a water sample, and the analyses are so complex and costly, that usually only a few major or index pollutants are determined. Most analyses summarize pollution levels with parameters such as *total dissolved solids (TDS)* or *biochemical oxygen demand (BOD)*. The following sections briefly describe, in general terms, some major classes of pollutants, their sources, human health effects, and impacts on aquatic ecology (Table 10.6).

Disease-Causing Organisms Of the many living things found in natural or polluted waters, only a small fraction can be regarded as important pollutants from a human standpoint. These are the bacteria, viruses, and parasites that cause disease in humans and livestock. The earliest awareness of water pollution as an important human problem came from the recognition that water, particularly drinking water, transmits many diseases. Among the infectious diseases communicated largely through drinking water are cholera, typhoid fever, hepatitis, and dysentery, but many other lesser known diseases are also transmitted in this manner. Most of these are transmitted through human or animal wastes; hence sewage pollution and livestock operations are their primary sources. Many different organisms are potentially dangerous, and just a few organisms in a large amount of water may be sufficient to cause infections.

The presence of *coliform bacteria* is used as an indicator of the possibility of contamination by infectious organisms. Coliform bacteria live in great numbers in human and animal digestive systems. They are not dangerous in themselves, but their presence indicates the possibility that disease-causing organisms could also inhabit the water. Chlorination of public water supplies has eliminated these diseases from common occurrence in the developed nations, although disease outbreaks occasionally occur. In areas without such water treatment, as in most developing nations, waterborne diseases are a major problem. For example, it is estimated that billions of cases of diarrhea occur annually in Africa, Asia, and Latin America, resulting in perhaps 3 to 5 million deaths each year, mostly of children (Meybeck et al. 1989; United Nations Environment Program 1997; World Resources Institute 2000). Many other diseases are transmitted via organisms such as snails or insects that live in water; schistosomiasis and malaria are well-known examples. However, infection results from insect bites, skin contact, and other means rather than ingestion; hence, these organisms are not usually considered components of water quality.

Plant Nutrients Although aquatic plants need many different substances for growth, algal growth requires just a few key substances, primarily nitrogen and phosphorus. Nitrogen is available to plants in the form of nitrate (NO_3), nitrite (NO_2), and ammonia (NH_4), while phosphorus is available mostly as phosphate (PO_3). In natural systems, nitrogen is derived primarily from the decay of plant matter. Phosphorus, on the other hand, is made available by the weathering of phosphorus-bearing rocks and enters streams either directly in groundwater or surface water or through decay of organic matter. Nitrogen and phosphorus are found in large quantities in sewage, and they enter waterways by the decay of organic particulates and by being dissolved in sewage treatment plant effluent. Runoff from urban and rural areas is also an important source. The close association between intensive agriculture and nitrogen in streams is clearly seen in Figure 10.10. Water in densely populated areas, such as the mid-Atlantic states, also has high nitrogen concentrations, which is derived from a combination of agricultural and urban sources.

When one or both of these nutrients are the factors limiting algal growth, their introduction stimulates rapid algal growth, also called blooms. The algae then die and decay, releasing still more nutrients and adding to BOD. In swift-flowing rivers, this extra BOD loading is a relatively minor problem, but in sluggish rivers and standing bodies of water serious problems can result. One effect of increased nutrients in surface water is accelerated *eutrophication*, which is the process whereby a water body ages over geologic time, with the water becoming progressively shallower and nutrient rich. In North America, eutrophic lakes typically support species such as carp and catfish, whereas geologically young *oligotrophic* lakes support pike, sturgeon, whitefish, and other species that require higher oxygen levels or cooler temperatures. In summer, lakes commonly develop a *stratification*, or layering, which prevents mixing of bottom and surface waters. If

Table 10.6 Major Water Pollutants, Their Sources, and Environmental Effects

Pollutant Type	Major Sources	Indicator Measurements	Major Effects on Aquatic Life	Major Effects on Human Use
Disease-causing agents	Sewage releases; municipal discharges, urban and agricultural runoff, feedlots	Coliform bacteria	Few	Health hazard for human consumption and swimming, shellfish contamination
Oxygen-demanding wastes	Sewage; municipal discharges, industrial discharges	Biochemical oxygen demand (BOD), chemical oxygen demand (COD)	Oxygen depletion. Loss of diversity, elimination of intolerant organisms, fish kills	Unpleasant odor if severe
Plant nutrients	Agricultural runoff, municipal and industrial discharges	Phosphorus, nitrogen, ammonia	Increased algal growth resulting in elevated turbidity, sedimentation of organic matter, oxygen depletion (eutrophication)	Ammonia toxicity in infants
Suspended particulates	Runoff and erosion, municipal and industrial discharges	Suspended solids, turbidity	Reduced light penetration and reduced photosynthesis, alteration of stream- and lake-bottom environment by sedimentation	Filtration required before consumption; damage to turbines and other machinery
Dissolved solids	Natural sources (rock weathering), municipal and industrial discharges, urban and agricultural runoff	Total dissolved solids, pH, alkalinity	Loss of diversity, elimination of intolerant organisms, fish kills due to altered pH or toxicity of effect unless concentrations are high	Health effects at elevated concentrations of some substances
Toxic substances	Pesticides, industrial activities, runoff from developed areas	Bioassays, analyses for specific chemicals	Bioaccumulation, toxicity for some organisms	Health hazards for consumption of water or organisms that concentrate toxins
Heat	Thermal electric power plants, industrial discharges, deforested stream banks	Temperature	Oxygen depletion causing loss of diversity, elimination of intolerant organisms, increase in warm-water species	None
Radioactivity	Natural sources, mining radioactive wastes	Radioactivity	None except in extreme situations	None except in extreme situations; fear of radiation may limit use

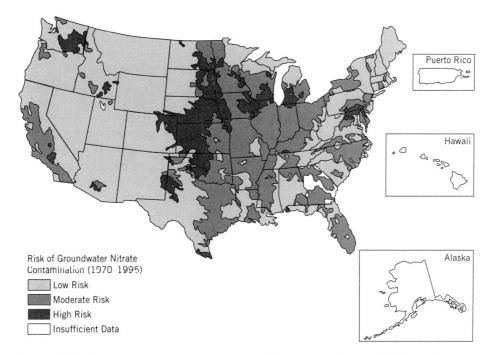

Figure 10.10 Nitrogen concentrations in continental U.S. streams. *Source:* USEPA, 2002.

algal blooms occur, the algae settle to deeper waters, where decay depletes oxygen and deep-water organisms suffocate. The absence of oxygen can also cause anaerobic decomposition of organic matter on the bottom, which produces unpleasant odors and may make water unsuitable for drinking or affect the aesthetic quality of a river or lake.

In drinking water, phosphorus is not a problem because it is an essential nutrient that humans require, and we generally ingest far more in food than in drinking water. Nitrate and nitrite, however, do present health hazards. When ingested in high concentrations, these lead to methemoglobinemia, in which the blood's ability to carry oxygen is impaired. In addition, ingestion of nitrate or nitrite may lead to the formation of compounds called nitrosamines, some of which have been found to cause cancer in animals, but the carcinogenic potential in humans is unknown.

In much of the central United States as well as parts of Canada and Europe, application of large amounts of nitrogen fertilizers to cropland has resulted in elevated nitrate concentrations in both surface and groundwater. This is particularly a problem for homes and communities that depend on this water for drinking. In the United States, around 14,000 community water systems incurred reported violations of nitrate standards in 2000, affecting the drinking water quality for more than 4 million people (USEPA 2001). The problem is less severe in most developing countries because fertilizer application rates are lower, but nitrate levels are climbing worldwide as use of nitrogen fertilizer increases.

Oxygen-Demanding Wastes Organic matter is the pollutant that places the greatest burden on a stream or lake as a pollution assimilator. Particulate organics are small bits of living or dead and decaying plant and animal matter. They are broken down by bacteria in the water, which use dissolved oxygen in the process. The most widely used measure of oxygen-demanding wastes is biochemical oxygen demand (BOD). BOD is a measure of the amount of dissolved oxygen that is required to decompose the organic matter. A stream with a high BOD loading will consequently have a low concentration of dissolved oxygen (DO). The depletion of oxygen and increased nutrient levels are primarily responsible for the ecological degradation of rivers and lakes.

Organic matter is derived from surface runoff, internal production by algae, agricultural wastes, various industries, especially food processing and paper pulp, and sewage. The relative contributions of these sources vary from one area to another, but historically large point sources have been responsible for the most severe cases of organic particulate pollution. These point sources include feedlots, pulp mills, sewage systems, and other major dischargers. In areas where treatment facilities have been installed and upgraded, such as most industrialized countries, these sources have been reduced in importance. In the United States, this most noticeable form of pollution has been significantly reduced since the 1950s. As a result, water quality has significantly improved in many streams both large and small, and fish populations have recovered dramatically.

Sediment By weight, sediment is the largest pollutant in our waters. It is measured along with organic particles as total suspended particulates in a water sample, and it consists of particles of soil and rock that are eroded from the land and from stream beds. Erosion is a natural process, and the movement of sediment through a river system helps to maintain the ecological integrity of that system. However, the accelerated erosion of agricultural lands and erosion associated with urban construction and similar activities have greatly increased the sediment loads of many streams.

For practical purposes, sediment is chemically inert and thus has little direct effect on the chemical quality of water. However, fine-textured sediment plays a role in the transport and deposition of trace substances in water. In this way, it can carry pesticides and nutrients from agricultural fields, as well as a wide range of harmful substances contained in urban runoff. Most sediment is easily filtered from water in drinking-water treatment plants, and thus health hazards associated with sediment pollution are minimal.

The major harmful effects of sediment are economic, including damage to turbines and pumps and reduction in reservoir capacity as sediment is deposited in impoundments. In extreme cases, sediment also may reduce stream channel capacity and contribute to flooding. Excessive sediment loads also modify stream habitats and restrict fish reproduction. High sediment loads associated with logging activities have contributed to the decline of salmon populations in the Pacific Northwest of North America, for example (see Issue 9.1). Some fish are sensitive to chronic high suspended-sediment loads, which clog gills, restrict vision, or otherwise interfere with normal activity. There are also a few examples, especially in the western United States, where reduction in sediment loads by reservoir construction has caused detrimental effects downstream, notably erosion in the Grand Canyon and along West Coast beaches.

Dissolved Solids Dissolved solids form a major part of the load of most rivers, and they include many different elements and compounds. Most of these are derived from rock weathering and soil leaching, and thus geographic variations in concentrations are often attributable to varying bedrock types. Hardness ($CaCO_3$ and related minerals) is a major indicator of dissolved minerals. Calcium carbonate is a good example of a substance derived from natural sources, primarily from marine sedimentary rocks. In areas of limestone bedrock, such as Florida and many areas of the central United States and Canada, hardness is commonly several hundred milligrams per liter, but in areas of calcium-poor rocks, such as New England and the Canadian Shield, values of 5 to 25 mg/liter are typical. For most trace minerals, regional variations are attributable to natural factors, whereas local "hot spots" are almost always human-made. Many different inorganic minerals are found in water, and it is impossible to make generalizations about their effects on humans or the environment. Many minerals are essential nutrients in trace quantities, but virtually all have detrimental effects at higher concentrations.

Toxic Substances A wide variety of substances that are harmful at very low concentrations (parts per million or less) are introduced to surface and groundwater. Although many of these substances occur naturally, we are most concerned about those that are introduced by human activity. Among the most troublesome are organic chemicals. They include herbicides, insecticides, and a wide variety of industrial organic chemicals such as benzene, carbon tetrachloride, polychlorinated biphenyls (PCBs), chloroform, and vinyl chloride. In addition, oils and grease can be included in this category, although they are usually found at higher concentrations than the other compounds.

These toxic substances are a major concern because many of them are toxic, carcinogenic, or both. They can be dangerous if present in only parts per billion or parts per trillion, particularly if they are accumulated in tissues or biomagnified in the food chain. Adverse health effects may not be observed until many years after exposure; consequently, there is great uncertainty as to what substances are dangerous and at what levels of exposure. Many more years of intensive research are needed to understand the hazards associated with these substances.

In surface waters, toxic substances are usually diluted so that they are present only in very low concentrations, generally parts per trillion. They are found in higher concentrations in fine-grained sediments in many waterways, with runoff sources being particularly important. In groundwater, dilution is very slow and much higher concentrations have been found than in most surface waters.

Many metals are toxic to plants and animals at relatively low concentrations and so are a concern in surface water and groundwater. Some are also essential nutrients to certain organisms. Among the metals that often are identified as important toxic pollutants in water are arsenic, barium, cadmium, chromium, cobalt, copper, lead, manganese, mercury, nickel, silver, and zinc. The degree of health hazard associated with these metals depends on whether they are chemically or physically available to organisms. Most metals are relatively insoluble in water and tend to become associated with particulates. Metals bound to particulate matter are easily removed by filtration and thus are less available to organisms than are dissolved forms. The solubility of most metals in water is affected by pH, with solubility increasing as pH decreases. Acid precipitation may therefore increase problems of metal pollution, as acidic water dissolves metals in soil and sediments and causes them to enter the food chain.

Heat Electric power generation, petroleum refining, and many other industrial processes depend on the production and dissipation of large amounts of thermal energy—heat. For example, typical efficiency levels in electric generation are 32 to 36 percent. This means that about a third of the energy produced at a power plant is converted to electricity, and the other two-thirds must be dissipated as heat, usually in condensing steam and exhaust gases. Any industrial process that requires heating and cooling will produce waste heat, and water is usually the most effective means of dissipating that heat. Depending on the amount of heat discharged and the rate at which it is dispersed by receiving waters, the temperature increase of the water may be as much as 10° to 20°C, though usually it is less. Another cause of *thermal pollution* in streams is the removal of vegetation that shades the water. This is particularly severe when an area is deforested. Stream corridors, where shade trees are left along the streambanks, are effective in preventing this pollution.

Heat in water has little direct effect on humans; warm water may be less pleasant to drink, but it is no less safe. The primary detrimental effects of thermal pollution are to fish, because most fish have critical temperature ranges required for survival, and these ranges differ among species. Spawning and egg development in lake trout, walleye, and northern pike, for example, are inhibited at temperatures above 48°F (9°C). Smallmouth bass and perch will not grow at temperatures above about 84°F (29°C), whereas growth of catfish is possible at temperatures as high as 93°F (34°C). In some cases, thermal discharges have benefited commercial fisheries by making otherwise cool water suitable for species that require warmer temperatures, but generally the effects are negative. Equally important is the effect of temperature on dissolved oxygen concentrations. The amount of oxygen that can be dissolved in water decreases with increasing temperature; water at 92°F (33°C) holds only about half the oxygen that water at 32°F (0°C) will hold. At high temperatures, then, increased rates of bacterial activity put more demand on oxygen supplies just when saturation concentrations are low. Many fish kills are caused by a combination of high BOD and high temperatures, particularly in summer.

Radioactivity *Radioactivity*, or the emission of particles by decay of certain radioactive substances, is a subject of public concern today. Ionizing radiation, consisting primarily of alpha, beta, and gamma radiation, is derived from many natural and human-made sources. The *sievert* is a unit that describes all ionizing radiation in terms of the biological damage it causes. On the average, Americans receive a dosage of about 100 millisieverts (0.1 sievert) per year from natural sources and another 80 millisieverts from artificial sources, primarily diagnostic X rays. The radiation from natural sources comes mostly from cosmic radiation (the sun) and from terrestrial materials (rocks, bricks, and concrete). An average of about 15 to 20 millisieverts come from radioactive potassium-40 found in bone tissue.

Radioactive substances in water are derived primarily from rock weathering, particularly by groundwater. The greatest amount of radioactivity in water is from potassium-40, but this source is probably only about 1/100 of the amount derived from food sources. However, some substances tend to become concentrated in bone tissues, particularly strontium-90, radium-226, and radium-228. In certain areas, these isotopes occur in groundwater, and if the concentration is high, an increase in the risk of bone cancer is possible. In areas of mining or industrial operations that process rocks with high radionuclide content, local radioactive water pollution may occur. In general, however, surface waters dilute these substances to the extent that concentrations are lower than those found in natural groundwater.

Groundwater Pollution Problems

Groundwater pollution is a serious problem in industrialized countries and stems from municipal and industrial sources as well as from agriculture. Groundwater represents a large storage of water that is replaced very slowly. Whereas typical flow velocities for rivers are measured in meters per second, groundwater is likely to flow at rates of meters per day to meters per year. In most cases, flow distances are quite large, and it takes decades to millennia to replace contaminated water in an underground reservoir, if it can be replaced at all. This has two important consequences. First, once an aquifer is contaminated, it is lost for an indefinite period of time, except for uses not affected by the contaminants. Second, the contamination being discovered in wells today may result from pollutant discharges that occurred years in the past, and chemicals dumped today may not show up in well water for years to come. Not only are flow rates low, but the purification processes that remove particulates and bacteria are not as effective against human-made chemicals such as chlorinated hydrocarbons. Such chemicals seep into an aquifer and are likely to remain there with little or no dilution or degradation.

Many different sources of groundwater contamination exist, including municipal and industrial landfills, industrial impoundments, household septic systems, and waste disposal wells (Fig. 10.11). Municipal and industrial landfills are used to dispose of nearly every kind of waste imaginable, most of it relatively harmless but some of it quite dangerous. Industrial landfills may receive much greater volumes of toxic materials, and most of the sites that pose the most immediate threat to human health are those in which industrial wastes have been discarded on the ground or in landfills. Municipal landfills, of which there are thousands in the United States, also receive hazardous wastes from household, commercial, and industrial sources, though generally in small quantities. In the past, landfills were often located on whatever land was available rather than in areas that were geologically suited for waste disposal. Until recently, little care has been taken to see that *leachate* (liquid seeping out of the base of a landfill) does not percolate down to an important aquifer.

Industrial impoundments such as storage lagoons and tailings ponds are another important

Figure 10.11 A toxic waste lagoon near the Shenandoah River, Virginia. Sources of groundwater contamination include septic tanks, landfills, lagoons, and waste-disposal wells.

Figure 10.12 Waste lagoon adjacent to a hog production facility in North Carolina. If properly managed, treatment of animal wastes in concentrated animal feeding operations may result in less pollution than occurs from nonpoint sources associated with relatively dispersed meat production in farm feed lots.

cause of groundwater pollution. Lagoons may be used to temporarily store liquid wastes prior to disposal, reprocessing, or other use (Fig. 10.12). If they are unlined, as most are, liquid wastes can percolate into groundwater. In still other cases, wastes may be intentionally pumped into the ground as a disposal method. In confined, unusable aquifers this can be a safe practice, but leakage may occur. Tailings ponds, or impoundments used to trap mining debris, sometimes cause severe contamination with acids or metals.

Household septic tanks with leach fields are used for sewage disposal in about one-quarter of all homes in the United States. Properly designed, constructed, and maintained septic systems are effective water purifiers, returning clean water to the ground and nutrients to the soil. They are generally used when population density is relatively low, such as in rural or low-density suburban areas.

WATER POLLUTION CONTROL

Because of the many different sources and kinds of water pollution, control is a complex and expensive problem. Wastewater discharged by point sources can be treated by a variety of methods, but nonpoint sources must be controlled through land management.

Wastewater Treatment

Sewage treatment methods include primary, secondary, and tertiary techniques. *Primary treatment* consists of removal of solids by sedimentation, flocculation, screening, and similar methods. Primary treatment may remove about 35 percent of BOD, 10 to 20 percent of plant nutrients, and none of the dissolved solids. *Secondary treatment* removes organic matter and nutrients by biological decomposition, using methods such as aeration, trickling filters, and activated sludge (Fig. 10.13). It became widely used in the United States during the 1960s. This treatment removes about 90 percent of BOD, 30 to 50 percent of nutrients, and perhaps 5 percent of dissolved solids. *Tertiary methods* have come into widespread use only in the past decade or so, and still only a small proportion of communities have tertiary treatment. There are many methods, and they vary considerably in their effectiveness, but generally they remove 50 to 90 percent of nutrients and dissolved solids. Treatment methods for industrial wastewater are usually specific to the type of wastes being considered. Many industries discharge into municipal sewage systems rather than treat wastes on site, although pretreatment is often required.

Prior to the early twentieth century, little was done to control water pollution in most of the world. The problem of contamination of water supplies by sewage was identified in Europe in the seventeenth century, but the solutions were to change drinking-water supplies rather than to try reducing pollution. Beginning around 1900 and growing rapidly by the 1940s, wastewater treatment was instituted in larger cities in the industrialized world. In the United States, a few states had water pollution control laws by the 1940s, and the 1948 Federal Water Pollution Control Act provided impetus for construction of treatment plants. As late as 1960, however, only about 36 percent of the population served by sewers had wastewater treatment, and this was almost exclusively primary

Figure 10.13 Activated sludge tanks at a sewage treatment plant in Wisconsin.

treatment. The remaining 64 percent were served by sewer systems with no treatment at all.

Awareness of water pollution problems increased dramatically in the 1960s, especially in the industrialized world. In the United States, for example, new federal laws in 1961 and 1965 greatly increased nationwide efforts at pollution control, mostly by providing funds for construction of treatment plants. By 1970, over 85 million Americans were served by treatment plants, or 52 percent of those with sewer systems. The most ambitious and comprehensive law to date, the Federal Water Pollution Control Act of 1972 and its amendments of 1977, 1980, and 1987 now form the basis of our nationwide pollution control efforts. By the early 2000s, over 62 percent of the population was served with secondary sewage treatment plants, double the portion with such treatment in 1980 (Issue 10.2).

In the industrialized world, virtually all urban residents are served by sewage collection and treatment systems of some kind. In Japan, Canada, the United States, and western Europe, secondary treatment systems are most common. But in the developing world, urban sewage systems are not universal. For example, in Brazil, about 80 percent of the urban population is served by sewage collection systems, in Vietnam about 43 percent have such service, and in Nigeria the figure is 50 percent (World Resources Institute 2000). Sewage collection does not necessarily mean treatment—in poor countries the sewage is typically piped to rivers or the sea untreated. In less industrialized regions of some countries, a large percentage of the population lives in rural areas, where sanitation systems are uncommon.

Nonpoint Pollution Control

Nonpoint sources of pollution are the most difficult to control. In rural areas, they consist primarily of suspended and dissolved solids, nutrients, and pesticides contained in runoff, either dissolved or in particulate form. In agricultural areas, control of overland flow can do much to limit these sources because soil eroded by water often contains harmful pollutants such as pesticides and nutrients. As you recall from Chapter 6, such management practices are often difficult to establish or enforce. In urban areas, runoff from streets, parking lots, and similar surfaces usually contains large amounts of suspended solids and BOD as well as many toxic substances. In cities with combined storm and sanitary sewers, runoff is routed through the treatment system, but during storms the treatment plant cannot handle the increased flow; thus, sewage and runoff are discharged in an untreated form. Sewage discharge has generally been regarded as the more serious problem, and most cities have converted or are converting to separate sanitary and storm sewer systems. This eliminates the problem of untreated sewage discharges but does little to solve the problem of urban runoff pollution, as stormwater is discharged directly without treatment. In newly developing areas, storage basins can be incorporated into stormwater systems to retain runoff temporarily or permanently, and these may be useful in reducing runoff pollution. But in developed areas the control of urban runoff is usually prohibitively expensive.

The need to reduce nonpoint water pollution in the most cost-effective ways is best achieved through watershed management. *Watershed management* is an approach that considers many different aspects of water quality at the scale of the watershed. Because many different people are involved in using and managing water and water pollutants across the landscape, effective watershed management requires participation by a cross-section of people, from farmers to industries to domestic water users and environmental interest groups. In the United States, watershed management is the primary approach being used today to reduce water pollution. The process focuses on calculation of *Total Maximum Daily Loads* (TMDLs), or the maximum amount of certain pollutants that can be discharged to lakes and streams on a daily basis without impairing water quality. Once the TMDLs are determined, then all polluters in a watershed—point and nonpoint alike—must agree on procedures to reduce their discharges to acceptable levels. The process requires cooperation among diverse interests, many of whom must cooperate voluntarily rather than under threat of legal action. TMDL allocation thus must include many different people in discussions rather than simply allowing a government official to impose rules.

Pollution Prevention

The cost of pollution control becomes a major problem as the amount of control increases. As a general rule, controlling the worst pollution is relatively cheap per unit of pollutants removed. But as pollution control requirements become more stringent, the costs of cleaner water increase. For example, consider the conventional technologies used in primary, secondary, and tertiary treatment described earlier. Tertiary treatment using carbon-absorption filters is much more expensive than primary treatment using a settling tank. In the industrialized countries, by the late 1980s demands for pollution control remained strong, but costs were beginning to escalate. In the United States, for example, the goal of "fishable and swimmable" water by 1985 remained in place, but the costs of achieving that goal were so great that it ceased to be practical with use of conventional technologies.

As a result of this and other factors, a new approach known as *pollution prevention* emerged. This approach focuses on activities that reduce pollutants in the first place, rather than on removing them from wastewater before it is discharged to the environment. The idea of pollution prevention first caught on in industries which, faced with requirements that they comply with strict pollution standards, sought ways to minimize the cost of compliance. When engineers and plant managers started to look at alternative ways of dealing with these compliance problems, they found that modifying their practices to stop generating pollutants in many cases cost much less than removing pollutants from their effluents.

As the pollution prevention idea caught on, it came to be recognized as the best and perhaps the only practical means for achieving higher levels of water quality than could be reached with conventional approaches. Government regulatory agencies began to promote voluntary pollution prevention as an alternative to more stringent regulations, and corporations learned that other benefits were to be gained from voluntary actions to reduce pollution (see Chapter 15). The result has been a significant number of environmental success stories in areas such as paper, chemicals, and general manufacturing.

Much of the impetus for pollution prevention came from forces that only affected the business sector strongly, such as consumer demand or government regulatory pressure. These forces are less significant for polluters in the government sector, and reduction of pollution from municipal sewage has been much less. Pollution prevention in municipal systems starts with the individual and thus has been limited to those individuals who have chosen to modify their practices on their own, regardless of what their neighbors do. Positive results have been much less visible.

QUALITY, QUANTITY, AND THE WATER-SUPPLY PROBLEM

Relations Between Quality and Quantity

Although many significant improvements have been made in water quality in the past few decades, especially in wealthier countries, water quality remains a critical problem worldwide. In wealthy countries, ecological impacts of water pollution are the focus of most public attention to water issues. Restoring degraded ecosystems, protecting endangered species, and comprehensive watershed management top the agendas for regulators (Issue

ISSUE 10.2: WATER POLLUTION LEGISLATION IN THE UNITED STATES

The Federal Water Pollution Control Act of 1972 (now called the Clean Water Act) established a federal goal to "restore and maintain the chemical, physical, and biological integrity of the Nation's waters." The original act had the goal of making all waters clean enough to fish and swim in by 1983 and to eliminate the discharge of pollutants into navigable waters by 1985. The Clean Water Act contains provisions for establishing effluent standards for industries and municipal treatment plants and for comprehensive local planning to reduce both point and nonpoint pollution. Municipal plants were required to achieve secondary treatment by 1977 and "best practicable" technology by 1983. Similarly, industries were required to use the best practicable technology by 1977 and the best available technology by 1983. All point dischargers are required to obtain discharge permits under a National Pollutant Discharge Elimination System (NPDES), which was originally administered by the Environmental Protection Agency (EPA), but today most states have taken over the permitting process. Permits allow discharges only within limits established by the permitting agency.

The actual conditions for issuance of permits are determined primarily by the permitting agency, and these conditions have changed with changing public opinion and availability of funds. During the 1970s, for example, the EPA was relatively rigorous in enforcing compliance with effluent standards, although deadlines for compliance were frequently postponed. During the 1980s, however, standards in some areas were relaxed, because some argued that water quality was already good enough or that improved treatment would not result in significant improvement of water quality. One example of this administrative modification of the law came in 1982, when the EPA announced that it would no longer require secondary treatment for certain cities (including New York) discharging wastes into coastal waters.

The importance of nonpoint sources (particularly when major point sources are controlled) is recognized by the Clean Water Act, which requires the establishment of sedimentation management plans to reduce nonpoint pollution. Plans vary from one area to another, depending on the nature of the sources and local needs. Most plans include provisions for runoff and sediment control at construction sites, as well as guidelines for nonpoint pollution control in new developments. In some urbanized areas, measures such as street sweeping were instituted. As with the measures for controlling point sources, local plans are subject to modification by the agencies concerned, depending on local needs and desires, because of the technical difficulty of controlling nonpoint pollution. As a result, actual implementation of the guidelines of the federal law is highly variable from place to place and time to time.

In 1977, the Clean Water Act Amendments were passed. One of the more important achievements of this law was to focus government regulatory efforts on toxic substances rather than on the more conventional pollutants such as BOD or nutrients. Under this law, the EPA has established industry-specific effluent limits for many common toxic substances and has developed a system of monitoring certain index contaminants as a means to reduce monitoring costs.

10.3). Drinking-water quality still serves to galvanize public opinion and draw attention to water pollution problems. In 1996 the U.S. Congress passed legislation (the Safe Drinking Water Act Amendments) that requires drinking-water purveyors to inform their customers, once each year, of the presence of regulated substances in their water even if concentrations were below levels considered hazardous. Such notices would stimulate more widespread public awareness of water-quality issues, and it is hoped that this awareness would lead to improvements in water quality. In 1993, for example, 79 percent of the community water systems in the United States met all existing health-based water quality standards. In 1997 (one year after the Safe Drinking Water Act was passed) the number increased to 87 percent, and by 2000 it was 91 percent (US EPA 2001).

Water-supply limitations in most wealthy countries are increasing. In the United States, for example, population is increasing at about 1 percent per year; thus, even if we did not increase in-

During the 1980s, most of the efforts at water pollution control were led by the states, as the Reagan administration sought to reduce the federal role in this area. The administration of pollutant discharge permit programs was turned over to state regulatory agencies, which in many cases enforce regulations that are more stringent than federal criteria. Efforts at reducing point-source pollution from sewage continued, and many new treatment facilities were built. In 1987, the U.S. Congress overrode a Reagan veto of a bill continuing federal subsidy of these efforts and passed the Water Quality Act.

The Water Quality Act recognizes and reiterates the need for strong regulatory control of nonpoint sources of toxics and other pollutants. The act continues the effluent standards based on designated use through the NPDES permit process. The legislation also tightens the control on point-source polluters and provides additional money for control of nonpoint sources through state grants. The Water Quality Act also strengthens the protection of specialized environments like estuaries and wetlands. Finally, the Water Quality Act of 1987 instructs states to establish clean water strategies and assessments of nonpoint sources of pollution. Subsequent amendments to the Clean Water Act have strengthened many of the existing provisions.

As a result of these and other legislative efforts, water quality has improved in many areas. By 1978, virtually all sewer systems had treatment, with most of these having secondary treatment or better. Water-quality violation rates for some pollutants have declined markedly in major rivers. The Cuyahoga River, near Cleveland, Ohio, once notorious for catching fire repeatedly in the 1950s and 1960s, is no longer flammable. Lake Erie, pronounced dead by environmentalists in the 1960s, is significantly cleaner. But generally progress has not been as dramatic as had been hoped. The regulations were effective in reducing industrial discharges in many areas, and industries made substantial investments in pollution control equipment. Municipal pollution control efforts depended on both local revenues and federal assistance, and often a lack of funds or political disputes delayed treatment plant construction. For example, a 1980 estimate by the EPA indicated that 63 percent of the major municipal treatment facilities were not in compliance with the 1977 deadline for secondary treatment (Council on Environmental Quality 1980). Difficulties experienced by municipalities in meeting federal requirements have led to some relaxation of the regulations.

Under the 1977 Clean Water Act Amendments, states classified their streams according to the uses they should support, including fish propagation, fish maintenance, drinking, swimming, and boating. Throughout the 1980s, construction of improved treatment facilities continued, resulting in locally improved water-quality conditions. The total population served by secondary treatment or better is now 165 million, while 8 million people are still relying on direct discharges of untreated sewage into the nation's lakes and streams (U.S. Census Bureau 2000). A recent survey of water quality found that only 28 of the more than 2100 watersheds in the nation were ranked as having extremely poor water quality.

dividual water use, our total withdrawals for domestic consumption would increase. More and more people are moving to the Sunbelt states, which are often more arid environments, yet they still want their dishwashers and green grass lawns. In developing nations, population increases and migration to urban areas are severely stressing water resources as well. Transboundary water resources and their use are disputed in many parts of the world, including the Aral Sea, Ganges River, Jordan River, Nile River, and Tigris-Euphrates River. In the absence of water-sharing agreements, projected population increases and demand for water will increase the tension between bordering nations (Postel 2000; Milich and Varady 1998).

How can we remove more water from rivers and thus reduce their capacity to dilute and remove wastes, while at the same time demanding lower pollutant concentrations? The answer, of course, is to reduce our output of pollution. But advanced sewage treatment is expensive, and publicly owned treatment works have been slow to respond to calls

ISSUE 10.3: SURF YOUR WATERSHED

The Internet has revolutionized the way in which environmental data are reported and made available to the public. Water resources and quality issues demonstrate the importance of the Internet in making real-time observations readily available with just a click of a mouse and good search engine. At the USGS, for example (www.water.usgs.gov), you can view current (real-time), daily, or monthly stream conditions. Through their interactive GIS system, you can see current and historic data on surface and groundwater resources and use, as well as water quality. You can also gather basic hydrological data on watersheds.

More detailed analyses of water quality can be obtained from the U.S. Environmental Protection Agency (www.epa.gov/surf) and their Surf your Watershed program. An Index of Watershed Indicators was created so that diverse watersheds could be compared to one another. One set of indicators monitors conditions in the watershed (uses, fish and wildlife consumption advisories, ambient water quality, wetlands loss, and many more) and the other focuses on vulnerability (aquatic species at risk, pollutant loads, urban runoff potential, increasing population). In addition, you can generate profiles of your watershed (again, using an interactive GIS) that reveal not only the use designations but also whether the rivers (and estuaries) met state water quality standards, as well as causes and sources of pollution.

---

for reduced discharges. Instead, industrial and agricultural water users have been forced to reduce their use of water. In the case of U.S. industry, pressure to reduce water pollution under the 1972 Clean Water Act ultimately led to industries reducing withdrawals by replacing once-through systems with processes to clean and recycle water within the plant or disposing of wastes in some other way such as through a municipal sewage treatment plant or conversion to solid form and disposal in a landfill. Wastewater reclamation and reuse is also used to supplement supplies of nonpotable (nondrinkable) water. In Japan, for example, 40 percent of the wastewater reuse is for industrial demands, with another 30 percent used to augment natural flows in rivers and streams. In the United States, California leads the nation in the use of reclaimed water, primarily for crop irrigation, groundwater recharge, and outdoor landscape irrigation (Gleick 2000). However, it is the sustainable use and management of watersheds that holds the most promise for improving both the quality and quantity of water resources in the future (Platt et al. 2000).

Water Quality in Developing Regions

Water-quality problems in most developing countries are a stark contrast to those in the wealthy world. While drinking water is a concern in wealthy countries, even the worst drinking-water problems there pale in comparison to those faced by the majority of residents in developing countries. In 1998, over 1.7 billion people, or nearly a third of the world's population at that time, did not have access to safe water (Fig. 10.14). Only about one-third of the population has access to sanitation. New sewage systems are being built, but the number of people served by these systems is growing slower than the

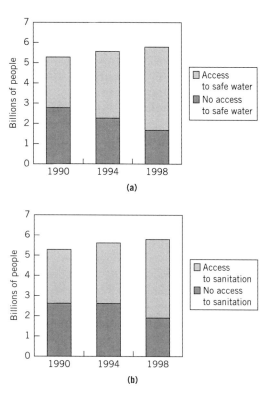

Figure 10.14 Access to safe water (a) and sanitation services (b). *Source:* World Bank, 1992:46; UNDP, 2000.

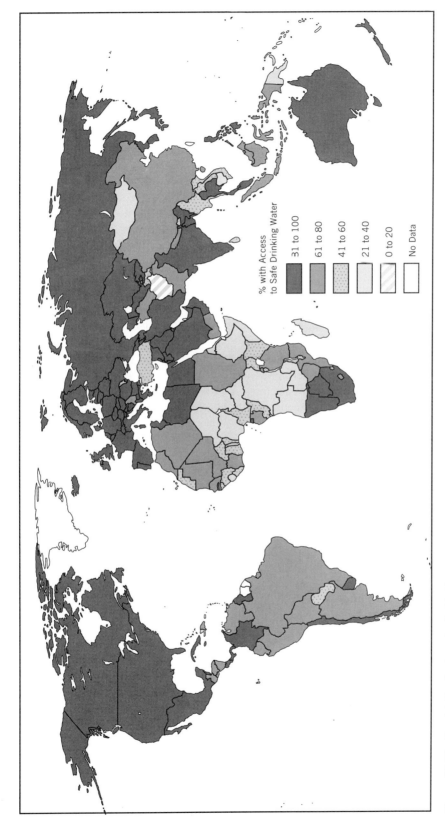

Figure 10.15 Spatial distribution of safe drinking water. Notice the large disparities in access to safe drinking water in Africa. *Source:* World Resources Institute, 2000.

population, so that the number without access to sanitation services is increasing. The problem is especially acute in rural areas. Even where sewage is collected, most is discharged to surface waters untreated. There is improvement in making safe drinking water available in developing countries, with about 71 percent of people in less industrialized countries now having access to safe drinking water (Fig. 10.15). However, there are strong regional disparities, with many sub-Saharan nations lagging behind most nations in the developing world. Despite these improvements, it is estimated that every year over 3 million children die of diarrhea caused by consuming contaminated water.

CONCLUSIONS

Clearly, the world's water problems are acute. Most of the world's available water supplies are already in use, and those that are not in use are found in places where population densities are very low. Water supplies in the world's megacities are a major source of concern. Poor sanitation resulting in waterborne diseases is still the leading cause of sickness and death worldwide. Water-supply scarcity and contamination are critically important in Africa, Asia, Latin America, and west Asia. Currently, about one-third of the world's population have moderate to severe problems with access to fresh water. It is estimated that in 2050, around 3 billion people will face severe water shortages (United Nations Environment Program 1997, 2002).

Opportunities for increasing the quantities of water available to people are few. The only means for making significant amounts of new water available on a renewable basis is construction of large dams to store runoff from wet seasons, making it available in dry seasons, but this option is limited by environmental, political, and financial constraints. As the world's population increases to nearly 9 billion in the next few decades, improving access to clean water can be achieved only through decreases in per capita consumption and increases in water reuse.

REFERENCES AND ADDITIONAL READING

Burkholder, J. M. 1999. The lurking perils of *Pfiesteria*. *Scientific American* August:42–49.

Council on Environmental Quality, 1980. *Environmental Quality, 1980*. Washington, D.C.: U.S. Government Printing Office.

Crutchfield, S. R., L. Hansen, and M. Ribaudo. 1993. *Agricultural and Water Quality Conflicts: Economic Dimensions of the Problem*. Washington, D.C.: U.S. Department of Agriculture, Economic Research Service.

Gilpin, A. 1995. *Environmental Impact Assessment*. Hong Kong: Cambridge University.

Gleick, P. H. 2000. *The World's Water 2000–2001: The Biennial Report on Freshwater Resources*. Washington, D.C.: Island Press.

_____. 2001a. Making every drop count. *Scientific American* February:40–45.

_____. 2001b. Global water: threats and challenges facing the United States. *Environment* 43(2):18–28.

Mallin, M. A. 2000. Impacts of industrial animal production on rivers and estuaries. *American Scientist* 88:26–37.

Meade, R. H. 1995. *Contaminants in the Mississippi River, 1987–1992*. U.S. Geological Survey Circular 1133.

Meybeck, M., D. Chapman, and R. Helmer. 1989. *Global Freshwater Quality: A First Assessment*. Oxford: Basil Blackwell.

Milich, L. and R. G. Varady. 1998. Managing transboundary resources: lessons from river-basin accords. *Environment* 40(8):10–15, 35–41.

National Research Council. 1991. *Toward Sustainability: Soil and Water Research Priorities for Developing Countries*. Washington, D.C.: National Academy Press.

_____. 1999. *New Strategies for America's Watersheds*. Washington, D.C.: National Academy Press.

Opie, J. 1993. *Ogallala: Water for a Dry Land*. Our Sustainable Future Series V.1. Lincoln: University of Nebraska Press.

Platt, R. H., P. K. Barten, and M. J. Pfeffer. 2000. A full, clean glass? Managing New York City's watersheds. *Environment* 42(5):8–22.

Postel, S. 1989. *Water for Agriculture: Facing the Limits*. Worldwatch Paper no. 93. Washington, D.C.: Worldwatch Institute.

_____. 1992. *Last Oasis: Facing Water Scarcity*. New York: W. W. Norton.

_____. 2000. Redesigning irrigated agriculture. In L. R. Brown et al., *State of the World 2000*. Washington, D.C.: Worldwatch Institute, pp. 39–58.

Reisner, M. 1986. *Cadillac Desert: The American West and Its Disappearing Water*. New York: Viking Penguin.

Reuss, M., ed. 1993. *Water Resources Administration in the United States: Policy, Practice and Emerging Issues*. East Lansing: Michigan State University Press; Bethesda, MD: American Water Resources Association.

Rogers, P. 1993. *America's Water: Federal Roles and Responsibilities*. Cambridge, MA: MIT Press.

Sexton, R. 1990. *Perspectives on the Middle East Water Crisis: Analysing Water Scarcity Problems in Jordan and Israel*. London: Overseas Development Institute; Sri Lanka: International Irrigation Management Institute.

Shah, T. 1990. *Sustainable Development of Groundwater Resources: Lessons from Amrapur and Husseinabad Villages, India*. London, England: Overseas Development Institute; Colombo, Sri Lanka: International Irrigation Management Institute.

Smith, R. A., R. B. Alexander, and K. J. Lanfear. 1993. *Stream Water Quality in the Coterminous United States—Status and Trends of Selected Indicators During the 1980's. National Water Summary 1990–91: Stream Water Quality*. U.S. Geological Survey Water-Supply Paper 2400.

Steinhart, P. 1990. The water profiteers. *Audubon* 92(2):38–51.

Stone, J. A. and D. E. Legg. 1992. Agriculture and the Everglades. *Journal of Soil and Water Conservation* 47(3):207–215.

United Nations Development Program (UNDP). 2000. *Human Development Report 2000*. New York: Oxford University Press.

United Nations Environment Program (UNEP). 1997. *Global Environmental Outlook*. New York: Oxford University Press.

_____. 2002. *Global Environmental Outlook 3*. New York: Oxford University Press.

U.S. Bureau of the Census. 2000. *Statistical Abstract of the United States*. www.census.gov.

U.S. Department of Agriculture. 1989. *The Second RCA Appraisal: Soil, Water, and Related Resources on Nonfederal Land in the United States, Analysis of Condition and Trends*. Washington, D.C.: U.S. Government Printing Office.

U.S. Environmental Protection Agency. 2002. *Watershed Information Network, U.S. Map: Risk of Groundwater Nitrate Contamination*. http:www.epa.gov/iwi/1999sept/iv21_usmap.html.

_____. 2001. *Factoids: Drinking Water and Ground Water Statistics for 2000*. EPA 816-K-01-004. www.epa.gov/safewater.

U.S. Geological Survey. 2001. *Offstream Use*. www.water.usgs.gov/wateruse/pdf1995/html.

_____. 2002. http://water.usgs.gov/hwis/peak.

U.S. General Accounting Office. 1986. *The Nation's Water: Key Unanswered Questions about the Quality of Rivers and Streams*. Washington, D.C.: U.S. Government Printing Office.

Valentine, J. and J. Carochi. 1993. Making a difference: agencies can, will, do work together to solve non-point source pollution problems. *Journal of Soil and Water Conservation* 48:401–406.

Water Resources Council. 1978. *The Nation's Water Resources, 1975–2000*. Washington, D.C.: U.S. Government Printing Office.

White, G. F. 2000. Water science and technology: some lessons from the 20th century. *Environment* 42(1):30–38.

World Bank. 1992. *World Development Report, 1992: Development and the Environment*. New York: Oxford University Press.

World Commission on Dams. 2000. www.dams.org/global.

World Resources Institute. 2000. *World Resources 2000–2001*. Washington, D.C.: World Resources Institute.

For more information, consult our web page at *http://www.wiley.com/college/cutter*.

STUDY QUESTIONS

1. What pollutants are most significant in agricultural areas? What pollutants are mostly derived from industrial sources? What pollutants are mostly derived from sewage?
2. Where does your drinking water come from? What are the most important sources, or potential sources, of pollution affecting that water? Where does your wastewater go? What pollutants are discharged as part of your wastewater? What are the likely effects of these pollutants on the receiving water body?
3. For a river near you, list all the significant uses of the waterway, including in-stream and off-stream uses. Which ones are most dependent on good water quality? Which ones are not affected by water quality?
4. Visit the home page of the U.S. Geological Survey (*http://www.usgs.gov*) and download information on streamflows for a stream near you. If possible, find a stream with at least a 30-year record. Make a graph of streamflow through time, and examine the fluctuations in water availability over time. How much is available in the driest year in the record, as compared to the average?

CHAPTER 11

THE AIR RESOURCE AND URBAN AIR QUALITY

INTRODUCTION

Although air quality may appear to be a recent issue to many people, some parts of the United States and Europe have been plagued with air pollution problems since the Industrial Revolution. *Air pollution* is normally defined as a human-caused addition of impurities to the air. However, many of the same substances that humans release to the atmosphere, such as dust, also come from natural sources. While naturally derived impurities are not really pollution, we must consider them as we attempt to manage environmental quality. Air pollution is a significant health hazard; acute episodes can cause death while lower, prolonged pollution levels also adversely affect health. Some of the more prolonged air pollution episodes can even be classified as disasters. For example, 20 people died in Donora, Pennsylvania, in 1948, and 4700 people lost their lives in London, England, in 1952 due to thick smog (Elsom 1987). In another example, nearly 6000 people were treated for "smog poisoning" in Tokyo during a 1970 oxidant and sulfate episode. All of these disasters were the result of a combination of meteorological conditions and excessive emissions of sulfur from coal burning. The situation today is not so dramatic, yet in some parts of the world we find that major pollution episodes require both industry and individuals to curtail their activities on a fairly regular basis.

On a global level, emissions of traditional air pollutants (sulfur dioxide and particulates) continue to rise, particularly in developing countries. Air pollution in many of the world's cities chronically plagues local residents from Auckland to Zagreb. On the basis of one estimate, Shenyang and Xian, in the People's Republic of China, and Lahore, Pakistan, have the highest concentrations of sulfur dioxide and particulates, respectively, in the world (World Resources Institute 1996). Lanzhou, China, exceeds the World Health Organization's health guidelines for SO_2 by a factor of eight! Central Europe's air pollution is more dangerous and more widespread than that in any other industrialized region (Fig. 11.1). Mexico City, however, still retains its title as the world's largest city with the worst air. Excessive amounts of ozone, lead, and other contaminants spew forth daily from the city's vehicles and factories, most of which have no pollution control equipment, affecting more than 18 million. In addition, emissions of carbon monoxide and other so-called greenhouse gases have far-reaching regional and global consequences, which will be discussed in Chapter 12.

AIR POLLUTION METEOROLOGY

Composition and Structure of the Atmosphere

The atmosphere is divided into several layers, based on temperature and gaseous content. The *homosphere*, or lower atmosphere, extends from sea level to an altitude of 50 miles (80 km) (Fig. 11.2). It is called the homosphere because the gases are highly diffused, so that they act as a single gas. These gases include nitrogen (78 percent), oxygen (21 percent), carbon dioxide (0.03 per-

Figure 11.1 Copsa Mica, Romania, a city where industrial air pollution contributes to an exceptionally high death rate.

cent), and inert gases, such as argon, neon, helium, and krypton (less than 1 percent), and trace concentrations of several other substances.

The homosphere is further divided into the troposphere, stratosphere, and mesosphere. The *troposphere* is the layer in which humans live and extends from sea level to approximately 8–9 miles (13–14 km). In this layer, temperature steadily decreases with altitude, at an average rate of 3.5°F/1000 feet (6.4°C/km). This rate is called the *environmental lapse rate*. The next layer is the *stratosphere*. Air temperatures gradually increase with altitude until they reach 32°F (0°C) at an altitude of about 30 miles (50 km). The protective ozone (O_3) layer is located in the stratosphere; this layer serves as a shield in protecting the earth's

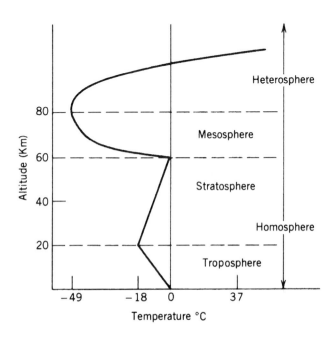

Figure 11.2 Temperature variation with altitude in the atmosphere. Most pollution is found in the troposphere, but some pollutants are carried to the stratosphere.

surface and the troposphere from harmful ultraviolet radiation. The third layer of the homosphere is the *mesosphere*. Here, temperatures decrease with altitude, reaching a low of −120°F (−83°C) at approximately 50 miles (80 km) altitude.

Air pollutants are not confined to the lower parts of the troposphere. Certain concentrations of contaminants have disastrous effects at higher altitudes by inducing global changes in climate. Similarly, ozone is considered a pollutant in the troposphere but becomes an essential gas necessary to protect human health in the stratosphere.

Role of Meteorology and Topography

Air pollution problems are the result of two factors: excessive emissions of pollutants and insufficient atmospheric dispersal. The first factor is the reason most cities have pollution problems and most rural areas do not. The second explains much of the variation in pollution problems from one city to another and why some very small cities have pollution problems as severe as those in major metropolitan regions.

Atmospheric dispersal of pollutants depends on air motion, both horizontal and vertical. Horizontal movements, or winds, carry pollutants away from cities. On windy days, the air in most cities is generally cleaner, and on calm days it is dirtier. Horizontal movements also contribute to vertical motions, which play a more direct role in air pollution. Despite the reputation that some cities have for being windy, average wind speeds do not vary much from place to place, and wind speed is not an important factor in explaining spatial variations in pollution.

Vertical movement in the atmosphere and low-pressure systems such as wave cyclones result from wind-generated turbulence and convection. Convection in turn is a result of differential heating of the lower layers of the atmosphere by sunlight, whereby the warmer layers become less dense and therefore rise, while cooler layers sink. Regional circulation patterns, characterized by areas of high and low pressure, can be seen as a larger-scale form of convection. The normal temperature pattern—cooler air at higher elevations—prevails when there is sufficient vertical mixing through the lower atmosphere. Sometimes, however, warmer air overlies cooler air, a condition called a *temperature inversion* (Fig. 11.3). An inversion keeps the atmosphere stable and thus inhibits vertical motions. Such inversions are the major meteorological factor in most air pollution problems.

Temperature inversions are caused by several different processes, including subsidence, radiation, and advection. A *subsidence inversion* develops when an air mass sinks slowly over a large area, as is common in a high-pressure cell. The atmosphere is compressed as the air mass sinks, and

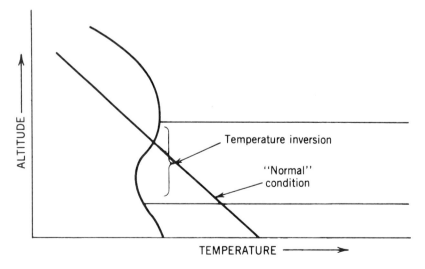

Figure 11.3 Temperature inversions. A temperature inversion consists of a layer in which temperature increases with altitude instead of decreasing. This temperature change prevents vertical circulation through the inverted layer.

higher layers are warmed more than lower layers, resulting in an inversion. Subsidence inversions are formed over large areas (thousands of square kilometers), usually at relatively high altitudes, but can occur as low as 3000 ft (900 m) above the surface. The weather that produces them (slow-moving high-pressure cells) also produces sunny conditions and gentle winds, which contribute to photochemical smog formation and poor dispersal of pollutants. Subsidence inversions are responsible for most of the severe pollution episodes in large cities east of the Rocky Mountains and also contribute to problems in mountainous areas.

Radiation inversions also develop in clear, relatively calm weather, but, unlike subsidence inversions, they are a diurnal phenomenon. On clear nights, the ground radiates heat upward, and the absence of clouds allows this radiation to escape to the upper atmosphere and into space. The result is that the ground cools more than the atmosphere, thus cooling the air near the ground so that it becomes cooler than air higher up. Radiation inversions are fairly thin and usually temporary, but cold air drainage can cause them to thicken and thus slow their dispersal in the morning. In hilly or mountainous areas, the dense, cooler air near the ground flows downhill, accumulating in valleys and producing a large pool of cool air. In hilly areas, most cities are situated in the valley bottoms, and thus the inversion traps the city's pollutants in the valleys. Valley walls prevent pollutants from dispersing horizontally, and the inversion keeps them from dispersing vertically. Valley inversions, often reinforced by subsidence inversions, are responsible for pollution problems in many cities in western North America, including Denver, Salt Lake City, Albuquerque, and Mexico City.

The third type of temperature inversion, the *advection inversion*, is a problem primarily on the West Coast of the United States, where local winds in the form of sea breezes blow off the Pacific Ocean. Before reaching land, air passes over the cold ocean current along the coast of California, and the lower layers of the air are cooled by contact with this water and thus become cooler than the air above. These inversions are usually of moderate thickness, from a few hundred to 5000 ft (1600 m) or more. Los Angeles, San Diego, and, to a lesser extent, the San Francisco Bay area are bordered on the east by mountains that prevent pollutants from being dispersed inland. The particularly severe pollution problems of these cities are essentially the result of the presence of the mountains, combined with very persistent advection inversions (Fig. 11.4).

In addition to dispersion, two other aspects of weather are important in understanding air pollution problems. These are sunlight and atmospheric humidity. Sunlight contributes to the formation of photochemical smog, and such smog is therefore more severe on sunny days than on cloudy ones. Cities that have a lot of sunshine have more photochemical smog than do those in cloudy areas. High-altitude cities, such as Denver and most other cities in the Mountain West, have particularly intense sunlight because of the thinner atmosphere, and this is an important factor in their pollution problems. In areas of high sulfur oxide emissions, atmospheric humidity is more of a

Figure 11.4 Smog in Los Angeles, California. Per capita emissions in Los Angeles are low, but weather conditions limit dispersal of pollutants.

problem, because water and oxygen combine with sulfur oxides to form sulfates and sulfuric acid. In areas of high humidity, high sulfur emissions and foggy days can be more dangerous than dry days.

The various combinations of these factors make the problems of each metropolitan region different. Some cities suffer mostly from photochemical smog, whereas others have the greatest problems with particulates or carbon monoxide. Some cities have pollution episodes that last only a day or two, and others have much longer ones. In the Northeast United States, for example, pollution is usually the most severe in summer and fall, because that is when emissions are highest (high electrical demand and increased automobile usage) and subsidence inversions most frequent. In the Mountain West, winter is usually the time of the most persistent inversions, and thus pollution is worse. Local variations in wind direction or speed also contribute to variations in pollution. These regional and local differences in weather conditions are the major factor in explaining differences in pollution problems. They also influence the management of these pollution problems where local, regional, national, and sometimes international approaches to air pollution control are needed.

MAJOR POLLUTANTS

Air pollutants can come from both natural and nonnatural sources, with the latter being the most important in the United States. Some natural sources of pollutants include smoke from forest fires, hydrocarbons from coniferous trees and shrubs, dust from a variety of sources, volcanic eruptions, and pollen. Natural sources can be quantitatively significant, even dominant in some places. But in those areas with severe pollution—most urban areas—human-made (*anthropogenic*) sources are much more important, especially in the United States.

Anthropogenic sources of air pollutants are either stationary or mobile. *Stationary sources* are site-specific and include stack emissions from refineries, smelters, electric power plants, and other manufacturing industries. *Mobile sources* are those that are not site-specific. They include automobiles, motorcycles, buses, trucks, airplanes, trains, ships, boats, and off-highway vehicles.

Particulate Matter (PM)

Particulate matter includes any solid or liquid particles, such as soot, fly ash, dust, pollen, various chemicals, and metals, such as arsenic, cadmium, and lead. Extremely small particles (less than 10 microns) are more likely to cause adverse health effects, such as respiratory distress and asthma. Other adverse health effects of particulates include the direct toxicity of some of the metals and chemicals and aggravation of cardiorespiratory diseases, such as bronchitis and asthma. Suspended particulates also have been linked to lung cancer. Aside from health, some of the negative effects involve the corrosion of metals and the soiling and discoloration of buildings and sculptures. More important, suspended particulates both scatter and absorb sunlight, thus reducing visibility. They also provide nuclei on which condensation can occur, which increases cloud formation.

Particulates are produced primarily by stationary sources, especially in those industries that use coal as a fuel source, such as power plants, steel mills, and fertilizer plants. Construction activities, quarry operations, and solid waste disposal (burning) also contribute minor percentages of particulate emissions. Natural sources of suspended particulates are volcanic eruptions, forest fires, and wind erosion.

Sulfur Dioxide (SO_2)

Sulfur dioxide is a colorless gas with a strong odor. It is highly reactive in the presence of oxygen and moisture and forms sulfuric acid, a corrosive chemical. SO_2 stings the eyes and burns the throat. More important, SO_2 contributes to respiratory diseases, including bronchitis, emphysema, and asthma; chronic exposures can permanently impair lung functions. SO_2 also corrodes metals, discolors textiles, and speeds the deterioration of building material, especially stone and metals. Perhaps the most significant effect of SO_2 is its role in the formation of acid rain and the resulting damage and decrease in plant growth (see Chapter 12). SO_2 emissions are a direct result of burning sulfur-bearing fossil fuels and smelting sulfur-bearing metal ores. Certain industrial processes, notably petroleum refining, also contribute SO_2 to the atmosphere. The most significant natural source of SO_2 is volcanic eruptions.

Nitrogen Oxides (NO_x)

Nitrogen oxide emissions include nitrogen monoxide (NO) and nitrogen dioxide (NO_2). Nitrogen dioxide is a reddish-brown gas that aggravates respiratory diseases and increases susceptibility to pneumonia and lung cancer. NO_2 also causes paints and dyes to fade. However, two effects of NO_x cause it to be considered a major pollutant. The first is its crucial role as an ultraviolet light absorber in the formation of photochemical smog, or ground-level ozone. Second, NO_x is a factor in the formation of acid rain. Gaseous nitrogen is usually inert, but it combines with O_2 at high temperatures in internal combustion engines and furnaces to form NO_x. Thus, the primary sources of NO_x are power plants and motor vehicle exhaust.

Carbon Monoxide (CO)

Carbon monoxide (CO) is a tasteless, odorless, colorless gas. It combines with hemoglobin in the blood, reducing its oxygen-carrying capacity and damaging some of the functions of the central nervous system. In small doses, CO impairs some mental functions as well, resulting in headaches and dizziness. In large doses, CO causes death.

Most CO pollution results from the incomplete combustion of carbon materials, including fossil fuels. There are some natural sources of CO, such as forest fires and decomposition of organic matter. Most of the anthropogenic CO emissions in urban areas are from transportation (mobile) sources, with an additional contribution from stationary sources, including industrial and power plants.

Ozone (O_3) and Volatile Organic Compounds (VOCs)

Ozone is a photochemical *oxidant* that is the most important component of photochemical smog. In combination with volatile organic compounds (VOCs), NO_x, and sunlight, oxidants comprise the now famous Los Angeles–type smog. In simplified form, the process is as follows: sunlight causes NO_2 to break down into NO and monatomic oxygen (O). This O atom combines with O_2 to form O_3. In addition, VOCs, O_2, NO, and NO_2 interact to form both ozone and a class of compounds called peroxyacetyl nitrates (PANs), which, like ozone, are harmful photochemical oxidants.

Photochemical oxidants are eye and respiratory irritants, and prolonged exposures will aggravate cardiovascular and respiratory illnesses. Other effects include deterioration of rubber, textiles, and paints and reduced visibility and vegetation growth. Leaves and fruit seem to be the most susceptible to oxidants, the effects of which result not only in injury but also in leaf drop and premature fruit. Since oxidants are produced in chemical reactions in the atmosphere, there is no direct source of emissions other than the sources of VOCs and NO_x.

Volatile organic compounds are released through the incomplete combustion of carbon-containing fuels and through the evaporation of fossil fuels from natural gas pipelines, gas tanks, and gas station pumps. Methane, propane, ethylene, and acetylene are some of the specific compounds generically called VOCs. Although many VOCs are suspected carcinogens, their most significant effect on air quality is their role in the formation of photochemical smog. Most of the anthropogenic VOC emissions are from stationary industrial and fuel-combustion sources, with the remainder emitted by transportation. There are some natural sources of VOCs such as coniferous forests, but they are relatively insignificant in their contribution to urban pollution problems.

Lead (Pb)

Lead is a nonferrous, heavy metal that occurs naturally. In the atmosphere, lead occurs in the form of a vapor, dust, or aerosol. Lead acts as a cumulative poison in the human body, causing general weakness and impaired functioning of the central nervous system. High lead levels in children contribute to neurological damage and learning disabilities. Ingestion can lead to severe anemia and even death.

Lead is often added to high-octane gasoline to reduce engine knock. However, this practice was banned in the United States in the 1970s, so all gasoline sold in the U.S. is now unleaded. This is not true in other world regions. The primary sources of lead in the atmosphere are vehicle exhaust from lead additives in gasoline, lead mining and smelting, and manufacturing of lead products, such as batteries. Volcanic dust, the major natural source of lead, contributes less than 1 percent of total emissions. Cigarette smoke is another source of lead in the air.

URBAN AIR POLLUTION: THE WORLD'S MEGACITIES

In evaluating pollution, we must distinguish between ambient concentrations (pollutants existing in the air) and emissions (pollutants discharged to the air). *Ambient data* are the concentrations of pollution in the air that are recorded at specific monitoring locations. On the basis of the appropriate health standards, ambient data indicate how close we are to achieving clean air. *Emissions data*, on the other hand, are estimates of the amount of pollutants released into the air from tailpipes or smokestacks. They illustrate how well the regulations on industrial and vehicular emissions are working. The distinction between ambient levels (the concentration of the pollutant in the air) and emissions levels (how much is coming out of tailpipes or smokestacks) is an important one, especially when we consider air pollution control measures.

The world's population is now more than 50 percent urban, and the proportion living in cities will continue to grow. Urbanization, though still greater in wealthy countries than in poor ones, is growing rapidly, especially in the developing world. In 2000, there were 19 cities in the world with populations over 10 million (Issue 5.2). The megacities are places of significant, if not severe, air pollution, although such pollution problems also occur in hundreds, perhaps thousands, of smaller cities as well.

Monitoring Network

In 1973, the World Health Organization (WHO) set up a global program to assist countries in monitoring air pollution. Shortly thereafter, the United Nations Environment Program (UNEP) set up its Global Environmental Monitoring System (GEMS), which now represents more than 175 sites in 75 cities in about 50 countries. More than one-third of the sites are in developing countries (Tolba et al. 1992). The six main pollutants of concern in urban areas are sulfur dioxide, nitrogen oxides, carbon monoxide, ozone, particulate matter, and lead. All of these are monitored by GEMS.

In many developing countries, however, indoor pollution is more of a health risk to millions of people who burn biomass fuels (wood, crop residues, dung) in stoves for cooking and heating. Women and children are most at risk from carbon monoxide and other contaminants (Cutter 1995). According to the WHO, biomass burning is the major indoor air pollution problem in the world today.

Air-Quality Patterns

Air pollution in urban areas is a local problem, although there are some transboundary considerations, as we will see in the next chapter. This means that global estimates of total emissions by pollutants are difficult to quantify. However, we do know that industrialized countries accounted for 40 percent of the SO_2, half of the NO_x, two-thirds of CO, and one-quarter of PM emissions. Ambient levels are a little easier to quantify, and estimates suggest that nearly 900 million people living in urban areas, worldwide, are exposed to unhealthy levels of sulfur oxides, while more than a billion are exposed to particulate levels that are so high they are termed a health hazard.

In one of the first detailed assessments of air quality in urban areas, PM was found to be the most prevalent air pollutant in 17 of the 20 largest world cities (megacities) (UNEP/WHO 1992). Mexico City exceeds WHO concentrations for all pollutants by more than a factor of two for four of the six (ozone, SO_2, PM, and CO). In the late 1990s, 17 cities exceeded WHO guidelines for particulates, sulfur dioxide, and nitrogen dioxide; most of these are located in the People's Republic of China (World Bank 2001). Comparing air quality among selected world megacities shows Beijing, Shanghai, and Mexico City consistently failing all WHO guidelines (Table 11.1). It is interesting to note that Delhi, India, has the worst problem with particulate matter among those cities listed, but meets health guidelines on the other indicators.

There is no systematic collection of information on the health risks and effects of air pollution in most of the megacities. Moreover, fewer than a handful have adequate assessment monitoring networks (United Nations Environment Program 1994). Suffice it to say that as these cities increase in size, consume more energy, and use more automobiles, the level of air quality will deteriorate to dangerous levels unless air pollution control measures are implemented quickly.

Economic Development and Air Pollution

Despite the fact that the greatest quantities of pollutants are emitted by relatively wealthy countries,

Table 11.1 Comparative Air Quality in Selected World Megacities, 1995–1998

City	Particulate Matter[a]	Sulfure Dioxide[b]	Nitrogen Dioxide[c]
	Micrograms/m^3 of Air		
Bangkok	223	11	23
Beijing	377	90	122
Mumbai (Bombay)	240	33	39
Cairo		69	
Calcutta	375	49	34
Delhi	416	24	41
Jakarta	271		
London		25	77
Los Angeles		9	74
Manila	200	33	
Mexico City	279	74	130
Moscow	100	109	
New York		26	79
Rio de Janeiro	139	129	
São Paulo	86	43	83
Seoul	84	44	60
Shanghai	246	53	73
Tokyo	49	18	68

[a]WHO guideline is 90 micrograms per cubic meter of air.
[b]WHO guideline is 50 micrograms per cubic meter of air.
[c]WHO guideline is 40–50 micrograms per cubic meter of air.
Source: World Bank, 2001 *(http://www.worldbank.org/data/databytopic/databytopic.html#ENVIRONMENT).*

some of the worst air pollution problems are found in the developing world. Concentrations of suspended particulate matter in major Chinese cities are an order of magnitude higher than those in similar-sized cities in North America. Severe air pollution problems occur in Bangkok, Lahore, Delhi, Cairo, and many other smaller cities. Respiratory problems are a major cause of death in many cities in the developing world, with 4–5 million new cases of chronic bronchitis each year.

Why do many cities in developing countries have severe air pollution problems, even though per capita energy consumption levels are much lower? Several factors are important. First, cities are centers of industrial activity, and many cities in developing countries are experiencing rapid industrialization. In such countries, industrial development is often a higher priority than environmental protection; thus, restrictions on industrial emissions are usually not as severe as in wealthy countries. In addition, the funds needed to install pollution control equipment are less readily available. Industrial emissions may therefore be substantial, even though the value of industrial output is usually not as great as in developed-world cities.

Second, even though cities in developing countries may have fewer automobiles per capita than the wealthy countries have, they still have quite a few. In most cases, this is a relatively recent phenomenon, and there has not been time or money to build urban highway networks like those in wealthier countries. As a result, traffic jams are nearly perpetual. Where it may take an hour or less to go 30 miles by auto in a U.S. city, much shorter journeys in developing-world cities often take two hours or more. In the process, cars are running continuously, emitting pollution in the stop-and-go traffic at much higher rates per mile than if they were moving faster. In addition, a higher proportion of vehicles are older, not equipped with the sophisticated pollution control devices required on cars in the United States, or not maintained as well.

URBAN AIR POLLUTION IN THE UNITED STATES

Air Pollution Monitoring in the United States

The U.S. government has identified several pollutants, called criteria pollutants, that are the focus of its air-quality management efforts. *Criteria pollutants* are those specific contaminants that adversely affect human health and welfare, for which the U.S. Environmental Protection Agency (EPA) has set ambient air-quality standards. *Primary standards* are designed to protect human health, and *secondary standards* are designed to protect human welfare (property and vegetation).

Legislative Mandates The original enabling legislation establishing air pollution control was the Clean Air Act, passed in 1963. Amendments to that legislation, the Air Quality Act of 1967 and the Clean Air Act Amendments of 1970, provided the framework for air resource decision-making at both regional and national levels. The Clean Air Act Amendments of 1970 established standards for ambient air quality for the five major pollutants and provided timetables for achieving those standards.

The Clean Air Act Amendments of 1977 further refined the monitoring of air pollutants and clarified previous legislation. The 1963, 1967, 1970, and 1977 acts are collectively known as the *Clean Air Act*. The 1977 amendments required standard monitoring of the criteria pollutants and standardized reporting methods. Under this legislation, the EPA was to review the standards for criteria pollutants and establish deadlines for compliance with the standards. States were to meet the primary standards for SO_2, NO_x, and PM by 1981 and the primary standards for O_3 and CO by 1987.

The Clean Air Act expired in 1982. However, all the rules and regulations in effect at that time were still valid. In 1982, Congress passed a continuing resolution that provided appropriations and legal authority to the EPA to continue the air-quality program under the 1977 amendments. In essence, this placed the legislation on a hold status while it was debated in Congress. The Reagan administration wanted to relax standards as well as limit provisions for transboundary pollution problems, such as acid rain. In the fall of 1990, Congress finally passed a revision of the Clean Air Act. The new law mandates a 50-percent reduction in sulfur dioxide emissions to help reduce acid rain. It also requires a phaseout of CFCs and other ozone-destroying chemicals in an effort to curtail stratospheric ozone depletion. To help alleviate urban air pollution problems, the new Clean Air Act requires lower vehicular emissions of nitrogen oxides (60 percent reduction) and hydrocarbons (40 percent reduction) and cleaner-burning gasoline, particularly in the country's smoggiest cities. The 1990 revisions call for a 90-percent reduction in the output of toxic emissions, particularly the 189 known toxic and cancer-causing chemicals. Provisions in the Clean Air Act on acid rain and other transboundary issues are discussed in the following chapter.

National Ambient Air-Quality Standards (NAAQS) Under the 1970 amendments, a national network of air-quality control regions (AQCRs) was established (Fig. 11.5). There are 247 AQCRs, with monitoring stations in 3000 counties in the United States (National Commission on Air Quality 1981). Data from each of these regions are stored in a national aerometric database, and monitoring is done on a county level. In 1979, problems with the frequency and accuracy of monitoring data led the EPA to standardize and regulate the monitoring network. State and local monitoring sites were thus incorporated into a national system, with consistent and uniform readings, including frequency, type of pollutant, and placement of monitoring stations (central city versus suburban location). Ambient air-quality data are submitted to the EPA's Aerometric Information Retrieval System (AIRS), where trends (10-year and 5-year) in air quality are monitored.

Primary and secondary standards were established under the Clean Air Act for criteria pollutants (Table 11.2). As stipulated by the 1977 amendments, these standards were subject to review and revision prior to the 1982 reauthorization. In July 1987, new standards were promulgated for particulate matter on the basis of size. These smaller particles (designated PM_{10} and so

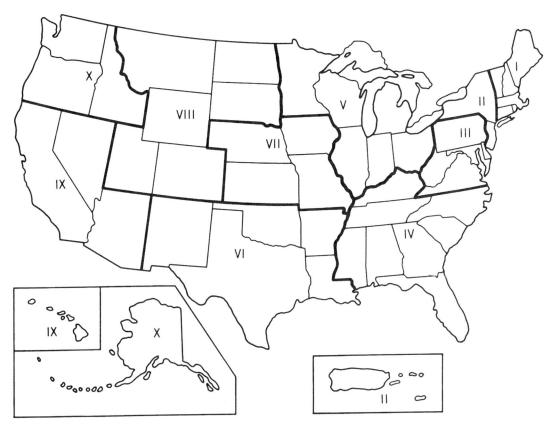

Figure 11.5 Federal Air Quality Control Regions. *Source:* U.S. Environmental Protection Agency, 1980.

Table 11.2 National Ambient Air-Quality Standards (NAAQS), 2001

Pollutant	Averaging Time	Primary	Secondary
PM_{10}	Annual arithmetic mean	50 µg/m^3	Same as primary
	24-hour	150 µg/m^3	Same as primary
$PM_{2.5}$[a]	Annual arithmetic mean	15 µg/m^3	
	24-hour	65 µg/m^3	
SO_2	Annual arithmetic mean	80 µg/m^3 (0.03 ppm)	Same as primary
	24-hour	365 µg/m^3 (0.14 ppm)	Same as primary
	3-hour		1300 µg/m^3 (0.50 ppm)
CO	8-hour	10 µg/m^3 (9 ppm)	No secondary standard
	1-hour	40 µg/m^3 (35 ppm)	No secondary standard
NO_2	Annual arithmetic mean	100 µg/m^3 (0.053 ppm)	Same as primary
O_3	Maximum daily 1-hour average	235 µg/m^3 (0.12 ppm)	Same as primary
	8-hour[a]	157 µg/m^3 (0.08 ppm)	
Pb	Maximum quarterly average	1.5 µg/m^3	Same as primary

[a]Proposed standard but not yet implemented.
Source: U.S. Environmental Protection Agency, 2001 *(http://www.epa.gov/airs/criteria.html)*.

on for sizes less than 10 microns in diameter) cause the most serious health threat since they become lodged in lung tissue and remain in the body for significant lengths of time. New standards for particulates $PM_{2.5}$ (2.5 microns in diameter) were adopted in 1997 but not fully implemented because of legal challenges.

The Nondegradation Issue Interesting quirks in the clean air legislation began to emerge in the mid-1970s, and these involved conflicts between economic development and air quality. The intent of the 1970 amendments was to keep clean air clean, while cleaning up dirty air. Primary standards for the criteria pollutants were to have been met by 1975. But there were no provisions or policies for those areas that were already clean in 1970. Industry noticed this and began to relocate into these relatively clean areas. The EPA did not take action on this issue, which became known as Prevention of Significant Deterioration (PSD), until the Sierra Club filed a legal suit over the Kaiparowitz energy facility in southern Utah.

In response to a court order, the EPA established its PSD policy, which effectively limits the extent to which clean air can be degraded by managing economic growth in various regions (National Research Council 1981). The entire United States was divided into three classes. Class I areas could not have any increases in particulate (PM) or SO_2 levels. All National Parks and National Wilderness Areas were designated mandatory Class I areas, a designation which limits industrial growth in the area. Most of the Class I areas are located in the western half of the country (Fig. 11.6). The primary goal of this designation is to protect the visibility in the national parks and wilderness areas (Issue 11.1).

Class II allows for moderate development and industrial growth. All areas of the country that were not mandatory Class I regions were assigned to this group. The states were then given the opportunity to change this designation to Class I or Class III. Class III permits significant industrial growth and residential development. Changing to a Class III designation, however, requires envi-

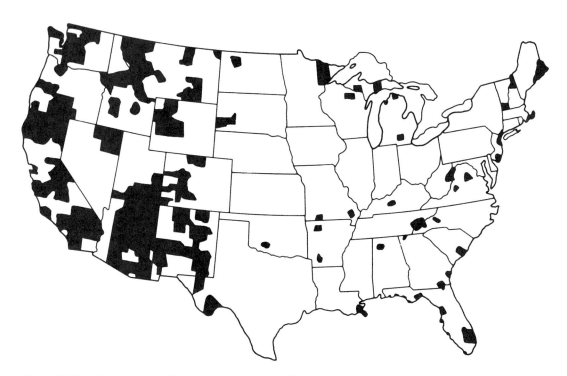

Figure 11.6 Class I counties. These counties have been designated as Class I areas because they have the best air quality. Therefore, new industrial growth in these areas is somewhat restricted. *Source:* U.S. Environmental Protection Agency, 1980.

Issue 11.1 On a Clear Day You Can See the Grand Canyon

One of the primary motivations for strengthening some of the provisions of the Clean Air Act in 1990 was the limited visibility in many of our national parks and national monuments. On "bad" days, it is virtually impossible to see across the Grand Canyon. On good days, the view extends for miles and miles. Visibility conditions are influenced by humidity, direct emissions of particulate matter, sulfur dioxide emissions, and nitrogen oxide emissions. Without pollution, visibility in the east would range from 45 to 90 miles, while in the west, greater vistas can be viewed, ranging from 120 to 180 miles. In the eastern half of the nation, the reduced visibility is largely due to higher humidity levels (which enhance the effect of pollution) and sulfates (formed by the combination of sulfur dioxide and nitrogen oxides, often located in distant areas and simply transported to the region). In the west, reduced visibility is most often caused by dust and emissions from burning wood and locally generated sources. Under the Clean Air Act, the visibility in national parks and wilderness areas is protected (they are listed as Class I areas).

Perhaps the most troublesome pollutant to affect national parks and monuments is ozone. Monitored since 1990, ozone concentrations have inched upward, affecting 40 percent more parks at the end of the decade than in the beginning. Ozone concentrations exceeded national ambient standards in nine national parks and monuments in 1990, and this number increased to 13 ten years later. Sequoia–Kings Canyon National Park exceeded the ozone standard 27 times in 1990, 58 times in 1994, and 23 times in 1999. Great Smoky Mountain Park, in the east, began the decade exceeding the standard only 4 times, but by 1999, this number had inched upward to 26 (USEPA 2001). Cape Cod National Seashore and Great Smoky Mountain National Park have the most serious and persistent problems with ozone. It is surprising that ozone concentrations in the Grand Canyon have remained relatively unchanged during the decade.

In 1999, the USEPA established a regional haze program, acknowledging that reduced visibility in many national parks is caused by the long-range transport of pollutants emitted elsewhere. This requires the cooperation and coordination of local, state, and other federal agencies working in partnership to improve the air quality and visibility in our national parks and wilderness areas. It is too soon to tell how effective this regional approach will be, but it is certainly a step in the right direction.

---

ronmental impact statements, public hearings, and EPA approval.

There was and still is considerable debate over the visual impairment program. In the west, visibility impairment (for the worst days) has remained relatively unchanged during the 1990s. These debates are particularly acute in the western half of the country, where issues over energy development, industrialization, and pristine areas are hotly contested.

National Trends

The United States has had air pollution control measures for more than three decades, during which time noticeable improvements in air quality have been made. However, some parts of the country still exceed the primary health standards, posing serious risks to human health and the environment. In 1993, slightly more than 79 million people lived in areas that did not meet air-quality standards. By 2000 121 million people were living in areas designated as nonattainment for one or more of the criteria pollutants. Overall, air quality in the United States has improved, but there are still some problem pollutants, most notably ozone, which accounts for the increased number of people living in nonattainment areas. Advances in pollution control have been offset by the presence of more people, creating increasing demand for heating/cooling and driving more miles.

Emissions Significant decreases in emissions of criteria pollutants have been recorded since the Clean Air Act first was implemented. Nitrogen oxide levels, however, have shown an increase during the past two decades (Fig. 11.7, Table 11.3). Total suspended particulate emissions have

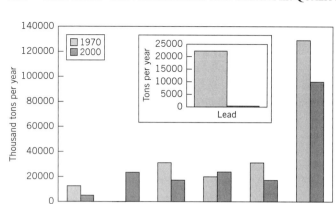

Figure 11.7 Change in U.S. emissions, 1970–2000. All criteria pollutants decreased in emissions, with the exception of nitrogen oxides. *Source:* U.S. Bureau of Census, 1997; USEPA, 2001.

declined by 66 percent since 1970 and by almost 5 percent since 1990. Fugitive dust from agricultural erosion, construction, and so forth increased in the first half of the decade, on the order of 1 percent per year. Although some improvement has been made in emissions from industrial sources due to the installation of control equipment, particulate emissions from mobile sources remain problematic. Emissions standards for diesel automobiles and tailpipe standards for diesel trucks and buses took effect in 1988, helping to reduce particulate emissions from transportation sources. However, during the past decade there was a 6 percent increase in these emissions. On the other hand, emissions from fugitive sources (Fig. 11.8), especially anthropogenic sources, declined during the decade.

The largest and most persistent source of SO_2 emissions is coal-burning electric power plants. Emissions have declined during the past decade, largely as a result of the switch from coal and high-sulfur oil to natural gas and low-sulfur oil.

The recent trend shows a 26 percent decrease in SO_2 emissions during the past decade, yet 17.5 million tons of SO_2 were emitted into the atmosphere. The increased use of emission control devices by industry (especially the chemical sector) and the recovery of sulfuric acid at smelters have contributed to the decline (Council on Environmental Quality 1995).

Emissions of nitrogen dioxide fluctuated during the last decade. While emissions were reduced from stationary sources (electrical utilities and industrial boilers), those from transportation sources have increased slightly, despite advances in emissions control on highway vehicles. There are simply more people driving more miles. The net result is a decline in emissions (around 2 percent) during the past decade.

Since oxidants are byproducts of chemical reactions, we have no direct emissions data for them. However, we do have emissions data for both precursors: nitrogen oxides (already described) and volatile organic compounds (Fig. 11.9). VOC emis-

Table 11.3 Percentage Change in U.S. Emissions of Selected Pollutants

Decade(s)	PM_{10}	SO_2	NO_x	VOC	CO	Lead
1940–1950	+7.3	+12.1	+36.9	+22.0	+9.6	NA
1950–1960	−9.2	−0.6	+40.1	+16.8	+6.9	NA
1960–1970	−16.2	+40.2	+45.9	+25.3	+16.7	NA
1970–1980	−45.9	−16.9	+12.9	−15.5	−9.7	−65.8
1980–1990	−44.9	−13.4	−1.0	−8.9	−13.0	−92.4
1990–2000	−12.2	−26.0	−2.2	−15.3	−2.7	−19.6
1970–2000	−65.9	−43.8	−12.4	−42.8	−26.0	−98.2

NA = not available
Source: U.S. Bureau of the Census, 1997; USEPA, 2001.

Figure 11.8 Blowing dust from unvegetated areas such as this site in Nebraska contributes to particulate pollution problems.

sions were down by 15 percent between 1990 and 2000, but 17.7 million tons of VOCs were still emitted. The majority of VOC emissions continue to come from highway and off-highway vehicles.

Two-thirds of the carbon monoxide emissions are from vehicles. Thus, emissions reductions in vehicles are offset by the increased number of vehicles and vehicle miles driven. However, between 1990 and 2000, there was a slight decrease in CO emissions. Pollution control and the retirement of older vehicles without catalytic converters have helped bring about the decline.

One of the greatest success stories in emissions reductions is lead, the use of which declined by 98 percent between 1970 and 2000. The phaseout of lead in motor fuels began in the early to mid-1970s and was virtually completed by the mid-1980s. As a result, ambient lead levels declined by over 75 percent during that 10-year period. Total emissions in 2000 were around 4000 tons, as compared to over 220,000 tons in 1970. Point sources now account for the major emissions, with metal processing representing about 40 percent of the total (U.S. Census Bureau 2001).

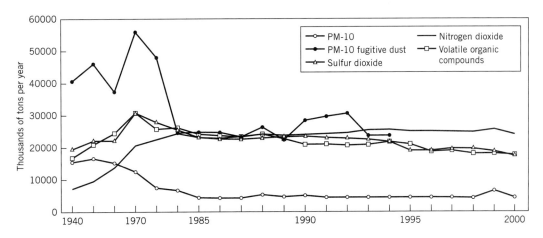

Figure 11.9 Emissions trends for criteria pollutants, 1940–2000. Note the differences in sources of pollutants. *Source:* U.S. Bureau of Census, 1997; USEPA, 2001.

Ambient Concentrations Overall, ambient concentrations of the criteria pollutants have improved since 1980. PM_{10} ambient concentrations decreased by 19 percent between 1990 and 2000, for example. In 1985, 290 areas did not achieve air-quality standards for one or more of the criteria pollutants, and by 1990 this was reduced to 274 metropolitan areas exceeding air-quality standards. A decade later, the number of nonattainment areas dropped to 135 (U.S. Environmental Protection Agency 2001), but 121.4 million were still affected by poor air quality. Regionally, these nonattainment areas are concentrated in the Atlantic coastal states, Ohio, and throughout California. Most of the nonattainment areas did not meet the ozone standards or had especially high particulate matter readings. The latter is the case in the arid West, where wind-blown agricultural dust continues to degrade the air quality.

The construction of tall stacks by industry in the 1970s dispersed sulfur dioxide emissions far from the local source. The result was that ambient conditions at the local level improved even though emissions increased. The pollutants were simply transported farther downwind to more remote areas. There were 60 nonattainment areas for sulfur dioxide in 1985, and this figure dropped to 43 in 1995 (U.S. Environmental Protection Agency 1995). By 2000, all metropolitan areas in the nation had achieved the sulfur dioxide standard. Despite the overall decline in sulfur dioxide ambient levels, the problem still exists in those areas that burn high-sulfur coal to generate electricity, where nonferrous smelters operate, and where steel and chemical plants and pulp and paper mills predominate, such as in the intermontane West and the Great Lakes states.

Ambient concentrations of NO_x have also improved. Ambient levels have remained relatively constant in the 1990s. Los Angeles–Long Beach, California, had the highest annual arithmetic mean for NO_x in the nation, but it was still below the standards. There are currently no recorded violations of the annual average nitrogen dioxide standard anywhere in the country.

The trends in ozone concentrations tell a different story. For example, high levels were found in 1983 and 1988 and were attributed largely to meteorological conditions that favored the formation of ozone. The number of areas where the ozone standard (one-hour concentration) was violated dropped from 368 in 1985 to 51 by 1990. In 1995, 43 metropolitan areas were in nonattainment, and this dropped to 30 five years later. However, if the 8-hour standard is included, another 72 areas were in nonattainment, most in suburban areas surrounding the central cities. Again, California has the worst ambient ozone levels in the country. Houston has the highest readings of any individual metropolitan area and is closely followed by Riverside–San Bernardino, California. Although stationary sources do contribute to the problem, EPA control efforts focus on auto emissions by stressing motor vehicle inspections and maintenance. In addition, major life-style changes may be required to significantly reduce ozone levels in some of the nation's harder-hit sunshine states.

Ambient concentrations of CO have also declined, again because of emissions controls on automobiles. In 1996, EPA designated 31 areas as nonattainment for CO, with only one, Los Angeles, classified as serious. In 1995, for example, Los Angeles had 14 days when it failed to meet the CO standard, compared to the next highest city, Fairbanks, Alaska, which had 9 such days. In 2000, only Los Angeles failed to meet the carbon monoxide standard.

Finally, lead concentrations in urban areas (where most of the monitoring stations are located) show significant improvement. There has been a steady decline in ambient lead levels as well. Some areas, however, have still not met the ambient standard. These urban areas include St. Louis, Missouri, and Tampa–St. Petersburg, Florida, both with localized point sources of lead emissions.

How Healthy Is the Air You Breathe?

In an effort to standardize monitoring efforts nationwide, the EPA adopted a uniform air-quality index in 1978. This index, the *Pollution Standards Index*, or PSI, is a health-related comparative measure based on the short-term national ambient air-quality (NAAQS) primary standards for criteria pollutants. It is widely used to report daily air-quality readings to the public.

The PSI integrates concentrations of nitrogen dioxide, sulfur dioxide, carbon monoxide, ozone, and particulate matter (PM_{10}) for an entire monitoring network into a single value, which ranges from 0 to 500 (Table 11.4). When the levels for all five of these pollutants are below NAAQS primary standards, the air is called good or moderately polluted (PSI values 0–99). When ambient

Table 11.4 Comparison of Pollution Standards Index (PSI) Values

PSI Value	PM^a	$SO_2{}^a$	CO^b	$O_3{}^c$	$NO_2{}^c$	Descriptor
400+	875+	2000+	46.0+	1000+	3000+	Very hazardous
300–399	625–874	1600–1999	34.0–45.9	900–1099	2260–2999	Hazardous
200–299	375–624	800–1599	17.0–33.9	480–899	1130–2259	Very unhealthful
100–199	260–374	365–799	10.0–16.9	240–479	NR^e	Unhealthful
50–99	75^d–259	80^d–364	5.0–9.9	120–239	NR	Moderate
0–49	0–74	0–79	0–4.9	0–119	NR	Good

^{a}24-hr, μg/m³.
^{b}8-hr, μg/m³.
^{c}1-hr, μg/m³.
^{d}Annual primary NAAQS.
^{e}NR = no index value reported at concentration levels below those specified by "alert level" criteria.
Source: Council on Environmental Quality, 1980:156–157.

concentrations of any of the criteria pollutants exceed their primary standard, the PSI reading is in the 100–500 range, depending on the concentration level. PSI values in the 100–200 range are labeled unhealthful; values from 200 to 300 are called very unhealthful; and values in excess of 300 are labeled hazardous.

Public warnings are issued when PSI values rise above the "good air" value of less than 100. An air-quality alert is called when PSI values range from 100 to 200; at this time, persons with heart or respiratory ailments should reduce physical exertion. An air pollution warning is given when the PSI ranges from 200 to 300. During a warning, elderly and other persons with heart and lung diseases should remain indoors. Industry is also asked to curtail emissions temporarily, until the warning is removed. An air pollution emergency is called when PSI readings exceed 300. Then the general population is advised to refrain from outdoor activities, and persons with heart and lung diseases are advised to remain indoors and minimize their physical activity. Industry and motorists are asked to curb emissions through lower production and less driving, respectively (Issue 11.2).

ISSUE 11.2: GREEN DAYS, RED DAYS

In the mid-1990s, the USEPA developed a user-friendly way of communicating the health risks of air pollution to communities. The Air Quality Index generates air quality forecasts and communicates the air quality information in a very user-friendly manner to the public. Based on its predecessor, the Pollution Standards Index, the *Air Quality Index (or AQI)* similarly ranges from 0 to 500. The index uses the five criteria pollutants (ground level ozone, particulate matter, carbon monoxide, sulfur dioxide, and nitrogen dioxides) and ambient air quality standards to assess the placement on the scale. For air to be rated unhealthy (for some sensitive groups) or greater than 100 on the scale, a carbon monoxide reading of more than 9 ppm, a particulate reading of 150 micrograms per cubic meter (for PM_{10}), or an 8-hour averaged ozone reading of .08 ppm is needed. The scales are color-coded and range from good (green, with values 0–50) to moderate (yellow, with values 51–100) to unhealthy for sensitive groups (orange, with values 101–150) to unhealthy (red, with values 151–200) to very unhealthy (purple, with values 201–300) and, last, hazardous (maroon, with values exceeding 301). The information is communicated in newspapers, on the radio and television, and even on the Internet (*www.epa.gov/airnow*). Health advisories are issued along with the readings when levels exceed 100. These include such warnings as "people with asthma should consider limiting outdoor activities" to more severe warnings that "people with respiratory problems should remain indoors." So, when you get up in the morning, be sure to check the local newspaper or television weather report to see what the air quality forecast is for the day.

Overall air quality in metropolitan regions is improving. Selected metropolitan regions had an average number of 37 days of unhealthful air (above 100 on the PSI) in 1984, and this dropped to 12 days by 1993; by 1995 it was down to 8 days (Council on Environmental Quality 1995; U.S. Environmental Protection Agency 1995). Although the national trend exhibits signs of improvement, three California metropolitan areas (San Bernardino, Bakersfield, and Fresno) had air-quality alerts (PSI > 100) for the equivalent of two months or more.

Southern California still has the most troublesome air quality. In 1984, Los Angeles had 204 days with PSI levels greater than 100, but by 2000, it had decreased to a mere 48 days (or a little over a month and a half). There are, however, year-to-year fluctuations that are often more reflective of local conditions than of any success in pollution abatement (Fig. 11.10). While portions of California have the distinction of the worst air pollution in the nation, it is also one of the most innovative regions for pollution reduction strategies (Issue 11.3).

Air-Quality Control and Planning

Economic Considerations In 1999, $198 billion was spent in the United States for pollution abatement and control (U.S. Bureau of the Census 2000). Nearly $16.2 billion was for air-quality control alone. The government (federal, state, and local) spent $500 million (less than 2 percent of all air pollution abatement expenditures). Industry spent nearly $21 billion (67 percent) for air pollution abatement, and consumers spent another $8 billion (27 percent). Governmental regulation, monitoring, and research cost another $1.7 billion.

ISSUE 11.3: SMOG CITY, USA

"I love L.A.," echoes the refrain from a popular song about the city of angels. But how much longer will Los Angeles residents be feeling this way if they must begin to wear gas masks before venturing outside in the warm California sun? People in Los Angeles are becoming less enamored of their city as increased air pollution continues to choke not only their city but all of Southern California and as the costs of curbing it escalate.

Good air quality has been a persistent problem in Southern California since Juan Rodriguez Cabrillo first discovered the Bahia de los Fumos (Bay of Smokes) in 1542. By 1877, air pollution, in the form of dust from the streets, was so bad it prompted one citizen to remark: "It does not allow invalids with lung disease to remain here" (Weaver 1980:197). By 1944, the term *smog* (smoke and *fog*) was coined to describe the brown haze that hung over the Los Angeles basin. With postwar urbanization and industrialization, the now famous Los Angeles smog worsened, and residents began to experience discomfort and adverse health effects. Smog alerts became commonplace, and as early as the 1950s, people were advised to curtail their outdoor physical activities.

Four decades later, the smog problem remains. In March 1989, the South Coast Air Quality Management District (AQMD) finally decided to confront the Southern California icon, the single-passenger automobile. The AQMD developed a 20-year strategy for cleaning up the air by proposing and passing an innovative and far-reaching plan. The plan covers 13,350 square miles (34,600 km^2) of Southern California and its 15 million inhabitants, as well as their 7 million cars and 2 million trucks and other vehicles.

The primary goal of the AQMD is to improve air quality by changing residents' everyday habits in the Los Angeles Basin. Restrictions have been placed on activities ranging from driving and parking to use of powered lawn-care tools and even backyard barbecues. Ultimately, this may change how cities are organized in the future.

The plan has 123 specific actions that involve not only residents but also industry. It placed a number of limits on sources of air emissions. Restrictions on car use included the elimination of free parking in some downtown locations and an increase in registration fees for motorists with more than one car. Emissions standards for diesel engines were tightened, and paints and solvents were reformulated to decrease emissions.

All over the city, businesses and industry began changing. The local dry cleaners had to in-

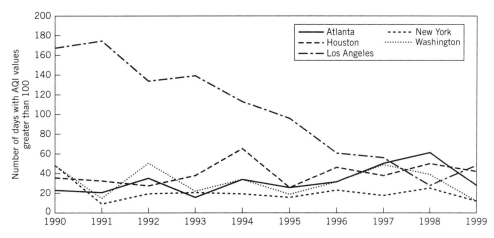

Figure 11.10 Air quality in selected metropolitan areas, 1991–2000. Overall air quality has improved in these cities, but Los Angeles just beats out Houston for the moniker, the worst air quality of any major metropolitan area in the United States.

stall a combined wash/dry system to reduce VOC emissions when clothes are moved between machines. Some paint shops switched to water-based paints with fewer solvents. Body shops began to use a lower-pressure spray to reduce the amount of paint in the air.

At the same time, the AQMD established its RECLAIM pollution trading program. Originally covering 300 firms, RECLAIM allows companies to buy pollution credits on the "smog exchange." Using annual emissions for 1989 or 1990 (the company chooses) as a base, industries can either cut their own annual emissions or buy the unused portion of emissions (credits) from another firm. Because the base year of 1989 was quite high, emissions could actually increase. This is precisely what happened between 1993 and 1996.

In 1993, Southern California had its lowest emissions in more than a decade, owing largely to the recession in the region caused by the downsizing of the aerospace industry. As the economy picked up, so did emissions. From 1993 to 1996, emissions rose from their 1993 low values (remember, the targets were set on 1989 emissions), so that the AQMD has now fallen 60 percent short of its emissions reduction goals. By 1997, the RECLAIM program was expanded to more than 1000 industries in an effort to increase compliance. Emissions targets were set at 38 tons of VOCs daily, compared to 31 tons in 1993. Under this new plan, the target emissions were slightly higher than those in the 1994 plan, but the rate of decline was faster since the timetable for emissions reductions was shortened from 2010 to 2005 in order to meet the federal ozone health standard.

As one might guess, the debate about the AQMD plan continues. Since Southern California already has the toughest air pollution legislation in the country, there was no alternative other than this complete overhaul. Legal challenges and business threats will no doubt fuel the fire over whether or not the costs exceed the benefits. The costs are astounding—around $1.7 billion annually (Cone 1997). Opponents claim that the benefits are not nearly that great, and proponents say it must be done to protect the health and welfare of residents. This far-reaching plan is the first of its kind in the nation and will surely serve as a model for other urban areas whose air pollution problems are worsening. A major change in life-style and the pattern of business is occurring in Southern California. Los Angeles has said yes to clean air, but at what cost? (Cone 1995; Reinhold 1989; Weisman 1989.)

Clean air is a costly business, and in an era of fiscal austerity, the return on investments in air pollution reduction must be demonstrated. EPA's Pollution Prevention Program (see Chapter 2) is one example of the need for the government to forge private partnerships and to make it cost-effective for industries to comply.

Another example is EPA's emissions trading program for stationary source control (Bearden 1999). Instead of considering each smokestack as an emitter, which was the previous policy, the EPA now views the entire plant as a point source. Thus, it allows emissions from one smokestack to exceed standards as long as another stack at a different location in the same plant has compensating reductions. As long as the emissions from the entire plant do not exceed the standards, then the plant is not in violation of the Clean Air Act and thus not subject to criminal prosecution. This policy allows the plant to average emissions from all stacks, allowing internal decision-making about what is most appropriate for the plant. Production levels and expenditures for control equipment are made by the plant management, as long as the total emissions are below federal limits.

Control Programs Until recently, it was thought that vehicle exhaust emissions offered the greatest potential for decreasing mobile source contributions to the nonattainment of NAAQS. However, the newest-model cars already remove up to 95 percent of the emissions. Real improvements in vehicle emissions will come from a switch from gasoline as a fuel and from reductions in the number of vehicles and miles driven.

Vehicle exhaust emissions have been federally regulated since 1968 and have become increasingly strict (Table 11.5). The emissions standards apply only to the newer-model cars and trucks; older models have less stringent controls. Increased fuel efficiency is another way to reduce emissions. Smaller, more fuel-efficient cars produce less pollution. During the 1980s, fuel efficiency standards were rolled back to 26 mpg to ease pressures on U.S. auto makers. The result was that Americans began driving bigger and less fuel efficient cars, thereby producing more air pollution. In early 1989, the fuel efficiency standard was increased to 27.5 mpg as a result of increased pressure to reduce smog. Today, Americans are still driving bigger cars such as minivans and sport utility vehicles with lower fuel efficiencies.

Table 11.5 Exhaust Emission Standards for Automobiles

Model Year	VOC	Standard (g/m^3) CO	NO$_x$
Pre–1968	8.2	90.0	3.4
1968–1971	4.1	34.0	NA[a]
1972–1974	3.0	28.0	3.1
1975–1976	1.5	15.0	3.1
1977–1979	1.5	15.0	2.0
1980	0.41	7.0	2.0
1981	0.41	3.4	1.0
1987	0.41	3.4	1.0
1991	0.25	3.4	0.7
1996	0.25	3.4	0.7
2000	0.25	3.4	0.4

[a]NA = not available.
Source: National Commission on Air Quality, 1981; Renner, 1988; U.S. Environmental Protection Agency, 1997; USDOE, 2001 *(http://www.afdc.doe/gov/pdfs/standrds.pdf).*

One area specifically addressed by the 1990 CAA is the requirement for cleaner burning fuels. For example, in 1992 oxygenated fuel was introduced as a method for reducing CO emissions and improving total fuel combustion, especially in colder areas. In 1993, limits on the sulfur content in diesel fuel came into effect. Also, regulations for cleaner, reformulated gasoline to reduce hydrocarbons and toxic emissions were put in place in the mid-1990s, especially in nonattainment areas. Last, California has required that a certain number of cars sold in the state have zero emissions. This has prompted automobile companies to increase production of these vehicles.

State inspection and maintenance programs for vehicles in order to monitor emissions on a yearly basis are one method to control mobile sources. Unfortunately, there are no uniform requirements for such programs, and as a result there is great variation from state to state, with California's being the most stringent. Under the new CAA, enhanced vehicle inspection and maintenance, including stricter tailpipe and evaporative emissions control, are now under development. Electric-powered vehicles are another option, particularly for localized commuting within metropolitan areas (Jensen and Ross 2000).

The most obvious way to reduce vehicle emissions is for people to drive less. However, this is an unpopular solution to the problem, given the increasing number of motor vehicles and trucks on the road and our attachment to personal transportation. During the past decade, the number of cars and buses increased by 25 percent, while the number of trucks using diesel engines increased by 40 percent. The number of miles driven, however, doubled; thus, any improvement in emissions control was more than offset by the increase in vehicle miles driven. Under the new legislation, states are required to rethink their planning and transportation services provision so as to consider them within the context of meeting air-quality goals. In 1998, for example, 22 cities were required to have a new fleet of public transportation vehicles (taxis, buses) that met more stringent tailpipe emissions standards than automobiles.

Stationary-source control involves installing mechanical devices on smokestacks and switching from high-sulfur to low-sulfur fuels. Fitting gasoline pumps with pollution control equipment is another method currently in use to prevent hydrocarbons from escaping at the gas station (Fig. 11.11). There are over 27,000 major stationary sources of air pollution in this country alone. The EPA considers "major" any plant that produces more than 100 tons of pollutants per year. Stack scrubbers, precipitators, and filters are costly capital investments for industry, especially for facilities with old, outdated plants. For example, industry spent $4 billion on capital expenditures for air pollution abatement and another $6 billion in operating costs in the mid-1990s (U.S. Bureau of the Census 1996). The chemical and petroleum industries had the largest air pollution control expenditures.

Compliance with federal standards is spotty. Unfortunately, no systematic assessment of compliance has been made at the national level since 1980. Reductions in enforcement actions as a result of cuts in the federal budget and decentralization of the federal role in air-quality control during the Reagan administration obviously have not improved air quality. Air pollution control is expensive and requires federal action and support from both industry and consumers.

TOXINS IN THE AIR

Over 70,000 synthetic chemicals are available in the world today. Although the effects of immediate (acute) exposures to human health and the environment may be known, scientific information on the effects of lower-level, longer-term exposures is often incomplete or missing. Disasters such as the 1984 release of methyl isocyanate in Bhopal, India, which killed more than 2000 people, do occur, though not frequently. Perhaps of greater concern are the daily emissions of airborne toxic substances that result in chronic exposures. Unfortunately, we have very little information on the number of deaths attributed to longer-term exposures to toxic chemicals in the air we breathe.

For years, the EPA has been grappling with the problem of toxic substances in the air, or "toxics" for short, that are emitted from a wide range of mobile and stationary sources, including incinerators, municipal waste sites, plastics and chemical manufacturing plants, and sewage treatment

Figure 11.11 The rubber boot on this gasoline pump nozzle captures hydrocarbon-rich vapors that are expelled from a fuel tank as it is filled with gasoline, reducing the amounts that are discharged to the atmosphere. Measures such as these have become increasingly important in capturing so-called fugitive emissions of hydrocarbons.

plants. The agency's first approach was to regulate the source emissions for a small number of toxic pollutants, even though hundreds of organic compounds, many with carcinogenic or mutagenic properties, are routinely emitted into the air. The National Emission Standards for Hazardous Air Pollutants (NESHAP), for example, monitors criteria pollutants and has set emissions standards for eight toxins: asbestos, beryllium, mercury, vinyl chloride, arsenic, radionuclides, benzene, and coke oven emissions. Under the 1990 CAA, emissions standards for 188 extremely hazardous substances are being phased in over the next decade. This should reduce "routine" toxic emissions into the air. Furthermore, the EPA has implemented toxic emissions reductions for certain industries—dry cleaners, coke ovens, industrial solvent users, and chromium electroplating factories—in an effort to reduce heavy metal and perchlorethylene emissions.

Underlying the current EPA policy was the assumption that emissions are directly related to ambient quality and thus human exposure. However, the air toxics problem is complex and requires an integrated approach in managing toxic substances and exposures from a variety of sources and media. As a result, the EPA is now using an exposure assessment methodology that measures all the exposures to toxics regardless of media (e.g., air, water, land). In this way, we can understand the total burden placed on the environment from toxic releases. The largest obstacle to widespread use of this method may be the myriad environmental laws that restrict an agency's ability to undertake integrated studies of human exposures from different media as well as institutional inertia.

Under Title III of the Superfund Amendments and Reauthorization Act (1986), industry was required to report on the quantities of toxic emissions for about 320 chemicals. In the first survey, the EPA found that 2.7 billion pounds of toxic chemicals were emitted into the air in 1987, significantly more than anyone thought (Cutter and Solecki 1989). Unfortunately, this is a conservative estimate, for it excludes those toxic emissions from autos, toxic waste dumps, and, most important, companies that produce less than 75,000 pounds of toxic materials. In 1998, roughly 2 billion pounds were released, of which 9.2 million pounds were hazardous air pollutants (or HAPs). The electric utility industry was the largest source of toxic emissions (784 million pounds), followed by the chemical industry (321 million) and the paper industry (186 million) (U.S. Census 2001).

Indoor Air Pollution

Indoor air pollution has become a major health issue in the United States. As houses and buildings become more energy-efficient, concentrations of pollutants build up because of lack of ventilation in both the winter and summer. Many people spend as much as 90 percent of their time indoors. The problems of indoor air pollution have received widespread attention during the past decade. Potential health effects range from short-term symptoms such as headaches, nausea, and throat irritations to longer-term health problems like lung disease and even cancer. There are a wide array of pollutants coming from a variety of sources such as tobacco smoke, building materials, gas ranges, cleaning agents, and drinking water.

Home, Work, and School The most serious pollutant in the home is tobacco smoke (especially the benzene it contains), followed by radon and particulates from wood-burning stoves. Other sources of pollutants include some consumer products such as paint thinners and wood conditioners. Formaldehyde, which is often used in furniture, foam insulation, and some wood products, is a major source of volatile organic compounds; hence, the manufacturers recommend use of these products in "well-ventilated areas." Other sources of VOCs are carpets and carpet adhesives, latex paint, and products made from particleboard, such as bookcases.

We sometimes hear complaints about "sick buildings" that cause their inhabitants to complain about eye and throat irritations, drowsiness, headaches, and so on. The likely sources of many of these ills are elevated pollution levels within the buildings, often caused by the same agents that create elevated levels at home. Heating, ventilating, and air conditioning systems can bring biological contaminants indoors and circulate them throughout buildings, causing allergic reactions to pollen and fungi, as well as promoting more serious bacterial and viral infections. Asbestos, once used widely for insulation and fireproofing, is a known carcinogen. Some of the highest concentrations of asbestos have been found in the nation's schools. In 1986, Congress passed the Asbestos Hazard Emergency Response Act, which required all schools to inspect for asbestos-containing materials and to develop plans to remove them. The role of tobacco smoke in increasing benzene levels, together with the increased cancers that result from "passive smoking," has resulted in restrictions on smoking in most public and private buildings and on airplanes. The recent California law to ban smoking in bars and clubs is a good example.

Radon Radon is a tasteless, colorless, odorless gas that is a natural byproduct of the decay of radium. Radium occurs naturally in many different types of soils and rocks, and radon enters buildings through cracks or openings in the foundations or basements. A study by the EPA suggests that radon may account for between 5000 and 20,000 lung cancer deaths each year (U.S. Environmental Protection Agency 1988). The EPA estimates that up to 8 million homes nationwide may have radon levels exceeding 4 picocuries per liter, their indicator of high risk. The radon problem is found everywhere, and high-risk areas can be identified with use of geological maps to identify uranium-rich rocks. Elevated levels of radon were once thought to occur only in winter but are now found during summer months, as buildings remain airtight to keep cool with air conditioning.

Despite the relatively high risk, radon can often be controlled at the individual-homeowner level. Once a radon test is done, exposure can be lessened by sealing cracks in the basement, installing home ventilation systems, removing radium-tainted soil, and, if all else fails, relocation. Educational materials and testing kits are often available free from individual state governments. The most comprehensive educational and testing programs are found in Florida, Maine, New Jersey, New York, and Pennsylvania (where the problem was first "discovered" in 1984). New Jersey has the most active educational and testing program, and Pennsylvania offers low-interest loans for home repairs to ameliorate radon exposure.

CONCLUSIONS

As we have seen, the quality of our air resource has improved in some regions of the country but has worsened in others. Instead of cleaning up pollutants, we often exacerbate the problem by shifting from one pollutant source to another or simply transferring the problem to greater distances. Internationally, the problem of urban air pollution is increasing as many of the world's developing countries become more urbanized.

Cleaning up the air resource is costly and will entail cooperation between industry, government, and citizens. Unless we are willing to don gas masks every time we venture outside, it is essential that we reduce our reliance on the automobile and decrease fossil fuel use. Industry must also do its part by making fuels burn more efficiently and reducing the amount of toxins that are routinely emitted into the air we breathe. Unfortunately, reducing emissions entails substantial costs for both industrial polluters and consumers. While tangible emission reductions have been achieved, further reductions will require substantial capital investments or changes in the way we live. It remains to be seen whether we have the political will to force these improvements in air quality. The goal of clean air has bipartisan political support, but the means for achieving it vary from one political administration to another. The Clean Air Act has always been contentious and will continue to be so.

REFERENCES AND ADDITIONAL READING

Bearden, D. M. 1999. *Air Quality and Emissions Trading: An Overview of Current Issues.* CRS Report for Congress 98–563. http://www.cnie.org/nle/crsreports/air/air-27.cfm.

Cone, Marla. 1997. Revised anti-smog plan is unveiled. *Los Angeles Times*, October 20, p. A3.

_____. 1995. AQMD's smog plan for L.A. Basin Ok'd. *Los Angeles Times*, January 24, p. A3.

Council on Environmental Quality. 1995. *Environmental Quality, 1993. 25th Anniversary Report.* Washington, D.C.: U.S. Government Printing Office.

Crawford, M. 1990. Scientists battle over Grand Canyon pollution. *Science* 247:911–912.

Cutter, S. L. 1995. The forgotten casualties: women, children, and environmental change. *Global Environmental Change* 5(3):181–194.

_____ and W. D. Solecki. 1989. The national pattern of airborne toxic releases. *Professional Geographer* 41(2):149–161.

Elsom, D. 1992. *Atmospheric Pollution: A Global Problem*. Cambridge, MA: Basil Blackwell.

Goldsmith, J. R. and L. T. Friberg. 1977. Effects of air pollution on human health. In Arthur C. Stern, ed. *The Effects of Air Pollution*. 3rd Ed. Chapter 7, pp. 457–610. New York: Academic Press.

Jensen, M. W. and M. Ross. 2000. The ultimate challenge: developing an infrastructure for fuel cell vehicles. *Environment* 42(7):10–22.

Liroff, R. A. 1986. *Reforming Air Pollution Regulation: The Toil and Trouble of EPA's Bubble*. Washington, D.C.: Resources for the Future.

Machado, S. and S. Ridley. 1988. *Eliminating Indoor Pollution*. Washington, D.C.: Renew America Project.

National Commission on Air Quality. 1981. *To Breathe Clean Air*. Washington, D.C.: U.S. Government Printing Office.

National Research Council. 1981. *Prevention of Significant Deterioration of Air Quality*. Washington, D.C.: National Academy Press.

Reinhold, R. 1989. Southern California takes steps to curb its urban air pollution. *New York Times*, March 18, p. A1.

Renner, M. 1988. *Rethinking the Role of the Automobile*. Worldwatch Paper no. 84. Washington, D.C.: Worldwatch Institute.

Tietenberg, T. 1985. *Emissions Trading: An Exercise in Reforming Pollution Policy*. Washington, D.C.: Resources for the Future.

Tolba, M. K. et al. 1992. *The World Environment 1972–1992*. London: Chapman & Hall.

United Nations Environment Program 1994. *Environmental Data Report 1993–94*. Cambridge, MA: Basil Blackwell.

United Nations Environment Program/World Health Organization. 1992. *Urban Air Pollution in Megacities of the World*. Cambridge, MA: Basil Blackwell.

U.S. Bureau of the Census. 1996. *Statistical Abstract of the United States, 1996*. Washington, D.C.: U.S. Government Printing Office.

_____. 1997. *Statistical Abstract of the United States, 1997*. Washington, D.C.: U.S. Government Printing Office.

_____. 2000. *Statistical Abstract of the United States, 2000*. http://www.census.gov.

_____. 2001. *Statistical Abstract of the United States, 2001*. http://www.census.gov.

U.S. Environmental Protection Agency. 1980. *Environmental Outlook*. EPA-600/8-80-003. Washington, D.C.: U.S. Government Printing Office.

_____. 1988. *Environmental Progress and Challenges: EPA's Update*. EPA-230-07-88-033. Washington, D.C.: U.S. Government Printing Office.

_____. 1995. *National Air Quality and Emissions Trends Report 1995*. http://www.epa.gov/oar/aqtrnd95tap.html.

_____. 1997. *Motor Vehicles and the 1990 Clean Air Act*. http://www.epa.gov/docs/omswww/11-vehs.htm.

_____. 2001. *Latest Findings on National Air Quality: 2000 Status and Trends*. EPA 454?K-01-002. Research Triangle Park, NC: Office of Air Quality. http://www.epa.gov/airtrends.

Weaver, J. D. 1980. *Los Angeles: The Enormous Village 1781–1981*. Santa Barbara, CA: Capra Press.

Weisman, A. 1989. L.A. fights for breath. *New York Times Magazine*, July 30, pp. 14–17, 30–33, 48.

World Bank, 2001. *World Development Indicators*. http://www.worldbank.org/data.

World Resources Institute. 1996. *World Resources 1996–97*. New York: Oxford University Press.

STUDY QUESTIONS

1. What is an inversion, and how does it affect air pollution? How do sunlight and atmospheric humidity affect air pollution?
2. What are the criteria pollutants? For each one, what are the major emission sources? What emissions cause photochemical smog?
3. What has the government done to prevent air-quality deterioration in relatively clean areas? In nonattainment areas?
4. Does your city meet all federal air-quality guidelines? What pollutants are the greatest problems? What are the major sources of pollution in your city?
5. Summarize the trends in emissions and ambient concentrations for each major pollutant over the past several years. What is the difference among them?
6. How do cities in developing countries and those in wealthy countries differ in the kinds of pollution problems they face and in their potential solutions?
7. Economic development leads to increased industrial activity and fuel use, but it also makes possible investments to reduce air pollution. Overall, do you think economic development will improve air quality in developing world cities, or worsen it?
8. Why are toxic substances in the air of particular concern? Are toxics a more important problem than conventional pollution?

CHAPTER 12

REGIONAL AND GLOBAL ATMOSPHERIC CHANGE

Traditionally, air pollution has been a local concern, but recently it has taken on global significance in light of increasing scientific evidence about climate change and the destruction of the protective stratospheric ozone layer by anthropogenic air pollutants. Although the pollutants may have a specific local or regional source, their effects are so widespread and there are so many sources that the emissions and their impacts often transcend national boundaries. This phenomenon is called *transboundary pollution*. In this chapter we discuss three major atmospheric transboundary issues facing the world today: acid deposition; stratospheric ozone depletion; and climate change.

ACID DEPOSITION

Acid deposition is a general term that refers to the deposition of acids and substances that contribute to soil acidification, carried by rainfall, snow, and dust particles falling from the atmosphere. Although normal rainfall is slightly acidic, air pollution, particularly emissions of sulfur and nitrogen oxides, has greatly increased the acidity of rainfall over wide areas in the past several decades. It is a regional rather than a local consequence of fossil fuel combustion, having been identified as a problem in the eastern United States, Canada, Europe, China, and southern Brazil (Fig. 12.1). Acid deposition has broad-ranging environmental effects, including damage to vegetation and structures and reduced surface-water quality. It will no doubt continue to be an important environmental issue in industrialized and industrializing regions of the world in the years to come (Hedin and Likens 1996).

Formation and Emissions Sources

Acids are substances that give up a proton (hydrogen nucleus) in a chemical reaction. The acidity of solutions is measured on the pH scale, which is the negative logarithm of the concentration of hydrogen nuclei in a solution. A pH of 7 is neutral, and numbers less than 7 are increasingly acidic. A pH greater than 7 indicates alkaline conditions. Rainwater is slightly acidic, even under natural conditions. Carbon dioxide dissolves in water to form carbonic acid, a weak acid of about pH 5.6. Other natural sources of acids exist in the atmosphere, including sulfur compounds emitted from volcanoes and nitric acids created by lightning passing through the atmosphere. The relative contributions of these other sources to the acidity of rainfall are not known, but the pH of natural rainfall is variable. The pH of natural rainfall is usually slightly lower than 5.6 and perhaps as low as 5.0. Acid precipitation, on the other hand, has much lower pH values, commonly in the low 4's and sometimes in the low 3's, about the same as vinegar. A drop of one unit on the logarithmic pH scale represents a tenfold increase in acidity, so these are substantial differences.

Two major pollutants are responsible for acid deposition: sulfur dioxide and nitrogen oxides. Sulfur dioxide (SO_2) combines with oxygen and water in the atmosphere to form sulfuric acid (H_2SO_4). Nitrogen oxides (NO_x) combine with water to form nitric acid (HNO_3). These acids are found in water droplets and on dust particles in the atmosphere, and they are deposited on the ground

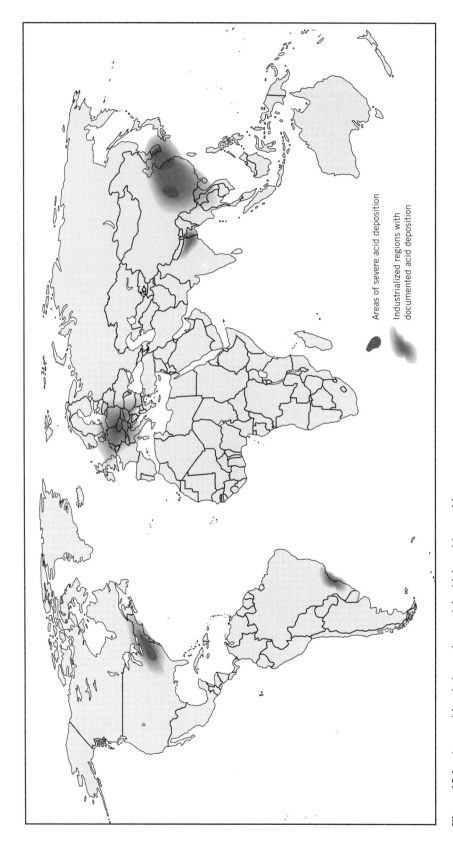

Figure 12.1 Areas with existing and potential acid deposition problems.

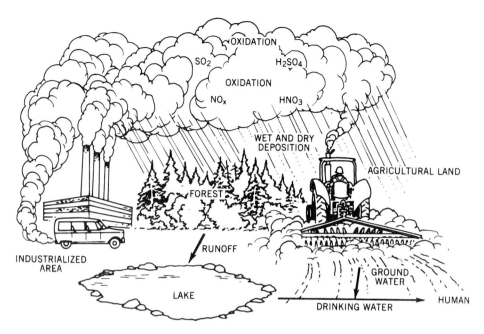

Figure 12.2 Formation and deposition of acid rain. Sulfur and nitrogen oxides emitted from industries, automobiles, and other sources are oxidized in the atmosphere to form nitric and sulfuric acids. These are deposited on the land, either in precipitation or in dry dustfall, and affect vegetation, soil, and water quality. *Source:* U.S. Environmental Protection Agency, 1980.

either in precipitation or in dry dust. The chemical processes that form these acids are not instantaneous; they take from minutes to days to occur, depending on atmospheric conditions (Fig. 12.2).

Sulfur dioxide emissions derive primarily from the combustion of impure fossil fuels, such as coal and fuel oil. Globally, anthropogenic sources of sulfur account for 55 to 80 percent of all emissions (UNEP 1993). Most of these emissions originate in the Northern Hemisphere. While sulfur and nitrogen oxide emissions are declining in the industrialized nations (Table 12.1), they are increasing noticeably in rapidly developing countries in Asia such as China and Pakistan (McDonald 1999).

In the United States, the areas with the largest sulfur emissions are in the urban industrial areas of the Midwest, particularly the region extending from Illinois to Pennsylvania (Fig. 12.3*a*). Most of the sulfur emissions are from coal-fired power plants, which are heavily concentrated in the region and account for the majority of all sulfur dioxide emissions nationally. These electricity-producing plants burn coal that is higher in sulfur content than the coal burned in the West. It recently was estimated that moderate-sulfur coal (which accounts for 40 percent of the total coal burned) is a prime source of emissions, accounting for the majority of all sulfur dioxide emissions. In addition, older, smaller electrical utilities emit larger amounts of sulfur dioxide than newer facilities. In Canada, smelting is a major contributor of sulfur. A nickel

Table 12.1 Anthropogenic Emissions of Sulfur Dioxide and Nitrogen Oxides in Selected Countries

Region/Country	Percent Change, 1990–1996	
	SO_2	NO_x
North America		
Canada	−16	−4
United States	−17	−2
Europe		
Czechoslovakia	−50	−42
France	−20	+4
Germany	−71	−30
Poland	−26	−9
United Kingdom	−46	−25
Russia	−40	−32
Ukraine	−54	−57

Source: World Resources Institute, 2000.

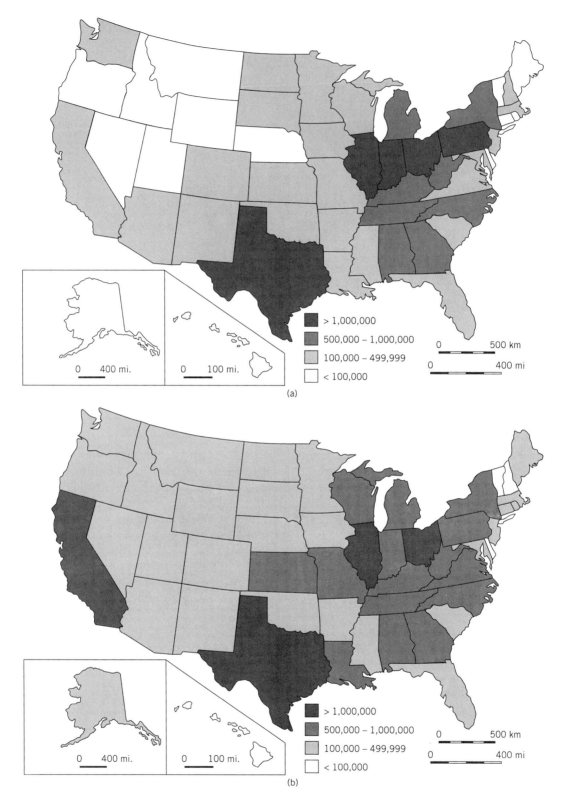

Figure 12.3 Sulfur dioxide and nitrogen oxide emissions, which contribute to the formation of acid precipitation. In 1999, sulfur dioxide emissions were greatest in the Ohio Valley states and Texas (*a*), while nitrogen oxide emissions were more dispersed, with Texas, Illinois, Ohio, and California having the most (*b*). *Source:* U.S. Environmental Protection Agency, 2001.

smelter at Sudbury, Ontario, was, for years, one of the largest single point sources of sulfur dioxide in the world, but it has found a way to reduce its emissions, reduce costs, and market the improved technology (Munton 1998). In Europe, sulfur emissions are greatest in the heavily industrialized regions of central Europe, although they have decreased dramatically since 1990.

The pattern of nitrogen oxide emissions is somewhat similar to that of sulfur dioxide (Fig. 12.3b), but the overall emissions are lower. Illinois, Ohio, Texas, and California have the highest levels. There is a lower concentration of nitrogen oxide emissions in New England and the Pacific Northwest.

Geographic Extent and Effects on the Environment

Acid deposition is a significant and widespread problem in several regions of the world—eastern North America, central and northern Europe, China, and Japan—and is locally important in many others. These major regions of acid deposition are all industrial areas with large amounts of coal use and major emissions of sulfur and nitrogen oxides. In North America, precipitation pH is in the low 4's over a large area extending from the mid-Atlantic coast to southern Ontario and from Ohio to New England.

The U.S. National Atmospheric Deposition Program monitors the chemistry of precipitation throughout the United States. Comparison of the eastern and western portions of the country shows a number of interesting trends (Fig. 12.4). We also know that the East has more precipitation than the West. When the greater amount of "wet" deposition is coupled with the concentration of acids, the East has a much more significant problem. While deposition and concentrations of sulfates in precipitation have been reduced in the past decade, especially in the east, the acidity of precipitation is still four to five times greater in the East than in the West (USEPA 2001). As seen on the map, the most acidic precipitation is found in eastern Pennsylvania.

In Europe the most severe acid deposition is found in Germany, Poland, and the Czech Republic, but significant problems occur over much of Europe. The situation has improved during the past decade with the closure of older factories and improvements in pollution control. Sulfur dioxide emissions from coal burning in Asia are almost equal to those in Europe and the United States combined in 2000 (UNEP 2000). While data are not readily available on acid deposition in China, the most severe problems are in the southeastern part of that country. Coal is used intensively in the northeast as in the southeast of China, but the acidity of precipitation is reduced in the northeast because of a higher input of neutralizing alkaline dust derived from the arid regions to the west. Chinese emissions contribute to acid deposition in Japan.

The most recognized effect of acid deposition is on lakes and aquatic life. Acid-neutralizing substances, such as calcium and magnesium compounds, are leached by water as it passes through the ground. In areas of calcium-rich rocks or soil, there are more than enough neutralizing substances to buffer most of the downstream effects of acid deposition. In areas of more acidic soils or in headwater areas where the water does not pass through much soil or rock before entering streams and lakes, the problem is more severe. In the Adirondacks of New York, for example, several high-altitude lakes have become so acidic that most fish cannot survive in the water. In some areas, lime has been added to lakes in an attempt to buffer the acids, but this has had limited success and is viewed as only a temporary solution to the problem. While sulfur levels in Adirondack lakes have declined in the 1990s, nitrogen levels have steadily increased, thereby increasing the acidity of the lakes (Dao 2000).

In addition to its effects on fish life, acid deposition contributes to slower growth and increased disease susceptibility in trees. Many mechanisms are involved, including mobilization of soluble metals in the soil, leaching of nutrients (especially calcium), and harm to soil biota that are vital to nutrient cycling. On Camels Hump Mountain in Vermont, spruce trees have been dying for several years. A drop in the pH tends to make minerals more soluble and thus increases their uptake by plants. Aluminum is of particular concern because of its toxic effects on plants. Analysis of cores from these trees suggests markedly increased accumulations of aluminum leached from the soil since 1950. Damage to spruce has also been documented in the Adirondacks. The main damage to red spruce, for example, is at 3000–4000 feet (910–1200 m) in the boreal zone of the Green

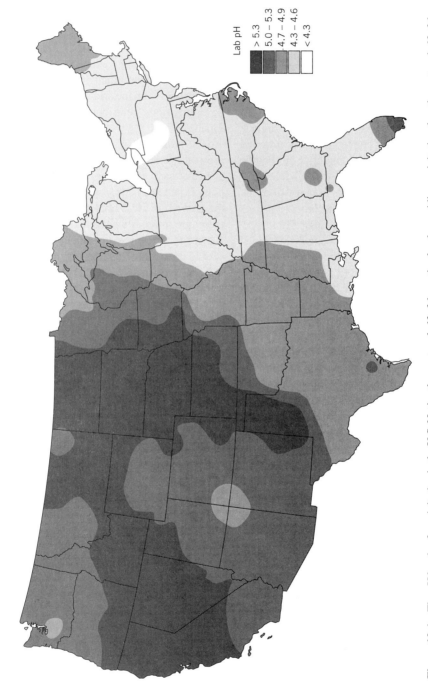

Figure 12.4 The pH levels of precipitation in the U.S. Notice the western half of the nation has less acidic precipitation than the eastern half of the country. Source: *National Atmospheric Deposition Program/National Trends Network*. http://nadp.sws.uiuc.edu/isopleths/maps2000/phlab.gif.

Figure 12.5 These trees in the Black Forest of southern Germany are dying because of the effects of air pollution, especially acid deposition. Documented damage to this famous forest helped galvanize public opinion, forcing action to reduce sulfur emissions in Germany.

Mountains (Vermont) and the Adirondacks (New York). In west Germany, trees have died over 150,000 acres (60,000 ha), and another 9.4 million acres (3.8 million ha) of forests are reported to be severely damaged (French 1990). In addition, about 35 percent of all European forests are now damaged as a result of acid deposition, totaling approximately 120 million acres (50 million ha) (French 1990; Little 1995) (Fig. 12.5).

Concerns have been voiced about acid rain and its effect on drinking water supplies and human health. Most metals are relatively insoluble in water at near-neutral pH, but as pH decreases, their solubility increases. Aluminum in soils, for example, may be leached out by acid rain and lead to an increase of aluminum concentrations in surface water. Lead, derived from both natural and human sources, is found in lake and stream sediments. As pH decreases, lead concentrations in water may increase. In addition, copper and other metals used in water-supply and distribution systems may be dissolved if the water becomes too acidic. Concern with these problems prompted officials in Massachusetts to add lime to Quabbin Reservoir, the major source of water for metropolitan Boston.

Finally, acid deposition contributes to the corrosion of building stone and exposed metal, including steel rails, unpainted metal surfaces on bridges and buildings, and so forth (Fig. 12.6). The economic costs of these effects are difficult to estimate, but they are believed to be substantial. The cost of pollution damage to European forests is estimated to be nearly $30 billion per year (World Resources Institute 1992). Damage estimates resulting from acid deposition are spotty, and their effect on crops, forests, human health, or buildings has not as yet been calculated.

Control and Management

Acid deposition can be controlled by the same methods used to control sulfur dioxide and nitrogen oxide concentrations in urban areas, including burning low-sulfur fuels, installing sulfur scrubbers, and reducing combustion temperatures. However, it is virtually impossible to link deposition in one area to emission sources in another because of the large number of sources broadly distributed over vast regions. In addition, NO_x, another component of acid rain, also derives from mobile sources (vehicles).

Figure 12.6 Destruction of sandstone sculpture caused by air pollution in Germany. The sculpture was installed in 1702; photo at left was taken in 1908, and photo at right was taken in 1969.

Control of acid deposition is costly. Coal combustion is preferred in many areas because it is relatively cheap; switching to other fuels or to low-sulfur coal increases costs. Emission controls like scrubbers or advanced technologies such as fluidized bed combustion allow higher-sulfur coal to be burned, but these measures also bear high costs.

Because acid deposition affects large regions rather than just isolated areas, the solutions to the problem must also be regional, and this has been one of the primary barriers to reducing acid deposition. Failure to recognize the regional dimension to the problem delayed corrective actions. In the United States, for example, the 1970 Clean Air Act was aimed at urban air pollution problems, and its goal was to reduce pollutant concentrations in those areas. One of the techniques used was tall smokestacks, which dispersed the pollutants over a larger area in lower concentrations. This reduced local pollutant concentrations but also helped to spread sulfur and nitrogen oxides over larger areas by placing pollutants into stronger upper winds, thus adding to the acid deposition problem at remote locations.

The U.S. government was slow to act on the problem, but the 1990 Clean Air Amendments set goals for emissions reductions. The law called for a 2-million-ton reduction in nitrogen oxide emissions and a 10-million-ton reduction for SO_2 by 2000, goals that were not achieved, but steady improvements were made (Table 12.2).

EPA's Acid Rain Program uses a variety of mechanisms to encourage energy efficiency and pollution prevention by industry and automobiles. To achieve SO_2 emissions reductions, the EPA has set a national cap of 8.95 million tons of sulfur emissions annually by fossil fuel power plants. The EPA allocates emissions allowances, and it is up to each individual plant to decide how to meet them. The concept of trading allowances between different plants owned by the same utility or trading allowances from one utility to another is generally regarded as a success. In 1993, for example, 150,000 allowances were traded on the Chicago

Table 12.2 Emissions Reductions Based on the U.S. Acid Rain Program

	Emissions (million tons)	
Year	SO_2	NO_x
1980	17.30	...
1990	15.73	6.66
1995	11.87	6.09
1996	12.51	5.91
1997	12.98	6.04
1998	13.13	5.97
1999	12.45	5.49
2000	11.20	5.11

Source: USEPA, 2001.

Board of Trade at prices of $122 to $450 per allowance.

Early in the program prices rose rapidly, but in the mid-1990s the market softened, and prices fell significantly in 1996 (see Fig. 2.5). Low prices for emission allowances mean that utilities can emit more sulfur and have less of an incentive to reduce pollution. In order for the price to rise, either the government must issue fewer permits for sale or environmental interests must be willing to purchase substantial numbers of permits and thus prevent sulfur from being emitted.

International agreements are needed to solve transnational pollution problems, because some countries cause pollution and others suffer its effects. The convention on Long-Range Transboundary Air Pollution (LRTAP) signed in 1979 by most European countries as well as the United States and Canada specifically addressed transboundary pollution issues. In 1987, a protocol to the 1979 convention, Reduction of Sulphur Emissions or Their Transboundary Fluxes by at Least 30 Percent (referred to as the 30 Percent Club), was ratified by 16 European countries and later joined by 5 others. The 30 Percent Club agreed to reduce sulfur dioxide emissions by at least 30 percent of 1980 levels in order to reduce transboundary pollution problems. Some of the largest sulfur dioxide emitters, notably the United States, the United Kingdom, and Poland, have not signed this protocol. Not all countries have achieved the promised reductions, but some have done better than a 30-percent reduction, resulting in an overall continental decline. In addition to the sulfur protocol, a similar agreement was reached in Sofia, Bulgaria, in 1988 to limit nitrogen oxide emissions by freezing them at their 1987 levels. The Sofia protocol was signed by 24 nations, including the United States. Finally, a VOC protocol was adopted in 1991 requiring nations to reduce their VOC emissions by 30 percent of 1984–1990 base levels (the individual year is up to the country) (McCormick 1998).

STRATOSPHERIC OZONE DEPLETION

Another global air pollution problem involves the accumulation of *ozone-depleting chemicals (ODCs)*, such as *chlorofluorocarbons (CFCs)*, which contribute to a loss of ozone in the stratosphere. Ozone intercepts ultraviolet radiation from the sun, and so one of the results of ODCs in the stratosphere is that more ultraviolet radiation reaches the earth's surface. This radiation is responsible for biological damage to plants and animals. Increased radiation may decrease photosynthesis, water use efficiency, leaf area, and ultimately crop yields by as much as 30 percent. Crops such as corn, soybeans, wheat, tomatoes, lettuce, and cotton seem to be the most affected in laboratory studies. Among nonagricultural environments, aquatic ecosystems may be most damaged, with estimates suggesting that a 25-percent reduction in ozone would result in a 35-percent reduction in the productivity of these ecosystems and their fish populations.

From a human standpoint, increased ultraviolet radiation would produce an increase in the number of skin cancers worldwide in the coming years. Every one percent decline in ozone is linked to a 4 to 6 percent increase in skin cancers. There could be 150 million new skin cancers, resulting in 3 million deaths, in the next 50 years in the United States alone. Some scientists also feel that ultraviolet radiation plays a role in depressing the human immune system, thus lowering the body's resistance to infections and disease. New satellite sensor systems now allow us to measure ultraviolet radiation at ground level (UV-b). Current measured levels are 4 to 5 percent higher than they were a decade and a half ago (USEPA 2001).

Chlorofluorocarbons also play a role in the troposphere similar to that of carbon dioxide; they absorb longwave radiation and thus enhance the greenhouse effect. Although they are present in much lower concentrations, they are much more effective than carbon dioxide per unit of concentration in raising atmospheric temperatures.

The worldwide response to the threat of ozone depletion represents a significant success story in international management of a commonly owned resource. The following sections describe the events contributing to that success and some problems that remain to be solved.

Ozone-Depleting Chemicals

Chlorofluorocarbons (CFCs) are the most important ozone-depleting chemicals. CFCs are a class of synthetic substances that were originally developed for use in refrigeration. First discovered in 1930, these chemicals became widely used because they are neither toxic nor flammable. Marketed by E.I. du Pont de Nemours & Company under the trademark Freon (CFC-12), these chemicals have amazing versatility and hence many industrial applications. Over time, new chemical formulations were discovered and new applications were found for CFCs. CFC-113, for example, was one of the most widely used members of the family. Its sole use is as a solvent, and it is used to clean everything from microchips in computers to dry-cleaned clothes.

CFCs were widely used in cooling and foam blowing. For example, before use was restricted by law in the United States, about a third of all CFCs were used as coolants for refrigeration and air conditioning, including the air conditioning in cars. Another third were used as blowing agents or synthetic rigid foams (insulation, styrofoam ice chests and cups, and fast-food containers) or flexible foams (furniture cushions and foam pillows). Other important uses included industrial solvents and aerosol propellants. The use of CFCs in aerosol sprays was discontinued in the United States in 1979.

Halons are a related family of compounds that contain bromine, a more potent destroyer of stratospheric ozone. It is speculated that per unit, emitted halons cause 20 times the damage to the ozone layer as CFCs. Because of their superb fire-retardant properties, halons (Halon-1211, -1301, -2402) are used primarily in fire extinguishers. Finally, carbon tetrachloride (an intermediate product in the production of CFCs) and methyl chloroform also affect ozone. Carbon tetrachloride is used in chemical and pharmaceutical applications, but its impact on ozone depletion is minor in comparison with that of the other ODCs. Methyl chloroform is used as an industrial degreasing agent and as a solvent in paints and adhesives.

Emissions of CFCs steadily increased over the years. In 1931, less than 220 million pounds (100 metric tons) were released; by 1976, worldwide CFC emissions peaked at 1.6 billion pounds (707,000 metric tons). A downturn in CFC emissions occurred between 1976 and 1983, but by 1986, CFC emissions were again quite high, averaging 1.5 billion pounds (671,600 metric tons) (World Resources Institute 1988). As discussed later in this chapter, emissions decreased dramatically in the 1990s, and it appears that stratospheric ozone concentrations will begin to recover early in the 2000s.

The Ozone Hole Is Discovered

One property of CFCs that makes them so useful is that they are chemically inert and therefore very stable. Unfortunately, this property also influences their adverse effect on the atmosphere, as they remain intact, accumulating rather than breaking down and being removed. The atmospheric lifetime of CFCs, for example, ranges from 65 to 400 years (Table 12.3). This stability allows them to reach the stratosphere, where they are broken down by intense solar radiation, and their components enter into other chemical reactions. One of these reactions reduces the amount of ozone in the stratosphere. The process is quite

Table 12.3 Ozone-Depleting Chemicals

Chemical	Atmospheric Lifetime (Years)	Contribution to Ozone Depletion (%)
CFC-11	65	26
CFC-12	130	45
CFC-13	400	
CFC-113	90	12
CFC-114	200	
CFC-115	400	
Carbon tetrachloride	50	8
Methyl chloroform	7	5
Halon-1301	110	4
Halon-1211	25	1
SF_5CF_3	1000	<1

Source: Tolba et al., 1992; Sturges et al., 2000.

simple, but a brief background on its scientific discovery is needed.

In 1985, a group of atmospheric scientists reported a 40-percent reduction in stratospheric ozone during the spring over Antarctica, where they were gathering data. Thus, the existence of the "Antarctic *ozone hole*" was established. During the sunless Antarctic winter (summer in the United States), the air mass is extremely cold, cold enough to freeze water vapor. Polar stratospheric clouds are formed, and chemical reactions on the ice crystals convert chlorine from nonreactive forms to the more reactive hydrogen chloride and chlorine nitrate, which are sensitive to sunlight. As the Southern Hemisphere spring approaches, sunlight appears and releases the chlorine, which starts a chemical chain reaction that transforms ozone (O_3) into oxygen (O_2). The chlorine remains in the stratosphere to initiate many more chain reactions.

In one of the worst years (1987), 50 percent of the ozone over Antarctica was destroyed. Recent evidence indicates that CFC-derived chlorine is also destroying the ozone layer in the Arctic region as well, with 2 to 8 percent of the stratospheric ozone depleted. With global air circulation patterns, the chlorine could easily be transported to midlatitudes, and this is what concerns scientists. In the late 1990s, springtime (October) averages of the ozone over Antarctica were 50 to 70 percent less than they were 30 years earlier. Compared to those in the 1970s springtime, ultraviolet radiation levels are now 130 percent higher in Antartica and 22 percent higher in the Arctic (World Resources Institute 2000).

Reducing ODCs: The Montreal Protocol

In 1974, a group of U.S. scientists postulated that CFCs added chlorine to the stratosphere and through a series of complex chemical reactions reduced the amount of ozone. As a partial response to this concern over potential ozone depletion, the United States banned CFCs as aerosol propellants in all nonessential applications (ranging from hair sprays to deodorants to furniture polish). U.S. production of CFCs dropped by 95 percent. Considering that the United States and the European Community accounted for 84 percent of all CFC output, this was significant. Canada, Sweden, Norway, Denmark, and Finland also banned the use of CFCs as propellants in spray cans. The rest of the European Community (EC) was slower to respond and delayed action until 1980, when it cut aerosol use by 30 percent of 1976 levels (Conservation Foundation 1989). Because of CFCs' stability, the drop in production and use did not significantly alter the amount of CFCs in the atmosphere. With the discovery of the so-called ozone hole in 1985 in Antarctica, the theoretical relationship between CFC use and ozone depletion became fact and widely accepted by scientists, policymakers, and the public. The time was ripe for international action (Table 12.4).

In 1985, 28 nations participated in the Vienna Convention on the Protection of the Ozone Layer. This meeting was designed as a legal instrument to protect the atmosphere as a resource by reducing the use of CFCs. The conferees were able to agree in principle to protect the atmosphere, but in practical terms they were unable to agree on CFC control. The United States, Canada, and Sweden, for example, wanted an immediate freeze on CFC production and a gradual phaseout of its use. European Community members wanted a freeze in production at existing levels. In early 1986, the Chemical Manufacturers Association, an industry group, supported production limits. du Pont, the developer of Freon, vowed to find a substitute for CFCs within five years.

By 1987, significant international concern had been raised, resulting in the negotiation of the *Montreal Protocol* (officially known as the Montreal Protocol on Substances That Deplete the Ozone Layer). The treaty, which has since been ratified by over 125 countries, became effective January 1, 1989. It froze the production of CFCs by European Community members at mid-1989 levels and called for 50-percent reductions in emissions globally by 1999. The treaty permitted developing countries to increase CFC use for ten more years, while allowing the former USSR to continue production through 1990. Finally, the Montreal Protocol limited the use of halons as fire-suppressant chemicals.

In early 1989, the European Community countries went a step further and agreed to totally eliminate the use of CFCs by the end of the century. The United States supported this move, resulting in the total ban of CFC production by 65 percent of the world's producing countries. In 1990, an amendment to the Montreal Protocol called for the complete elimination of CFC use by 2000. Two years later, another amendment to the Montreal Protocol (the Copenhagen Amendment) accelerated the

Table 12.4 Policy Milestones in Ozone-Depleting Substances

Year	Event
1978	UNEP Coordinating Committee for the Ozone Layer established
1985	Vienna Convention for the Protection of the Ozone Layer held
1987	Vienna Convention enters into force; 19 nations ratify
1987	Montreal Protocol on Substances That Deplete the Ozone Layer established; Regulates consumption and production in the following manner: 1. CFCs frozen at 1986 levels in 1990; 20% reduction in 1993; 50% reduction in 2000 2. Halons frozen at 1986 levels by 2005 3. Developing countries given 10-year exclusionary period
1989	Montreal Protocol enters into force; 23 nations ratify
1990	London Amendment to Montreal Protocol ratified Accelerates production and consumption phaseouts: 1. CFCs reduced by 20% in 1993; 50% in 1995; 85% in 1997; phaseout in 2000 2. Halons reduced by 50% by 1995; phaseout (except for essential uses) by 2000 3. Carbon tetrachloride emissions reduced by 85% in 1997; phaseout in 2000 4. Methyl chloroform emissions frozen in 1993; reduction by 30% in 1995; 70% in 2000; phaseout in 2005 5. Other fully halogenated CFCs reduced by 20% in 1993; 85% in 1997; phaseout in 2000
1991	Interim Multilateral Fund established to provide financial assistance to developing countries to meet timetable for phaseout
1992	Vienna Convention now ratified by 82 nations Montreal Protocol now ratified by 76 nations London Amendment to the Montreal Protocol ratified by 19 nations
1992	London Amendment to the Montreal Protocol enters into force; 20 nations ratify
1994	Copenhagen Amendment to the Montreal Protocol ratified Accelerates complete phaseout of CFCs by January 1, 1996
1995	Nobel Prize in chemistry awarded to Paul Crutzen, F. Sherwood Rowland, and Mario Molina for discovery that trace amounts of gases can profoundly change the upper atmosphere Copenhagen Amendment to the Montreal Protocol enters into force
1996	Production of CFCs in developed countries ceased
1997	Montreal Amendment to the Montreal Protocol ratified Requires a system for licensing the import and export of new, used, recycled, or reclaimed controlled substances
1999	Beijing Amendment to the Montreal Protocol ratified Bans the import and export of controlled substances

Source: Tolba et al., 1992:37; United Nations official Montreal Protocol website *(http://www.montrealprotocol.org).*

CFC phaseout to January 1, 1996. Finally, by January 1, 1996, all developed nations ceased their production of CFCs. By the end of 2001, 183 nations had ratified the Montreal Protocol, and 136 ratified the London Amendments. Two additional amendments have been added. In 1997, the Montreal Amendment was ratified, which banned the import and controlled the export of used, recycled, or reclaimed controlled substances (CFCs, halons, carbon tetrachloride, methyl chloroform, HCFCs, and methyl bromide). It required signatory nations to develop licensing systems to control export and import. Two years later, the Beijing Amendment developed a monitoring system to record production and consumption levels of the controlled substances. At the end of 2001, 73 nations had ratified the Montreal Amendment, while 22 had ratified the Beijing Amendment (UNEP 2001).

The result of these agreements has been dramatic. In the late 1980s, it was widely believed that the concentrations of ODCs in the stratosphere would increase well into the twenty-first century. A decade later, concentrations of ODCs are declining; there is evidence that the ozone hole may have reached its worst level, and improvement is beginning (Figure 12.7). Projections indicate that ozone levels could recover to their 1979 levels by 2050, although continued release of

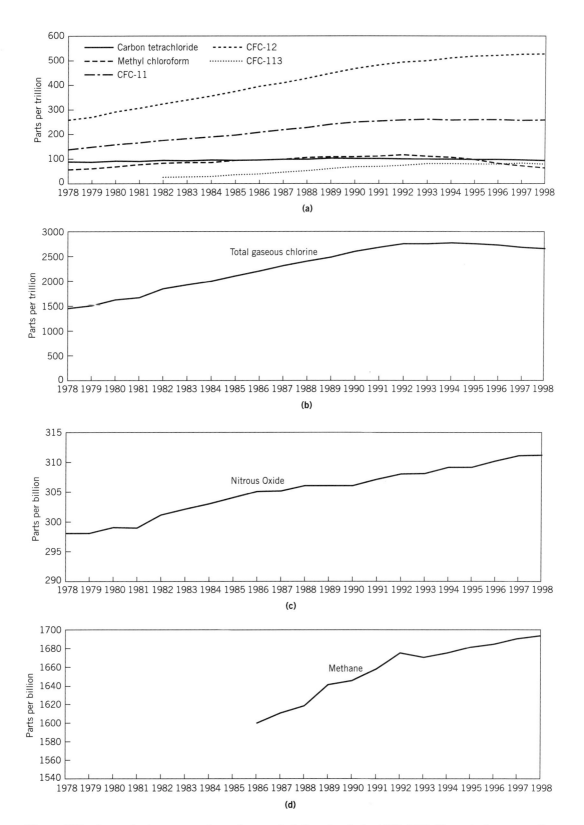

Figure 12.7 Atmospheric concentrations of ozone-depleting chemicals, 1978–1998. Concentrations generally rose rapidly in the 1980s, but the rate of increase declined significantly in the early 1990s. (*a*) Concentrations of methyl chloroform have declined since 1992. Chlorine (*b*) and nitrous oxides (*c*) show steady increases and a leveling off, while methane concentrations (*d*) appear to be increasing. *Source:* World Resources Institute, 2000.

ISSUE 12.1: BLACK MARKET FREON

Freon is the du Pont trade name for chlorofluorocarbons (CFCs). Since January 1, 1996, CFCs, halons, carbon tetrachloride, and methyl chloroform (also called ozone-depleting substances) have been outlawed in the United States as part of the implementation of the Montreal Protocol. When the Montreal Protocol was ratified in 1987, it set a timetable for phasing out ozone-depleting chemicals by the year 2000 for the developed countries and the year 2010 for developing countries. However, in the early 1990s, developed nations agreed to a January 1, 1996, deadline.

CFCs were widely used coolants in refrigerators, freezers, room air conditioners, and car air conditioning units. With the ban, new coolants are now required for these consumer products. However, what do you do with the old appliances? More important, how do you get your older car serviced?

Recycled Freon is available from discarded freezers, refrigerators, and unit air conditioners. The government also stockpiled Freon (and other CFCs) in anticipation of potential shortages once the ban took effect. In 1989, for example, one pound of Freon cost around $1. Now, with the recent governmental excise tax ($5.35 per pound), the average price has jumped from $8 to $20 per pound.

Since 1991, all cars and trucks have been required to use a more environmentally friendly product. While older cars can be retrofitted to use the new refrigerants, most have not been converted, despite the relatively low cost ($200–300). This means that in 1996 there were about 140 million cars in the United States that still used Freon.

Unfortunately, the demand for Freon exceeded the legal supply. In addition, the legal supplies are costly: more than 20 times the price on the black market. A canister (30 pounds) costing $160 in Tijuana was worth $600 in Santa Ana less than 100 miles north. The result was a thriving black market in Freon. Some governmental officials estimated that nearly $200 million in tax revenues were lost because of illegal smuggling.

Smugglers mostly operate in India, Russia, and China, where the CFCs are still legal. They ship the contraband through Canada or Mexico into the United States. In 1994, for example, customs officials seized 1.5 million pounds of Freon worth an estimated $18 million. In 1996, authorities figured that $500 million of Freon was smuggled into the United States, making it a bigger trade commodity than illegal guns. Obviously, much more is coming into the country undetected.

Consumers have no way of knowing the source of Freon when they have their car serviced. Often the mechanics might not know either. Black marketeers will continue to supply the banned substance until older car owners finally realize it is both economical and environmentally sound to spend the $200 to $300 needed to convert their cars than to continue to support the illegal smuggling operations. It just might be cheaper and more environmentally responsible to simply roll down your windows (Goldberg 1996; *San Diego Daily Transcript* 1996).

---

halons from fire extinguishing systems may slow recovery (Kerr 1996). If these trends continue, significant international cooperation in managing a global common resource will have been achieved.

Problems remain, however. In the mid-1990s production actually rose in China and other countries not required to eliminate CFC production, raising concerns that the global phaseout of CFCs might take a little longer than expected. Some of this increased production may have supplied a growing black market in CFCs smuggled into countries in which use has been curtailed (see Issue 12.1).

GLOBAL CLIMATE CHANGE

Climate is naturally variable, from year to year and over longer periods of time. While natural climate variations cause problems, we are normally content to live with them—we have no choice. Human-caused climate change is another matter, however. Concern about the climatic impacts of increasing carbon dioxide (CO_2) in the atmosphere began decades ago but has been especially vocal since the early 1970s. In the 1970s, considerable controversy arose over whether elevated CO_2 would cause warming, cooling, or neither. By the late 1990s, most scientists agreed that climatic warming would result from increased CO_2, in the atmosphere and many believed that the warming had already begun.

The Greenhouse Effect

The earth's atmosphere is a partially and differentially transparent medium with respect to energy and regulates flows of energy between space and the earth's surface. Since it is relatively transparent to the wavelengths of most solar radiation, sunlight passes through the atmosphere relatively unimpeded. The energy returned to space by the earth has longer wavelengths than that of sunlight. The atmosphere is only partly transparent to these wavelengths, and much of the outgoing radiation is temporarily trapped in the atmosphere, which keeps the atmosphere warmer than it would otherwise be. Several atmospheric components are responsible for this action, among them water vapor, methane, ozone, CFCs, nitrous oxides, trace gases, and carbon dioxide. These are collectively known as *greenhouse gases,* and they have natural as well as human sources. The ability of carbon dioxide to admit shortwave solar radiation but absorb outgoing longwave terrestrial radiation causes atmospheric warming and has been called the *greenhouse effect*. In terms of its impact on atmospheric temperature, water vapor is the most important greenhouse gas. The warming of the atmosphere as a result of human-caused increases in greenhouse gas concentrations is known as *global warming*.

Greenhouse Gases

The amount of heat stored in the atmosphere and its distribution within different layers and regions of the atmosphere play an important role in regulating atmospheric circulation and ultimately climate. The carbon dioxide content of the atmosphere (about 0.037 percent of the atmosphere in 1998) has changed considerably throughout the history of the earth. A billion or more years ago, it was probably much higher than today. During the Carboniferous Era, about 280 to 345 million years ago, much of the land surface of the globe was covered with vast swampy forests. The fossil fuels we burn today are derived from carbon taken from the atmosphere and stored in those forests and other ecosystems in the past.

Atmospheric carbon dioxide content also fluctuates on a seasonal basis. In the Northern Hemisphere, the concentration of carbon dioxide varies about 5 ppm within a given year. The maximum occurs in April; as plants photosynthesize and store carbon throughout the summer, the carbon dioxide content steadily decreases. It reaches a minimum in October and then climbs back up as more plant matter decays than grows. The annual cycle shows the close relation between carbon dioxide in the atmosphere and processes at the earth's surface. The global carbon cycle is the biogeochemical cycle that human activity has affected the most. This will result in significant climatic changes within the next several decades and may cause a profound impact on the global environment. The long-term effects of increased atmospheric CO_2 are highly uncertain, however, because we still have an inadequate understanding of atmospheric dynamics. For example, global warming could cause water vapor concentrations to rise and thus increase the greenhouse effect, but it could also create more cloud cover, which could cause cooling.

Sources and Sinks Greenhouse gases have a wide variety of sources and sinks, depending on biogeochemical cycling and human activities. Some of the major natural sources of greenhouse gases include decomposition in wetlands (methane) and microbial processes in soil and water (nitrous oxides). Anthropogenic sources include fossil fuel consumption and land-use change. The oceans, soil, and biomass are important sinks of CO_2.

Since the early nineteenth century, we have steadily increased our extraction and combustion of fossil fuels, returning stored carbon to the atmosphere as carbon dioxide (see Fig. 4.6). Not all the carbon dioxide emitted stays in the atmosphere; a substantial amount enters the oceans, and some is stored in living biomass. Not all fuels contribute the same amount of carbon dioxide when burned. Coal contains about 75 percent more carbon per unit of energy than natural gas, whereas oil contains about 44 percent more. Coal and oil combustion thus release more CO_2 to the atmosphere than natural gas. Carbon dioxide represents about 81 percent of all the greenhouse gases released in the United States.

Globally, more carbon is currently being released from the biota than is absorbed by it as a result of land-use activities. It is estimated that about 20 percent of carbon dioxide emissions come from deforestation and the loss of soil carbon due to soil degradation. Tropical deforestation is the largest source related to land-use change. In some parts of the world, forest growth is causing net removal of carbon from the atmosphere and storage in the biota.

Most atmospheric methane comes from natural processes, particularly anoxic decomposition of organic matter. Natural methane sources include decomposition in rice paddies and swamps, the action of anaerobic bacteria on plant material in the intestines of ruminants (cows, sheep, and camels), and wood digestion by termites. Anthropogenic sources contribute significantly less and include the combustion of fossil fuels (natural gas in particular), biomass (clearing of forests), and landfills.

Nitrous oxide (N_2O) is another greenhouse gas that contributes to the greenhouse effect. In addition, nitrous oxide acts as a catalyst for stratospheric ozone removal. Nitrous oxide is derived primarily from anthropogenic sources, including the combustion of fossil fuels, the use of nitrogen-rich fertilizer, and deforestation.

Very recently, scientists discovered a new greenhouse gas, triflouromethyl sulfur pentafluoride (SF_5CF_3) (Sturges et al. 2000). While its concentration in the atmosphere is very low (0.12 ppt), it has increased from zero in the late 1960s to its current level. Scientists estimate that the concentration is growing around 6 percent year. The concern is over the ability of the gas to warm the atmosphere. Currently SF_5CF_3 has the greatest warming potential of all the greenhouse gases (nearly 18,000 times greater than carbon dioxide) (Revkin 2000).

The primary concern is its warming potential, the rapid increase in concentration in the atmosphere, and its longevity (estimated at around 1000 years).

Atmospheric Concentrations Since the nineteenth century, the atmospheric content of carbon dioxide has been steadily increasing (Fig. 12.8). The rate of increase has itself escalated as a result of the growth in fossil fuel use worldwide. Prior to the Industrial Revolution it was about 280 ppm, and in 1980 it was 340 ppm. By 1990 the concentration was 354 ppm, and in 1998 it was 367 ppm (Table 12.5) (World Resources Institute 2000). Between 1970 and 1995, global concentrations of carbon dioxide rose by 10 percent, or about 0.4 percent per year (World Resources Institute 1996). Nitrous oxides and methane concentrations also increased. Ambient concentrations of methane have more than doubled in the past 300 years, from 665 to around 1693, and are currently increasing at about 1.6 percent per year. The rate of increase of nitrous oxide is 0.6 percent per year.

Emission Trends Carbon dioxide emissions from fossil fuel combustion were 23.1 billion metric tons in 1996. The overwhelming majority of these were from industrial activity. In fact, emissions from industrial activity have increased by 40 percent over the past 20 years. The vast majority of the CO_2 emissions are from the burning of fossil fuels (97 percent), especially petroleum and coal. On a continental basis, Asia has the most CO_2 emis-

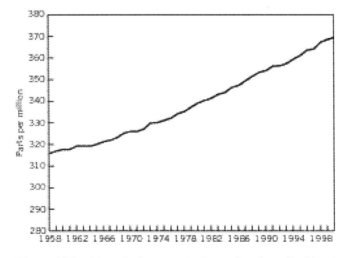

Figure 12.8 Atmospheric concentrations of carbon dioxide at Mauna Loa, 1958–2000. *Source:* Keeling and Whorf, 2001.

Table 12.5 Trends in Ambient Levels of Key Greenhouse Gas Emissions

Variable	CO_2 (ppm)	Methane (ppb)	NO_x (ppb)	CFC-11 (ppt)	CFC-12 (ppt)
Preindustrial	280	700	285	0	0
1990	354	1645	307	249	469
1994	359	1666	309	261	509
1998	367	1693	311	259	530
Current rate of change	+2.2	+1.6	0.6	−0.8	+4.1
Atmospheric lifetime	50–200	10	150	65	130

Source: Tolba et al., 1992; World Resources Institute, 1996, 2000.

sions (nearly a third of the world total), followed by Europe and then North America (Table 12.6). These percentages will surely change in the next decade, as Asia has shown a 44 percent increase in CO_2 emissions from 1990 to 1996, while those in the Middle East and North Africa have increased by 33 percent and those in South America by 29 percent. The United States is the largest source of CO_2 emissions (22 percent of the world's total), followed by China (14 percent), Russia (7 percent), and Japan (5 percent). Among the major global emitters, the United States still had the highest per capita rate (21.6 tons per person in 1996), more than four times the world average (4.6 tons per person in 1996). Globally, the United Arab Emirates has the highest per capita dioxide emissions (40 tons per person) as a consequence of petroleum production, including gas flaring, and a relatively small population (2.4 million). The regional variation in per capita emissions (Fig. 12.9) highlights the contributions of the industrialized nations.

Methane emissions averaged 270 million metric tons in 1992, the majority coming from industrial activity, followed by livestock. Regionally, China and India are the largest methane emitters, followed by the United States and Russia. In China, rice cultivation and coal mining produce more than 80 percent of the methane emissions. For India, it is a combination of livestock and wet rice cultivation. The methane emissions in the United States are more mixed, with solid waste dominating, followed by oil and gas production, coal mining, and livestock. A similar situation obtains for Russia, where oil and gas and livestock production dominate.

Impacts

Greenhouse gas emissions could cause a rise in global temperatures during the next century. Though seemingly small, this rise could have many significant impacts, including inundation of coastal areas caused by sea-level rise, increased variability in weather, and extreme weather events. Because of the climate's inherent variability it is difficult to tell whether observed climate change is natural or human in origin. We do know that global average temperature has risen about

Table 12.6 Regional Contributions to Greenhouse Gas Emissions

Region	% of World's Total, 1996	% Change in CO_2 Emissions, 1990–1996
Asia	31.2	43.5
Europe	25.6	NA
Middle East/North Africa	5.9	33.2
Sub-Saharan Africa	2.2	11.3
North America	23.9	9.1
Central America	2.1	21.1
South America	3.1	28.6
Oceania	1.4	15.9

NA = not available.
Source: World Resources Institute, 1996, 2000.

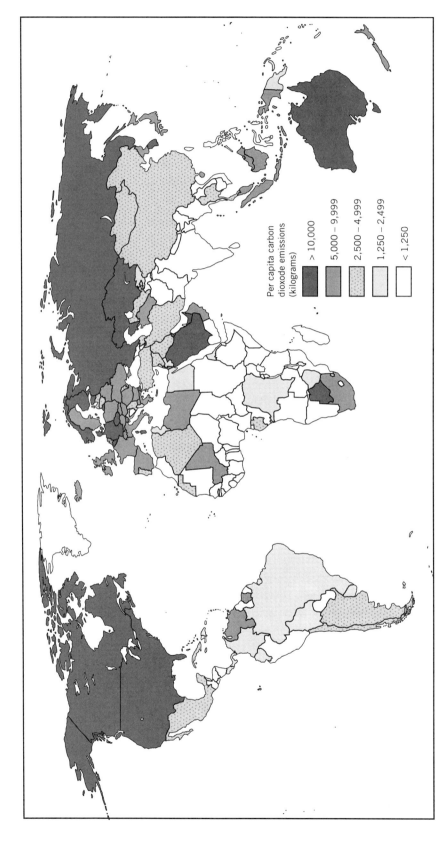

Figure 12.9 Per capita CO_2 emissions. With the exception of a few energy-producing states in the Middle East, the highest per capita emissions are in wealthy industrial countries. U.S. per capita emissions are about 5 times the world average and about 50 times those in sub-Saharan Africa.

1°C in the past century, and many believe that this is a result of increased CO_2 in the atmosphere.

A key question regarding the historic rise in temperature is as follows: Is the warming part of natural climatic variability, or is it a result of human activities? Those who argue that the warming may be natural note that the period from about 1500 to 1750 was distinctly cooler than the preceding centuries. This cool spell, known as the Little Ice Age, still may be coming to an end as a result of unknown natural processes, with temperatures returning to levels that existed several centuries ago. On the other hand, the recent warming coincides with the increase in atmospheric CO_2 since the Industrial Revolution, and it is logical to see that correlation in cause-and-effect terms.

In the 1990s, significant advances were made in climatic modeling that help to answer the question of causality. Modelers' analyses began to include changes in the atmospheric concentrations of sulfur oxides. These substances, emitted primarily through coal combustion, have a cooling effect on the atmosphere. When the effects of sulfur oxides are included, the models are able to produce both spatial and temporal patterns of change that are very similar to those observed in historical data. The similarities suggest a "fingerprint" of human impact, supporting the hypothesis that the observed warming is indeed a result of increased CO_2. By 2000 there remained little controversy in the scientific community over the basic conclusion that the increase in CO_2 was causing warming, and in 2002 the Bush administration finally acknowledged this fact.

Predicting future warming depends on predictions of future fossil fuel use. These are, of course, highly uncertain. In the 1960s and 1970s, fossil fuel use was rising rapidly, and projections indicated that the atmospheric concentration of CO_2 would double by sometime in the twenty-first century. But by the 1990s the rate of increase in atmospheric CO_2 slowed, largely as a result of the economic collapse in the former Soviet bloc. The 2001 report of the Intergovernmental Panel on Climatic Change makes projections based on a range of assumptions and suggests that earth's climate will warm between 1.4°C and 5.8°C by 2100 (Houghton et al. 2001), a significant increase from the 1995 projections of 1.0°C 1 to 3.5°C.

What will be the impacts of such a climate change? The answer to that question is even more uncertain than the temperature change itself. Perhaps the most worrisome potential impact is related to people living in vulnerable areas such as coasts and semiarid regions (Schneider 1997; Fischetti 2001). A rise in sea level could displace millions of people from their current homes. The amount and rate of rise are highly uncertain but could be as much as 3 to 7 feet (1 to 2 m). The combined effects of these changes would inundate low-lying coastal areas, submerge coastal wetlands, increase coastal erosion, increase flood and storm damage from coastal storms, and increase the salinity of groundwater in low-lying areas, thus threatening drinking-water supplies. Bangladesh and Egypt are among the most vulnerable countries, where a 7 foot (2 m) rise in sea level would result in a 20-percent reduction in inhabitable land area in each country. China is also vulnerable, as are several small island nations such as the Maldives, Kirabati, and Tuvalu, which are threatened with near-complete inundation (McCarthy et al. 2001). In the United States a similar rise would destroy 50 to 80 percent of our coastal wetlands and between 10 and 28 percent of the land area in Louisiana; Florida; Delaware; Washington, D.C.; Maryland; and New Jersey. Major changes in vegetation and land use would occur and have significant impacts on human society. If the predicted sea level changes are greater than 3 to 7 feet (1 to 2 m), then these effects will be even more pronounced and cover more of the world's low-lying areas (see Issue 12.2).

Another area of concern is the impact on agriculture. This impact will depend on changes both in temperature and precipitation, as well as the direct effects of CO_2 itself. Increases in atmospheric carbon dioxide cause more rapid plant growth, but some plants respond more dramatically than others. For example, soybeans respond much more than corn, and silver maple responds more than sycamore. These differences may lead to significant agricultural and ecological impacts. For example, some concern has been expressed about shifts in the range of forest species, increased forest fires, and the economic consequences of these factors on the forest industry. The impacts on agriculture also include shifts in the ranges of crops, increases in agricultural pests and livestock diseases, and shifts in the demands for irrigation.

Changes in regional climates resulting from warming are very hard to predict (Karl et al. 1997). Circulation changes might cause today's extreme weather to become the norm 50 years from now.

ISSUE 12.2: THE COSTS OF GLOBAL WARMING

In 1992, the countries of the world met in Rio de Janeiro and committed themselves to talking about doing something about CO_2 emissions. In 1995 and 1996 in Berlin, and in 1997 in Kyoto, the countries of the world again committed themselves to trying to reduce CO_2 emissions. But global CO_2 emissions continued their rise, which had been temporarily halted in the early 1990s by the economic collapse of eastern Europe. Despite the fact that the Intergovernmental Panel on Climate Change concluded in 1995 that there was strong evidence that human-induced global warming was occurring, and despite the fact that climate changes are probably the most profound and far-reaching impacts we have ever had on our environments, as of 2000 little of substance had been done to reduce CO_2 emissions. The reason for this lack of action is, of course, the enormous cost—perhaps as much as 4 percent of gross world product—involved in converting the world economy from one dependent on fossil fuels to one relying on less polluting sources of energy (Stevens 1995).

But what are the economic costs of not reducing CO_2 emissions? No one knows, and no one has even a very good guess, because no one knows exactly what climate changes will occur. Climate is made up of weather—storms and high pressure centers, clear skies and clouds, north winds and southern breezes. These circulation patterns are, in turn, driven by the distribution of heat in the atmosphere. Common sense tells us that because CO_2 is affecting heat in the atmosphere, circulation, and hence weather, should change. But our knowledge of the atmosphere is still too crude to be able to predict what areas will actually get warmer or cooler, wetter or drier. Thus, estimates of the impacts on agriculture, for example, are impossible. Some places that are on the dry margins of the arable earth may get wetter and thus more productive, or they might get drier and become unusable for farming. We just don't know.

One industry that is widely expected to be hurt by global warming is the insurance industry, and many believe these costs are already appearing. In the early 1990s, several very expensive natural disasters occurred. The biggest of these in terms of monetary value was Hurricane Andrew, which struck southern Florida in 1992. Andrew's damages were estimated at $25 billion, of which $16 billion were covered by insurance. Had Andrew hit Miami directly instead of the southern margins of the city, the insured losses might have been as high as $50 to $60 billion. Other major storms in this period included floods in China in 1991 and 1994, in Italy in 1994, and on the Mississippi River in 1993, as well as severe windstorms in

Summers in the midcontinental regions of North America, southern Europe, and Siberia, which are now some of the world's most productive agricultural regions, may become so dry that they will be unable to grow wheat. On the other hand, areas that are now marginal for agriculture may become more favorable. Droughts, floods, and similar problems could become commonplace rather than rare.

Finally, the possible health impacts of global warming are becoming more apparent (Epstein 2000). In addition to increases in respiratory diseases in industrialized nations, there are also changes in the distribution of vector-borne diseases such as malaria, dengue fever, and the West Nile virus and in the distribution of diseases related to contaminated water, such as cholera. It is suggested that climate warming could be a factor in these changes.

Greenhouse Politics and Emissions Stabilization

The United States is the leading source of carbon dioxide emissions, yet it was slow to acknowledge the importance of global warming and thus develop a comprehensive climate-change strategy. It was also slow in joining the greenhouse politics parade.

In June 1988, a number of nations met in Toronto under the auspices of the United Nations to develop a far-reaching plan to protect the atmosphere. The Intergovernmental Panel on Climate Change (IPCC) recognizes that in order to merely stabilize carbon dioxide levels, a 50- to 80-percent reduction in carbon dioxide emissions is needed immediately. The panel's goal is to reduce carbon dioxide emissions by 20 percent by 2005, with an equitable distribution of reductions

northern Europe in 1990. In just seven years, from 1990 to 1996, the insurance industry worldwide racked up losses from weather-related natural disasters totaling over $65 billion—over three times the total losses of the previous 30-year period (Abramovitz 2001). Many insurance companies were put in serious financial trouble, and some were wiped out altogether.

We don't know whether these disasters are a result of global warming or whether this is simply a string of bad luck. Some believe that an increase in atmospheric temperature causes an increase in the intensity of circulation and the severity of storms, but this is only a hypothesis. But whether or not they were caused by global warming, there are good reasons for the insurance industry, in particular, to be concerned. First, global warming is expected to cause a rise in sea level. If sea level rises only a few tenths of a meter, the damage caused by coastal storms can increase exponentially. A combination of increased storm frequency and slightly higher sea level is deadly. Second, insurance rates are based on experience. The history of storms and the damages they cause give insurers an idea of how much they are likely to lose in a given year, and they charge rates that, over a period of time, will provide them with sufficient assets to cover those losses. But if climate is changing, then history is not a good predictor of the future, and that means that the insurance companies' rates may be too low. Finally, we might think that while some kinds of disasters could get more common, others might become less so. This is partly true, but insurance companies worry most about the big events that cause most of their losses. A few extra years without major disasters helps insurers a little, but a few extra years with major disasters can bankrupt a corporation. And because insurance companies are themselves major investors and also sell some of their contracts to other companies (called reinsurers), major disasters can send shock waves through financial businesses worldwide.

As a consequence of these disasters, the insurance industry has become a significant player in discussions on controlling CO_2 emissions. It lobbied intensively at the Berlin conferences, siding with a coalition known as the Alliance of Small Island States in proposing an international insurance pool that would help cover the costs of disasters related to climate change. Contributions to the fund would be based on GNP and CO_2 emissions. In this way, insurance companies would shed some of the risk of the disasters, with the costs at least partly borne by the countries responsible for rising CO_2 levels in the atmosphere.

---

per country. This recommendation was obviously a compromise, since many countries wanted an immediate freeze on emissions, whereas others (the United States, Japan, and eastern and central European countries) oppose restrictions because they are viewed as too costly.

The Framework Convention on Climate Change (FCCC) was adopted in May 1992 and signed by over 150 countries, which agreed to stabilize emissions of greenhouse gases. The treaty entered into force in March 1994, at which time 166 nations had signed on, and by the end of 2000, 186 nations had ratified the treaty. The FCCC calls for reducing greenhouse gas emissions to 1990 levels. The treaty also has other elements that promote our understanding of the emissions as well as actions we can take to reduce them. All countries must provide a periodic greenhouse gas inventory that takes into account the global warming potential of various gases and their production within that country. The Framework also called for countries to produce an action plan that outlines their proposals for reducing emissions.

In 1993, the United States adopted its Climate Change Action Plan. Among the actions planned were forest protection and industry/government partnerships for emissions reductions in electric utilities. Under the 1990 Clean Air Act, greenhouse gas emissions from transportation sources are also restricted. However, the U.S. plan relied heavily on voluntary measures, which may or may not work.

Work on a treaty to control carbon emissions continued between 1992 and 1997. Preliminary negotiations were carried out in a series of meetings held in Berlin in 1994 and 1995. In these sessions several positions became clear. The United States, a few other major emitters, and the oil-producing nations strongly opposed significant reductions in emissions. Most European nations favored significant reductions in emissions to below 1990 levels. Such reductions were easier for the Europeans to achieve because a decline in coal use was already under way there. An alliance of small island nations concerned about rising sea level also favored significant reductions. Most of the poorer countries argued that the major emitters—especially the United States—bore responsibility for controlling carbon dioxide. These poorer countries were not willing to reduce emissions and potentially restrict their economic growth unless major reductions occurred in the United States and other rich countries first. Among these poorer countries is China, where emissions are growing rapidly. If current trends continue, China will become the world's largest emitter of CO_2 in the first decade of the twenty-first century (Drennen and Erickson 1998).

These negotiations culminated in Kyoto, Japan, in December 1997, where an agreement known as the Kyoto Protocol was signed. The Kyoto Protocol set binding reduction targets and assigned emissions caps based on 1990 greenhouse gas emissions. Nations were given a five-year window (2008–2012) in which to achieve the targets. The United States agreed to reduce emissions to 7 percent below 1990 levels by 2010, but only on the condition that credit for emissions reductions could be traded internationally, much as sulfur emissions permits are traded in the United States. The Kyoto Protocol requires that 55 nations ratify or accede to the treaty, including the major emitting nations (at least those that account for more than 55 percent of the world's total emissions) in order for it to go into effect. Since the United States emits about 35 percent of the world's total, it would require near unanimity among the remaining industrialized nations in the world on targets and implementation plans in order to implement the Kyoto Protocol. At the end of 2001, 46 parties had ratified or acceded to the Kyoto Protocol, none of whom who are major emitters (United Nations Framework Convention 2002). While the Kyoto conference failed (thus far) to reach agreements on the difficult issues of exactly how such trades would be carried out, negotiations continue on the development of a climate change regime (Depledge 1999).

Once agreed to in principle, the Kyoto Protocol would have to be ratified by signatory nations before it became binding. The political debates surrounding climate change in the United States are intense (Skolnikoff 1999) and often cite the uncertainty in estimates of global warming as the reason to do nothing (National Research Council 2001). Despite changes in adminstrations, political resistance to the treaty in the Republican-controlled U.S. Senate is intense, and ratification of the treaty is highly unlikely in the near future unless significant modifications are made.

Reducing greenhouse emissions will require governmental intervention and personal responsibility. Stabilizing emissions will be costly and probably borne by most of the industrialized nations. However, they are not the only major emissions sources. Global reductions in greenhouse gas emissions require personal responsibility as well as equitable strategies to achieve the goals in the Framework without deleteriously affecting any one nation or population subgroup.

CONCLUSIONS

No one knows the geographic extent of the predicted climatic changes or the exact timing. Most scientists agree that some changes are likely, perhaps within the next several decades. These effects will be slow to come; to prevent them, we would have to act now or, better yet, have acted a few decades ago. For the present, the best that can be done is to become more aware of transboundary pollution problems and the difficulty they pose for environmental management.

Worldwide, both tropospheric and stratospheric pollution levels have worsened and will rise as developing countries become more urban and industrialized. The long-term consequences of altering the earth's atmospheric chemistry are already generating global concern. Tough economic and political action must be undertaken, and international cooperation is critical if we are to protect this common property resource for future generations.

Success in managing problems associated with international air pollution has so far been spotty. By 1997, the regional problem of acid deposition had been effectively addressed in Europe, but less so in the United States, and work is just beginning in eastern Asia. Even in Europe, where the commitment to reduce emissions is strongest, acid deposition continues to be a severe problem. The international agreements to control ozone-depleting chemicals have been quite effective, to the point of reducing emissions and their damaging effects even faster than was envisioned in the 1980s. But negotiations to reduce carbon emissions produced only commitments to *attempt* reductions, which so far have had few meaningful results. The future of these international efforts remains cloudy.

REFERENCES AND ADDITIONAL READING

Abrahamson, D. E., ed. 1989. *The Challenge of Global Warming*. Covelo, CA: Island Press.

Abramovitz, J. 2001. *Unnatural Disasters. World Watch Paper #158*. Washington, D.C.: Worldwatch Institute.

Cifuentes, L., V. H. Borja-Aburto, N. Gouveia, G. Thurston, and D. L. Davis. 2001. Hidden health benefits of greenhouse gas mitigation. *Science* 293:1257–1259.

Consortium for International Earth Science Information Network (CIESIN). 1997. *Environmental Treaties and Resource Indicators (ENTRI)*. http://sedac.ciesin.org/entri/.

Dao, J. 2000. Acid rain law found to fail in Adirondacks. *The New York Times* March 27:A1, A23.

Depledge, J. 1999. Coming of age at Buenos Aires. *Environment* 41(7):15–20.

Dovland, H. 1987. Monitoring European transboundary air pollution. *Environment* 29(10):10–20, 27–29.

Drennen, T. E. and J. D. Erickson. 1998. Who will fuel China? *Science* 279:1483.

Epstein, P. R. 2000. Is global warming harmful to health? *Scientific American* August:50–57.

Fischetti, M. 2001. Drowning New Orleans. *Scientific American* October:77–85.

Flavin, C. 1989. *Slowing Global Warming: A Worldwide Strategy*. Worldwatch Paper no. 91. Washington, D.C.: Worldwatch Institute.

_____. 1997. Climate change and storm damage: the insurance costs keep rising. *WorldWatch Magazine* 10(1):10–11.

French, H. 1990. *Clearing the Air: A Global Agenda*. Worldwatch Paper no. 94. Washington, D.C.: Worldwatch Institute.

Goldberg, C. 1996. A chilling change in the contraband being seized at borders. *The New York Times* November 10, p. 34.

Hansen, J. et al. 1981. Climate impact of increasing atmospheric carbon dioxide. *Science* 213:957–966.

Hedin, L. O. and G. E. Likens. 1996. Atmospheric dust and acid rain. *Scientific American* 275(6):88–92.

Houghton, J. T., L. G. Meira Filho, B. A. Callendar, and N. Harris, eds. 1996. *Climate Change 1995: The Science of Climate Change*. Cambridge: Cambridge University Press.

Houghton, J. T., Y. Ding, D. J. Griggs, M. Noguer, P. J. van der Linden, and D. Xiaosu, eds. *Climate Change 2001: The Scientific Basis*. Cambridge: Cambridge University Press. Summary for *Policy Makers and Technical Summary:* http://ipcc.ch/reports.htm.

Karl, T. R., N. Nicholls, and J. Gregory. 1997. The coming climate. *Scientific American* 276(5):78–83.

Karl, T. R. and K. E. Trenberth. 1999. The human impact on climate. *Scientific American* December:100–105.

Keeling, C. D. and T. P. Whorf. 2001. Atmospheric CO_2 records from sites in the SIO air sampling network. In: *Trends: A Compendium of Data on Global Change*. Oak Ridge, TN: Carbon Dioxide Information Analysis Center, Oak Ridge National Laboratory, U.S. Department of Energy. http://cdiac.esd.ornl.gov/trends/co2/graphics/sio-mlgr.gif.

Kerr, R. A. 1995. It's official: first glimmer of greenhouse warming seen. *Science* 270:1565–1567.

_____. 1996. Ozone-destroying chlorine tops out. *Science* 271:32.

Little, C. E. 1995. *The Dying of the Trees: The Pandemic in America's Forests*. New York: Viking/Penguin.

Luoma, J. 1988. The human cost of acid rain. *Audubon* 90(4):16–29.

McCarthy, J. J., O. F. Canziani, N. A. Leary, D. J. Dokken, and K. W. White, eds. 2001. *Climate Change 2001: Impacts, Adaptation & Vulnerability*. Cambridge: Cambridge University Press. Summary for *Policy Makers and Technical Summary:* http://www.ipcc.ch/reports.htm.

McCormick, J. 1998. Acid pollution: the international community's continuing struggle. *Environment* 40(3):16–20, 41–45.

McDonald, A. 1999. Combating acid deposition and climate change: priorities for Asia. *Environment* 41(3):4–11, 34–41.

Munton, D. 1998. Dispelling the myths of the acid rain story. *Environment* 40(6):4–7, 27–34.

National Research Council. 1983. *Acid Deposition: Atmospheric Processes in Eastern North America*. Washington, D.C.: National Academy Press.

_____. 2001. Climate change science: an analysis of some key questions. Washington, D.C.: National Academy Press. http://books.nap.edu/books/0309075742.html.

Ott, H. E. 1998. The Kyoto Protocol: Unfinished business. *Environment* 40(6):16–20, 41–45.

Patrick, R., et al. 1981. Acid lakes from natural and anthropogenic causes. *Science* 211:446–448.

Revkin, A. C. 2000. New greenhouse gas identified, potent and rare (but expanding). *The New York Times* July 28, 2000:A1.

San Diego Daily Transcript. 1996. Smugglers make hot case from air conditioner chemical. October 14. http://sddt.com/files/li...0_14/DN96_10_14_cad.html.

Schneider, D. 1997. The rising seas. *Scientific American* 276 (3):112–117.

Schneider, S. H. 1989a. The greenhouse effect: science and policy. *Science* 243:771–782.

_____. 1989b. *Global Warming: Are We Entering the Greenhouse Century?* San Francisco: Sierra Club Books.

Scholle, S. R. 1983. Acid deposition and the materials damage question. *Environment* 25(8):25–32.

Skolnikoff, E. B. 1999. The role of science in policy: the climate change debate in the United States. *Environment* 41(5):16–20, 42–45.

Soroos, M. S. 1998. The thin blue line: preserving the atmosphere as a global commons. *Environment* 40(2):6–13, 32–35.

Stevens, W. K. 1995. Price of global warming? Debate weighs dollars and cents. *The New York Times*, October 10, 1995, p. B6.

Stone, R. 1994. Most nations miss the mark on emission-control plans. *Science* 266:1939.

Sturges, W. T., J. J. Wallington, M. D. Huyrley, K. P. Shine, K. Sihra, A. Engel, D. E. Oram, S. A. Penkett, R. Mulvaney, and C. A. M. Brenninkmeijer. 2000. A potent greenhouse gas identified in the atmosphere: SF_5CF_3. *Science* 289:611–613.

Tolba, M. K., et al., eds. 1992. *The World Environment 1972–1992.* New York: Chapman & Hall.

United Nations Environment Programs. 1993. *Environmental Data Report 1993–94.* Cambridge, MA: Basil Blackwell.

_____. 2000. *Global Environmental Outlook 2000.* http://www.unep.org/geo2000/english.

_____. 2001. *Status of Ratification/Accession/Acceptance/Approval of the Agreements on the Protection of the Stratospheric Ozone Layer.* http://www.unep.org/ozone/ratif.shtml.

United Nations Framework Convention on Climate Change, 2002. http://www.unfcc.de/resource/convkp.html.

U.S. Environmental Protection Agency. 1980. *Acid Rain.* EPA-600-79-036. Washington, D.C.: U.S. Government Printing Office.

_____. 1998. http://www.epa.gov:6703/airwdcd/owa/afs.count.

_____. 2001. *Latest Findings on National Air Quality: 2000 Status and Trends.* Research Triangle Park, NC: USEPA EPA 454-K-01-002. www.epa.gov/airtrends.

World Resources Institute. 1987. *World Resources 1987.* New York: Basic Books.

_____. 1988. *World Resources 1988–89.* New York: Basic Books.

_____. 1992. *World Resources 1992–93.* New York: Oxford University Press.

_____. 1994. *World Resources 1994–95.* New York: Oxford University Press.

_____. 1996. *World Resources: A Guide to the Global Environment 1996–97.* New York: Oxford University Press.

_____. 2000. *World Resources 2000–2001.* Washington, D.C.: World Resources Institute.

For more information, consult our web page at ***http://www.wiley.com/college/cutter.***

STUDY QUESTIONS

1. International negotiations to control emissions of ozone-depleting chemicals were quite successful, even leading to acceleration of the timetable for control. In contrast, efforts to control CO_2 emissions have been an almost complete failure. What are the main reasons for the differences in these two outcomes?

2. Who will be the winners and who will be the losers if global climate changes significantly over the next 50 years?

3. Examine Chapters 6–10 in this book, and make a list of all the important resource issues that might be significantly affected by global warming.

4. What changes in energy policy would be necessary for a significant reduction in acid deposition in the eastern United States?

5. What changes in energy policy would be necessary for a significant reduction in CO_2 emissions in the United States?

6. What countries should be asked to make the greatest sacrifices to control global CO_2 emissions? Should all countries bear equal responsibility, or should responsibility be proportionate to emissions?

CHAPTER 13

NONFUEL MINERALS

INTRODUCTION

In Chapter 1 we stated that resources are defined by their use. They fluctuate in response to changing human evaluations of resources as commodities and our abilities to use them. This is particularly true for minerals. For most minerals there is nothing inherent in the substance that makes it valuable. The value of a mineral derives from its usefulness in a given situation, which is largely a function of technology and cost relative to other materials. Our use of minerals therefore changes frequently as technology improves and as economic and political conditions vary. The physical characteristics of resources, including supply, are relatively unimportant in determining use.

Minerals have many different definitions. For the purposes of this book, minerals are substances that come from the earth, either from solid rocks or from soils and other deposits. Minerals include fossil fuels, such as coal, oil, and natural gas, but these are discussed separately in Chapter 14. The nonfuel minerals include metal ores, phosphate rock, asbestos, salt, precious stones, clay, gravel, building stone, and similar materials. Minerals are valued for their physical properties, such as strength, malleability, corrosion resistance, electrical conductivity, and insulating or sealing capacity, and they are fundamental to any industrial system. Table 13.1 lists major minerals, some of their uses, and major world producers.

An important feature of mineral resources is that most are traded internationally. Usually, a relatively small number of countries dominate the production of a particular mineral, other countries are consumers, and a few are self-sufficient. A country may have substantial quantities of a mineral available, the capacity to extract it, and the need to use it, but this does not mean that it will produce the mineral domestically. If sufficient quantities are available on the world market at prices below domestic production costs, then the mineral will be imported. Social or economic conditions in major producing or consuming nations significantly affect world markets and have far-flung impacts on the trading status of mineral resources. We must therefore examine mineral resources on a global scale.

RESERVES AND RESOURCES

Reserves is an important term used to describe the availability of minerals to any production system. Reserves of a mineral are the supplies available for use at the present time. Their location and physical characteristics are known, and they can be extracted with present technology and under prevailing economic conditions. Reserves are a subset of resources, which also include deposits that are unavailable for use at present because of poor geologic knowledge or unfavorable economic or technological conditions but that might become available in the future. *Unidentified resources* have not been discovered yet, whereas *subeconomic resources* may have been discovered but cannot be extracted at a profit at current prices.

The distinction between reserves and resources is illustrated in Figure 13.1. The horizontal axis of the diagram represents varying degrees of certainty about the existence or nature of deposits of a particular mineral. The vertical axis represents differing economic values of the deposits, that is,

Table 13.1 Uses, Major Producers, and World Reserves and Resources Relative to Demand for Selected Minerals

Mineral	Major Uses	Major Producers — Countries	Percent of World Production Accounted for	World Reserve/Production Ratio (Years)	U.S. Net Import Reliance (%)	Substitute(s)
Bauxite (aluminum ore)	Packaging, building, transportation, electrical	Australia, Brazil, Guinea, Jamaica, China	76	197	100	Iron, magnesium, plastics, ceramics
Chromium	Metallurgical and chemical industry, refractory industry	South Africa, Kazakhstan, Turkey, India, Zimbabwe	85	263	78	None for ferrochromium
Cobalt	Superalloys, catalysts, paint driers, magnetic alloys, cemented carbides	Congo, Australia, Canada, Russia, Zambia	80	146	74	Barium, nickel, manganese
Copper	Building construction, electric and electronic products	Chile, Zambia, United States, Indonesia, Australia	71	26	37	Aluminum, titanium, steel
Gold	Jewelry manufacturing, electronics	South Africa, United States, Australia, China, Canada	57	20	(Net exporter)	Palladium, platinum, silver
Iron Ore	Steel	China, Brazil, Australia, Russia, India	70	70	19	Ferrous alloys, scrap iron
Lead	Batteries, fuel additives	Australia, China, United States, Peru, Canada	70	21	24	Plastics, aluminum, tin
Magnesium	Metal refractories, aerospace, auto components	China, Turkey, United States, North Korea, Russia	67	714	45	Aluminum, zinc, calcium carbide
Nickel	Metal alloys, stainless steel	Russia, Canada, Australia, New Caledonia, Indonesia	68	40	58	Aluminum, plastics, titanium
Platinum group	Automotive, electronic, chemical	Russia, South Africa, Canada, United States	97	200	87	Palladium

Source: U.S. Geological Survey, 2002.

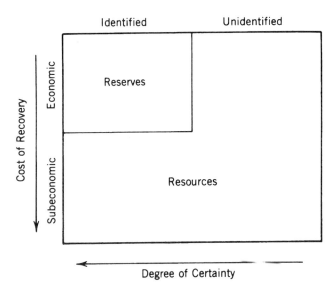

Figure 13.1 Reserves and resources. The boundary between reserves and resources is not fixed. It shifts with new discoveries, price changes, and other technological and socioeconomic factors. *Source*: Modified from Brobst and Pratt, 1973.

the varying economic profitability of extraction. All occurrences of a given mineral are contained symbolically within the boundaries of the diagram. For nonfuel minerals, this includes already-used and discarded resources, such as those in landfills. Use of nonfuel minerals does not destroy them but simply converts them to a less usable form.

The boundaries between reserves and other resources in Figure 13.1 shift over time. These boundaries change according to three factors: (1) economic conditions, (2) technology of extraction and use, and (3) geologic information. The economic profitability of extraction varies considerably, usually depending on the price of the commodity. As price goes up, more and more deposits, either of lower grade or those costing more to extract, become profitable to mine. Similarly, if price falls, only high-quality or cheaply extracted deposits are mined at a profit, and the quantity of reserves shrinks. Prices for minerals fluctuate widely, as discussed in Chapter 2. Variability in the level of production in the economy as a whole affects demand and price for raw materials, and competition between substitutable materials also causes prices to shift.

Technology is another factor that is important, especially as it affects the costs of mineral extraction and processing. For example, during the nineteenth and early twentieth centuries, high-quality iron ore was shipped in essentially raw form from Lake Superior ports to steel mills on the lower Great Lakes. In time the high-grade ores were depleted, and transportation costs per unit of steel became prohibitive. The development of a means for concentrating a lower-grade ore, called taconite, into enriched pellets at the mining site reduced transportation costs and allowed renewed mining activities. This technological development made lower-grade ores economically extractable, which they weren't in the past. Technological and social changes also cause increases in the costs of extraction. Environmental and safety regulations required changes in mining techniques to minimize environmental disruption and to protect miners. These changes increased the cost of extraction, thus tending to decrease production. For example, stricter safety standards for miners increased the costs of mining coal underground, and in some places this contributed to the decision to shift from subsurface to surface mining of coal.

Geologic exploration does not affect the profitability of extracting minerals directly but instead brings new deposits to light. If these are extractable at current prices and with current technology, they become reserves. Changes in geologic information are more important in long-term trends in reserves rather than in the short term. For most minerals there are identified deposits that will last at least a few years, and price fluctuations cause much more variation in reserves than do increases in information. It takes time to acquire new geologic information. Easily located surficial deposits are identified already; thus, any new deposits must be found beneath the surface or in very remote areas. Geologic structures and rock types must be mapped carefully, possible mineral associations checked with geochemical surveys, gravity surveys conducted, and so forth. When a potential deposit finally is located, it is necessary to drill test boreholes and analyze many rock samples before there is any certainty of the nature and extent of a deposit. The sophisticated exploration techniques available today help us to see the subsurface in greater detail and to make more cost-effective decisions on extraction.

AVAILABILITY OF MAJOR MINERALS

Geology of Mineral Deposits

There are so many different important minerals, and the geology of their deposits is so varied, that it is impossible to describe the specific conditions for all mineral occurrences. However, it is useful to discuss some general principles and examples of several major types of mineral deposits.

Minerals differ greatly in their crustal abundance, that is, in the percentage of the earth's crust composed of particular minerals. Iron, for example, makes up about 5 percent of the earth's crust and aluminum about 8 percent. Gold, on the other hand, is only about one-billionth of the earth's crust, and copper and zinc are each about one ten-thousandth. These fractions obviously affect the frequency with which we find usable deposits of these elements. Some minerals are valuable because of the particular chemical or molecular structure where they are found, and the frequency of such occurrences may be high or low. Carbon is a relatively plentiful element, for example, but diamonds, a crystalline form of carbon, are quite rare.

Many minerals, especially metal ores, are formed by similar geologic processes. For example, if rocks are fractured heavily by stresses in the earth's crust at high temperatures, then *hydrothermal mineralization* takes place. In this process, various elements dissolved in subsurface water flow below the surface, within the crust. If the chemistry and temperature of the water change, then certain minerals are deposited in surrounding rocks, creating concentrations of those minerals. Many valuable ores are created in this manner and are found near each other in mineralized districts. Mountain building often includes hydrothermal activity; thus, many of our important mineralized districts also are found there.

Shields are areas of very old igneous and metamorphic rocks that form the ancient cores of continents, and they are another geologic environment that yields large amounts of valuable minerals. In most of the world shields are buried under other rocks, but large-surface shield areas occur in Canada, Africa, Australia, and elsewhere. Because of their age, these areas contain different mineral assemblages than do most younger rocks, and many shield areas are very rich in metal ores. Important deposits of iron, nickel, copper, zinc, and other metals are found in the Canadian shield in Ontario and Quebec, for example.

Whereas concentrated metal ores are usually found in metamorphic rocks, many areas of sedimentary rocks also yield important minerals. Lead, zinc, and uranium are found sometimes in commercially extractable quantities in sedimentary rocks. Some substances, most notably gold and diamonds, are found in *placer deposits*. These are deposits of sand or gravel in which denser particles have been concentrated by the action of running water. Mining them usually requires excavating large volumes of sediment to recover relatively small quantities of minerals.

Another process that concentrates minerals at the earth's surface is *weathering*, or the gradual breakdown of rocks by mechanical and chemical processes. Weathering selectively removes some elements while leaving others in the soil. Bauxite, the most important ore of aluminum, is formed in this way in humid tropical environments.

These few examples illustrate the range of environments in which minerals are found. Some minerals show up in several different types of deposits, but many are located only in restricted geologic circumstances. Common minerals, such as iron, are found all over the globe, and the quality of those deposits varies greatly. As a rule, very high-grade deposits are relatively rare, but if industry is willing to accept slightly lower-grade deposits, more are available. This is certainly true for common elements such as iron and aluminum. The United States has little high-quality aluminum ore, and most of our best iron ore has been used. As a result, the United States imports most of its iron and aluminum ore from other countries. But should those foreign sources become too expensive, substantial domestic low-grade deposits are available. Rarer minerals, such as chromium, platinum, molybdenum, and vanadium, are found in commercial concentrations in fairly restricted locales. Only a few countries dominate world production and marketing for those minerals, and in some cases even low-grade domestic deposits are unavailable.

Variations in Reserves and Resources

Minerals can be regarded as stock or nonrenewable resources in that the amount available in rocks is finite. Thus, it is possible to examine how much is in the ground relative to present and projected future demand. But this is not as easy as it might seem. First, the amounts of mineral resources in the ground are extremely large, and the real question is how large the reserves are relative to demand. The amount of a mineral reserve available, however, is heavily dependent on price. A doubling in price, for example, may produce a 5- or 10-fold increase in reserves. Price increases also stimulate exploration, which further increases reserves through new discoveries. In this way, minerals act as flow resources rather than stock resources.

Second, an increasing number of minerals are recycled, and thus, from the point of view of human production systems, are essentially renewable resources. Over 40 percent of U.S. iron and steel supplies are derived from recycled material, and more than half the nation's lead consumption is met from recycled material. When the price of minerals rises, so does the economic attraction of recycling, and the contribution of recycled materials will increase undoubtedly in the future.

Third, there is a high degree of *substitutability* for most minerals. If we run short of steel, we could use more aluminum. If we run short of aluminum, we might use magnesium or synthetic materials. Substitution is not possible for all uses of a material, but partial substitution usually alleviates supply problems and keeps the material available for necessary uses. This substitutability is one factor causing demand for minerals to be highly elastic, changing significantly with variations in price.

Taken together, these factors mean that future demands for minerals are difficult to predict and may be significantly affected by small changes in prices and technologies. In addition, a distinction must be made between short- and long-term availability. In the short term, perhaps over periods of less than 5 to 10 years, sharp fluctuations in mineral prices are common (Fig. 13.2). These fluctuations both reflect and cause variations in supply or demand, presenting very real problems when they occur (Fig. 13.3). We often misunderstand the nature of these crises, however. They are not the result of the world running out of a mineral; rather, they are a consequence of the vagaries of the world economic and political system. In the long term, however, we will never "run out" of any minerals. We may find that a particular mineral has become too expensive to justify using it, and, in that sense, it may become unavailable. For groups of minerals, worldwide demand may cause a long-term trend of increasing prices, which would reduce demand for those minerals.

World Reserves and Resources

In the long run, mineral availability is determined by the level of economic activity, technological changes, and geologic considerations. Although these factors cannot be predicted with certainty, the current status of reserves relative to demand is indicative of future mineral availability and use. Table 13.1 includes a listing of current world reserves of selected minerals in relation to world consumption. Also included are world reserve/production ratios. These represent the number of years that current reserves would last at the year 2000 rates of production. The final column in the table lists some of the materials that could be substituted for several of the uses of the minerals if supplies were limited or prices climbed significantly. Substitution is possible only for some uses; for other uses substitution is not currently feasible. In most cases the use of these substitute commodities results in either increased costs or reduced quality of product, and in some cases new product development is needed. After all, if these substitutes were suitable under present economic and technological conditions, we would be using them.

Examination of current reserves indicates that for most minerals, currently identified deposits of economically recoverable materials are sufficient to meet world demand for several decades. If demand increases, as is likely for most minerals, the adequacy of current supplies will be somewhat less. On the other hand, changes in economic and technological conditions and geologic exploration certainly will result in increases in reserves.

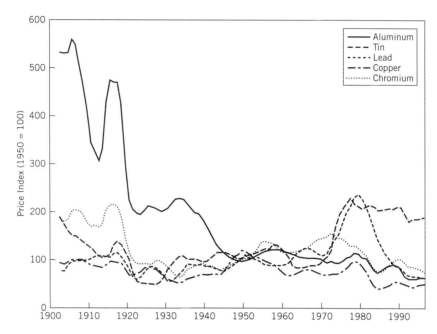

Figure 13.2 Indices of prices for selected minerals, 1900–2000 (1950 = 100). Data are based on 5-year running means, adjusted for inflation. Mineral prices are highly variable; in general, prices for these minerals climbed in the 1970s but fell dramatically in the 1980s. If minerals were becoming more scarce in relative terms we would expect prices to rise. The lack of any consistent trend of increasing prices is cited by some as evidence that the minerals are not being significantly depleted. *Source*: U.S. Geological Survey, 2002.

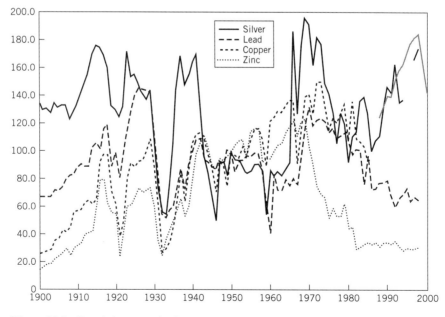

Figure 13.3 Trends in ore production at domestic mines, 1900–2000 (1950 = 100). Production has varied greatly over time. Unlike many other resources, our production of many minerals is not dramatically higher today than it was a century ago. *Source*: U.S. Geological Survey, 2002.

Finally, should shortages of any of these minerals develop, higher prices will stimulate substitution of other substances or increased recycling, resulting in reduced demand. At the world level, then, it seems likely that most of these minerals will continue to be available for the foreseeable future.

This is not to say that localized, short-duration shortages will not take place. As discussed earlier, in many cases a few countries, or a few mines, may dominate world production. For example, 5 countries control about 76 percent of the world's bauxite production, 6 countries produce 80 percent of all manganese, and 68 percent of phosphate rock comes from 4 countries. Nearly 90 percent of the platinum group metals are produced by two countries—South Africa and Russia. Moreover, one firm, Norilsk Nickel in northern Siberia, produced 62 percent of the world's palladium, 27 percent of the world's nickel, and 22 percent of the world's platinum in 1994, enough to influence the world price of these commodities at that time (Kotov and Nikitina 1996). Closure of a few mines or short-term restrictions in trade easily could result in market shortages and dramatic price changes.

One factor that sometimes leads to these short-term supply fluctuations is the *cartels* that artificially control supplies of minerals. While cartels exist, notably those in tin, copper, and bauxite, they are unsuccessful in manipulating world markets (see the discussion of OPEC in Chapter 14). Their lack of success is due to several factors. In some cases, important producers refused to join because of political differences with other producers or lack of cooperation with other cartel members. In addition, many countries have large stockpiles of minerals that were accumulated for military uses in times of war that are now being used to manipulate supplies and prices. Most minerals are simply not in short enough supply nor critical enough to world economies to allow a group of producers to put significant pressure on consuming countries. The result is a continuation of a relatively unregulated market at the global level. For those developing nations that are mineral-rich and derive substantial foreign exchange from mineral exports (Congo, Gabon, Zimbabwe, and Malaysia are examples), this situation has thwarted plans to substantially increase national income through market manipulation of metals and minerals.

The diamond cartel provides a good example. One company, De Beers of South Africa, controls more than 80 percent of the world's rough diamond trade. Starting in 1990, diamonds from Angola and the former Soviet Union republics began flooding the market. The glut of illicit diamonds had two causes: illegal mining in the Lunda Norte province in Angola by garimpeiros, or prospectors, who flooded to the region in search of riches; and increased smuggling in Russia once its economic crisis worsened during the early 1990s, after the breakup of the Soviet Union. This upsurge in supply forced De Beers to consider either reducing the price of their diamonds or buying the illicit diamonds themselves. De Beers spent nearly $200 million to initially purchase diamonds from Russia and Angola in order to maintain its control of the rough diamond market (MacLeod 1992). Since then, they have spent more than $4.8 billion to control production from Angola and Russia. At the same time, other production was slowed, yet DeBeers continued to increase its inventory to keep prices high. As a consequence, DeBeers is buying diamonds faster (to control supply) than it can sell them (at higher prices), creating huge economic losses for the company.

U.S. Production and Consumption

Although most minerals are essentially global commodities, inasmuch as supply and demand are dominated by worldwide markets rather than national policies, it is useful to examine the domestic situation in the United States with respect to mineral supply and consumption. The United States is the largest single consumer of many materials, and conditions here are of considerable importance in the world market. Domestic conditions are sometimes the focus of policy debates relating to employment in mining or the environmental impacts of mining. Because nonfuel mineral production adds significant income to many states (Fig. 13.4), these policy debates become significant.

Table 13.1 (page 296) includes information on the U.S net import reliance, or the percentage of our consumption that was supplied by imports. The United States consumes a substantial portion of world production of many minerals. Typically, the United States, with about 5 percent of the world's population, consumes between 10 and 30 percent of world production. This is roughly the same proportion as world

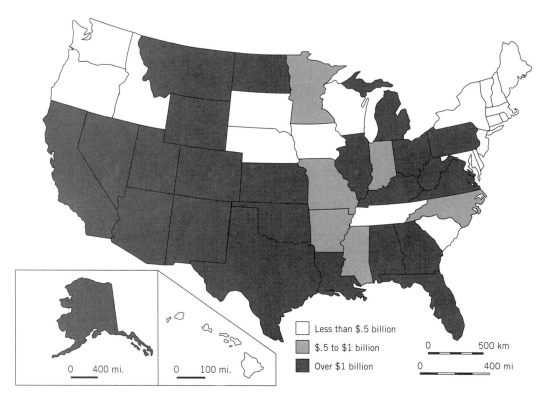

Figure 13.4 Total value added in nonfuel mineral production, 1997. The mining states in the west produce high-value minerals like gold and silver; in Minnesota and Michigan, iron ore; and in Florida, phosphate. Source: U.S. Bureau of the Census, 2002.

energy production consumed by Americans and simply reflects the United States' large population, advanced level of industrialization, and high standard of living. Much of our large consumption is associated with minerals used in technologically advanced industries like aerospace and electronics. On the other hand, our consumption of iron ore and bauxite is at levels more appropriate to our population size.

U.S. domestic reserves of many minerals vary considerably; some are virtually nonexistent, while others are very large. Important minerals for which our domestic reserves are quite small include bauxite, chromium, cobalt, manganese, nickel, and the platinum group metals. The United States has abundant reserves of copper, iron ore, magnesium, molybdenum, rare earths, thorium, vanadium, and numerous other minerals.

Finally, the United States is a major trader of minerals and is dependent on imports for such important commodities as bauxite, chromium, cobalt, graphite, iron ore, manganese, tin, tungsten, and zinc. The United States is a net exporter of boron, industrial diamonds, magnesium, and phosphate rock, for example. Note that for some minerals there is essentially no domestic production, but our import reliance for these minerals is well below 100 percent. For example, we have no domestic chromium production, but we imported only 78 percent of our needs in 2000. The balance of our consumption was met with recycled chromium. Recycling was also important for iron, aluminum, lead, copper, and tin.

STRATEGIC MINERALS AND STOCKPILING

Although effective mineral cartels or other political organizations have yet to significantly limit supplies of minerals, the possibility of such restriction is of some concern to the U.S. government. This is particularly true in the case of minerals that are significant to military/industrial production and that are imported from nations with unstable or unfriendly governments.

This dependency issue has been recognized for many years. The U.S. government therefore defined certain minerals to be of strategic or critical importance to the welfare of the country and developed policies to prevent shortages. *Strategic minerals* are those essential for defense purposes for which the United States is totally dependent on foreign sources. Examples of strategic minerals are cobalt, chromium, manganese, and platinum. *Critical minerals* also are necessary for national defense, but the United States can meet some of its demand through domestic sources and supplies from friendly nations. Examples of critical minerals are copper, nickel, and vanadium. It is an interesting quirk of geography that much of the U.S. supply of strategic minerals comes from two nations: South Africa and Russia. Over 30 minerals have been identified as having strategic or critical importance. The following examples illustrate that importance.

Chromium is used to harden steel and make it resistant to corrosion. It is an essential component of stainless steel and is found in ball bearings, surgical equipment, mufflers, and tailpipes. It is used in the defense industry in armor plating and weapons and for many parts of piston and jet engines. The leading world producers of chromium are South Africa and Kazakhstan. The United States imports about 78 percent of its chromium, mainly from South Africa, Turkey, and Zimbabwe.

Cobalt is a metal used in the aerospace industry. It is a high-temperature alloy used in the manufacture of jet engines, cutting tools, magnets, and drill bits. It is used in electronics equipment, especially in computers, television receivers, and transmitters. The primary producers of cobalt are Congo, Zambia, Canada, Australia, and Russia. The United States is heavily reliant on imported cobalt, mostly from Zambia, Norway, and Canada.

Other strategic minerals are those in the platinum group, which includes six different minerals with similar properties. These metals—platinum, iridium, palladium, oridium, rhodium, and ruthenium—are resistant to corrosion and are used to catalyze chemical reactions. Applications include catalytic converters in automobiles, petroleum refining, electroplating, electronics, and fertilizer manufacture. The major world producers of platinum group metals are Russia, South Africa, and Canada. The United States imports about 90 percent of its platinum, primarily from South Africa, the United Kingdom, and Belgium.

Manganese is essential in steel manufacture and thus has many industrial and military applications. It is added to molten steel to remove oxygen and sulfur, thus hardening the steel. Ukraine, South Africa, Gabon, and Brazil are the world's major producers, providing about 57 percent of the world's manganese supply in 2001. The United States imports most of its manganese from South Africa, Gabon, and France. The presence of large quantities of presently subeconomic manganese in deep-ocean beds was an important part of the United States' reluctance to sign the Law of the Sea Treaty, as we saw in Chapter 9.

Stockpiling, or maintaining large storages of commodities, is one method the United States uses to protect itself from restrictions in supplies of strategic minerals. The danger of minerals dependency was recognized at the turn of the century, but actual stockpiling did not begin until shortly before World War II. In 1949, the U.S. Strategic Stockpile was created to avert shortages of minerals during wartime. The intent was to purchase and store materials in sufficient quantities to meet defense and national security needs for a three-year period, which is presumably long enough for alternative supplies to be developed in the event of a cutoff of supply from other nations. Ninety-three substances, not all of them minerals, were on the list. Congress did not appropriate funds for procurement, however, and the stockpile was well below the official goals for many years. More recently, procurement increased, and the stockpile reached 100 percent for many materials, prompting the government to sell some of the stockpiled minerals. For example, in 1995 the government sold excess minerals from the stockpile, valued at $414 million, and occasionally used the stockpile to influence mineral prices through trading or threatening to trade in stockpiled commodities.

Stockpiling is an important minerals resource policy because it protects a nation from the short-term interruptions of supply and price increases. Stockpiling in the United States historically implied preparedness for war. It is unlikely that limited regional wars would cut off supplies from all major mineral suppliers, and in the case of all-out world war it is improbable that the United States would exhaust the three-year supply of industrial minerals. Nonetheless, the strategic stockpiles are very good guarantees against short-term reductions in supplies caused by an industrial collapse or prohibition of trade in one or two countries. From the standpoint of national economic health, it makes sense to guard against such events.

In summary, the market for most minerals is a world market in which a relatively small number of suppliers sell to a large number of customers. In the long term, the availability of some minerals is essentially infinite, but for others there will be substantial increases in price, leading to conservation, recycling, development of substitute materials, and other means for reducing demand for expensive materials. In the short term, prices fluctuate widely, stimulating changes in sources and uses of minerals, and creating some economic distress in sensitive industries. Even though the production of many minerals is concentrated in a small number of countries, because those nations lack political agreement and depend on exports for foreign exchange and because of the substitutability of materials, the formation of effective cartels is unlikely. The volatility of mineral prices makes stockpiling useful for insulating national production systems from catastrophic changes in the world market.

MINING IMPACTS AND POLICY

Environmental Considerations

Mineral extraction has significant environmental and social effects. Individual mines often are very large, as economies of scale are important in maintaining profitability. Few mineral ores are more than 30 percent pure, and some are less than 1 percent pure. Thus, large quantities of ore must be processed to obtain relatively small amounts of finished product. If the mine is at the surface, such as the open-pit mine shown in Figure 13.5, the area disrupted is quite large. In addition, it is often necessary to remove large amounts of undesired rock to get at an ore body, further increasing the area disturbed.

Unused rock and the waste from processing operations require some type of disposal (Fig. 13.6). These materials, called *tailings*, are processed with large amounts of water and are deposited in tailings ponds. Often these ponds are themselves quite large. The water pumped from mines after use in extraction or in ore processing is usually of low quality. In many cases it is highly acidic, and it is usually contaminated with the minerals being mined. The resultant pollution of receiving waters is severe. The huge volume of ore-bearing rock to be processed requires that this activity take place

Figure 13.5 The Bingham Canyon copper mine in Utah, one of the world's largest open pit mines.

Figure 13.6 Spoil piles from hydraulic mining for gold in California.

near the mine; thus, many mine sites are also locations for smelting or other methods of purifying minerals. Most of these processes produce large emissions of air pollutants, particularly sulfur oxides and metals. These emissions severely damage vegetation and soils in the surrounding area (Fig. 13.7). Sudbury and Wawa, Ontario, and Palmerton, Pennsylvania, are examples of areas where vegetation destruction is so severe that it is clearly visible on satellite photographs.

Mine sites are generally dictated by the presence of ore, not by the environmental suitability of the location for a large-scale industrial operation. Many of the mines in the Rockies and Sierra Nevada are in areas of considerable natural beauty, in or near important resort areas. In a region with extensive mining activities, the environmental consequences are widespread (Issue 13.1). Although long-abandoned mines and ghost towns are appealing to tourists, modern mines and associated support facilities generally are not.

Social Impacts

In addition to environmental effects, mining has many important social and economic effects in the areas where it occurs. Most mining towns are isolated, away from major urban centers, and frequently in mountains (Fig. 13.8). The populations of mining towns are dependent almost entirely on the mines for employment and income. When the mine is operating, they prosper, but when it is not operating, they become impoverished. Mine operations are governed by the prices of the materials mined, which, as we have shown, tend to fluctuate widely. When demand for a mineral drops, causing a reduction in price, production is reduced by closing whole mines rather than reducing production a little at all mines. This is because economies of scale are important in mining, and it is often more economical to close a mine entirely than to run it at a reduced level.

A good example is the San Manuel, Arizona, copper mine. Throughout the 1990s domestic production of copper was stable, but new mines were opened in Asia and Latin America (Indonesia, Peru, Chile, Argentina). These new mines had a higher-quality ore, lower labor costs, and fewer (and thus less costly) environmental regulations. As a consequence, global production levels increased. Unfortunately, consumption levels did not keep pace, and thus the price of copper fell. In the mid-1990s, for example, the price of copper was $1.20 a pound, but by 1998 it had dropped to $0.61 per pound. Given the global competition and the declining price for copper, the San Manuel mine was closed by its owner, Broken Hill Proprietary (an Australian company). Nearly 2200 of the town's 4000 people were laid off (Kopytoff

Figure 13.7 Destruction of forests and associated wildlife is intense in the vicinity of Sudbury, Ontario, where acid emissions from nickel smelting have fallen on soils lacking adequate buffering capacity.

ISSUE 13.1: THE NEW GOLD RUSH: PROSPECTING IS POISON

Gold has a special allure to mining prospectors. For centuries, it has been associated with wealth and affluence. The Spanish conquest of the Americas was largely driven by the quest for gold. In the Andes, the Inca and Chibcha Indians were quite distinguished artisans in their use of the element. The Gold Museum, in Bogotá, Colombia, houses one of the most impressive collections of pre-Columbian gold artifacts in the world.

Gold rushes have come and gone, resulting in many boom-and-bust local economies and environmentally ravaged lands. The latest gold rush is in Venezuela. As an attempt to revitalize its mining sector, the Venezuelan government permitted gold mining in one of the country's most environmentally sensitive regions, Bolívar. The Caroní River runs through this region, supplying most of the drinking water for the state. The river feeds the Gurí Dam, which produces more than 70 percent of the country's electricity. Canaima National Park is located in the southern section of Bolívar. More than 40 percent of the mining in Bolívar, according to government estimates, occurs in the Imataca Forest Reserve.

The mining process involves blasting the soil with powerful water jets (*hydraulic mining*). The sediments are carried into the tributaries and downstream. Increased sedimentation could threaten the power production potential of the Gurí Dam. In addition, the pools of water left behind become wonderful habitats for mosquitoes, many of which are malaria carriers. To separate the gold from other minerals, mining companies often use cyanide, mercury, and other toxic substances. These substances are dumped into holding ponds (and seep into the ground) or are washed into the nearest stream. High levels of mercury have already been detected in the lower Caroní basin. Local people avoid the fish, which was once a staple in their diet. The incidence of Down's syndrome and the number of birth defects in Bolívar are among the highest in the entire country.

Venezuela's largest mine, Minerven, run by the state-owned mining company, operates no differently than dozens of smaller scale operations. All use hydraulic mining, and the larger facilities tend to use cyanide rather than mercury in the separation process. The relaxation of environmental laws to promote large-scale mining was an attempt by the government to focus on economic development in southern Bolívar state, while ignoring some of the adverse environmental consequences of that development (Schemo 1996).

Halfway around the world, the town of Bergama, Turkey, has been the site of confrontations for over 3000 years, starting with the ancient Greeks and Romans and moving on to Alexander the Great and, much later, the Crusades. In 1989, gold was found. Eurogold, a French-based company backed by the Turkish government, proceeded to open the country's first modern gold mine. The pit mine is just eight miles south of Bergama, a major tourist spot. About 300,000 people live in the area around the mine, which also happens to be located in a seismically active area. An earthquake could rupture the liner beneath the waste pool, thereby releasing cyanide. In 1938, the region suffered a major earthquake in which many villages were totally destroyed.

The mine was built according to the most modern of standards. There is a clay and polyethylene sealer beneath the waste pool. However, residents are fearful of the vast amounts of water (diverting it from agriculture) and the ever-present cyanide. Nevertheless, the mine began operations in late 1997 (Kinzer 1997). In 2002, Turkish courts closed the mine due to the hazards of mining (Degirmenci 2002).

As gold prices fluctuate ($36 per ounce in 1970, $613 per ounce in 1980, $385 in 1990, $324 in 2002), prospecting will increase. Whether the return on the investment is achieved may be less important than the environmental disruption caused by the current mining practices. The get-rich-quick scheme to assist developing economies may in the end bring about their economic demise.

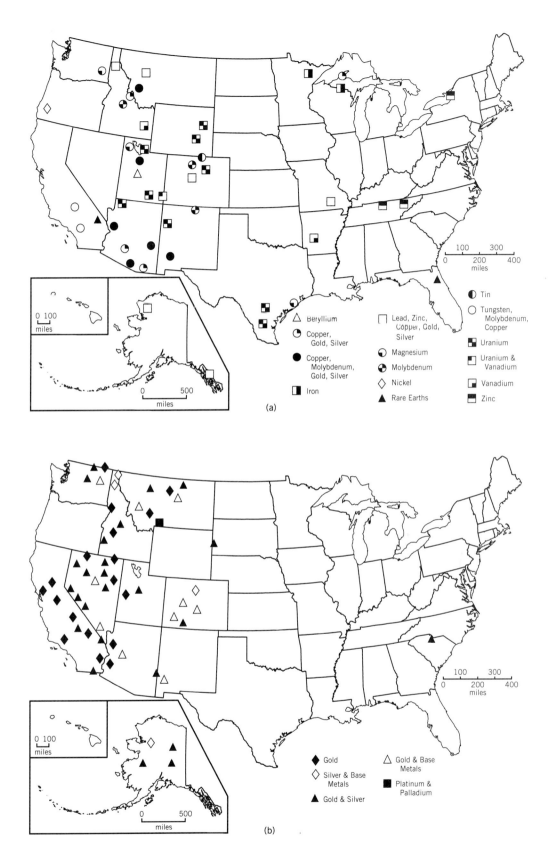

Figure 13.8 Major areas of production of domestic metal (*a*) and precious metals (*b*). Notice the concentration of mines in the west. In the east, most of the metal mining is for zinc and rare earths. South Carolina stands out as the only eastern state to mine gold and silver. *Source*: U.S. Geological Survey, 1997, 2002a.

1999). Mining towns thus go through repeated boom-and-bust cycles, alternating between full employment and extreme unemployment (Issue 13.2). The toll this takes in the lives of the mine workers and other residents is understandably severe.

Nonfuel Minerals Policy

Most of the minerals policy in the United States is guided by the 1872 Mining Act, which provided unlimited access to public lands for mineral exploration. Specifically, the Mining Law gives U.S. citizens and corporations the right to prospect for certain minerals on federal lands and allows them to file claims to mine and sell whatever minerals they find. The law does not provide for any payment to the federal government for minerals mined on federal lands. Despite repeated attempts to reform the 1872 Mining Act, most notably in 1997, this law still guides much of the United States' nonfuel minerals policy. Since the Mining Law was passed more than a century ago, the regulation of mining activities and environmental impacts continues to be done on an ad hoc basis. Reforms are needed to (1) ensure that mining companies are charged royalties for the minerals they extract from federal lands and (2) establish a federal program to minimize the environmental impacts of hard-rock mining after the mining operations cease. Finally, as part of the 1990 Clean Air Act Amendments, the mining and minerals processing industries are increasingly monitored for air emissions, especially lead compounds from smelters.

CONSERVING MINERALS: REUSE, RECOVERY, RECYCLING

Recycling is the process whereby a material is recovered from the consumer product and then becomes a raw material in the production of a similar or different product. A significant amount of material can be recovered during the manufacture, use, and disposal of goods to decrease our need for raw ores (Table 13.2). Aluminum recycling is a good example. Aluminum cans that are returned to redemption centers are recycled, melted down, and used in the manufacture of new aluminum cans and other aluminum products (Fig. 13.9). In 2000, about one-third of total aluminum production was from recycled aluminum. Because of its

Table 13.2 Mineral Recycling in the United States as a Percentage of Apparent Consumption, 2000

Mineral	%
Aluminum	36
Chromium	24
Copper	32
Iron and steel	55
Lead	63
Magnesium	39
Nickel	36
Tin	29
Titanium	50
Zinc	27

Source: U.S. Geological Survey, 2002.

high price and availability, aluminum is consistently recycled and is often the target for scavengers. Nothing appears to be sacred. Aluminum siding and gutters from New York City homes, swing sets from a Miami playground, and even cans from local recycling centers have been swiped by vandals desperate for money.

The ease of minerals recycling depends on several factors, including the technology of conversion to a desired product, the level of difficulty in obtaining sufficient quantities of used material, and the relative costs to manufacture a good from recycled as opposed to raw materials (Issue 13.3). This last factor is partly affected by differences in quality between products made from raw materials and those made from recycled materials. For most minerals, the cost of converting an ore to pure form is significantly greater than the cost of remanufacture from recycled material, if we exclude the cost of mining ore and the cost of accumulating recycled materials. This is true in monetary terms but also applies to energy use and environmental pollution. For example, manufacture of a given amount of aluminum from recycled scrap uses 90 to 95 percent less energy and generates 95 percent less pollution than refining aluminum from bauxite.

Accumulating sufficient material to recycle and maintaining the purity of recyclable materials are major problems. These depend on the use and handling of materials. One of the biggest uses of lead, for example, is in storage batteries, such as those in automobiles. Old batteries are found at specific locations, primarily automobile junkyards

ISSUE 13.2: LIVING WITH BOOM AND BUST

Boom and bust, a cycle of rapid growth and catastrophic decline, occurs wherever mining dominates a local or regional economy. When national or world production of minerals drops slightly, mines that are marginally profitable close operation, while those with larger profit margins stay open. When prices are high and production increases, new mines open and old ones are reopened. If a mine is located in a remote area, as is often the case, the communities around it depend on the mine for nearly all their income. Most local residents work for the mine, for businesses that serve the mine, or for businesses that serve the people who work in the mine. As the mine goes, so goes the community. The following scenario is a typical one.

Both boom and bust are stressful in a mining town. Boom times usually bring an increase in population, a shortage of housing, a rise in rents and property values, and a shortage of public services. When immigrants come in large numbers, they bring problems to small towns. Everyday life for the permanent residents is disrupted, shops are more crowded, housing costs rise, and streets are blocked by construction. Unemployment and crime may also be problems if more immigrants arrive than are hired.

Boom times are not all bad, however, for nearly everyone has a job, property values increase, and business is good. The new money in town brings in new businesses, perhaps a second barber shop, a few gas stations, a discount department store, a bigger supermarket, a few restaurants, and several bars. Those lucky enough to get into business at the right time do very well and may amass small fortunes. Those who came in when the boom was already under way meet stiff competition and high costs but still manage to earn a living.

Then comes the bust. Maybe competition from cheaper imports is too rough, the higher grade ore is played out, or the world price may have fallen because of a recession. Maybe new materials are replacing the ore in a few critical industries. Or perhaps environmental requirements are making the mine less profitable than one in another state or another country. The actual cause doesn't really matter to those affected, because whatever the cause, it is beyond their control. At any rate, the mine is shut down, and perhaps 40 percent of the town's workforce is laid off with two weeks' notice and two months' severance pay. A few stay on for several months to help with removing the machinery that can be used elsewhere.

Some people leave town immediately, particularly those who had arrived most recently and hadn't established themselves yet. Unemployment compensation and savings keep the rest going for the better part of a year, but eventually that runs out, so most of them leave, too. Those with large debts go bankrupt. Many of the rest can't sell their houses for as much as they owe on the mortgage, and their mortgages are foreclosed.

Two years after the mine is closed, the town's population may have dropped by 50 percent. Before the boom, there was one barber shop, three gas stations, and one supermarket, all doing well. Now there are two barber shops, both having trouble making rent payments, one supermarket open and one closed, a closed department store, and six gas stations, only one of which is doing well. Most of the bars that opened during the boom are still there, but now they get more business during the day than at night.

Although the specific town just described is fictional, these things have happened in many towns in the United States and elsewhere, and more than once in most of them. Today, concern about the social impacts of mining is increasing, and many states have taken steps to lessen the blows. Severance taxes, charged to the mining companies when minerals are removed, are often used to pay for improvements in town services and infrastructure during growth times and for relocation or job training afterward. Mining companies are becoming more willing to contribute to these expenses in their attempts to maintain public goodwill. These changes will ease the ups and downs to some extent, but they will still occur and recur (Gulliford 1989).

Figure 13.9 Recycling aluminum beverage cans saves both materials and large amounts of energy.

and repair shops. The lead in them is valuable enough to warrant recycling, as is much of the lead used in other applications. As a result, in 1998, nearly 80 percent of the U.S. lead production was from recycled materials (Fig. 13.10). Titanium, on the other hand, is not as recylable. Over 95 percent of titanium used in the United States is for pigments in paints, and thus it winds up on the walls of houses, buildings, and so forth. It is very difficult to collect used paint in sufficient quantities to recycle the titanium that it contains. Titanium recycling is therefore insignificant.

Over time, the changing use of metals in manufacturing also affects their recycling potential. Until the 1970s, the platinum group metals were recycled with efficiencies around 85 percent (Frosch and Gallopoulos 1989). However, in the mid-1970s the introduction of catalytic converters for automobiles to reduce exhaust emissions changed the recycling rates. Automotive use of platinum group metals now accounts for most of the permanent consumption of these metals. Although the industrial applications of the platinum group continue to recycle about 85 percent of the metals, poor recycling rates (less than 12 percent) characterize the automotive applications. The reason is quite simple. The poor recycling rates are a function of the limited means for collecting the discarded catalytic converters, which are found in scrap yards along with the discarded automobile. The technology for recycling exists, but the cost of locating, collecting, and emptying all the discarded converters and then transporting the platinum metals for recycling is too high for most recovery operations at current prices. If the price of platinum rises to more than $500 per ounce, then recycling of catalytic converters will become more pronounced.

Accumulation of material for recycling is easiest in industry, where large amounts of material tend to be found in a few locations. In many industrial processes, scrap is generated as part of the manufacturing process, and this is recovered easily. For example, steel that does not meet specifications is recovered at the steel plant and immediately is returned to the production process as a raw material. This recycling of so-called new scrap is

ISSUE 13.3: COMPUTERS AS SOLID WASTE

The spread and evolution of personal computers is probably the most dramatic technological development in the past three decades. In the early 1980s, personal computers started appearing in many U.S. homes. Those who purchased computers at that time found that within a few years the newer machines were much faster than the older ones and the software they used was similarly more advanced, to the extent that even though the older machines were still capable of doing everything they did when they were new, it was necessary to replace them.

Some people use computers for a longer period of time, while others find it necessary to buy the most up-to-date machines relatively often. It is common to replace a computer when it is 3 to 5 years old. Many of those who bought their first computers in the early 1980s are today on their fourth, fifth, or even sixth computer. By the end of 2001, over 300 million PCs had been sold in the United States, and by 2005 that number is expected to exceed 600 million. In American families it was (and still is) common that when a parent purchased a new computer, the older one was passed along to a child. Whereas in the early days of personal computers a household would have only one computer, now it is common for each person of school age or older to have his or her own machine. In the first two decades of PC use in the United States, many of the machines that had become obsolete for their first owners were still in use. Today, however, people who are likely to have computers already do, so when a new computer is purchased there is nowhere for the old one to go. Even though the machine is still fully functional, it becomes waste.

What is a computer good for if not computing? What is it made of? Unlike simple objects like books or plastic bags, computers consist of a wide variety of materials. By weight the largest amount is plastic, followed by copper and steel. But they also contain significant amounts of lead and tin (primarily in solder), nickel, antimony, zinc, silver, gold, cadmium, tantalum, palladium, and many other metals. Many of those materials are quite valuable if they are concentrated in significant quantities. In computers they are found in minute amounts (typically less than 1/10,000th of the weight of the computer) comingled with many other materials, and thus are hard to separate. As a result there is very little value in a used computer. In fact, it may have negative value—you have to pay someone to take it.

Where do the old computers go? Can and will they be recycled? The answers to these questions are still uncertain. Hundreds of PC recycling companies have sprung up in the last few years. In most cases these companies do some dismantling of at least the newer computers and resell the few components that are still useful. Circuit boards, which contain the highest concentrations of especially valuable materials, are separated out for recovery of gold, platinum, copper, and other metals. Some plastics are separated and recycled, and some other components are sent to landfills. The degree to which computers can be dismantled and recycled varies from machine to machine and is changing over time. There is pressure on the computer industry to make computers that are more easily broken down for recycling, but at the same time the computer manufacturers are under tremendous pressure to keep costs down in an intensely competitive market.

just a way of making the manufacturing process more efficient. Recycling of new scrap is not included in the data in Table 13.2.

CONCLUSIONS

Although we often think of minerals as nonrenewable resources, we are not going to run out of any of them. Variations in price are far too important in generating new supplies and in controlling demand to permit geologic considerations alone to determine use. Demand fluctuations occur primarily through substitution, but increased efficiency of use is also important. In the future, recycling, reuse, and recovery will become important, as increasing demand raises prices and makes this source of materials economically more feasible.

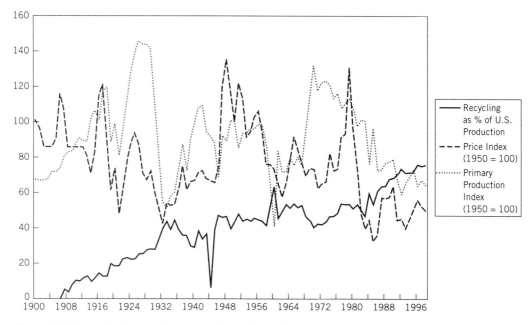

Figure 13.10 Lead prices, production, and recycling in the United States, 1900–1998. Price and production have fluctuated widely, with a tendency for production to be high when prices are high, reflecting the influence of demand and price on production. Recycling rates have increased steadily. *Source:* U.S. Geological Survey, 2002.

On the other hand, in many applications a shift has been made already from mineral-based products to plastics. Many parts of new automobiles, for example, are now made of plastic rather than metals. The same is true for children's toys like trucks and cars. If that shift continues, then demand for some minerals may decline, reducing prices and recycling rates.

REFERENCES AND ADDITIONAL READING

Anderson, E. W. and L. D. Anderson. 1997. *Strategic Minerals: Resource Geopolitics and Global Geoeconomics.* New York: John Wiley & Sons.

Bridge, G. 2000. The social regulation of resource access and environmental impact: Production, nature and contradiction in the U.S. copper industry. *Geoforum* 31:237–256.

Brobst, P. A. and W. P. Pratt, eds. 1973. *United States Mineral Resources.* Washington, D.C.: U.S. Geological Survey. Professional Paper 820. Washington, D.C.: U.S. Geological Survey.

Degirmenci, E. 2002. *Izmir third administrative court decides the Bergama mine should close.* http://www.minesandcommunities.org.

Frosch, R. A. and N. E. Gallopoulos. 1989. Strategies for Manufacturing. *Scientific American* 261:144–152.

Gulliford, A. 1989. *Boomtown Blues: Colorado Oil Shale 1885–1985.* Boulder: University Press of Colorado.

Kinzer, S. 1997. Turks fight gold mine, saying prospect is poison. *The New York Times*, April 16, p. A4.

Kopytoff, V. G. 1999. Tough times in the copper pits. *The New York Times*, September 11, p. B1.

Kotov, V. and E. Nikitina. 1996. Norilsk nickel: Russia wrestles with an old polluter. *Environment* 38(9):6–11, 32–37.

McKelvey, V. E. 1972. Mineral resource estimates and public policy. *Scientific American* 60:32–40.

MacLeod, S. 1992. Diamonds aren't forever. *Time* October 12, p. 73.

Roth, S. 2002. Of mines and men. *The Ecologist* 32(6):26–27.

Schemo, D. J. 1996. Legally now, Venezuelans to mine fragile lands. *The New York Times*, December 8, p. 6.

Sinclair, J. E. and R. Parker. 1983. *The Strategic Minerals War.* New York: Arlington House.

Solnit, R. 2000. The new gold rush. *Sierra* July/Aug:50–57.

U.S. Bureau of Mines. 1994. *Minerals Yearbook.* Washington, D.C.: U.S. Government Printing Office.

U.S. Congress, Office of Technology Assessment. 1979. *Materials and Energy from Municipal Waste.* Washington, D.C.: U.S. Government Printing Office.

U.S. Geological Survey. 1975. *Mineral Resource Perspectives*, 1975. USGS Professional Paper 940. Washington, D.C.: U.S. Geological Survey.

―――. Office of Minerals Information. 2002. *Mineral Commodity Summaries.* Washington, D.C.: USGS, http://minerals.er.usgs/pubs/, and http://minerals.usgs.gov/minerals.

―――. 1997. USGS, http://minerals.er.usgs/pubs/mcs/base.gif.

―――. 2002. http://minerals.usgs.gov/minerals/pubs/mcs/1996/.

For more information, consult our web page at ***http://www.wiley.com/college/cutter.***

STUDY QUESTIONS

1. What are three reasons for the shifting boundary between reserves and other resources?
2. While we usually think of minerals such as ores as nonrenewable resources, a broader view of all resources of a given element would see this substance as a renewable resource, in which all occurrences of that element are potentially usable. What would it take to make a given mineral, such as lead, aluminum, or iron, a renewable resource?
3. Some economists maintain that price is the best measure of the scarcity of resources. The prices of many minerals fell significantly between 1990 and 2000. Does that mean that they became more plentiful?
4. Why are mineral resources important to national security? What individual actions and governmental policies are used to strengthen the supplies of strategic and critical minerals?

CHAPTER 14

ENERGY RESOURCES

Of all demands placed on our world's natural resource base, the need for energy is perhaps the most far-reaching and basic. Energy provides heat for living and cooking, is used for lighting and refrigeration, and turns motors and wheels that power machines and move people and goods. Whether the energy is nuclear-generated electricity that lights an office building or is from wood burned in a cooking stove, everyone depends on it daily for physical health and economic prosperity.

Since the early 1970s, the world has experienced dramatic shifts in the availability and price of energy commodities, particularly petroleum, and widespread political tensions over their control. Because of supply disruptions such as embargoes and wars, temporary shortages of oil occurred repeatedly in the 1970s. This led to sharp increases in the prices of fuel oil, gasoline, chemicals, and other petroleum-derived products. During this period energy policy was a critical issue in many nations, and people came to view natural resources in a different way, seeing them as immediately exhaustible instead of indefinitely renewable. Since the early 1980s that situation has been dramatically reversed, at least on the surface, and we once again enjoy low energy prices. In 1990 a war was fought in Kuwait and Iraq over control of Kuwait's vast oil reserves. The region continues to be a battleground, with oil the focus of conflict. In the late 1990s, the real (inflation-adjusted) retail price of gasoline in the United States was lower than at any other time since World War II, and yet the underlying realities of a limited resource base and growing demand that precipitated the energy crises of the 1970s have not disappeared. Oil remains an integral part of international politics and world economies.

Energy exists in many different forms and is used for many purposes. It includes the kinetic energy of a speeding train or of the wind, the radiant energy of the sun or of a warm building, and the stored energy of water in a reservoir above a hydroelectric plant or in fossil fuels. This chapter is concerned primarily with those forms of energy that are used to do work in the production of goods and services and that form the basis of our national and world economies. It examines the complex nature of energy: where it comes from, how it is used, and how supplies and prices have varied in the past few decades.

ENERGY USE IN THE INDUSTRIAL AGE

The pattern of energy use in the past few hundred years shows a series of transitions from one fuel to another: from wood and other crude biofuels through coal, oil, and natural gas to nuclear and solar power. Each of these transitions was driven by a combination of two complementary forces: reduced supplies of a particular fuel (and thus higher prices) and the development of new technologies that use new energy sources. In some cases the new technologies were the main driving forces, independent of any restrictions in energy supplies. The first of these transitions, for example, was driven by the Industrial Revolution. Today we are seeing a movement away from direct use of fossil fuels toward use of electricity. This change is closely associated with the conversion from industrial- to service- and information-based economies.

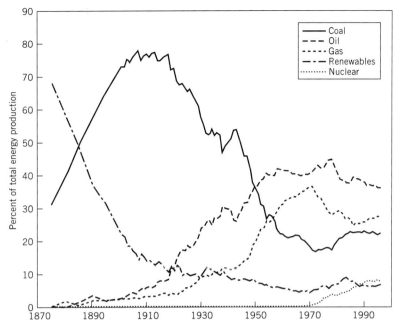

Figure 14.1 United States energy production, by source. In the late nineteenth century wood and coal supplied almost all United States energy. Today, a more diverse mixture is used, including oil, gas, and nuclear power, in addition to coal and wood. *Source:* EIA, 2002a

Wood, Coal, and the Industrial Revolution

Intensive energy use is a feature of the industrial age, which began in England and spread through northwestern Europe in the eighteenth century, North America and Japan in the nineteenth century, and the rest of the world in the twentieth century. Prior to the Industrial Revolution, the people of the world relied on renewable energy sources, primarily wood and charcoal but also dung, peat, and animal fats. But when modern steel-making developed, a new fuel was used: coal. Coal and steel use developed together—coal providing the energy source that made large-scale steel production possible, and steel providing the demand that led to the use of coal to power steam engines and factories. In the 1870s in the United States, three-fourths of energy was derived from renewable sources, primarily wood and mechanical hydropower. Thirty years later coal produced three-quarters of U.S. energy (Fig. 14.1).

In North America, as in Europe, the development of coal also came at a time when wood supplies were dwindling. The areas surrounding rapidly growing cities could not possibly supply all the wood needed to heat homes and cook food in those cities. The exploitation of coal thus grew through a combination of dwindling supply of an alternative fuel and development of a new technology that could make use of it.

Oil and the Internal Combustion Engine

Another transition in energy use took place in the early twentieth century. This time there was no shortage of the existing energy supply; rather, the development of a new technology was the dominant factor. The new energy source was petroleum, and the technology that led the way was the internal combustion engine.

Although oil development began in Europe and North America in the 1850s, relatively little was used. Oil was used mainly as a lubricant, with some used for lighting. In the 1860s and 1870s, however, the internal combustion engine was developed, and by the 1890s the first automobiles powered by this engine were in production. At first, automobiles were expensive and impractical, but by 1910 a large number of automobiles were in use, and society was adapting to this new form of transportation. Roads were improved for

Table 14.1 Energy Equivalents

One petajoule (10^{15} joules) is the same as 0.0009478 quads (10^{15} British thermal units)
One quadrillion British thermal units (quad) equal approximately:

- 50 million metric tons of coal
- 66 million metric tons of oven-dried hardwood
- 1 trillion cubic feet of dry natural gas
- 170 million barrels of crude oil
- 293 billion kilowatt-hours of electricity

automobile use, gasoline stations opened, and repair services became more available. When the Ford Motor Company adopted mass-production techniques that reduced the price of the automobile, demand and production increased further.

As the use of oil increased, the petroleum industry itself grew and developed. Geologists explored for oil, drilled wells, and built refineries. The development of an infrastructure of refineries, pipelines, and distribution facilities reduced the cost of oil, encouraging its use. This helped make oil competitive with coal for other uses, such as home heating or electric energy generation, further hastening the transition from a coal-based to oil-based economy. Today oil is the largest single source of energy in the world, accounting for almost 40 percent of total energy use.

Energy Use in the Late Twentieth Century

World energy use expanded rapidly in the second half of the twentieth century. In 1950, world consumption of energy was about 100,000 *quads* (Table 14.1); by the mid-1990s, it was approaching 400,000 (Fig. 14.2). The average growth rate between 1950 and 2000 was about 3 percent per year, faster than the population growth rate. This increase in energy use was driven by several factors, most important the expansion of the world industrial economy in the post–World War II boom. Petroleum was the major fuel supporting this growth, and the greatest increases in oil use occurred in the wealthy nations of Western Europe, North America, and East Asia. The automobile was at the center of this trend, but the growth was broad-based, including the residential and commercial sectors as well as industry and transport.

The growth has not been constant, however. Growth was most rapid in the 1950s and 1960s

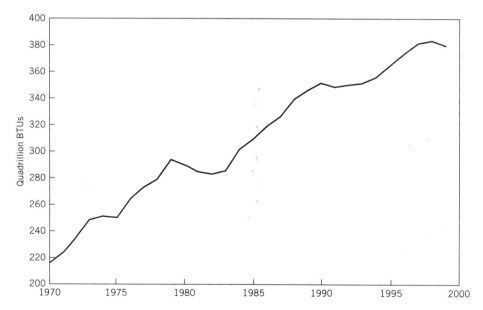

Figure 14.2 World energy production, 1970–2000. Energy production has increased substantially, but four distinct periods of decline are evident. The first two of these, the 1973 and 1979 energy crises, are associated with political tensions in the Middle East, while those in 1990–1992 and 1998–1999 are a result of economic contraction in the former Soviet Union and East Asia, respectively. *Source:* EIA, 2002b

and slowed significantly since then in four distinct episodes: 1973–1975, 1979–1982, 1990–1993, and 1997–1999. The first two of these episodes were periods of rapid increases in oil prices, precipitated by political events in the Middle East, and the second two resulted from regional economic declines (Fig. 14.3). The largest increase in oil prices came during (and were a partial cause of) a worldwide contraction in industrial production in the early 1980s, which further inhibited growth in energy use.

In the early 1970s, rapid growth in oil use increased demand for oil, much of which was derived from the Middle East. Oil production in countries such as Saudi Arabia, Iraq, Iran, Libya, and the Persian Gulf states was controlled by European and North American oil companies, which initally led and financed development of these resources. Nationalism was growing in the region, and many leaders in the area pressed for greater control of oil production and a greater share of the profits. In 1973 war broke out between Israel and its neighboring Arab countries, and Arab resentment of European and North American nations that were supportive of Israel increased further. For six months Arab oil-exporting nations refused to ship oil to Israel's supporters. Prices rose sharply on the world market, and consumption dropped in the United States.

A second energy crisis was precipitated by a revolution in Iran in 1979. The revolution caused another sharp increase in prices. But this time the world's industrial economies were more vulnerable. Many energy-intensive industries, especially the steel industry, were relatively inefficient, and there was excess productive capacity worldwide. When energy prices rose, the less efficient producers, especially those in older industrial regions such as the northeast United States and northern Europe, simply could not compete. Economic recession resulted, with a consequent decrease in energy use that lasted into the mid-1980s.

The third of the four episodes of low growth, the early 1990s, was a result of the economic collapse in eastern Europe that followed the breakup of the Soviet Union. That country and its Warsaw Pact allies in eastern Europe were heavy users of energy, especially coal, in their industrial sectors. Industrial production was controlled by the state, with only limited regard for profitability. But when those economies collapsed in the early 1990s, consumers could not afford to buy the industrial output of these countries, and factories shut down.

Then, in 1997, a financial crisis developed in several Asian economies, beginning in Thailand and spreading to many other countries in East Asia, notably Indonesia and Japan. Several rapidly industrializing countries, including Thailand, Malaysia, Indonesia, and the Phillipines, had currencies and stock markets that were significantly overvalued. Those markets collapsed in 1997, and the economic repercussions spread throughout much of East Asia, except China. The economic slowdown was dramatic enough to cause a reduction in world total

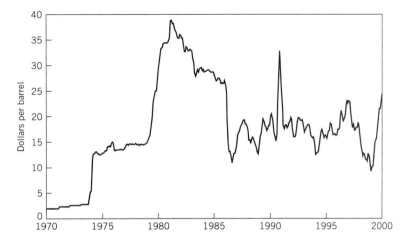

Figure 14.3 World oil prices, 1970–2000. The dramatic rises associated with the 1973 and 1979 energy crises are evident. The brief spike in oil prices in 1990 is a result of the Gulf War. *Source:* EIA, 2002a.

energy consumption. More recently, the contraction in the U.S. economy in late 2001, combined with the dramatic decrease in airline traffic following the September 11th attacks, contributed to a significant drop in oil consumption in the United States.

While world energy statistics are dominated by trends in petroleum use, the late twentieth century was a period of rapid growth in use of electricity worldwide, and this growth has continued into the twenty-first century. As electric production and distribution networks expand, so does demand for electricity. Much of the growth in electric energy demand has been met by expansion of nuclear and coal-fired electric generation. As we look ahead to the next few decades, we can see trends that will increase our use of electricity (the growing interest in electric automobiles, for example) as well as the potential development of new energy supplies, such as photoelectric cells.

The history of energy use is thus one of shifting from one energy source to another. The shifts are driven by two main factors: the development of technology that can use a given energy source and the relative availability of one energy source over another. The history of recent trends in energy use shows that technological developments and the social and economic changes that go with them are usually much more significant than the availability of natural energy resources in determining what fuel we use and how much.

Current energy consumption in the United States is dominated by the industrial (39 percent) and transportation (27 percent) sectors (Fig. 14.4). The preferred source of energy remains oil (primarily for transportation), followed by natural gas and coal. Regionally, nuclear sources are more dominant in the northeastern states, whereas hydro sources are more pronounced in the west. Fossil fuel use is fairly constant across all regions of the United States. Coal is the dominant fuel in the eastern north-central, southern north-central, and mountain states.

ENERGY SOURCES

Oil and Natural Gas

Petroleum is the most important fuel in the world today, supplying about 36 percent of total commercial energy production (World Resources Institute 2000). Most of the world's oil use is in the

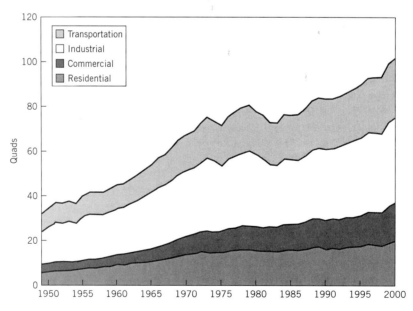

Figure 14.4 U.S. energy consumption trends from 1949 to 2000. Energy use in industry dropped dramatically in the mid-1980s as a consequence of increasing energy prices and economic recession. Use in transportation and residential and commercial sectors has stabilized, with slight increases since the mid-1980s. Industry, however, remains the dominant user of energy in the United States. Source: EIA, 2002a.

transport sector as fuel for internal combustion engines. This is especially true in the United States, where over 60 percent of oil use is in the form of gasoline and diesel fuel for automobiles and trucks and in aviation fuels. Industrial uses of oil are also important, mostly for heat production but also for petrochemicals. In the United States relatively little oil is used in electricity production, but elsewhere in the world this is an important use of oil.

Oil and natural gas are produced by the accumulation of organic matter in sedimentary rocks and the later alteration of that organic matter by the heat of burial in the earth's crust. Both oil and gas are composed of hydrocarbons, but they differ in their boiling temperatures. Oil is liquid at normal air temperatures, but gas is not. Oil and gas usually occur together, but sometimes they can become separated if gas migrates through a rock that is not permeable enough for oil to flow. *Crude oil* may contain natural gas in solution, and gas sometimes contains liquid petroleum. Because they coexist and are readily substitutable in most end uses, we discuss them together.

Sediments rich in organic matter are relatively common, but the geologic circumstances necessary to accumulate oil or gas in commercially useful quantities are not. For this to occur there must be a reservoir rock that is permeable enough for oil to flow through. Oil is less dense than water, and it is forced upward by density differences as well as pressures in the earth's crust. Another requirement is a trap, in which a reservoir rock is overlain by an impermeable rock that prevents the oil from escaping to the surface. Many different kinds of traps exist, and exploration geologists usually look for such structures in their initial searches for oil. Accumulations of oil, or fields, vary considerably in size, with the number of fields inversely related to their size. There are very few giant fields. The largest known field is the Ghawar in Saudi Arabia (about 75 billion barrels). Oil also comes in many different forms, and the quality of crude oil depends on the mix of different hydrocarbons in the oil. Crude oil with a high proportion of high-boiling-point hydrocarbons is called heavy oil. Tar and asphalt are major components of heavy oil. Light oil is rich in hydrocarbons with low boiling points, such as naphtha and kerosene. The proportion of various hydrocarbons affects both the ease of extraction and the price of the crude oil. Natural gas is made up of several different gases, including methane, propane, and butane.

Production and Consumption Oil-bearing formations are found all over the world, but the largest deposits are concentrated in a few areas. The 10 countries with the largest proven recoverable reserves are listed in Table 14.2. Most of the

Table 14.2 Known Recoverable Reserves of Fossil Fuels: Top 10 Nations

Coal: World total, 1,083,259 million metric tons		Oil: World total, 1028.1[a] billion barrels		Natural Gas: World total, 5288.5[a] trillion cubic feet	
State	% of World Total	State	% of World Total	State	% of World Total
United States	25	Saudi Arabia	26	Russia	32
Russia	16	Iraq	11	Iran	16
China	12	United Arab Emirates	10	Qatar	7
Australia	9	Kuwait	10	United Arab Emirates	4
India	8	Iran	9	Saudi Arabia	4
Germany	7	Venezuela	7	United States	3
South Africa	6	Russia	5	Algeria	3
Ukraine	4	Libya	3	Venezuela	3
Kazakhstan	3	Mexico	3	Nigeria	2
Poland	2	China	2	Iraq	2
All others	10	All others	15	All others	24

[a]Based on estimates from *Oil and Gas Journal*.
Source: EIA, 2002b.

world's reserves are located in the Persian Gulf area, with Saudi Arabia, Iraq, Kuwait, Iran, and the United Arab Emirates holding about 65 percent of world reserves. This concentration of deposits of the world's primary energy source in a few countries is a major source of world tension, with the United States particularly concerned because of its dependence on imported oil.

Global patterns of oil production and consumption changed dramatically over the past several decades and were the focus of many key economic and political issues. The United States was self-sufficient in oil for most of the first half of the twentieth century. There were shortages during World War I, and during World War II gasoline was rationed. Throughout this period, the real price of oil in the United States (adjusted for inflation) steadily declined, reaching a low in the late 1960s. By the mid-1950s, domestic oil began to get harder to find, and exploration in other regions of the world picked up. Domestic exploration, as measured by the number of wells drilled, peaked in 1956 and has not reached that level since, although by other measures exploration in 1981 was higher than ever before. American oil companies in the 1950s and 1960s placed more emphasis on exploration abroad, particularly in South America and the Middle East. Those areas held large untapped fields that could be developed more cheaply than the smaller fields in the United States. Even considering transportation costs, this oil was cheaper than domestic oil.

As a result of this exploitation of foreign oil, in the early 1970s the United States found itself a major importer of oil, with imports amounting to about 25 percent of consumption. Domestic oil production peaked in 1970, and natural gas production peaked in 1974. Domestic demand continued to climb, and the difference was made up by imports. In 1981, 53 percent of our oil consumption was supplied by domestic production, with 47 percent supplied by imports (Fig. 14.5). However, the dramatic price increases that accompanied the oil shortages of the 1970s stimulated conservation, and by 1981 demand was reduced. This was partly a result of the cumulative effect of conservation measures such as reduction in size of automobiles, but it was also caused by the economic recession of 1981–1983. At the same time, domestic production was spurred by high prices, and supply increased. Similar trends occurred in other countries, and a worldwide surplus developed,

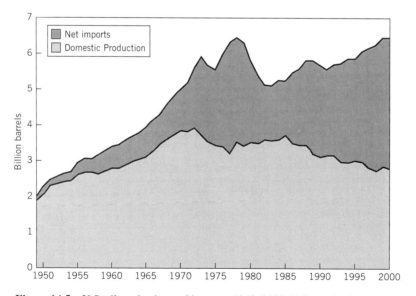

Figure 14.5 U.S. oil production and imports, 1949–2000. U.S. production peaked in 1972, rose slightly in the late 1970s and early 1980s as a result of high prices in that period, and has declined steadily since 1985. Low oil prices since the early 1980s have stimulated increased demand; this increase is met entirely by increases in imports. *Source:* EIA, 2002a.

with inventories at high levels and demand low. Oil prices peaked in 1981 and declined throughout most of the next two decades. The decline in oil prices in the 1980s and 1990s encouraged increasing consumption, and by the mid-1990s oil use (especially gasoline) in the United States had returned to mid-1970s levels (Fig. 14.6). In 2001 imports supplied about 47 percent of the United States' consumption (EIA 2002a).

One of the lasting consequences of the oil price-swings of the 1970s and 1980s was a desire on the part of importing and exporting nations alike to maintain relatively stable oil prices. Oil-exporting nations became fabulously wealthy almost overnight when prices rose in 1973 and 1979, but just as quickly they found themselves dependent on the revenues their oil exports brought. When prices sagged in the 1980s (primarily because of recession and conservation triggered by oil price increases), these countries found it difficult to meet the financial commitments they had made based on expected income from high-priced oil. The result was that they were forced to sell more oil, and OPEC's power was diluted.

When terrorists struck the United States on September 11, 2001, many consumers feared a new energy crisis and rushed out to buy gasoline. Lines formed at gas stations and retail prices spiked in some areas, if only for a few hours. Soon it was apparent that these events would encourage a drop in oil prices rather than an increase, as U.S. energy consumption fell with a weakening economy and oil exporters friendly to the United States (particularly Saudi Arabia) promised to maintain production. Even if Saudi Arabia curtailed production or refused to export oil to the United States, the effect on oil prices would probably not be a lasting one. In the short term prices would rise and other exporters would rush to fill the gap, and in the long term energy conservation and substitution with alternative energy sources would lead to reduced imports and resulting downward pressure on oil prices. Clearly it is in the long-term interests of oil exporting nations, as well as U.S. oil companies, to prevent this from happening. Their incomes depend on the United States and other consuming nations maintaining their high rates of oil consumption. Price increases undermine U.S. dependency on oil.

Oil Futures The oil shortages of the 1970s, like those of earlier years, generated much discussion about how much oil is available and how long it is likely to be before we run out. First, it should be recognized that, as with nonfuel minerals, we will never run out of oil. Rather, it will become so expensive that we will replace it with other sources, or we will reduce our demand for it. Nonetheless, it is important to ask just when oil and natural gas are going to become so hard to find

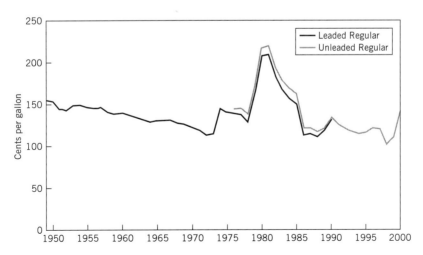

Figure 14.6 Retail gasoline prices in the United States, adjusted for inflation, 1949–2000. In the late 1990s gasoline was cheaper than it had ever been before. *Source:* EIA, 2002a.

and recover that the price must rise substantially, for at that time we will effectively cease to use these fuels for all but the most essential uses.

We can use two basic approaches to estimate how much oil we are likely to produce in the future: the geologic estimate and the performance-based estimate. The *geologic estimate* examines how much oil is in the ground in technically recoverable quantities, without regard to the larger economic forces affecting exploration and production. Most geologic estimates classify sedimentary basins with regard to petroleum-producing potential and use the well-known basins to extrapolate to the quantities likely to exist in poorly known areas. Thus, these are estimates of how much oil (or gas) is likely to be in an area, without regard to whether we have the desire or ability (other than technical ability) to get it out.

Geologic estimates varied greatly over the years, generally increasing with time. In the 1940s, for example, ultimately recoverable oil resources were estimated at about 125 billion tons. Recent world estimates are generally in the range of 300 to 700 billion metric tons, with most of them clustering nearer the low end of that range. About 125 billion metric tons had already been used as of the mid-1990s. In 2000 the world's proven recoverable reserves of oil were about 135 billion tons, or 42 years' supply at 2000 consumption rates (EIA 2002b). If these numbers are correct, and if oil consumption were to continue to grow as expected in the next few decades, then we would exhaust these supplies by the middle of the twenty-first century.

But such projections are based on many assumptions, not the least of which are the extrapolations of present conditions into the future. We know very well that conditions change. Consumption rates will change, and new deposits will probably be found. Perhaps more important, both our use of oil and the amount we are willing to find and extract from the ground depend on our desire to have that oil, which is measured ultimately by its price.

Performance-based estimates of oil resources focus on the economic factors that drive oil production and consumption. They examine the pattern of exploration and production through time, considering all the economic forces that determine whether or not an oil company supplies oil and whether or not a consumer buys it. This technique was pioneered by M. K. Hubbert, who argued that for any stock resource there is a pattern of exponential growth in use of that resource, until the amount used begins to approach the approximate amount in the earth. Then production will level off and decline exponentially (Fig. 14.7). The decline is related as much to the fact that substitute commodities are found and adopted as it is to actual difficulty in obtaining supplies. Performance-based estimates have the advantage of being able to consider the true economic factors affecting oil production but have the disadvantage of being subject to questionable assumptions about future economic conditions. Most performance-based estimates of remaining oil suggest that world oil production will peak sometime in the second or third decade of the twenty-first century (World Resources Institute 1996), but others argue it may be sooner than that (Deffeyes 2001; Campbell and Laherrere 1998). In the United States, production peaked in 1970 and will probably dwindle to a small fraction of consumption in the next decade.

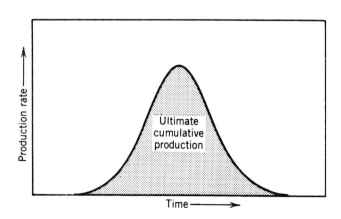

Figure 14.7 Performance-based estimates of oil recoverability. In oil production, a pattern of exponential growth first occurs. Production levels off when the amount used approaches the amount naturally found in the earth. This is followed by an exponential decline in production and use. *Source:* Hubbert, 1969.

As we ponder the future use of oil in comparison to other energy sources, we should also consider the environmental impacts of this activity. The most significant environmental concerns involve the transportation, refining, and burning of oil and natural gas. Because of the geographic locations of the major world producers and consumers, crude oil must be moved by ship (supertanker, tanker, or barge), pipeline, tank truck, or tank car. This is predominantly high-seas traffic, with shipments of oil from the producing nations in the Middle East to European, Japanese, and North American markets. As a result of shipping these massive amounts so frequently, accidents are likely to happen—and they do. Pollution of the oceans from tanker accidents has both local and global impacts (see Chapter 9). Another environmental concern is the refining and burning of petroleum, which releases hydrocarbons and carbon dioxide to the atmosphere. This was discussed in Chapter 12.

A number of land-based problems are involved in petroleum exploration, production, and natural gas storage. The withdrawal of both oil and natural gas has caused ground subsidence in such places as Long Beach, California, and Houston, Texas. Another problem with natural gas involves the transportation of *liquid natural gas (LNG)* by tanker and its storage and support at land-based facilities. Large volumes of volatile gas are vulnerable to fires, with potentially disastrous consequences. Fires have occurred, but they were less devastating than feared. However, the possibility of monumental conflagrations causes concern over the location of LNG facilities near urban areas. Finally, conflicts between exploration and drilling and preservation of pristine natural areas, especially in the United States, will continue (see Chapter 3).

Coal

Coal is the most abundant fossil fuel in the world, with reserves far exceeding those of oil or natural gas. The United States is particularly well supplied with coal, with one-fourth of the world's reserves. Coal is also a dirty fuel, and the environmental impacts associated with its extraction and combustion are a matter of considerable concern.

Coal is the partially decomposed and consolidated remains of plants that were deposited in ancient swamps and lagoons. The original material was modified by heat and the weight of overlying materials from the original plant matter to a substance that is much harder, drier, and chemically different. The degree to which modification of coal occurred varies greatly from one deposit to another, so that there are several different kinds, or ranks, of coal. The least modified form is *peat*. It has a very high moisture content and is being deposited in many areas today. In order, the remaining ranks of coal are *lignite, subbituminous, bituminous*, and finally *anthracite*, which is the most completely converted rank. Typical moisture contents vary from over 40 percent for lignite to less than 10 percent for anthracite. As moisture content is reduced and hydrogen is lost, the heat content per unit weight increases. In addition to rank, ash and sulfur contents are important to the value of coal deposits. Ash is derived primarily from mineral sediments deposited along with the plants. In some areas, ash content is very high, but generally coal with ash greater than 15 percent is uneconomic. Sulfur accumulates in most sediments deposited in swampy conditions, and in coal sulfur contents typically vary from less than 1 percent to more than 3 percent.

Production and Consumption Unlike oil and natural gas, there is so much coal in the world that there is little concern about remaining resources. The largest deposits of coal are found in the United States, Russia, China, and central Europe, though significant deposits also occur in South Africa, India, and elsewhere. Table 14.3 includes estimates of the ratio of reserves to production (R/P) for the top 10 coal-using countries. This ratio expresses the number of years a country's coal reserves would last at present rates of use. Nearly all the world's major coal users have abundant resources. Because of high transportation costs and the ease with which other fuels can be used, only those countries that have abundant coal are likely to use it extensively. Nonetheless, in those countries, R/P ratios frequently are in hundreds of years, and the ratio for the world as a whole is over 200 years. Thus, there is plenty of coal to meet both the United States' and the world's demands for at least several decades, if not hundreds of years, even if we consider increasing per capita energy consumption and conversion from oil and natural gas to coal for some uses.

The United States has the largest share of coal reserves, with more than one-third of the world's

Table 14.3 The World's Top 10 Solid Fuel Users

Country	A: Energy Production from Coal, 2000 (Quadrillion BTU)	B: Amount of Coal Required to Produce Amount of Energy in Column A (Million Short Tons)	C: Proven Recoverable Coal Reserves (All Types) (Million Short Tons)	D: Reserves to Production (R/P) Ratio (Years)
China	24.33	1338	126,215	94
United States	22.62	1244	273,656	220
South Africa	6.95	382	54,586	143
Australia	6.63	365	90,489	248
India	6.10	336	93,031	277
Russia	5.18	285	173,074	607
Poland	2.85	157	24,427	156
Germany	2.37	130	72,753	560
North Korea	2.36	130	661	5
Indonesia	1.72	95	5919	62
WORLD	89.70	4823.5	1,083,259	225

Source: EIA, 2002b.

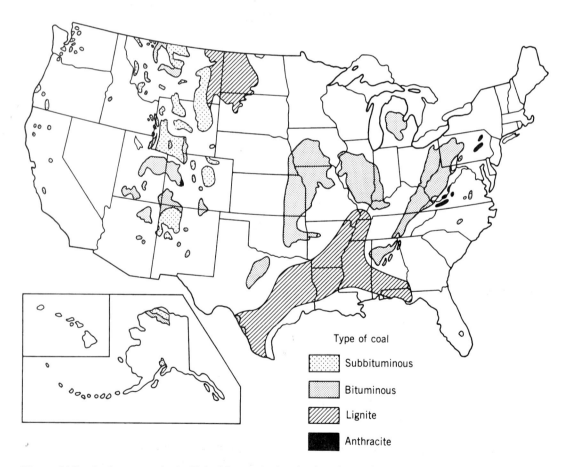

Figure 14.8 Coal resources in the United States. Anthracite deposits are found in eastern Pennsylvania and the Appalachian region of Virginia and West Virginia. Bituminous deposits are found throughout the United States, whereas lignite deposits are largely confined to the West. *Source:* Council on Environmental Quality, 1989.

proven recoverable reserves. Bituminous coal is the most abundant type of coal, and this is in the Appalachian Mountains in the East and in sections of the Midwest, particularly Illinois, Iowa, and Missouri (Fig. 14.8). Bituminous deposits are also found throughout the Rocky Mountains and in northern Alaska. Anthracite deposits are quite localized and are found in eastern Pennsylvania and in the Appalachian regions of West Virginia and Virginia. Subbituminous deposits are found throughout the Rocky Mountain region. Lignite deposits are found in North and South Dakota and Montana.

China and the United States together account for about half the world's coal use; Europe is the other major coal-using region. Most of the world's coal is used for industrial purposes, especially electric generation. In China coal is not only the primary industrial fuel but also widely used for space heating and cooking. In Europe and the United States little coal is used today for home heating, but it still is used in industry, primarily in producing iron and steel. In steel-making, coal is first converted to coke by driving off water and volatile matter through a distillation process, and then it is burned in blast furnaces.

Since the energy crises of the 1970s, U.S. coal production has risen by about 60 percent over 1970 levels. In 1900, coal contributed about 70 percent of U.S. energy supplies, but today its percentage is much lower. Its contribution reached a low of 17 percent in 1971 and has risen since then, contributing about 23 percent of total U.S. energy consumption and about 52 percent of U.S. electricity generation today.

Extraction and Environmental Impacts

Coal is mined in three ways: underground, in surface strips, and with augurs. In the nineteenth and early twentieth centuries, most coal mining took place in the eastern United States, where most of the coal seams are in hilly areas and many are steeply inclined relative to the surface. In such situations most of the coal is well below the surface, and underground mining is necessary. *Underground mining* involves drilling, blasting, or otherwise excavating tunnels and chambers underground, from which coal is removed and transported to the surface. Much of the coal must be left behind to support overlying rocks, and generally no more than 50 percent of the coal is removed. Underground mining is the most dangerous form of mining. Indeed, it is among the most dangerous occupations in the world. The greatest hazards are from explosions of methane gas, which is an important component of many coal deposits, and from rock bursts that result from removal of pressure on rocks deep underground.

Strip-mining or *surface mining* is conducted in areas where the coal is near the surface (Fig. 14.9). The *overburden*, or rock and soil overlying the coal, is first removed, and then the coal is taken out. In areas of relatively flat terrain where the coal seams are horizontal, strip-mining removes about 90 percent of the coal from a large area and is thus much more efficient than underground mining. In hilly areas, coal outcrops along hillsides are sometimes mined by stripping, leaving deeper coal unmined. Alternatively, coal in such areas may be mined by *auguring*, in which a large augur, or drill, bores into a coal seam, removing the coal in the process.

Until recently, most coal mined in the United States was mined underground in the East. In the past three decades, strip-mining has grown substantially, especially in the West. In 1974, surface mining surpassed underground mining, and as of 2000 about 64 percent of U.S. coal was surface-mined (EIA 2002a).

Coal mining and combustion are the most destructive methods in use today for obtaining energy. The environmental impacts are many and far-reaching. Underground mining has two major environmental effects: acid drainage and subsidence. *Acid drainage* results from air and water coming in contact with sulfur-bearing rocks and coal. The sulfur is oxidized to sulfuric acid, and groundwater flow carries this acid to streams. This is a particularly severe problem in mining regions of the Appalachians, where some streams are too acidic to support fish life. Acidity of streams also affects the solution of metals in water, making it less suitable for human consumption. *Subsidence* is the sinking of the land as underground voids collapse. It results in structural damage to buildings overlying mined areas, and it is widespread in Pennsylvania and other underground mining areas.

Strip-mining is generally much more disruptive of the land than underground mining. Overburden must be removed; thus, the soil and topography of the area underlain by coal are completely altered in

Figure 14.9 Area strip-mining of coal in Missouri. The active face is along the left; areas to the right are being graded prior to reclamation.

the mining process. While the mine is in operation and until the land is reclaimed, the overburden is exposed to air and rain, resulting in accelerated runoff and oxidation of newly exposed rocks. Runoff from mined areas results in increased sediment loads of streams, oxidation of sulfur-rich rocks leads to the formation of sulfuric acid, and runoff water from these areas is usually very acidic. These problems are particularly severe in areas with steep slopes, where runoff and erosion are more rapid. In addition, steeply sloping spoils piles are sometimes prone to landsliding.

In 1977, Congress passed the Surface Mining Reclamation and Control Act, which requires reclamation on surface-mined lands. This includes removal of topsoil separately from lower overburden layers. When mining is completed, the overburden is replaced with topsoil above it, graded to its approximate original contour, and replanted with vegetation similar to that present before mining. The Act also places a tax on surface mining activities that is used to pay for reclamation of areas mined and abandoned before 1977.

Although mining impacts are locally significant, air pollution from coal combustion has much more far-reaching impacts. Concerns about acid deposition increased demands for low-sulfur coal. Of U.S. coal reserves, about 20 percent is high-sulfur coal, 15 percent is medium-sulfur coal, and 65 percent is low-sulfur coal. Eastern coal is generally higher in sulfur than that west of the Mississippi River, which is unfortunate because the East is generally more susceptible to problems of acid precipitation than the West.

Carbon dioxide emissions, on the other hand, are not as controllable. CO_2 is produced as the end product of efficient combustion, and there is no known way to prevent this emission. As discussed in Chapter 12, coal produces proportionately more CO_2 than oil and natural gas, and if coal combustion is increased, then CO_2 emissions will increase. The nature of the environmental impact of those emissions is still open to question, but if we intend to prevent CO_2 buildup in the atmosphere, then the only alternative is to reduce coal combustion.

Other Fossil Fuels

In addition to oil, natural gas, and solid coal, vast deposits of potentially usable fossil fuels exist. The most significant of these fuels are tar sands and oil shale. These rocks can be mined, and burnable hydrocarbons can be extracted from them. Coal also can be converted to a gaseous form for easier transport and combustion. Fuels manufactured from these deposits are known as synthetic fuels, or *synfuels*. In some cases, the technology to convert these substances to usable fuels has been available for several decades, and in other cases it is yet to be commercially developed.

Coal gasification or *liquefaction* means the conversion of coal to a gas or liquid that can be transported via pipeline and burned much as we burn conventional fuels today. The most common gasification processes make gas that has a lower heat content than conventional fuels, though it can be further processed into high-quality gas and liquid fuels. The technology of coal gasification and liquefaction is fairly well known, but as yet, it is not economically competitive with conventional oil and gas.

Tar sands are deposits of sand that are high in heavy oil, or tar content. This tar is too viscous to be pumped from the ground as oil is, but if it is heated it will liquefy and can then be pumped and refined much as heavy crude oil is refined. In Alberta, Canada, extensive areas of tar sands near or at the surface are being commercially mined today.

Shale oil is not true oil; rather, it is a waxy hydrocarbon called *kerogen* that is found in shale, a fine-textured sedimentary rock. Shale oil is extracted, or retorted, by first crushing and then heating the rock. This liquefies the kerogen, which seeps out of the rock and can then be piped away.

One of the most attractive aspects of synfuels is that the raw materials are available in vast quantities. But despite the vast amounts of fuel present in these deposits, it is unlikely that any of it will be exploited in the near future because of the environmental impacts of extraction. All the impacts associated with mining coal also apply to coal gasification—unless the conversion is made underground. Gasified or liquefied coal probably would be cleaned of most sulfur in processing, and so sulfur emissions are not likely to be a major barrier to the use of these fuels. However, use of coal in any form, particularly one for which energy is lost in conversion, means that there will be substantial emissions of carbon dioxide.

Shale oil also has significant problems. First, to obtain a substantial amount of oil, large areas must be mined. If strip-mining is used, then all the problems associated with strip-mining of other minerals must be considered. Even with *in situ* retorting, about 25 percent of the rock's volume must be removed to make room for retorting. Second, retorting involves burning kerogen under conditions that are not conducive to clean combustion. In the United States the areas where retorting would take place are areas of good air quality, and retorting would almost certainly cause some deterioration.

In addition to these environmental problems, economic conditions have not allowed exploitation of this resource. Commercial interest began during oil shortages in the 1920s, but to date there has been no successful commercial-scale extraction of synfuels in the United States. This is due to the relatively low price of conventional fuels, uncertainties about the extraction technology, and, more recently, environmental concerns. It is more likely that clean, renewable energy sources will be developed instead of synfuels.

Nuclear Power

The first self-sustaining nuclear reaction took place in December 1942 at the University of Chicago. Although the first applications of nuclear energy were in weapons, atomic power also was used for peaceful purposes after World War II. The dichotomy between weapons applications and energy production still exists today, although we discuss only the commercial aspects of nuclear energy.

The first commercial use of nuclear power was to generate electricity in 1957 in Shippingport,

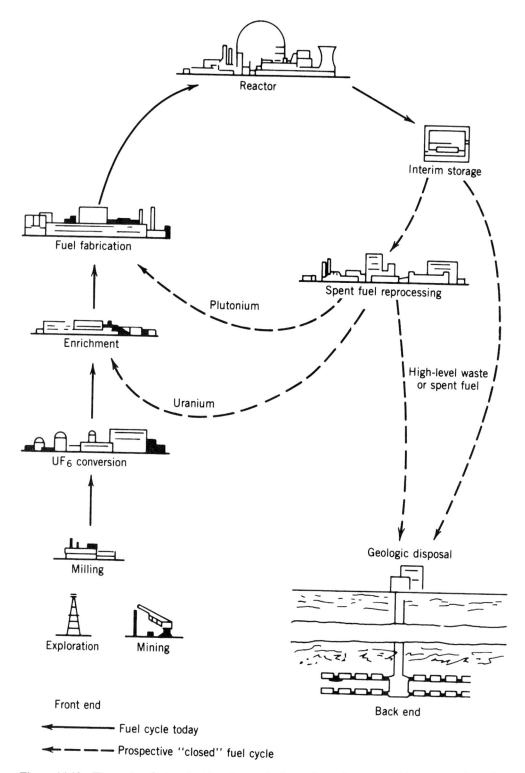

Figure 14.10 The nuclear fuel cycle. At each stage in the cycle, nuclear material is transported varying distances from mining regions to conversion and enrichment facilities, fabrication plants, and nuclear reactors. After use, spent fuel is reprocessed and/or disposed of.

Pennsylvania. Westinghouse Electric, in conjunction with the Atomic Energy Commission (now the Nuclear Regulatory Commission), opened the first full-scale nuclear electrical power plant to be operated by a public utility. At 60 mW, the plant was small by today's standards. Nuclear generation of electricity has grown steadily since then, and in 2000 nuclear power supplied about 23 percent of U.S. electricity and 8 percent of total energy consumption (Energy Information Administration 2002a). On a worldwide basis, nuclear energy production accounted for 7 percent of our total commercial energy production in 1999 (U.S. Bureau of Census 2001).

Nuclear power is based on the *fission* process, in which the nuclei of heavy atoms of enriched uranium-235 (^{235}U) or plutonium-239 (^{239}Pu) (the latter a byproduct of the fission process) are split into lighter elements, thereby releasing energy in a chain reaction. The energy released is used to heat water into steam. The steam is then used to drive a turbine, which turns an electric generator.

Nuclear Fuel Cycle The nuclear fuel cycle consists of eight stages (Fig. 14.10). Unlike some of the more conventional fuel sources, which can be used with a minimum of processing, nuclear power requires several processing steps, with transportation linkages between them. Uranium is the primary fuel for nuclear power plants. Most U.S. uranium resources are found in the Rocky Mountain states and are mined by both open-pit and underground techniques. Nearly two-thirds of the uranium is mined in New Mexico and Wyoming. Of the naturally occurring uranium, which is a mixture of the isotopes ^{235}U and ^{238}U, less than 1 percent, that is, only the ^{235}U, is highly fissionable. Thus, most of the uranium ore must be enriched or converted to make it sustain the fission process. Once mined, the uranium ore is milled to produce a purer concentrate called yellowcake. Yellowcake is about 85 percent natural uranium oxide (U_3O_8).

The third stage in the fuel cycle is the chemical purification and conversion of the yellowcake to uranium hexafluoride (UF_6). The conversion process prepares the uranium for enrichment, the next stage in the fuel cycle. At the enrichment plant the concentration of ^{235}U is increased from 0.7 percent to about 4 percent, to meet the requirements of the reactors. Once the UF_6 has been enriched, it is ready to be made into fuel rods. The UF_6 is converted into uranium dioxide (UO_2), which is then formed into small pellets and placed in alloy tubes. These tubes are then made into fuel rods and assembled into bundles, called fuel rod assemblies. The fuel rod assemblies are then shipped to individual reactors, where they are used to produce electricity.

Several different types of reactors are in use around the world. They differ in the medium used to extract heat from the reactor core and in the means for moderating the reaction. Most of the reactors in use in the United States are *light-water reactors* (Fig. 14.11), which means that ordinary water is used as the cooling agent. In some reactors, the water that circulates through the reactor vessel is used to generate steam directly, while in others pressurized water absorbs heat in the reactor, and this energy is transferred to the turbines via a separate loop that does not pass through the reactor itself. In a *high-temperature gas-cooled reactor*, helium gas is used as the coolant, transferring the heat from the core to the steam generator. *Heavy-water reactors* use water containing a higher-than-usual proportion of deuterium as the moderator of the fission process. Regular water passes through the core and carries heat to the secondary water/steam loop. Heavy-water reactors are not used in the United States, but they dominate the Canadian reactor program. Canada is also a large exporter of this technology and of the heavy water that is used in it. The reactor at Chernobyl, Ukraine, that exploded in 1986 was of a type that uses graphite to moderate the reaction.

One additional type of reactor bears mention because it is capable of creating more fissionable fuel than it uses. This is the *liquid metal fast-breeder reactor*, in which fissionable material is surrounded by nonfissionable material (^{238}U) in the core. Sodium is used as the moderating substance and heat exchanger. During the fission process, some of the nonfissionable material is converted to fissionable ^{239}Pu. The reactor produces more fuel than it consumes, hence the name *breeder reactor*. There are no breeder reactors in the United States, although they have been used in other countries for some time. Breeder reactors are a central feature of France's nuclear energy program, for example.

Spent fuel is removed from the reactor core and stored on site for several months to permit the

Figure 14.11 Power plant using a pressurized water reactor, Three Mile Island, Pennsylvania. The reactor experienced a loss-of-coolant accident in 1979.

levels of radioactivity to decline. Optimally, the spent fuel is then shipped to a reprocessing plant, where the unused portions of uranium and plutonium are separated from the fission wastes. The unused uranium is recycled back to the enrichment plant, and the unused plutonium is refabricated into new fuel pellets.

The last stage in the nuclear fuel cycle is disposal of the radioactive waste. *Low-level wastes* (only half of which are generated by nuclear power plants) are buried in metal or concrete containers. In 1998, for example, 800,000 cubic feet of low-level waste was buried in three locations; Barnwell, South Carolina; Hanford, Washington; and Clive, Utah, although not all of this was generated by nuclear power plants. A permanent disposal facility for 70,000 metric tons of high-level waste is being developed at Yucca Mountain, Nevada. Currently, all high-level waste (38,000 metric tons in 1998) is stored in pools at the reactor sites. The repository project has been plagued with technical and management problems, however, and it is unlikely that a permanent high-level waste storage facility will be available in the United States before about 2010 (Flynn et al. 1997; Ewing and Macfarlane 2002). Yet, estimates suggest that by 2003, these commercial reactors will produce an additional 48,000 metric tons of high-level waste.

The primary concern about the use of nuclear energy is the potential for human exposures to radioactivity, resulting in cancer, birth defects, and other health problems. Accidental releases of radioactivity occur at all stages in the fuel cycle, from the mining of uranium, which produces radon gas, to occupational exposures in fabricating fuel rods, to accidental releases at nuclear power plants. Releases of radioactivity can also occur as a result of sabotage and simple mishandling of materials. There is considerable debate on the effects of low levels of exposure to ionizing radiation.

There are also other environmental considerations at each stage in the fuel cycle. Land disturbance and radioactive mine tailings are the primary impacts of mining and milling. In the production of power, waste heat is produced, causing thermal pollution. This occurs with any type of steam-powered electricity plant. The most important environmental aspects of nuclear power production involve the accidental release of airborne or water-borne radioactivity. For example, in the Chernobyl incident (Issue 14.1), radioactive fallout was spread over much of Europe, resulting in levels of contamination high enough to prevent the consumption of numerous food products produced in contaminated areas for at least several years following the accident (Gould 1991; Hohenemser 1996; Hohenemser et al. 1986; Hohenemser and Renn 1988). These safety issues cause great public concern about large-scale use of nuclear energy. The technology is hazardous, and the future extent of nuclear energy production will be a function of the risks that society is willing to tolerate in exchange for additional electric power.

Power Production Trends Nuclear power is used only to generate electricity. In the United States, production of electricity using nuclear power increased substantially during the 1980s as plants under construction since the 1970s were completed and began operation. This growth in the United States nuclear industry ended in the 1980s, however, because nuclear power became much more expensive than other heat sources used to generate electricity and because the 1979 accident at Three Mile Island hindered public acceptance of nuclear power. The 1986 accident at Chernobyl (in the Ukraine) reconfirmed the public's fear about the safety of nuclear power plants. In 2001, there were 441 commercial nuclear units operating worldwide, producing 17 percent of the world's electricity; 103 of these were in the United States, where they produced about 19 percent of our electricity.

Because very few additional plants will be operating, United States' generation capacity will not increase substantially in the next decade. Within the next 20 years, nearly half of the commercial nuclear power plants will reach the end of their operating licenses and be decommissioned, resulting in a decline in generating capacity. Worldwide nuclear capacity is expected to increase by about 5 percent in the next few years. It remains to be seen whether confidence in nuclear energy will be restored and growth will continue or whether we already have reached the peak of nuclear energy generation. If nuclear energy is to become viable again, it will require the development of new, safer technology to restore public confidence in this energy source. Otherwise, the high costs of overcoming political opposition and meeting safety requirements will continue to make construction of nuclear plants prohibitively expensive.

In other parts of the world similar problems have arisen, though not everywhere. In much of western Europe, public concern about reactor safety increased dramatically after the Chernobyl incident. In 1989, the British government acknowledged that its nuclear industry (which now generates about 25 percent of that nation's electricity) was not economical in comparison to conventional generation technology and abandoned plans for any additional power plants. France, however, which has had a very aggressive nuclear program since the 1950s, generates 75 percent of its electricity from 56 nuclear power reactors. France has remained firm in its commitment to this energy source.

Fuel availability is not a major deterrent to the future use of nuclear power. Public concern over safety, however, poses the biggest challenge. On the other hand, public concern over global climate change may force us to reconsider the nuclear option. Coal is the primary alternative to nuclear power in many U.S. regions. Nuclear energy is seen as relatively clean from an environmental perspective, and it offers the potential to increase electric generation capacity without contributing to global warming or acid rain, although long-term waste storage remains a problem. As we saw earlier, almost 40 percent of the energy demand in the United States is in the transportation sector, which cannot be supplied by nuclear sources at the current time.

Renewable Energy

Several renewable sources of energy are available today, and many more will probably become available in the future. *Renewable energy* is a term used to describe energy sources that either are continuously available, like heat and light from the sun, or are replaced relatively rapidly and therefore can be depleted only temporarily, like wood. Renewable energy sources are quite varied, such as radiant energy (the sun), chemical energy (biomass), potential energy (water stored in reservoirs), and kinetic energy (wind). Some, like wood fuel, have been used for thousands of years, while others are relatively recent developments (hydroelectric generation) or future options (ocean thermal energy conversion, or OTEC) (USDOE 2002a).

Although many different technologies are available for renewable energy sources, a few technologies are likely to be most important in the

ISSUE 14.1: THE LEGACY OF CHERNOBYL

Although nuclear energy is very useful and reported accidents are rare, the fear of another accident like the one at Chernobyl is one of the largest problems facing the nuclear industry. On April 26, 1986, a nuclear reactor at Chernobyl, near the Ukraine–Belarus border, became unstable and exploded. The explosion and resulting fire released nearly 200 times the amount of radiation released by the Hiroshima and Nagasaki bombs combined and spread radioactive contaminants across Europe (Specter 1996). In the days immediately following, over 200,000 people were evacuated from areas exposed to the fallout (Stone 1996; Williams and Balter 1996). The ecological and medical impacts in the immediate area were enormous and long-lasting. The repercussions for the nuclear power industry were no less significant.

When the Soviet authorities finally acknowledged that the Chernobyl incident was major, they evacuated all the children in the immediate area, because young bodies are more vulnerable to the effects of fallout. It was too late for many. When they returned from their unexpected summer "vacation," many children were already ill with leukemia, thyroid cancer, and other radiation-induced diseases. The Soviet authorities paid a heavy price for their long-term policy of insisting that they could handle their own problems without help from the outside. In 1986 the Soviet government maintained that it could easily handle this local problem, that only 200 people had been contaminated, and that life and farming would soon return to normal. But the reality turned out to be much worse than the early official statements, and the Soviet government and later the governments of Ukraine and Belarus became much more open about the extent of the problem.

The effects of Chernobyl were felt internationally, and we still do not know the full impact of the drifting fallout on downwind areas in eastern Europe, the British Isles, and Scandinavia, where the Lapps' lives have been changed forever. Their reindeer, consuming contaminated lichens, are no longer permitted in livestock markets. In parts of Wales, England, and Scotland, heavy rainstorms occurring just as the cloud of fallout was passing overhead brought contamination to upland areas. For years following the accident, sheep from these areas had to be screened before being allowed on the market. Sheep pastured on contaminated hills were moved to uncontaminated areas for a period of several weeks to allow radiation levels to decline. These sheep were specially marked and are tested regularly until the levels reached are deemed safe for human consumption (Atherton and Atherton 1990).

With emergencies come technological breakthroughs, and it is now possible to detect irradiated areas from a helicopter equipped with radiation sensors. The maps resulting from this surveillance and from research on soils and runoff indicate that the cesium is becoming more environmentally active. The cesium is no longer simply lying on the vegetation but is weathering into the soil to be absorbed more efficiently by grass, the sheep's major food supply. The gradual loss of radiation in livestock in north Wales and other areas has leveled off and may begin to climb once more. North Wales farmers are unable to farm under these conditions; government compensation, where available, is no solace. Throughout the United Kingdom, 757 farms on 400,000 acres (160,000 ha) were affected by restrictions (Ghazi 1990).

Fifteen years after the accident, a vast region of southern Belarus and northern Ukraine is virtually uninhabitable. The area within a 30-km radius of the plant is officially closed to the public. Outside this are extensive areas in which no one is permitted to live or enter without a pass, and beyond these are zones in which residence is discouraged, health monitoring is required, and food production is prohibited. A thyroid disease epidemic is occurring among children in Belarus and Ukraine, with childhood thyroid cancer rates in Belarus increasing from fewer than 5 cases per year before the accident to over 80 by 1994 (Balter 1995). Those who inhabit the contaminated area fear more disease or another radioactive disaster. Fifteen of the 28 Chernobyl-type reactors in use in 1986 are still running, including two at Chernobyl. They have been modified to reduce the likelihood of another disaster, but many experts feel that the danger remains.

next few decades. Some of these are produced in large, concentrated facilities (as most of our electricity is produced today) which we will call *centralized renewables*. In other cases, energy production is in small facilities, like the photovoltaic cells that are used to power satellites, calculators, and remote communications facilities. These are called *decentralized renewables*.

Centralized renewables require large capital investments in single facilities, and most are used to generate electricity that is fed into the same grids that transport energy generated from fossil fuels or nuclear power. The primary centralized renewable energy sources in use today are hydroelectric generation, geothermal energy, tidal power, and biomass. The most promising technologies for future centralized renewable power include OTEC, low-temperature geothermal, and large-scale solar power facilities.

Decentralized renewables are energy sources that will be important for small facilities not connected to large-scale energy systems. The most significant of these is solar thermal energy for space or water heating. Other important decentralized energy sources for the future include photovoltaics and wind.

Hydroelectric Power On the basis of both economic and environmental considerations, hydroelectric power is the best source of electricity available to us today. It is generated by impounding water in a location where a substantial vertical drop is available, such as near a waterfall or in a steeply sloping and/or narrow portion of a river valley, and then passing the water through turbines that drive generators. Water power has been used for several centuries to drive various industrial facilities, but only in the past 100 or so years has it been used to generate electricity. In that time many large dams were built, and in 1998 about 10 percent of United States electricity, or 4 percent of total United States energy consumption, was supplied by hydroelectric power. Worldwide, hydroelectric generation supplies less than 3 percent of total commercial energy production (Fig. 14.12).

Hydroelectric production is concentrated at the most desirable sites for this technology, specifically locations with large volumes of available water, narrow and deep valleys, and, in some cases, low population density. At these sites the cost of generating electricity is quite low in comparison to other methods. A rather large capital cost is involved in dam construction and land

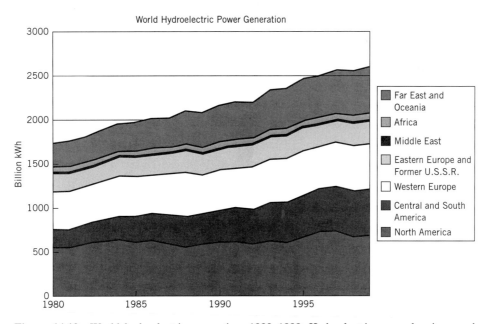

Figure 14.12 World hydroelectric generation, 1980–1999. Hydroelectric power has increased slowly but steadily in most of the world; in North America production was well established by the mid-twentieth century and has not increased very much since then. On a per capita basis, Central and South America have the highest hydroelectric production. *Source:* EIA, 2002b.

acquisition, but once these are paid for, the operating costs are extremely low. Most of the large dams in North America were built either by governments or with large government subsidies, and this contributed to the low cost of hydroelectric power in many areas. In the northwestern United States, for example, electricity prices are substantially lower than in the rest of the country because of the large contribution of hydroelectric power generated at federally built dams along the Columbia River and its tributaries.

Worldwide, there is considerable potential for increasing hydroelectric generation, especially in less industrialized countries (Fig. 14.13). World hydroelectric-generating capacity has grown steadily since the 1950s, at about 15,000 mW per year, and will probably continue to grow for some time. The greatest potential for constructing new dams for hydroelectric generation is in the less developed areas of the world, particularly in Asia, Africa, and South America, where many large rivers remain relatively unmanaged. The amount of untapped potential is vast: more than 15 times the current installed generating capacity. Tapping that potential, however, means overcoming substantial obstacles (see Issue 14.2).

There is relatively little potential for expansion of hydroelectric generation in the United States. Most of the best dam sites are already in use, and those that are not are generally unavailable because the river valleys are committed to other uses, such as agriculture or wilderness preservation. There may be some opportunity for increasing power production at smaller dams, and there are a few examples of existing dams that were unused until recently, when higher energy prices made hydroelectric power generation on a small scale more feasible. In New England, for example, hundreds of dams were constructed for hydroelectric power or industrial uses in the past but are no longer in use. In Canada, there is considerable potential for increased hydroelectric generation. Canada already exports substantial amounts of electricity to the United States, primarily from

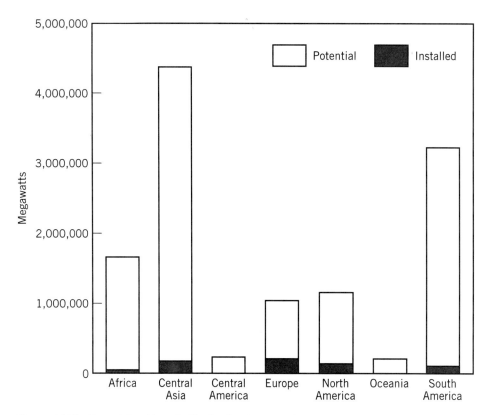

Figure 14.13 Potential and installed hydroelectric generating capacity. The potential for new hydroelectric development is vast, but political, economic, and environmental constraints will prevent most of this potential from being realized. *Source:* World Resources Institute, 1996.

Quebec to New York. This trade is facilitated partly because the peak demand for electricity in Quebec occurs in winter, whereas the peak in most of the United States is in summer. Several additional dam sites are available in northern Quebec to increase this production, although capital costs, transmission distances, and opposition from native populations are significant barriers.

While hydroelectric generation is the cleanest source of electricity we have, it is not without its environmental problems. One of these is the loss of the land and aquatic habitat that is submerged when a reservoir is filled. The construction of dams and subsequent regulation of flow patterns also have adverse effects on stream channels and aquatic life downstream. There have been problems of erosion in the Grand Canyon, for example, as a result of the construction of Glen Canyon Dam upstream. This led to an unprecedented experiment in the late 1990s in which water was released from the dam to create flood conditions. In arid areas, such as the Colorado basin, reservoirs contribute to evaporative water losses by increasing water surface areas. Many valleys contain fertile agricultural land, and reservoir development was a contributor to loss of agricultural land in the United States. There may also be effects on water quality as a result of impoundment, as nutrient-laden water stimulates algal blooms, leading to eutrophication. Despite these negatives, our existing hydroelectric facilities are important energy sources. Construction of new facilities, however, is a much more contentious issue, for it almost always entails the loss of some other valued resource.

Geothermal Energy Geothermal energy is derived from the internal heat of the earth. The core of the Earth is hot, and this heat is convected and conducted outward toward the surface. In many parts of the Earth's crust, sufficiently large amounts of heat are delivered to the surface to make the use of that heat feasible. Geothermal energy is used in many parts of the world today, for space and water heating, for industrial processing, and for electricity generation.

Three major types of geothermal resources are hydrothermal, *geopressurized*, and *dry rock*. Hydrothermal resources occur in locations where there is a source of heat relatively near the surface, and the overlying rocks are fractured enough to allow water to circulate. In addition, the water must be trapped in the ground to prevent heat from escaping rapidly to the surface. This results in the formation of hot springs and geysers. If this steam or hot water can be tapped, it can be used to generate electricity or heat buildings. Geopressurized resources are deeper pockets of water trapped in sedimentary formations in much the same way as oil and gas are trapped. Hot dry rock is simply rock in areas of high heat flow but without large quantities of water contained in the rock. To tap this heat, the rocks must be fractured and wells drilled so that water can be injected into the rocks to draw the heat out. At present, geopressurized and hot dry rock resources are not in use.

Only one commercial facility in the United States is currently generating electricity from geothermal energy—The Geysers, in California. It presently produces about 1000 mW at peak output, or about the same as a large conventional power plant. The output from this plant amounts to less than 0.2 percent of total U.S. electricity generation. One possibility for increased geothermal production is in the development of hot dry rock facilities (USDOE 2002a, b). There are numerous areas of the United States, mostly in the West, where subsurface temperatures are particularly high. The economics of such ventures are still uncertain, however, so it is unlikely that geothermal energy will make large contributions to total electric generation in the near future.

Other Centralized Renewables In addition to hydroelectric and geothermal power, a few technologies offer limited potential for future development. One of these is the incineration of solid wastes. Heat from this burning can be used directly for industrial processes or for generating electricity. This process is often called *resource recovery*. At present several dozen resource recovery plants in the United States use solid waste as fuels, with most of these producing steam for industrial processes. In a few cases, wood or agricultural wastes are used for fuel, but in most plants municipal wastes are burned. The major incentive for burning solid waste at the present time is the high cost of disposal rather than the value of the steam produced, and the disposers pay to have the fuel burned rather than the plants buying fuel. In 1995, combustion of wood and municipal solid waste together with the production of alcohol from sugar and grain provided approximately

ISSUE 14.2: THE THREE GORGES DAM

As world population increases and demands on earth's water resources increase, the need for dams is certain to grow. Dams offer two important benefits that will be greatly needed in the coming decades: an increase in the amount of renewable water that is available for human use and the production of clean energy in the form of hydroelectricity. But dams are also expensive, and they fundamentally transform the landscapes in which they are placed, with immense impacts on humans and nature. They inevitably become the focus of controversy, and the Three Gorges Dam is no exception.

The Three Gorges Dam, currently under construction, is located in central China, more than 1600 km upstream from the river's mouth. The dam will be among the world's largest: about 185 m high, 1.9 km long, and impounding a lake 600 km long. The capacity of the reservoir will be 22 billion m^3, or about 24 times the mean annual discharge of the Yangtze at its mouth. Cost estimates range from about $20 billion to $75 billion. Completion is scheduled for 2009.

Like any big dam, the Three Gorges project has multiple benefits. In this case, the principal justification is flood control. The Yangtze has a long history of floods that cause vast damage to communities and farms along the river's floodplain. One of the most devastating recent floods occurred in 1954, killing 30,000 people and displacing 19 million. Following that flood, Mao Zedong vowed to proceed with the project that would lessen this flood risk. Forty years later construction finally began.

In addition to flood control, the Three Gorges Dam will generate 18,000 mW of electricity, boosting China's electric production by 10 percent. China's energy use is growing rapidly, and this added hydroelectric generation capacity could substitute for coal-generated electricity. The project also opens up possibilities for flow regulation that could be used to make more water available for irrigation and other uses, although to some extent this could conflict with the primary flood-control objective.

Opposition to the project focuses on its considerable negative environmental impacts and major engineering uncertainties. The greatest environmental impacts are associated with the loss of land along the Yangtze Valley. The area to be inundated includes 190 km^2 of farmland and the homes of roughly 1.2 million people, who will be relocated to other areas. The loss of agricultural land is very small in comparison with China's 950,000 km^2 of cropland. But the potential loss of cultural resources in the valley is large. The area to be flooded includes Shibaozhai, a tower of rock overlooking the river on which a pagoda was built in

3 percent of the nation's energy needs, with alcohol being the largest part of this total (EIA 2002a).

Another centralized renewable energy source is *tidal power*. Tidal fluctuations produce strong currents in coastal embayments, and these currents can be harnessed to produce electricity. A large tidal range is required, and a dam must be constructed across the bay to create the hydraulic head necessary to drive turbines. Power can be generated only during a portion of the tidal cycle, and diurnal or semidiurnal tidal fluctuations do not always correspond to the fluctuations in demand. Thus, tidal power can be viewed as only a supplement to other sources of energy. At present, there is one large tidal power plant in operation in France and a smaller one in Nova Scotia.

Large-scale solar thermal and photovoltaic collectors, plants that harness wave energy, and large-scale wind-powered generating facilities have also been proposed. There has been relatively little activity in these areas since the 1970s because of the low costs of conventional energy and the weak growth in demand for electricity. If prices rise significantly in the future, these methods may again be considered. Centralized renewable sources in general will probably continue to provide only small portions of our total energy supply, with only biomass (primarily municipal solid waste) and hydroelectric generation making a large contribution to total electricity generation. New power generation from all centralized renewable sources together will probably not contribute

the eighteenth century and a village that contains some of the best remaining examples of Ming Dynasty (1368–1644) structural art (Tyler 1996). And, of course, the lake will forever transform one of the most spectacular river valleys of the world from a natural-flowing stream into a reservoir, with unknown impacts on the river's biota.

Because the dam is by far the largest in China and among the largest in the world, the technical uncertainties are many. In projects of this scale, engineers simply don't have the experience on which to base their plans. One of the critical problems is sedimentation. The Yangtze carries an enormous load of sediment that would be deposited on the floor of the reservoir, diminishing its capacity. A much smaller dam in China, Sanmenxia, experiences a much higher sedimentation rate than that anticipated by its designers. As a result, the dam had to be modified and reservoir operation altered after just two years of operation (Leopold 1996). Even more important questions are those relating to the dam's flood-prevention potential. The Yangtze is fed by several major tributaries that enter the river downstream of the dam. Although the flow from the upstream portions of the Yangtze could be reduced, the flow from these tributaries would not be controlled by the dam. In addition, although the lake's volume is immense, most of this would already be filled with water at the beginning of a flood, and the amount of water that could be stored may be insufficient (Mufson 1996). These and other technical problems raise questions about not only the engineering feasibility of the project but its financial aspects as well.

Like any major environmental issue, many of the debates are intertwined with other political differences. Domestically, opposition from academic and scientific circles has been strong, as have been the objections of those whose communities will be destroyed. In the late 1980s, internal opposition was sufficient to delay the project, but after the Tiananmen Square demonstrations in 1991, the government used the opportunity to forbid public opposition to the project and went ahead with construction of the dam. Internationally, strong opposition to the project has come from many points, and China has had a hard time borrowing money to pay for the dam. The World Bank, which includes environmental considerations in its policies, withheld its support for years. Despite pressure from some large engineering and construction-equipment manufacturers, in 1996 the U.S. Export-Import Bank also decided not to support it. As a result, virtually all the financing is coming from domestic Chinese sources, placing a considerable strain on that country's limited budgets for capital development.

more than a small percentage of the total U.S. energy supply for a few decades.

Solar Energy Among decentralized renewable energy sources, solar energy offers enormous potential for heating buildings and water and thereby to replace more conventional energy sources such as fossil fuels and electricity. Solar energy is plentiful, though it is not uniformly available in space and time. The technology is reasonably well developed and is economically feasible in most of the United States. The major barrier to its use at present is capital costs.

Solar energy reaches the earth's surface in the form of radiant energy, and it is converted to heat when sunlight is absorbed on an exposed surface. The amount of radiant energy received varies with the time of day, season, weather conditions, and location on earth. For example, there is no sunlight at night, and during cloudy weather, sunlight is less intense, though it is still present. Seasonality influences the length of day, so that there are fewer hours of daylight in winter than in summer. Also, the angle of the sun changes during the year, reducing the intensity of solar radiation per unit of horizontal surface area in the winter. Because of these large fluctuations in solar energy availability, only a limited portion of the world can rely on solar energy as a single consistent source. In other regions, backup energy systems are required.

Solar energy is often discussed in terms of passive and active technologies. Passive solar heating and cooling involve neither mechanical devices nor the production and storage of electricity. *Passive solar power* employs proper design of structures, building materials that insulate or store energy, correct orientation of structures, and careful landscaping to provide heating and cooling. A house with many windows on the southern exposure allows for maximum sun during the winter. Planting deciduous trees on the south side of the house protects it from the hot summer sun's rays yet allows the sunlight to penetrate during the winter months, when the trees have dropped their leaves. The use of adobe as a building material and the design of houses with shaded arcades and small windows, such as those of Spanish-style haciendas in the Southwest, are other examples of passive approaches to solar heating and cooling.

Active solar power uses mechanical devices to collect and store solar radiation for heating and cooling. The solar collector is the basic unit of a solar space heating system and can vary greatly in size and complexity. The simplest type of collector consists of a flat plate, painted black for maximum absorption, encased in an insulated, glass-covered box. The plate absorbs light, while the glass impedes energy loss, so that temperatures rise to about 200°F (93°C). This heat is then used to warm rooms and water as air moves across the black plate or water circulates through a pipe attached to the plate. Higher temperatures (up to 1000°F or 540°C) are obtained by concentrating the light with one or more curved mirrors, which rotate along with the sun's movement across the sky. Temperatures of up to 4000°F (2200°C) can be attained by focusing a bank of mirrors on a central point, such as a tower containing a water boiler for operating a steam turbine.

The amount of solar energy received by a collector varies with the time of day, season, and weather. These variations, as well as obstructions by trees and buildings, still allow for the use of solar energy for space heating. The heating and hot water needs of a one- or two-story building can be met by using available roof surfaces, southern walls, and other areas for the installation of collectors.

The use of solar collectors nationwide rose substantially in the late 1970s and early 1980s but slowed in the 1990s. Low-temperature collectors are used almost exclusively for swimming pool heating. Medium-temperature collectors are used for both space heating and cooling and for domestic water heating. Regionally, more collectors are in use in the western, high-sunshine states than in the East, but advocates maintain that solar collectors are capable of a quick return on investment even in the cloudiest of climates.

Among the more promising solar technologies for the future are *photovoltaic cells*—thin silicon wafers that convert sunlight directly to electricity. When first produced in the 1950s they were extremely expensive, but with the rapid growth in production technologies and concurrent increases in demand, the prices have fallen dramatically in the past decade. In 1982, the capital cost of purchasing solar cells was about $10 per peak watt, and by 1994 it had reached $4 per peak watt. By 1997 costs for entire electrical systems had dropped to $8.50 per watt for large-scale (70-kW) systems—large enough to power several homes. These prices are still too expensive for large-scale generation, but there are many applications for these cells today, primarily in running small machines in remote locations where installing conventional electric service is expensive. Further price reductions may greatly expand the usefulness of photovoltaic cells (Fig. 14.14), and large-scale photovoltaic electricity production could become viable in the first decade of the twenty-first century.

Although many reports claim that solar energy is the way of the future, obstacles will have to be overcome before it makes a truly significant contribution to our energy supplies. The biggest challenge to be overcome in adopting solar energy technology is the need to store energy for use at night, on cloudy days, and in cooler seasons. Considerable research and development activities are currently under way in battery technology, both for stationary systems such as buildings and for vehicles. The high capital costs and relatively long payback period for an investment are another economic barrier to residential heating and cooling. This is particularly troublesome for retrofitting, but less so for new installations. As a result of the slowdown in solar energy installation in the mid-1980s, substantial infrastructure in the form of manufacturers, dealers, and installers was lost (Rembert 1997). The economics of solar energy will have to improve substantially relative to conventional energy sources before there is another wave of installation of these systems.

Figure 14.14 Photovoltaic cells generating electricity for lighting a road sign in New Mexico.

Wind Power Windmills have been used worldwide since ancient times. We perhaps know them best in the United States as a symbol of the nineteenth-century farm, where wind energy was converted to mechanical energy to pump water from the farm well. Although windmills have largely been replaced with other pumping devices, the windmill is again feasible as a method of generating electricity. The industry is small but has considerable potential. Growth in this area was encouraged by a 1978 federal law requiring public utilities to purchase electric power offered to them by small generating companies. The price is determined by the avoidance costs of producing equivalent amounts of energy by conventional sources.

The first utility company to incorporate windmill-generated power into its power grid was Southern California Edison Company in 1980 (Pasqualetti 2001). A privately financed 200-foot-tall wind turbine installed in the desert near Palm Springs, California, is capable of generating enough electricity for about a thousand homes. The blades are driven by reliable winds, which average 17 mph with gusts up to 40 mph. The amount of electricity generated by this turbine is equivalent to an annual saving of about 10,000 barrels of oil. Another example is the Wind Farm project operated by Pacific Gas & Electric in the Altamont Pass area in northern California. Each of the 407 windmills there turns out about 50 kW (Fig. 14.15).

In the late 1970s and early 1980s, the U.S. Department of Energy put millions of dollars into developing prototype wind machines that were rated from 200 to 2500 kW at peak power output. The largest of these has a blade span of 300 feet. Early 1980s projections estimated that wind energy would provide 2 to 4 percent of U.S. electricity by 2000, but lower oil prices in the middle and late 1980s slowed development of this and other alternative sources. In 2001 wind generation amounted to only 0.01 percent of total electric generation in the United States and less than one percent of the world's electricity.

Wind power is recognized as the first renewable energy source since hydroelectricity to move beyond government sponsorship into control by traditional public utilities. Its application to large-scale electricity generation holds fewer uncertainties than other renewable sources. Wind power is also more easily integrated into existing utility power grids, which are necessary to provide backup power when wind velocities are low.

Figure 14.15 A wind energy-generating facility at Altamont Pass, California.

Energy Efficiency and Energy Conservation

One way to measure the efficiency of energy use is in strict physical terms: the amount of productive work accomplished for a given expenditure of energy or fuel, such as the number of miles a car travels on a gallon of gas. Modern automobiles are much more efficient than those made a generation ago, as a result of more precisely controlled engines, less weight, and improved aerodynamics. Similarly, fluorescent lights provide much more light per amount of electricity consumed than do incandescent lights, because they emit less heat. Many improvements in physical energy efficiency already were achieved in recent years, and many more are possible. We discuss some of these in more detail later in this chapter.

Another way to measure the efficiency of energy use is in terms of economic gain per unit of energy expended. Throughout much of recent U.S. history, energy was plentiful enough that we did not consider the efficiency of energy use in most of our economic decisions. Today, however, we are very conscious of the contrasting patterns of energy use and efficiency between different nations and economic systems.

Traditional agricultural economies produce goods and services at a relatively low level, using virtually no energy except that provided by the sun through heat and photosynthesis. In contrast, mechanized agriculture uses large amounts of nonrenewable fossil fuel in every step of the production system, from building the tractor to manufacturing fertilizer to fueling farm machinery. This system produces much more food per unit of human labor but much less per unit of energy input.

One indicator of energy efficiency is a comparison of per capita gross national product (GNP) with per capita energy consumption (Fig. 14.16). As a rule, the higher the GNP per capita, the higher the energy consumption, and vice versa. A nation with a higher income can spend more money on machines and raw materials to increase production and quality of life. This in turn leads to more energy consumption, which if expended in production, leads to more income. Industrialized nations consume much more energy per capita than less industrialized nations. The United States, for example, with about 5 percent of the world's population, consumes 25 percent of the world's energy, whereas China, with 21 percent of the world's population, consumes only 9 percent of the world's energy. Furthermore, there are great variations in energy efficiency among industrialized nations. Most of the western European nations, for example, use less energy per unit of economic return than the United States. This is be-

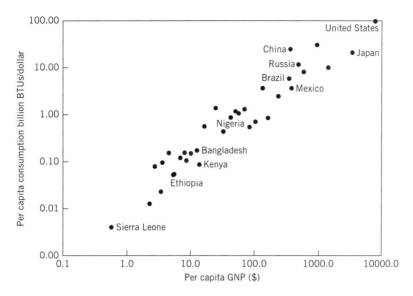

Figure 14.16 Per capita GNP and per capita energy consumption for selected countries, 1999. Countries that plot above and to the left of the trend are relatively inefficient; countries that plot on the lower right of the trend are more efficient. Note the logarithmic axes. *Source:* EIA, 2002b.

cause energy prices have been much higher there for a long time, and Europeans have adjusted by driving smaller cars, driving shorter distances, heating buildings to lower temperatures, and so on. In contrast, Russia and other eastern European nations use large amounts of energy but have much lower incomes than their neighbors to the west.

Recent trends in the overall economic efficiency of energy use are illustrated in Figure 14.17. The graph shows *energy intensity*, or the amount of energy used to produce a unit of income. Considerable gains in efficiency occurred in the United States and other industrial nations in the 1980s, spurred by increases in prices during the energy crises in the 1970s. At that time energy users at all levels, from government officials and corporate executives to individual consumers, realized that substantial savings were possible through energy conservation. Accordingly, a wide range of energy-saving technologies came into wider use. Also, during the 1980s production in energy-intensive heavy industries stagnated, while most of the economic growth took place in the service sector, which uses much less energy per unit of income generated. Some of the increase in energy efficiency seen in Figure 14.17 is a result of economic restructuring from the industrial sector to the service sector. Since the 1980s, changes in efficiency have been less dramatic.

It is also clear that considerable differences in energy efficiency exist among nations. In Brazil, for example, increases in the number of automobiles and relatively slow economic growth contributed to an decrease in energy efficiency. The most dramatic change in energy use in the world since 1980 has been in China, which used only about one-third as much energy to produce a dollar of income in 1999 as it did in 1980. Most of that increase in efficiency was the result of dramatic economic growth as China entered the world marketplace, combined with elimination of some of its most inefficient energy-using activities, particularly in heavy industries.

Energy conservation has been the industrialized world's largest single source of "new energy" since the early 1980s. Conservation in this sense simply means using less or using what you have with more efficiency. As we have seen earlier in this chapter, total energy consumption in the United States peaked at 81.0 quadrillion BTUs in 1979 and declined until 1988. It has since risen to nearly 98.5 quadrillion BTUs in 2000. Total output was greater in 2000 than in 1979, so that the

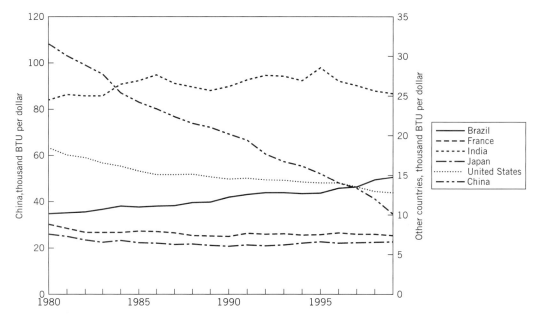

Figure 14.17 Energy intensity for selected countries, 1980–1999. Energy intensity is calculated as the amount of commercial energy consumed divided by GDP. In some nations energy intensity has decreased, meaning increased energy efficiency, while in other countries energy intensity is increasing. China has made very dramatic improvements in energy efficiency, mostly as a result of very rapid economic growth as its economy has become increasingly market-oriented. Source: EIA, 2002b.

amount of economic activity per unit of energy was about 68 percent higher in 2000 than in 1979. Put another way, the amount of energy required to produce a dollar's worth of goods and services (adjusted for inflation) decreased by a third between 1979 and 2000 in the United States.

Energy conservation developed gradually during this period. Initially, less fuel did less work; for example, people drove more slowly and reduced heat and light use. By the early to mid-1980s less fuel was used to do the same amount of work. This required minor design and investment decisions, not major overhaul of facilities or reorientation of an industry. Ideally, we would like to reach a condition in which less fuel is used to do more work. This would provide the greatest savings, yet it would require investment in applied research efforts in order to be effective. This applied research requires substantial technological innovations and usually takes at least several years to realize.

For example, in the mid-1970s consumers and auto manufacturers in the United States responded to higher gasoline prices by shifting from larger to smaller cars, smaller engines, and manual transmissions. Consumers also responded by driving less. Miles per vehicle decreased over 4 percent, and total motor fuel consumption dropped by 16 percent between 1978 and 1982. By the mid-1980s, most people were driving much more fuel-efficient cars. In 1976 the average fuel consumption by United States cars was 12.1 miles per gallon; by 1987 this had increased by 25 percent to 15.1 miles per gallon. In 1985 more people were driving more cars more miles per car but using less fuel than had been consumed in 1978.

But by the late 1980s gasoline prices were again low, and larger vehicles grew in popularity. In the 1980s the minivan replaced station wagons and sedans as the primary family vehicle. In the 1990s this was followed by a dramatic increase in the popularity of pickup trucks and four-wheel-drive sport utility vehicles. The greater height and lack of streamlining of these vehicles, as well as the added friction and weight of four-wheel drive, reduced fuel efficiency significantly, but low fuel prices meant that this was not a concern. Families bought the larger models, while those who needed only smaller cars bought two-seaters instead of the small sedans that were popular in the early 1980s.

In 1973, only about 10 percent of new light-duty vehicles sold in the United States were trucks, vans, or sport utility vehicles, compared to around half now. The average fuel efficiency of vehicles rose from about 12.2 mpg in 1975 to over 20.2 mpg in 1991 but declined in 1999 to 16.9 mpg.

The trends in less industrialized countries in the past few decades have been very different from those in the industrial world. To begin with, the less industrialized countries use vastly less energy per person and proportionately more traditional fuels (primarily wood and dung). The commercial fuels found in these countries tend to be used in older, less efficient machines and factories. In India, for example, per capita use of commercial energy is about one-twentieth the level in the United States. Between 1980 and 1998 India's population increased by 58 percent, its GNP increased, and energy consumption went up by 190 percent. The economic efficiency of that energy use, in dollars of GNP per kilogram consumed, increased by almost 30 percent. A different trend was evident in Brazil in that same period, when population increased 53 percent, GNP increased, and energy consumption increased 160 percent, but the efficiency decreased, so that it took 10 percent more energy to produce the same GNP (UNDP 2000).

Why is energy efficiency decreasing in some poorer countries but increasing in others? The answer to this question differs from country to country, depending on the particular circumstances of economic growth and change. Probably the most common explanation is the growth of manufacturing in less developed countries. The economic restructuring in the wealthy countries that resulted in a shift from manufacturing to service industries also involved a transfer of manufacturing overseas, primarily to poorer countries with lower labor costs and, in some instances, with modernized factories that were more efficient. The rapid growth of the maquiladora manufacturing centers just south of the Mexico–U.S. border is an excellent example. Consumer goods are manufactured in Mexico by U.S.-owned companies using low-cost labor, and the products are shipped north to be consumed in the United States. Thus, the benefits of the energy used in manufacturing are still reaped in the United States, but statistically the energy consumption is recorded as having taken place in Mexico.

ENERGY FUTURES

As we consider the future of energy use in the world, it is clear that the range of possibilities is large. There is a high degree of substitutability between the various sources of energy, and a wide range of alternatives exists for aggregate energy supply and demand in the future. One thing is clear: electricity will form an inceasingly important form of energy end-use. The choices for sources of energy and which to use for heating and which to use for electricity are important policy decisions with far-reaching impacts. Let us consider some of the possibilities.

High-Energy Options

Most of the world is following an energy-intensive course. Economic growth is focused on industrial production of consumer goods, on international trade and the transport it requires, and on personal mobility (both on an everyday basis and in tourism). If we continue on such a course and include a greater portion of earth's growing population in economic development, then large increases in total energy use are necessary. Where will this energy come from?

Fossil Fuels One obvious possibility is continued and increased exploitation of fossil fuels. In the short term—the next 30 years or so—this means continued reliance on oil. The price of oil would have to rise somewhat to stimulate new exploration and development, but use would continue much as in the present. Coal would grow as the primary fuel for electricity generation, and long-distance trade in either coal or the electricity generated from it would grow. Increasing oil use would lead to significant depletion of this resource, however; thus, in the longer term—30 to 60 years—significant development of synthetic fuels would be necessary. In such a world, the United States, with its vast deposits of coal and oil shale, would be fuel-rich and could conceivably become an energy exporter, as would other coal-rich nations.

Such a scenario would mean significant and continuing environmental impacts of energy use. A doubling of the CO_2 content of the atmosphere above pre-industrial levels would happen by the mid-twenty-first century, and in all likelihood global warming would continue or accelerate.

Unless new technology for significantly reducing emissions of SO_x and NO_x from internal combustion engines, power plants, and other sources were developed, the extent of damage to forests and other ecosystems would expand accordingly. Finally, large areas would be subject to disruption by surface mining, some of them in areas that presently have relatively low levels of environmental impact.

Nuclear and Renewable Energy Alternatively, a high-energy future is possible without heavy dependence on fossil fuels. Instead, we could increase our use of nuclear power and renewable energy, both centralized and decentralized. First, we would have to convert many activities that currently use fossil fuels to electricity or solar thermal power. Gasoline-powered automobiles would be replaced with hydrogen-powered or electric ones. Freight-hauling with diesel-powered trucks would be limited, and further growth in long-distance land-transport would focus on electric-powered railroads. Home heating in midlatitude climates, which today is accomplished mostly with oil and gas, would rely much more heavily on solar thermal systems.

This would require large increases in electric generation, not to mention replacing existing electric production from fossil fuels. Nuclear power easily could be expanded and generate large amounts of electricity. Despite the low standing of nuclear power today, the major barriers to expansion of this technology are social and political, not technical. Public resistance is based more on fear and on the potential for accidents, despite their rarity. If these fears could be overcome, then nuclear power would again have great promise.

Another major new source of electricity would be hydroelectric generation. As discussed earlier in this chapter, vast untapped hydroelectric power potential exists throughout the world. Hydroelectric power could make a much greater contribution to total energy production, without carbon emissions or radioactive waste. However, it would require both a very large capital investment in capital-poor parts of the world and the loss of many ecosystems dependent on free-flowing rivers.

Finally, a significant investment would have to be made in installing solar energy systems in new homes and buildings and retrofitting older structures to use solar energy instead of fossil fuels. Such an effort would also involve large investments in energy efficiency, such as increased insulation and heat storage systems.

A high-energy, low-fossil-fuel future requires a commitment to both development of nuclear and hydroelectric power and significant new investment in energy relatively soon. Such commitments are technically and economically feasible but not necessarily realistic in the current social/political climate. It is not clear whether climate would change quickly enough to allow a conversion to such an energy system in the near future.

Low-Energy Options

It is not a foregone conclusion that the world will continue to increase the amount of commercially produced energy it uses in making and distributing goods and services. Many ways of reducing the amount of energy we use are available, and some of these also allow for continued economic growth and reduction of economic disparities between rich and poor.

Conservation-Intensive As discussed earlier, conservation of energy has been the largest single source of new energy in the United States since the 1970s. In 1999 we produced 46 percent more wealth per unit of energy consumed than in 1980. At 1980 economic efficiencies, the amount of wealth produced in the United States in 1999 would have required 142 quadrillion BTUs (quads) instead of the 97 actually consumed. In these terms, conservation *produced* 45 quads in the United States in 1999, an amount greater than the energy consumption of 1.2 billion people in China and greater than the entire 1999 energy consumption of Africa and South America combined! Most of this increase in efficiency was achieved in a brief period during and immediately after the energy crises of the 1970s. Imagine the amount of energy efficiency improvement that would have occurred if gasoline prices had remained at their 1981–1982 levels (adjusted for inflation), instead of falling below 1972 prices in the early 1990s, as they did.

The increases in energy efficiency that occurred in the United States and some other countries were not entirely painless. They required new investments, and with the new investments came some social dislocations, such as the decline of the heavy-industrial sector and the communities that depended on it. It must also be remembered that much of the increase in economic energy efficiency was a consequence of economic restructuring, with a replacement of energy-intensive manufacturing with service industries that use less energy. We still consume manufactured goods and we still need energy to make them, but much of this energy consumption has been transferred overseas.

Nonetheless, many opportunities for energy conservation remain. Suppose the entire world had the economic energy efficiency of Japan. Based on recent data, the GNP of the world could be doubled, and we would still use only three-fourths of the energy used in that year. For those of us who live in the United States, this would involve some sacrifices, such as living in high-density cities, using public transport for most of our everyday trips, and using automobiles only occasionally. But imagine the benefits to human welfare if the GNP of the world were doubled and if most of that increase took place in the world's poorest countries!

Changes of this magnitude will not occur in the near future. To achieve them would require overcoming immense economic and political barriers and changes of life-style among the energy-guzzling countries of the world like the United States. But thinking about the possibilities should help us understand the enormous potential of reducing environmental damage without loss of wealth, which is possible through energy conservation.

Low-Growth Finally, we recognize that we could continue to use energy much as we do today, but simply use less (or stop increasing our use of energy). Without increases in efficiency, this would mean an end to economic growth in both the industrialized and less industrialized world. Such an alternative would be undesirable for a large portion of earth's human population. Unless significant changes in the way we use energy take place in the next few decades, a low-growth scenario will be a necessity, particularly given the increase in world population expected during this time. A low-growth future would have at least some advantages for the earth's natural systems, to the extent that it would not lead to increased pollution, for example. But it can be argued that some of the worst environmental degradation is a consequence of poverty, rather than of wealth, as people clear forests for fuelwood or cause overgrazing in the desperate search for adequate food supplies.

Energy Policies for the Future

The scenarios described above represent extremes, and none of them is either entirely desirable or particularly likely. But all are physically possible. These scenarios demonstrate the great range of energy-use options that are available to us in the next few decades and our considerable choices with regard to future conditions. What policies can we pursue today to affect these future outcomes? How can governments, businesses, and individuals make such choices?

At the present time, most of the governments of the world are relying primarily on market forces to determine energy policy. In the United States, for example, the choice of fuels for generating electricity is made by private corporations, largely on the basis of cost (Issue 14.3). Environmental regulations somewhat alter these choices by favoring low-sulfur coal over high-sulfur coal or increasing the cost of nuclear power through stringent safety rules, but these have relatively minor impact. The fact remains that coal is still being used to generate most of our electricity, and the lack of new nuclear power plant construction is as much a function of flat demand as it is of the higher cost of that energy form. Similarly, larger, less fuel-efficient vehicles were much more popular in the 1990s than they were in the early 1980s because of a decade of low fuel prices. If the higher prices of the late 1970s and early 1980s had persisted, we would not be driving sport utility vehicles in such numbers as we are today.

Exceptions to these market-based policies do exist. In Europe, fuel taxes have long been used as a tool for altering energy-use choices by making motor fuels artificially expensive. This encourages people to drive more fuel-efficient vehicles (producing less pollution), drive less (reducing road congestion and construction costs), and take public transport (maintaining the viability of this mode of transport). For the most part, relatively short-term cost determines what fuel we use. We have even gone to war to maintain free trade in energy commodities, as when Iraq's invasion of Kuwait in 1991 threatened to change the balance of control over oil, and again in 2003 to affirm U.S. control of the region.

Some argue that the governments of the world's energy-consuming industrial nations should be taking more aggressive steps to reduce dependence on fossil fuels and speed the transition to renewable energy and energy conservation.

ISSUE 14.3: ELECTRIC ENERGY DEREGULATION AND THE CALIFORNIA ENERGY CRISIS

Electricity will be the most important form of energy in the twenty-first century, and the ways in which we generate electricity will need to change. For most of the twentieth century, the electric power industry in the United States consisted of highly regulated regional monopolies. Monopolies made sense because of the high cost of installing and maintaining transmission lines—there was no reason to have two or more competing sets of transmission lines serving the same area. But because they were monopolies it was necessary to regulate prices so that consumers would not be unfairly exploited. Following the 1974 energy crisis it became apparent that the regulated monopoly structure inhibited efficient and innovative alternative-energy technologies. For example, a large factory or a complex of buildings might combine electricity production and space heating in one efficient process called cogeneration. Homeowners might install wind generators to supply some of their own electricity. These systems can be made more cost-effective if they are permitted to sell surplus energy back to the power company, but existing rules did not require utilities to accept that energy.

The first step in modernizing the industry was the 1978 Public Utilities Regulatory Policies Act, which made it possible for nonutility power generators to enter the wholesale electricity market. It

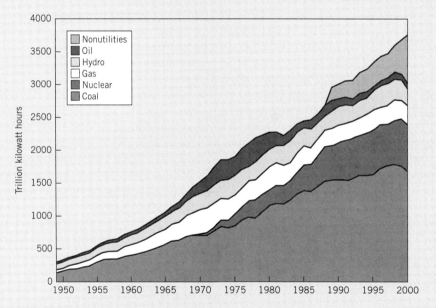

Electric generation in the United States, 1949–2000. Coal has always been our major electric generation fuel, but the mix of electric generation technologies is changing. In the 1990s regulatory changes in the electricity industry created a new set of generating entities separate from the traditional private and government-owned electric utilities that had dominated previously. These are companies that are not regulated as strictly and generate and sell electricity on the wholesale market to meet changing needs. Most of the electricity generated by nonutilities is generated with natural gas, though other fuels, including renewables, are important. *Source:* EIA, 2002a.

Such steps could include investing in renewable energy technology such as photovoltaics, loosening restrictions on energy-saving or energy-producing innovations, increasing taxes on fossil fuels to encourage conservation, and providing tax incentives for development and installation of energy-conserving technologies. With the exception of increasing fuel taxes, all of these measures were used in the United States and in many other nations during the past two decades.

took some time for these new players to contribute significantly to the power market, but by 1988 they were generating about 7 percent of electricity produced nationwide. Then, in 1992 the Federal Energy Policy Act made it possible for electric utilities to participate in the less regulated wholesale electricity market in addition to generating electricity and selling it to retail customers at regulated rates. In 1996 the Federal Energy Regulatory Commission issued rules that required utilities to make their transmission systems available for sale of electricity by third parties. Thus, a company in Texas, for example, could use transmission lines owned by an Arizona company to sell electricity to California.

In addition to this activity at the federal level, most states have been moving to deregulate electricity. In the traditional regulatory environment a customer had no choice regarding what company supplied his or her electricity, and prices were set by state regulatory agencies. The principal motivation for state-level deregulation was to increase competition between power producers, with the expectation that this would drive prices down. Another motivation was to allow consumers the opportunity to demand electricity produced in nonpolluting or renewable ways (even when such electricity is more costly than electricity produced from other sources).

California was one of the first states to move toward deregulation. One reason deregulation was particularly attractive to Californians was that in all the surrounding states electricity was much cheaper. In 1998 the average price of electricity in California was 9 cents per kilowatt hour (kWh), while in Oregon, Nevada, and Arizona it was 4.9 cents, 5.6 cents, and 7.3 cents, respectively. It was expected that deregulation would open up the California market to this cheaper electricity. In 1996 California passed the Electric Utility Restructuring Act. This law opened up the electric markets in California, allowing customers to choose electric suppliers and facilitating sales of electricity within the state and across state lines. Because it was expected that deregulation would reduce electricity prices, prices were capped at or below their 1996 levels.

Deregulation went fairly smoothly for the first few years, but in 2000 several different factors precipitated a crisis. First, between 1990 and 1999, electric demand in California rose 11 percent while generating capacity only rose 1.7 percent. Second, dry conditions in the Pacific Northwest, which supplied much of California's electricity, led to reduced hydroelectric capacity there. Third, during 2000 a few key power plants were temporarily out of service for maintenance and upgrades. Finally, transmission lines were unable to provide the capacity needed to move electricity from place to place within the state.

Wholesale prices rose dramatically in 2000, and some utilities were required to buy electricity at high prices but were prevented from passing those costs on to their customers. Three major utilities faced major financial problems, and some wholesale electricity providers were reluctant to sell to them for fear the bills would not be paid. Rolling blackouts plagued the state in early 2001. In the spring of 2001 both the federal and state governments responded to the crisis with a variety of measures designed to speed the construction of new generation and transmission facilities, relieve the financial burden on public utilities, and facilitate sales of electricity to California from other states. By summer 2001 the crisis had passed, and the power shortages ended.

The California experience certainly cooled interest in deregulation nationwide, but probably only temporarily. Electric demand is rising rapidly in the United States, and it will be necessary to increase both generating and distribution capacity. It is still widely believed that market-based competition is a fundamentally sound strategy for increasing flexibility in electric generation.

How should the lessons of these decades be interpreted? We can see that a small increase in the price of one energy commodity, oil, can stimulate enormous savings through conservation as well as moderate increases in supply. On the one hand, this can be interpreted as a demonstration of the power of the market and the flexibility of market economies. If the need arises in the future, these economies can respond again and shift to other energy sources relatively quickly. On the other hand, by demonstrating the possibilities of energy conservation and development of renewable technologies, we might argue that we are foolish not to take advantage of the opportunity of converting to renewable technology sooner, both avoiding future oil shortages and protecting the environment through reduction of pollution.

REFERENCES AND ADDITIONAL READING

Atherton, C. and D. Atherton. 1990. Disaster that fell with the rain on a bleak hill. *Guardian*, April 27, p. 25.

Balter, M. 1995. Chernobyl's thyroid cancer toll. *Science* 270:1758–1759.

———. 1996. Children become the first victims of fallout. *Science* 272:357–360.

Campbell, C. J. and J. H. Laherrere. 1998. The end of cheap oil. *Scientific American* 278:78–83.

Committee on Nuclear and Alternative Energy Sources (CONAES). 1978. U.S. energy demand: Some low energy futures. *Science* 200:142–152.

Commoner, B. 1977. *The Poverty of Power*. New York: Bantam Books.

Council on Environmental Quality. 1989. *Environmental Trends*. Washington, D.C.: U.S. Government Printing Office.

Deffeyes, K. S. 2001. *Hubbert's Peak: The Impending World Oil Shortage*. Princeton: Princeton University Press.

Dunn, S. 2001. *Hydrogen Failures: Toward a Sustainable Energy System*. Worldwatch Paper no. 157. Washington, D.C.: Worldwatch Institute.

———. 2001. Decarbonizing the energy economy in L. R. Brown, et al., eds. *State of the World 2001*. Washington, D.C.: Worldwatch Institute, pp. 83–102.

Energy Information Administration. 1989. *Commercial Nuclear Power 1989: Prospects for the United States and the World*. Washington, D.C.: Energy Information Administration.

———. 2002a. *Annual Energy Review, 2002*. Energy Information Administration. http://www.eia.doe.gov/aer/.

———. 2002b. *International Energy Annual, 2002*. Energy Information Administration. http://www.eia.doe.gov/iea/.

Ewing, R. C. and A. Macfarlane. 2002. Yucca Mountain. *Science* 296:659–660.

Feldman, D. L. 1995. Revisiting the energy crisis: how far have we come? *Environment* 37(4):16–20, 42–44.

Fischetti, M. 2002. Windmills: Turn, turn, turn. *Scientific American* July:86–88.

Flavin, C. and A. B. Durning. 1988. *Building on Success: The Age of Energy Efficiency*. Worldwatch Paper no. 82. Washington, D.C.: Worldwatch Institute.

Flavin, C. and N. Lenssen. 1994. *Powering the Future: Blueprint for a Sustainable Electricity Industry*. Worldwatch Paper no. 119. Washington, D.C.: Worldwatch Institute.

Flynn, J., R. E. Kasperson, H. Kunreuther, and P. Slovic. 1997. Overcoming tunnel vision: redirecting the United States high-level nuclear waste program. *Environment* 39(3):6–11, 25–30.

Ghazi, P. 1990. Chernobyl fallout may affect British farmland for decades. *Sunday Observer*, April 29, p. 9.

Gibbons, J. H., P. D. Blair, and H. L. Gwin. 1989. Strategies for energy use. *Scientific American* 261(3):136–143.

Gould, P. 1991. *Fire in the Rain: The Democratic Consequences of Chernobyl*. Baltimore, MD: Johns Hopkins University Press.

Hoffert, M. I., K. Caldeira, G. Benford, D. R. Criswell, C. Green, H. Herzog, A. K. Jain, H. S. Khesghi, K. S. Lackner, J. S. Lewis, H. D. Lightfoot, W. Manheimer, J. C. Mankins, M. E. Mauel, L. J. Perkins, M. E. Schlesinger, T. Volk, and T. M. L. Wigley. 2002. Advanced technology paths to global climate stability: Energy for a greenhouse planet. *Science* 298:981–987.

Hohenemser, C., M. Deicher, A. Ernst, H. Hofsass, G. Linder, and E. Recknagel. 1986. Chernobyl: an early report. *Environment* 28(5):6–13, 30–43.

Hohenemser, C. and O. Renn. 1988. Shifting public perceptions of nuclear risk: Chernobyl's other legacy. *Environment* 30(3):4–11, 40–45.

Hohenemser, C. 1996. Chernobyl: 10 years later. *Environment* 38(3):3.

Hubbert, M. K. 1969. Energy resources. In NAS, *Resources and Man*. San Francisco: W. H. Freeman, pp. 157–242.

Iranpour, R., M. Stenstrom, G. Tchobanoglous, D. Miller, J. Wright, and M. Vossoughi. 1999. Environmental engineering: energy value of replacing waste disposal with resource recovery. *Science* 285:706–710.

Lake, J. A., R. G. Bennett, and J. F. Kotek. 2002. Nuclear power. *Scientific American* 286(1):73–81.

Leopold, L. B. 1996. *Sediment Problems at Three Gorges Dam*. International Rivers Network. http://www.irn.org/im/programs/3g/leopold.html.

Lovins, A. B. 1977. *Soft Energy Paths*. New York: Harper Colophon Books.

Mufson, S. 1996. Floods leave Beijing eager for new dam. *Washington Post*, August 6, p. A11.

Pasqualetti, M. J. and K. D. Pijawka, eds. 1984. *Nuclear Power: Assessing and Managing Hazardous Technology*. Boulder, CO: Westview Press.

———. 2001. Wind energy landscapes: society and technology in the California desert. *Society and Natural Resources* 14:689–699.

Rembert, T. C. 1997. Electric currents. *E Magazine* 8(6):28–35.

Righter, R. W. 1996. *Wind Energy in America; A History*. Norman, OK: University of Oklahoma Press.

Sawyer, S. W. 1986. *Renewable Energy: Progress, Prospects*. Washington, D.C.: Association of American Geographers.

Shea, C. P. 1988. *Renewable Energy: Today's Contributions, Tomorrow's Promise*. Worldwatch Paper no. 81. Washington, D.C.: Worldwatch Institute.

Specter, M. 1996. Ten years later, through fear, Chernobyl still kills in Belarus. *The New York Times*, March 31, 1996, p. A1.

Stone, R., 1996. The explosions that shook the world. *Science* 272:352–354.

Taylor, J. J. 1989. Improved and safer nuclear power. *Science* 244:318–325.

Tyler, P. E. 1996. Chinese dam's inexorable future dooms rich past. *The New York Times*, October 6, 1996, p. A1.

Turner, J. A. 1999. A realizable renewable energy future. *Science* 285:687–689.

United Nations Development Programme 2000. *Human Development Report 2000*. New York: Oxford University Press.

U.S. Bureau of the Census. 2001. *Statistical Abstract 1997*. Washington, D.C.: U.S. Government Printing Office. www.census.gov.

U.S. Congress, Office of Technology Assessment. 1980a. *An Assessment of Oil Shale Technologies*. Vol. II: *A History and Analysis of the Federal Prototype Oil Shale Leasing Program*. Washington, D.C.: U.S. Government Printing Office, OTA-M-119.

———. 1980b. *World Petroleum Availability 1980–2000*. Washington, D.C.: United States Government Printing Office, OGA-TM-E-5.

U.S. Department of Energy. 1987. *Energy Security*. Washington, D.C.: U.S. Department of Energy.

———. 2002a. National Renewable Energy Lab. www.nrel.gov.

———. 2002b. Office of Energy Efficiency and Renewable Energy. www.eren.doe.gov.

Wilbanks, T. J. 1988. The impacts of energy development and use. In H. deBlij, ed. *Earth '88: Changing Geographic Perspectives*. Washington, D.C.: National Geographic Society, pp. 96–114.

Williams, N. and M. Balter. 1996. Chernobyl research becomes international growth industry. *Science* 272:355–356.

World Resources Institute, 1996. *World Resources 1996–97*. New York: Oxford University Press.

———. 2000. *World Resources 2000–2001*. Washington, D.C.: World Resources Institute.

Yergin, D. 1991. *The Prize: The Epic Quest for Oil, Money, and Power*. New York: Simon and Schuster.

For more information, consult our web page at ***http://www.wiley.com/college/cutter***.

Study Questions

1. In the past, human energy use has focused for the most part on one fuel at a time: wood, then coal, then oil and gas. Today, we may be entering an era in which we use a diversity of fuels, with different fuels for different uses. How does this energy diversity affect overall energy efficiency? How does it affect energy security?

2. What would be the effect on energy use if the United States followed the lead of most other wealthy countries and instituted relatively high fuel taxes (to raise the price of gasoline to, say, $2 or $3 a gallon)? What would be the effect on the environment? Would such an action eliminate or reduce any environmental externalities?

3. In the history of oil prices between 1970 and 1990, we can see evidence of scarcity and evidence of plentiful supplies. Do oil prices accurately reflect the relative scarcity of oil in the world? What do you think the future of oil prices will be in the next decade? In the next 50 years?

4. Should the United States encourage the construction of dams for hydroelectric generation? Why or why not?

5. Are electric vehicles the answer to our energy problems in the next 50 years? Why or why not?

CHAPTER 15

THE TRANSITION TO A GLOBAL SUSTAINABLE SOCIETY

The history of economic growth and resource use since the Industrial Revolution is one of a direct relationship between development and environmental degradation. As technology has advanced, we have grown wealthier and demanded more goods, thereby using more resources and producing more pollution. This trend, in combination with population growth, resulted in very rapid increases in raw material consumption and pollution output, with the consequences described in the preceding chapters.

In the late 1980s, an old word took on a new meaning in the debates about resource use, economic development, and the future. As we discussed in Chapter 3, at that time a new global awareness of the environment and human society emerged, and that awareness needed a word to focus debates. *Sustainability*, a well-established concept in renewable resources management in general and forestry in particular, began to be applied to the entire global environment and the human societies that rely on it and have profoundly altered it.

In this chapter we explore the medium-term future of resource use: the next century or so. Our discussion begins with a look at some predictions for the future that stimulated concern for sustainability. We then examine the different visions of sustainability being discussed today, indicating considerable divergence in how the concept is expressed. The final sections of the chapter describe technologies that may help us move toward sustainability and suggest some social actions that may encourage their adoption.

LIMITS TO GROWTH?

Growth in population and resource use during the 1950s and 1960s was unprecedented, and many alarms were sounded about future disasters. Books with titles like *The Hungry Planet* (Borgstrom 1965), *Famine, 1975!* (Paddock and Paddock 1967) and *The Population Bomb* (Ehrlich 1968) were prominent in bookstores and academic discussions. Then in 1972, at the height of the environmental movement, an influential book entitled *Limits to Growth* was published (Meadows et al. 1972). This book reported on the results of a computer-based modeling study that extrapolated into the future then-recent trends in population growth and resource use. It concluded:

> If the present growth trends in world population, industrialization, pollution, food production, and resource depletion continue unchanged, the limits to growth on this planet will be reached sometime within the next 100 years. The most probable result will be a sudden and uncontrollable decline in both population and industrial capacity (Meadows et al. 1972:24).

The trends that were extrapolated to reach these conclusions were those of the 1950s and 1960s, an era of rapid growth in both population and resource use, with few if any checks on that growth. It was therefore impossible for the authors of *Limits to Growth* to see, for example, the energy and industrial crises that began in 1973, which profoundly changed the nature of economic growth and its relation to resource use. Instead, they saw only exponential growth in resource consumption and pollution output, inevitably leading to an "overshoot and collapse" outcome (Fig. 15.1).

A similar study, called *Global 2000*, was carried out by the U.S. government in the late 1970s (Council on Environmental Quality 1981). Like *Limits to Growth*, it was based on detailed global models, including such conditions as population growth, economic development, food production, environmental pollution, and energy use. Like *Limits to Growth*, it made projections of the future based on recent (1960s and 1970s) trends. *Global 2000* concluded:

> If present trends continue, the world in 2000 will be more crowded, more polluted, less stable ecologically, and more vulnerable to disruption than the world we live in now. Serious stresses involving population, resources, and environment are clearly visible ahead. Despite greater material output, the world's people will be poorer in many ways than they are today. (Council on Environmental Quality 1981:1).

Had *Global 2000* been published in the mid-1970s it would have had more impact on government policy. Instead, it emerged at the beginning of the Reagan administration, when the prevailing views were less concerned about environmental conditions and much more focused on the ability of business to solve the world's problems. Many of the reactions to *Global 2000* were strongly critical, concentrating on the dangers inherent in extrapolating past trends into the future and highlighting occasional factual errors (Weinberg 1984). These criticisms were strengthened by the fact that in the 1970s two very significant changes in these trends occurred. Population growth and industrial resource use both began to slow considerably. In short, the doomsday predictions of the 1960s and 1970s that had been so convincing when they were first published weren't very credible a decade or two later.

Despite this reassessment and reaction to the concerns expressed in these studies, the basic fact of significant population growth and the desire for more economic growth in poorer countries still pose troubling questions for the future of the earth's environment and resources. When the awareness of global environmental change, centered on issues such as climatic warming, ozone depletion, and deforestation, grew again in the late

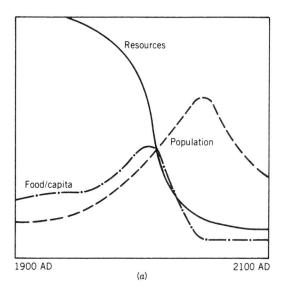

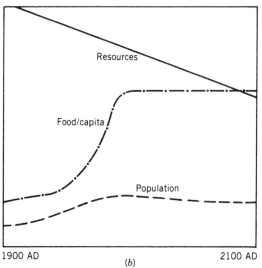

Figure 15.1 Results of the Limits to Growth model. (*a*) These curves were based on simple extrapolation of current (c. 1970) trends into the future. (*b*) These curves are based on early achievement of stability in population and resource use. *Source:* U.S. Congress, 1982.

1980s, the population/resources question once again came under discussion. The most significant document of this period is *Our Common Future*, also known as the Brundtland Report, prepared by the United Nations Commission on Environment and Development (WCED) in 1987. One of the central arguments of *Our Common Future* is that economic development and environmental protection are not mutually exclusive. Instead, it argued that economic development is necessary for good resource management. Unless people's basic needs are met and they have choices about how they will live, they will not have the opportunity to protect Earth's resources for future generations. This idea is at the core of the movement for sustainable development.

At the Earth Summit in 1992, participating nations adopted a blueprint for sustainable development called *Agenda 21* (see Chapter 3). This action plan included 27 different principles for achieving sustainability that include a common vision for the protection of natural resources, equity, and economic growth. Special consideration was given to reducing the disparities in standards of living between rich and poor nations, particularly as it leads to increased poverty and further environmental degradation. The goal is to improve the human condition without harming the environment as the world moves toward a sustainable future (Annan 2002).

What Is Sustainable Development?

In the late 1980s the terms sustainability and sustainable development entered widespread use in the context of Earth's natural resource future. These terms mean many things to many people, and a definition of sustainability is a continuing problem. One thing is clear: sustainable development must be fundamentally different from the kind of economic development that dominated the planet in the past few hundred years, because that development, if projected into the future, is clearly not sustainable. The central issue is: What is to be sustained? The problem is not a simple one, so it is important that we define our concept of sustainability.

Environmental Versus Economic Sustainability

If there are indeed real limits to growth, and if we are approaching them in the foreseeable future, then now is the time to change the way we use resources. One solution is to reduce economic growth or halt it altogether. Many people believe that we have become addicted to economic growth, and so we demand growth for growth's sake. This is certainly true of our financial markets, where a company that doesn't grow is much less valuable than one that grows steadily.

But demanding an end to economic growth has three basic problems. One is population growth. As we learned in Chapter 5, the world's human population is almost certain to reach 9 billion sometime in the twenty-first century. Simply maintaining current standards of living would require economic growth equal to population growth. A second problem is that much of the world has a standard of living that many observers regard as undesirable, if not unacceptably low. Improving the standard of living in the developing world means increasing industrial and agricultural output to provide food, housing, safe water and energy supplies, and consumer goods for large numbers of people. Finally, from the standpoint of political and social action, few people are likely to support a no-growth policy. Many people have desires or expectations of a better life for their children than for themselves, or at least the prospect of future improvement. To abandon that hope would be difficult.

In Chapter 1, in the context of environmental ethics, we described two different views of natural resources: a human-centered view and a nature-centered one. These two views illustrate the problems in setting priorities for sustainability. One way to approach the problem is to distinguish between environmental sustainability, economic sustainability, and social sustainability (Goodland and Daly 1994; National Research Council 1999).

Environmental sustainability means maintaining the physical condition of the environment, especially those aspects of natural systems on which we depend for our health and welfare but also including other living things. This means not only maintaining the restorative capacity of renewable resources but also limiting our output of waste so as not to exceed the capacity of the environment to absorb and reprocess it. The primary objectives of environmental sustainability are to maintain ecosystem integrity, carrying capacity, and biodiversity.

Economic sustainability means maintaining sufficient capital to support human needs indefinitely by providing the basis of production. This capital may include human-made capital such as machines, natural capital (forests), and social and human capital (the ability of workers and societies to organize themselves in productive ways). Its objectives are continued economic growth, equity, and efficiency.

Some authors also distinguish a third type, social sustainability, which means using resources in ways that increase social justice and social equity, while decreasing social disruptions. It requires direct community involvement in resource management decisions, as well as a significant commitment to restructuring social relations (Goodland and Daly 1994; Serageldin 1993).

Some would argue that the capacity of the earth's ecological systems to support life—human and otherwise—with a reasonable quality of life should be sustained. Others take sustainable development to mean continuing economic development and economic growth. Still others would consider maintaining and improving the health and well-being of the human population as the most important goal.

A Working Definition of Sustainability

The sustainability concept implies a much longer time-frame than what commonly is used in most day-to-day resource-use decisions. We recognize our obligation not only to those living today but also to future generations hundreds or even thousands of years hence. We must use resources in such a way as to leave for future inhabitants the same or better opportunities than we enjoy, rather than depending on future technological innovations or migration to other planets as a means to support future populations. In the words of the Brundtland Report, sustainable development means "development that meets the needs and aspirations of the present without compromising the ability of future generations to meet their own needs" (World Commission on Environment and Development 1987:45). For the purposes of this book, however, we need a working definition, and one that is more specific than that provided by the Brundtland Report.

We recognize that achieving sustainability is carried out by societies rather than by individuals, and thus defining goals should be left to political processes. Whether one takes an environmental, economic, or social perspective, it seems clear that the world faces several fundamental problems today. Foremost among these are large disparities in wealth between rich and poor, with many of the world's people suffering unacceptably poor conditions of health and welfare; and severe and increasing stresses on the world's natural resource base. Either of these situations could be improved by making changes in resource-use systems that increase the environmental efficiency of resource use. By this we mean *increasing the amount of material wealth generated per amount of natural resource degradation*. At present we believe that our economic, social, and resource-use systems require improvement, both in increasing wealth (especially for the poor of the world) and decreasing resource degradation (especially in industrial societies responsible for the majority of resource use). Thus, we can describe actions that move *toward* sustainability without specifying in detail the ultimate goals of sustainability.

How Does Sustainability Work?

Increasing the environmental efficiency of resource use means breaking the link between economic wealth and environmental degradation. We already know that this link is not universal. In Chapter 14, for example, we saw that increases in energy prices in the 1970s were followed by a period of economic growth without increased energy use. In Chapter 10 you read that discharges of biochemical oxygen demand (BOD) and other important pollutants to rivers declined in the 1980s and 1990s, while Chapter 11 described the significant decrease in sulfur oxide emissions during the same period. Although these environmental improvements had an economic price tag, we were able to bear such costs without suffering a decrease in our standard of living. At the same time, we have not been able to significantly reduce either our output of CO_2 or our consumption of forest products. One reason for the lack of improvement in these areas is the economic costs that would be incurred.

Nonetheless, the possibility exists that we can decouple environmental degradation and economic activity (Fig. 15.2). Such a decoupling will be necessary to improve the standards of living of developing countries without overwhelming the earth's capacity to support a growing population. Realizing this change is at the heart of the transition to a sustainable society.

Sulfur dioxide concentrations in cities provide an interesting illustration of the problem. In general, cities in poor countries have lower concentrations

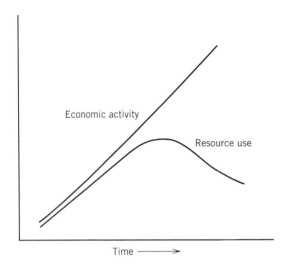

Figure 15.2 Decoupling of economic activity and resource use. In the past, increasing wealth has always meant increasing use of natural resources and pollution output. Can we effectively separate the two and continue to grow economically while improving the environment? *Source:* Redrawn from World Bank, 1992.

tion controls at this time, but reductions also occurred in the poorest nations. In other words, you don't have to be wealthy to reduce pollution. SO_2 levels have declined since the 1990s in China, where significant achievements have been made in reducing coal use and the pollution that it causes.

Thus, an alternative to stopping growth altogether is to change the nature of economic growth. Most of our present resource-use systems are destructive and inefficient, and waste products are poorly managed. Thermal electric energy production is an obvious example. When coal is used to make electricity, less than 40 percent of the heat is actually converted, and the rest is discharged to the environment as waste. At the same time, we burn other fuels, or perhaps even use electricity, to heat buildings. If we could use the waste heat from electricity generation to heat buildings, we could use much less coal without losing either electricity or warm spaces to work and sleep.

Some of the decrease in SO_2 emissions that occurred in the 1980s resulted from energy conservation through reduction of waste. Some resulted from emissions controls, which forced changes in the way fuel is burned. Another part of the decrease may have come from replacing fossil fuels with renewable energy such as hydroelectric power. A significant part came from changing the types of things we do to generate income: decline in the industrial sector and growth in the service sector.

In the following paragraphs, we will describe four kinds of changes in resource use that will lead to sustainability: waste recycling; waste reduction; design for reuse and recycling; and changing the resource-use structure. These changes are feasible today, and many are already occurring. We believe, however, that these and other changes must

of SO_2 than cities in countries of intermediate levels of wealth, while cities in very wealthy countries have the lowest SO_2 levels (Chapter 12) (Fig. 15.3). The disparity is due to the poor countries' relatively light use of fossil fuel. Cities in newly industrializing countries use substantial amounts of fossil fuel but without effective emission controls. The wealthiest cities can afford pollution controls; the poorer ones generally cannot.

Does this mean that the poorest cities will have to endure a period of high pollution before they can afford to control emissions? Not necessarily. SO_2 levels declined between 1976 and 1985, the period of rapid energy price increases, in cities of all wealth levels. The declines were larger in cities in wealthy nations, which were installing pollu-

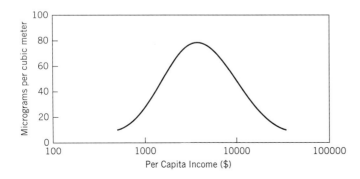

Figure 15.3 Sulfur dioxide concentrations in relation to GDP per capita. *Source:* Redrawn from Shafik, 1994.

be implemented much more rapidly if significant progress is to be made in the next few decades.

Waste Recycling

To achieve sustainability without continually switching from one set of resources to another requires that all resource use be renewable. Furthermore, the stocks from which those renewable resources are drawn should be maintained at a constant, if not improving, quality. As discussed in Chapter 13, recycling is the process whereby a material is recovered from the consumer product and then becomes a new material in the production of a similar or different product.

Efforts to improve recycling are widespread, especially in areas that are facing difficulty in disposing of wastes. In the United States, these efforts initially took the form of encouraging waste-generators to collect materials for recycling rather than disposing of them. This is sometimes called supply-side recycling. Among the early efforts were laws enacted in the 1970s requiring that a deposit be collected on the sale of beverages in bottles or cans. Part of the impetus for bottle-deposit laws came from a problem of roadside litter. The consumer gets his or her deposit back when the container is returned to the point of sale. The deposit is merely a behavioral incentive to get the consumer to save and return the container.

When the landfill crisis struck in the United States in the mid-1980s, a new reason for recycling emerged: saving precious space in landfills. A variety of programs were instituted in communities large and small; some were voluntary and some were mandatory, but nearly all encouraged consumers to separate materials into different types to facilitate recycling. The programs were very successful in generating recyclable material—large amounts of paper, plastic, aluminum, and other materials were collected (Fig. 15.4). New York City recycles around 1100 tons of metal, glass, and plastic daily. But problems quickly emerged. The supply of recyclable material was suddenly much greater than the amount that could be absorbed as raw materials. Without adequate outlets for used paper, stocks accumulated at collection points, creating nuisances and sometimes fire hazards.

The problem was particularly acute for municipalities that had instituted mandatory recycling.

Figure 15.4 Recycling programs instituted in response to the landfill crisis in the 1980s dramatically increased supplies of recycled material, in many cases beyond the ability of processors to use the material. Nonetheless, the vast majority of solid waste generated in the United States is still deposited in landfills.

They made substantial investments in recycling equipment and bore the cost of collecting recyclable materials. They had planned to recoup these expenses through the sale of these materials. But the rapid increase in the supply of recyclables drove down the prices of these materials, which cut revenues for recycling programs. Most recycling programs ran at substantial deficits, and often the deficits were so large that it became cheaper to dispose of these materials in landfills than to recycle them, even with landfill space at a premium.

One result of this problem was a shift in recycling policy toward so-called demand-side programs. These programs were intended to increase the demand for recyclable materials, usually by encouraging or requiring manufacturers to increase their use of recycled materials. Germany was a leader in this area, in part because of its high landfill costs. In 1991, Germany enacted a program that required manufacturers to take full responsibility for their packaging. In response, a consortium of about 600 manufacturers developed a system for joint collection and recycling of materials. Packages produced by member companies bore a green dot, indicating that they were eligible to be collected and processed by the consortium.

Mandatory recycling programs tend to be expensive. The increased costs are of two types. First, collecting and processing recyclable materials is often more expensive than disposing of these materials in landfills. In the case of the German green dot program, the cost of collection and processing materials in 1994 was about double the cost of collecting and disposing other household waste. For example, if New York City abandoned its recycling program it would save more than $102,000 per day, or $56 million over 18 months (Anonymous 2002). Second, the quality of recycled materials is lower than that of virgin materials. In some cases this difference in quality is not important, but in other cases it can be significant, prohibiting the use of recycled materials. Even if recycled materials are acceptable, they may be more expensive to use than virgin materials because of added processing necessary to remove impurities. These higher costs, of course, are passed on to the consumer.

Although these costs may be high in the short run, if the law stays in place and if companies work hard to adapt, they can modify their products and packaging in ways that will reduce these costs, perhaps to levels below the cost of using virgin materials. The high costs of processing waste materials can be a powerful incentive to reduce material requirements and waste generation. But such reductions may mean complete redesign of packaging and sometimes of the products themselves. Following redesign, new manufacturing, handling, and retailing facilities must be developed, and consumers have to learn to accept the new products. Shifting toward sustainability in materials use is not a simple matter.

The paper industry provides an example of the shift toward increased recycling in the long run. Paper recycling seems relatively simple—just substitute used paper for trees in the pulping process and make old paper into new paper. But it isn't that easy. Like most recycling situations, the biggest problems result from recycled materials that aren't clean or uniform. Used newspapers, for example, include inks, coatings on papers, metal staples, plastic wrappers, and miscellaneous materials mixed with the paper fibers used to make the original paper. As a result, these materials must go through special processing to remove unacceptable substances before they can be used to make new paper.

In the short run, these problems have limited the amount of paper that could be recycled, contributing to the problems with recycling programs. The situation was aggravated by relatively low paper prices, which made it difficult for paper companies to invest in new plants. But the price of paper rose substantially in the early 1990s, and by the mid-1990s several companies had installed new facilities capable of processing larger quantities of recycled paper. This resulted in a significant rise in the volume of paper recycled, as well as an increase in the price of used paper. Worldwide, recycled paper averages about 43 percent of the inputs into new paper production, with ranges from a low of 27 percent in China to72 percent in Germany (Abramovitz and Mattoon 1999). The United States averages around 46 percent.

Waste Reduction

As discussed in Chapter 10, pollution control can be approached in two fundamentally different ways. One is to remove pollutants from the waste stream just before it is discharged to the environment; a sewage-treatment plant is a good example.

This is known as *end-of-pipe pollution control*. The other approach is to not generate waste materials in the first place. This is known as *waste reduction*. Progress has been made in waste reduction, through a combination of governmental programs (USEPA's Pollution Prevention and local "pay as you throw" programs) and voluntary actions by major industries.. Among the leaders in waste reduction in the United States are chemical companies that were major emitters of pollutants (Dow, duPont, and 3M, for example), but they were joined by a large range of manufacturing concerns anxious to reap the benefits of a cleaner way of doing business.

What are the benefits of waste reduction to a manufacturing company that would lead it to reduce pollution voluntarily? There are many benefits, though not all of them are directly economic. Let us consider a few.

- **Customer demand** Consumers of manufactured goods, at both the retail and wholesale level, are interested in buying "environmentally friendly" products. If two products are side by side in the supermarket and one appears to have been produced with less pollution or has been produced by a company with a good environmental reputation, the consumer will be more likely to pick that product. This applies not only to individual decisions but also to purchases by institutions and other manufacturers. Environmentally conscious firms prefer to buy clean products, so they can tell their consumers that the products they sell are made with clean materials. Students and faculty ask universities to buy recycled paper. Publishers demand recycled paper so that they can say their books are printed on recycled stock. Usually, this means that the environmentally friendly product can command a higher price, though perhaps not very much higher. Procurement of environmentally friendly products is an untapped resource. For example, the 3700 or so American colleges and universities spend about $300 billion per year on supplies (Motavalli and Harkinson 2002). What if that purchasing power was targeted to buy only environmentally responsible products? Similarly, what if the U.S. government actually mandated that all federal agencies develop "green purchasing" policies? No doubt we would see rapid movement toward sustainability as industries would compete for part of the $385 billion that the government expends yearly on goods, services, and supplies!
- **Reduced waste disposal costs** It is expensive to dispose of wastes. Tightening environmental regulations greatly increased costs for manufacturers' waste disposal. These costs are especially high for toxic and hazardous substances, which must be taken by specially licensed waste-handling companies, which in turn have high costs associated with special equipment, employee training, and liability insurance. If wastes are disposed of in air or in water, a permit is usually needed. This kind of disposal requires that the polluter meet special conditions in addition to going through the costs of obtaining a permit. But even if the wastes are ordinary materials that can be taken to a landfill, those costs also have increased dramatically in recent years.
- **Reduced liability exposure** Many companies produce toxic or hazardous wastes as part of their manufacturing processes. In the past these wastes were often disposed of in the most convenient and inexpensive manner available, such as in landfills. When the Comprehensive Environmental Response, Compensation, and Liability Act (Superfund) took force in 1980, companies were suddenly made liable for their previously disposed wastes. If those wastes threatened human health, thereby creating a hazard to people or the environment, then they had to be cleaned up. Often a company thought it was disposing of wastes properly, only to learn later that it had created a major contamination problem (Cutter 1993). Even if a company produced very little waste, if it was among the parties that had deposited wastes in a given site it could be held responsible for a share of the cleanup costs. These costs grew rapidly in the 1980s and 1990s, sometimes running into billions of dollars for a single site. Under the Superfund program, responsible parties cleaned up 70 percent of the sites the USEPA placed on its National Priority List, spending $18 billion in the process (USEPA 2002).

    A manufacturing company could therefore be held liable for substantial cleanup costs associated with past waste disposal. Furthermore, if a company purchased another company, it also purchased that company's potential liabilities. In many cases, companies didn't even know what wastes they might be held responsible for, let alone the potential costs they might be asked to bear. This presents a major risk for the company.

    How can such risks be avoided? One way, as we have said, is to not generate the wastes in the first place. If we don't use or create toxic materials in manufacturing, then there can't be a disposal problem. Eliminating the use of potentially hazardous materials also eliminates the possibility that practices believed to be safe today will later be found to be dangerous.
- **Reduced material costs** In some cases, finding ways to manufacture a product using nonpolluting materials also means reducing the costs of inputs to a process. This may be especially true for hazardous substances, which often are more expensive to manufacture and handle than safer substances. Manufacturing processes that create less pollution do not

always cost less, but cost savings are realized surprisingly often. A new field of study called *industrial ecology* examines materials and energy flows in products and manufacturing processes, industrial sectors, and local, state, and national economies, with an eye toward more sustainable approaches to materials use and energy (Graedel and Allenby 1995). College courses in industrial ecology are becoming the norm in many schools, and a new journal, the *Journal of Industrial Ecology*, helps promote research and practice in this field.

- **Enhanced public relations** Businesses need to forge good relations with the communities where they are located. Government officials and agencies, from the national government down to the municipality, have regulatory authority over many business activities. Usually, decisions made under this authority involve a significant amount of discretion on the part of the decision-makers. It helps if these decision-makers have a positive attitude toward a company and its practices.

  For example, suppose a manufacturer wants to expand its facilities at an existing plant and this expansion involves a zoning change, a building permit, an emissions permit, or some other approval from the local government. Suppose also that the company has a history of being secretive about its operations, or perhaps it experienced a chemical spill or significant industrial accident recently. The authorities responsible for approving the expansion plans may look very carefully at the proposal and may attach difficult conditions to their approval to protect the safety of the community. On the other hand, if the manufacturer has a good reputation for operating a clean facility, for handling its wastes properly, and not using significant amounts of hazardous materials, the local authorities are likely to be much more willing to approve the proposed expansion with a minimum of red tape.

- **Employee morale and loyalty** Good relations with one's employees are just as important as good relations with government regulators. Good employee morale is fundamental to a smooth-running organization. Employees are likely to feel better about working for a company that has a good environmental reputation than for one that is a known polluter. Recruitment of new employees is made easier for the same reasons. In addition, employees not only work for the company but also usually live in the same community and thus are exposed to its emissions. For employees, the environmental impacts of their work are even more important than they are for other members of the community.

Design for Reuse and Recycling

As mentioned in Chapter 13, perhaps the greatest barrier to recycling is the mixture of materials found in waste. Separating different materials from each other is expensive, time-consuming, and difficult to mechanize. Asking consumers to separate glass from paper from plastic is one thing. But if different materials are joined together in a single object, as in drink boxes or foil-covered wraps, they become nearly impossible to recycle. What can be done to make materials more recyclable?

Fortunately, there are many opportunities to make recycling easier. Consider the changes in manufacturing practices described in the previous section. Nearly all of them brought changes in the way products are manufactured, distributed, sold, used, and disposed of. Such changes occur routinely as part of the process of developing new products, more efficient manufacturing techniques, and marketing strategies. These changes in product design and manufacturing are opportunities to move toward sustainability, if we choose to take them.

Automobiles are an excellent example. Autos are a source of much environmental degradation, because of their manufacture, use, and disposal. They are also complex machines containing a wide variety of materials, including metals (primarily steel and aluminum), glass and fiberglass, rubber, oils and other petroleum products, and a wide array of plastics. A typical auto consists of about 70 percent ferrous metals (iron, steel), 5 percent nonferrous metals (copper, aluminum, zinc, magnesium), and about 25 percent nonmetal materials. These materials are mixed together, attached to each other, and not easy to separate. Yet nearly all of them are reusable or recyclable in some way.

Most of the world's automobiles are at least partially recycled when their useful lives end. Valuable parts from newer wrecked vehicles and some older vehicles are removed and sold as spare parts at auto-salvage yards. Easily removed and valuable components such as tires and batteries are removed, and fluids are drained. The vehicle is then flattened and shipped to a shredding facility, where it is cut up into many small pieces. The iron and steel are magnetically separated, and the remaining material (fabric, rubber, glass, etc.) is disposed of in landfills. About 44 percent of the ferrous metals in a typical family vehicle made in the United States are recycled iron and steel. Relatively little of the nonferrous content of such vehicles is recycled (Steel Recycling Institute 1996).

Recognizing the need to improve automobile recycling, in the mid-1980s the BMW Company of

Germany embarked on a program to design and produce automobiles that would be much more easily recycled than those presently produced (Management Institute for Environment and Business, no date). The plastic content of automobiles is increasing, and most of this plastic is nonrecyclable. In Germany nonrecyclable parts of automobiles make up about 1 to 2 percent of municipal solid waste. With demands for recycling increasing, BMW hoped get a headstart in responding to the problem and gain a competitive advantage over other companies that would be forced by government regulations to adopt recycling requirements later. BMW continues to conduct research and development activities and is sharing its knowledge through partnerships with other companies.

One way in which industries can improve material recycling is to develop closer coordination among companies involved in different parts of the materials cycle. Waste from one operation can be a raw material for another, if the material is handled properly and costs are kept to a minimum. For example, industrial parks could be developed in which a group of companies that use related materials locate near each other, sharing waste-handling facilities and exchanging materials between them. Plans are being developed for such facilities in several U.S. cities.

Clearly, there is much potential for movement toward sustainability in manufacturing and waste handling. But long lead times are needed for activities such as the complete redesign of an automobile with recycling in mind or the coordination of independent companies in a single industrial park.

Changing Consumption Patterns

Changes in industrial practices designed to improve recycling and reduce waste are significant factors in moving toward sustainability (Kates 2000; Myers 2000). But some argue that these are only superficial changes and that real progress requires major changes in the kinds of goods we use and how we lead our day-to-day lives. This is a difficult challenge, especially for societies like the United States, which has a long history of mass consumption. But opportunities exist to change consumer habits, both in wealthy industrial societies and in the developing world. Some of these opportunities involve sophisticated technology that is probably accessible only to consumers in wealthy countries, while others are relatively inexpensive and use simple, low-cost technology.

From the standpoint of environmental impact, among the most damaging activities of industrial societies are those involving energy use, especially combustion of fossil fuels. Together, the United States and the European Union emit about 9 billion metric tons of CO_2 per year, or 35 percent of total world emissions. On a per capita basis, the United States emits about 4 times the world average and 22 times India's level. A significant improvement in energy efficiency would reduce emissions not only of CO_2 but also of the hydrocarbons, nitrogen oxides, and other pollutants produced at the same time.

In Chapter 14 we discussed technology that could help automobiles and houses expend less energy. But what would it take for people to use less? One way might be for them to travel less, for transportation is the largest single user of energy in the United States. Much of this travel consists of commuting to and from work. Efforts are being made to encourage commuters to car-pool or use public transit, but thus far these measures have not substantially reduced the use of individual automobiles for commuting. In the meantime, relatively inexpensive fuel has encouraged the development of housing in remote suburban and rural areas far from urban workplaces. The rapid development of communications technology in recent years has made the virtual office possible, and increasing numbers of people are working at sites remote from their employers' offices.

Another way in which computer technology can improve energy efficiency is through use of "smart machines." Computers can be used to automatically monitor the environmental conditions and performance of machines as well as to regulate those machines to improve their efficiency. For example, much of the improvement in automobile performance and emission reductions since the mid-1980s was achieved with use of computers to control ignition timing and fuel mixtures in engines. Modern office buildings are equipped with sensors that monitor temperature, humidity, and ventilation, automatically controlling the flow of heat and fresh air to maintain comfortable conditions without excessive energy use. Tiny computer chips are being studied for possible use in micro-flaps that can actively respond to small variations in air movement and control the air flow across the surface of an airplane, reducing drag and improving fuel efficiency. The development of a Sustainable Universities Initiative has helped integrate formal and informal learning and practice in fostering the wise use of resources. For example, in South Carolina, resource efficient "Green Buildings" are now being built as student dormitories. Not only are front-loading washing machines saving more than 2 million gallons of water per year (a 30 percent reduction in water use), but the university is saving more than $20,000 annually in water bills. Furthermore, these washing machines communicate with dorm residents to let them know when their wash is done.

Technology such as computerized buildings could be commonplace fairly soon in wealthy countries, but it will probably be too costly for most people in developing countries. Yet most of the economic growth taking place in the world is in relatively poor countries. Does this mean that new, environmentally friendly technology will not be available? Or should we view the rapid economic growth taking place in China, Latin America, and other developing areas as an opportunity to develop in ways that will be more sustainable?

The bicycle is an example of inexpensive technology that provides effective transportation without appreciable environmental impacts. It is a sustainable transport solution that can be used in poor as well as rich countries. In cities and suburban areas, where travel distances are generally less than a few miles, bicycles provide free transport that is often faster than motorized transport. In some countries, bicycles are a vital part of the transport system. In Denmark, for example, one in five journeys is made by bicycle. In the Netherlands, two-thirds of children ride to school on bicycles (*The Economist* 1995). In addition to being emission-free, bicycles take up much less space when parked.

Unfortunately, however, bicycles don't mix well with either pedestrian or motorized traffic. Safety is an issue here: on a per-mile basis, serious injuries are 20 times more likely on bikes than in cars. This means that roadways need to be designed in ways that will keep bikes separate from other traffic. In some cases, separate bicycle lanes can be designated on streets, while in other cases entire separate roadways must be constructed. Such designated bicycle routes are common in Denmark and the Netherlands, where they clearly have a positive effect.

Energy use is not the only area in which inexpensive sustainable technologies are readily available, but it provides a good illustration of the potential. Many existing sustainable technologies are more labor-intensive than their counterparts in industrialized countries. When cultural attitudes include a bias against labor-intensive activities, it will be difficult to encourage them. But if the exercise and involvement of riding a bicycle instead of a bus, walking instead of driving, or donning a sweater instead of turning up the thermostat can be seen as desirable ways to live, then these approaches offer enormous potential in improving the efficiency of resource use.

Science and Technology for Sustainability

The drive to sustainability must be supported by science and technology. The development of computers is an example of a technological advance that can contribute to sustainable development in a major way and is of great significance. But there also needs to be a more focused effort to create the tools needed to move our social, economic, and political systems toward sustainability. That focus in turn depends on clear scientific knowledge that will increase awareness of the need for change and inform policymakers as they guide that change.

Kates and others (2001) argue that the science that is needed is different from most science taking place today. Whereas the research that led to development of computers (and emerging nanotechnology, which may be as significant a development as computers) is compartmentalized into very focused and specialized work, sustainability science must be broad, integrative, and interdisciplinary. Understanding problems like the degradation of agricultural lands or managing water resources to meet diverse needs, even at the regional or local level, requires bringing together scientists with a wide range of abilities and knowledge.

TIPPING THE BALANCE

If the technologies that will lead to sustainability either are available today or can be created in a decade or two, why aren't we moving more rapidly to develop and adopt them? If relatively minor changes in how consumers use resources will have a significant positive impact on environmental quality at relatively low cost, why aren't we making these changes today?

Individual Action

The answer is that individually and collectively we lack the desire or the motivation to change. It is too easy to continue doing things the way we do. Gasoline is cheap, and we can afford to use as much as we want. Life in the suburbs or in the country is more desirable than life in the city, the land is available, and roads will take us to work quickly even if it is many miles away. Products come in packages designed to entice us to buy, and we have no convenient way of disposing of these packages except in the landfill.

Changing these habits requires a conscious decision to do so, and the decision has to come from the grass roots. On an individual basis, we have to decide that we are willing to pay slightly more for environmentally friendly products than for conventional products. We have to decide to do a little extra work to recycle our waste or to use less energy. These desires must be conveyed to the product designers, the manufacturing engineers, and the advertising agencies so that the companies that produce and sell commodities realize that making them in ways that conserve resources is good not only for business, but for society as well. Most important, our government leaders—politicians and administrative staff—need to understand that the people want change and that they want government to play an active role in that change.

Grass-roots support for environmental protection and conservation remains strong and enjoys bipartisan political support. Numerous public opinion polls show that the environment is a major concern for Americans. People want government and business leaders to do more, not less. When the Republican Party embarked on a major campaign in 1995 to scale back environmental protection and transfer control of natural resources to private interests, there was a groundswell of public opposition to these changes, which forced them to withdraw or scale back their initiatives. In much of Europe, public opinion in favor of environmental protection is even stronger than in the United States, leading the European Union to adopt strict controls on many forms of pollution and to argue strenuously for significant reductions in CO_2 emissions. In South Korea, a 1996 government-funded poll found that 85 percent of those surveyed felt that pollution was getting worse every year and that the environment was more important than economic development. Nearly three-quarters of them indicated that they would support increased taxes on water and fees for mass transit to help control water and air pollution.

One way this public opinion is translated into action is through increased demand for environmentally friendly products, and corporations are responding to this demand. Major consumer products companies, for example, are producing more goods using recycled materials, and they are labeling these goods prominently to that effect. This movement extends to a wide range of industries. For example, motels and hotels are installing recycling bins in rooms and are offering options to extend towel and linen use beyond a day. Furniture made with wood from forests certified as "sustainably managed" is growing in popularity, as are products made from "simulated" tropical woods. Although relatively few consumers are willing to pay large premiums for these products, simple attention to the difference can be enough to encourage businesses to act sustainably in pursuit of competitive advantage in the marketplace.

Corporate Action

Consumers' willingness to reward responsible businesses with higher prices and customer loyalty is just one factor motivating a change in corporate behavior. The realization that waste reduction can improve profits, discussed earlier, is another. In addition, a few industries are beginning to realize that excessive resource use and environmental damage affects them directly.

The insurance industry is an interesting example of an industry that is directly affected by environmental problems. In the past decade several large storms struck Europe and North America, most notably Hurricane Andrew, which devastated parts of south Florida in 1992, and the recent floods in central and eastern Europe. Economic losses especially from weather-related disasters rose substantially in this period. In 1999, for example, global losses from natural disasters exceeded $100 billion, the second highest figure on record (UNEP 2002). At the beginning of the United Nations' International Decade for Natural Disaster Reduction (1990s), worldwide disaster losses worldwide averaged about 0.11 percent of gross world product. The average for the period 1960 to 1989 was less than 0.02 percent (Flavin 1994; Worldwatch Institute 1996). By the end of the decade, losses were escalating and represented 0.35 percent of gross world product. This increase came at a time when there was also growing concern about global warming and much speculation that at least some of these hydrometeorological disasters might be associated with climate change. If global warming caused a rise in sea level, then the frequency of disasters like Hurricane Andrew or expansive flooding could be expected to increase.

An increase in the number and severity of hurricanes could be a real problem for the insurance agency, which bases its premiums on the historic occurrences of such storms. Such concerns were at the forefront when insurance interests began to lobby in favor of actions to reduce CO_2 emission and of risk and vulnerability reduction programs at federal, state, and local levels.

Tourism is another important economic activity that depends on maintaining environmental quality. Most tourism focuses on outdoor activities, especially in coastal areas. Beach resorts have been critical to economic development in many tropical countries, especially in the Caribbean but also in the Indian Ocean and the Pacific. In many such countries, tourism is the major source of foreign exchange in addition to providing employment for large numbers of workers. Development of these resorts brings population growth and prosperity to the surrounding regions, but this same growth also puts pressure on resources in the coastal zone. Often the coastal waters are the only available place to dispose of sewage. At the same time, good water quality is obviously critical for attracting bathers to the beach. The tourist industry thus has a powerful influence on environmental policy in these areas.

Perhaps the greatest incentive for companies to change their behavior comes from the financial industry—large banks and financiers and occasionally the stock market itself. Bankers in particular are careful to see that the collateral that secures their loans is worthy. But pollution can become a big financial risk if, for example, a company can be required to clean up its wastes. Concern about this risk has been an important factor forcing companies to reduce pollution and to adopt environmental auditing procedures that will verify compliance with environmental laws. At the same time, businesses, creditors, and insurers have been lobbying hard for government-imposed limits on this liability.

Although we can point to many examples of improved business performance with regard to natural resources, many problems remain. The simple fact is that in many situations it is still more profitable to exploit resources quickly or dispose of wastes carelessly without regard for the environmental quality or future resource needs. For example, Shell Oil was criticized heavily by environmentalists for pollution associated with its operations in Nigeria. Although some U.S. paper manufacturers were proactive in controlling pollution in advance of new regulations, others have staunchly resisted pressure to reduce waste discharges. Timber companies continue to argue for unrestricted logging of old-growth forests in the tropics and the midlatitudes. Without the *genuine* support of the majority of businesses (as opposed to support from the marketing and public relations departments only), progress toward sustainability will be limited.

Government Action

What can governments do to help and encourage individuals and businesses in the drive to sustainability? Fortunately, many things can be done, provided they have the political incentives to act. Opportunities abound in three major areas: considering the full economic consequences of resource-use decisions; enforcing stiff penalties for illegal resource degradation; and using the powers of taxing and spending to promote sustainability rather than waste.

Natural Resource Accounting Perhaps the most fundamental requirement in measuring sustainability and in moving toward it is improved environmental accounting. As we discussed in Chapter 2, environmental degradation can be considered a consequence of externalities. Externalities are transfers of commodities such as air pollution without the consent of all parties involved—specifically, those who bear the consequences of pollution without gaining directly from the activities that generated it. If the receivers of these wastes were compensated for this service, then those who generate the pollution would probably pollute less. Similarly, if those who exploit natural resources were required to pay the full social costs of this exploitation, then they would probably use resources more carefully.

What do we mean by improved environmental accounting? It means considering the value of all the resources we use to produce wealth, and not just the money earned from the sale of goods and services. For example, if you wanted to evaluate your own economic performance in a given year, you would consider changes in the amount of money you have in the bank as well as your income. Suppose you began the year with $10,000 in the bank. You invest $5000 of this in a business. During the year the business generates $25,000 in net income, and at the end of the year the initial $5000 investment has depreciated to zero. You spend $22,000 of your income on living expenses, put $3000 in the bank, and end the year with $8000 in the bank. The total amount of money you earned in the year would thus be $23,000: the $25,000 net business income minus the $2000 net drawdown of capital in the bank.

At present when economists calculate the net income of a country—its gross national product (GNP)—they include changes in the amount of capital stored in bank accounts, but they do not include changes in supplies of *natural capital*. Natural capital includes the value of standing trees in the forest, oil in rocks, and fertility of the soil. Conventional calculations of GNP do not include the value of social capital: the education, health, and skills of a country's residents. Yet, if we clear-cut a forest and sell the timber, that is equivalent to spending money saved in the bank—money (or forests) that could be used to generate income in future years.

Economists argue that if we include such depletion of natural capital in GNP calculations, we will find that we are not generating as much wealth as we might think, because we are depleting natural capital faster than it is renewed. For example, Daly and Cobb (1989) calculated an index of sustainable economic welfare (ISEW) that attempted to include environmental values in GNP calculations. Although the conventionally calculated GNP of the United States rose more than 50 percent between about 1970 and 1985, the ISEW remained relatively unchanged. If we were more aware of dwindling stocks of natural capital, we would be more likely to take action to protect them.

Mandated Environmental Responsibility

Earlier in this chapter, we described the move away from government-mandated environmental protection toward private-sector initiatives, a movement that gained momentum in the early 1990s. Businesses were taking responsibility for improving their environmental performance by reducing waste and use of hazardous chemicals, increasing recycling, and opening dialogues on environmental issues with the communities in which they operate.

These initiatives are motivated by several factors, some of which relate directly to the cost of doing business and some of which are intended to avoid liabilities and confrontations that have resulted from government regulations. To the extent that this improved business behavior is a response to government pressure, we may wish to maintain that pressure.

In what ways has government pressure led to voluntary change in business practices with regard to resources? One significant force has been the assignment of financial responsibility for environmental cleanup to parties contributing to pollution. The Superfund law is an excellent example, as we discussed earlier. Similarly, accidental releases of toxic substances within or outside a manufacturing facility can easily bring expensive lawsuits. To reduce the possibility of litigation, many companies have instituted programs of periodic environmental audits to ensure that they are in compliance with the law and are handling hazardous materials safely.

Legislation such as the Superfund law and damage suits filed by parties exposed to pollutants have at times created very large penalties for polluters. Some argue that these penalties or cleanup costs are sometimes excessive; for example, restoring polluted land to a near-clean condition can cost much more than the value of the land after it has been cleaned. Corporations and insurance companies that provide businesses with liability coverage have pressured government to pass laws limiting corporate liability in such cases.

The annual publication of the Toxic Release Inventory (TRI) is another government program that has been effective in encouraging businesses to reduce pollution. When the lists of emissions are published, newspaper headlines immediately announce who are the biggest polluters in town. Few corporations want to look like the dirty kids on the block, and whether or not the bad publicity associated with the TRI is truly deserved, it helps to show businesses the value of reducing their emissions. More important, right-to-know advocates have ensured that residents retain access to this information through Web sites such as Environmental Defense's Scorecard.

The TRI has been criticized on many grounds, including its tendency to add amounts of different pollutants together, like apples and oranges, without full regard for their differences. Similarly, the costs of cleaning up toxic wastes and the penalties imposed in civil lawsuits are sometimes excessive. But the fear of bearing such costs has provided a powerful incentive for businesses to reduce pollution. Requiring continued public accountability for corporate behavior can be important in maintaining and strengthening the growing record of good corporate citizenship.

Using the Power of Government Governments are organizations set up by groups of people for the purpose of doing things that they cannot do as individuals. Despite widespread cynicism (at least in the United States) about the ability of government to accomplish meaningful goals and the popularity of criticizing excessive government regulation, governments can do many positive things to encourage responsible resource use.

One of these things is to use the necessary act of collecting taxes to promote sustainability. Most governments use tax policy routinely as a means of achieving various social ends, such as taxing the wealthy at higher rates than the poor so as to redistribute income more evenly. Perhaps the most effective way taxes can be used to improve the environment is through increasing the costs of resource degradation to those who contribute to it. If individuals are held responsible for at least some of the social/environmental costs of their activities, then the environmental externalities can be reduced.

For example, driving an automobile benefits the individual while generating many social and environmental costs. It discharges pollutants to the air, increases congestion on roads, and helps wear down the roads themselves. It also helps deplete future supplies of petroleum, increasing future costs of that fuel. Most countries charge taxes for operating motor vehicles, usually in the form of license fees and fuel taxes.

In most European countries, for example, taxes are assessed for both licensing an automobile and purchasing fuel. European fuel costs in the late 1990s were typically $1 to $1.50 per liter ($4 to $6 per gallon). As a result, people buy more fuel-efficient cars, drive less, and take public transit more than they would otherwise. But in the United States, these taxes are very low, and fuel costs much less: about $0.35 per liter ($1.33 per gallon). As a result people in the United States drive more frequently and greater distances in less efficient cars, and CO_2 emission rates per person are two to three times as much as typical European rates.

Not only does the United States fail to charge individual polluters such as automobile drivers for the social costs of their actions, but in many cases its tax system actually encourages inefficient environmental behavior. For example, sport utility vehicles are heavier and less fuel-efficient than conventional sedans, and some of them are classed as trucks. Trucks are subject to less-stringent efficiency rules than autos, and some taxes are lower on trucks than on autos.

Governments encourage wasteful practices in other ways, often through subsidies. In Chapter 10 we saw how government subsidies on water in the western United States encouraged use of this water to grow low-value crops such as hay, instead of conserving the water for other uses. The U.S. Forest Service subsidizes timber harvesting on its lands through road construction and other activities, sometimes even to the point of spending more money to encourage logging than it earns in the sale of the timber. Such subsidies are common in many countries.

Instead of using government funds to encourage waste, they could be used to encourage the development of new sustainable technologies. For example, photoelectric cells offer much promise for reducing demand for fossil-generated electricity. That potential is not yet realized because the cost of electricity that is generated using photoelectric cells is still relatively high in comparison to the cost of other sources. Their high cost, in turn, is in part caused by the small numbers of cells produced and the lack of economies of scale in manufacturing.

What is needed is for someone to make a large initial investment in photoelectric cells, perhaps in the form of a commitment to buy a given number of cells. This would prime the pump and would begin the positive feedback of more demand, more production, lower unit costs, more purchases, and so forth. In 1996, plans were announced to build a 10-mw photovoltaic power plant in Nevada, and a 50-mw plant in Rajastan, India. In both cases, the plants were made possible in part by long-term government commitments to purchase the power generated. Although these plants are small in comparison to modern, conventional electric-generating plants, they are significant because they will increase the production of photovoltaic cells, leading to reductions in prices.

Action at Many Levels In addition to action by national and state governments, there must be cooperative action at the international level. The Convention on Long Range Transport of Air Pollutants and the Montreal and Kyoto Protocols developed under it are excellent examples, provided they are adopted and enforced by sovereign national governments. The United Nations has struggled to act on the principles set forth in Agenda 21 but has been hampered by lack of funds and limited willingness of individual governments to implement those principles. In particular,

> "A fragmented approach . . . has seen policies and programmes address economic, social and environmental concerns, but not in integrated manner; the world continues to use far more resources than ecosystems can support; there is a lack of coherent policies in areas of finance, trade, investment and technology and policies that take a long-term view; a lack of resources [has been] dedicated to implementing Agenda 21; developing countries have had difficulties obtaining new technologies and private investment from developed countries, and official development assistance has fallen over the last decade." (UN Social and Economic Council, 2001).

In 2002 the World Summit on Sustainable Development took place in Johannesburg, to focus the world's attention and action on improving people's lives and conserving natural resources in a world that is experiencing increasing demands for food, water, shelter, sanitation, energy, health services, and economic security. Unfortunately, little substantive progress was made at the summit in the implementation of sustainability goals. The United States, in particular, resisted establishment of binding commitments with regard to either environmental improvement or measures to reduce poverty, arguing instead for voluntary projects and programs. The unwillingness of the United States to make such commitments, let alone take a leadership role in alleviating these problems, caused considerable resentment in the international community (Fig. 15.5).

In Chapter 3 we described the elements of the decision-making process and the roles of social agents and managers—individuals and groups working within political structures—in creating natural resource policy. Clearly the move toward sustainability requires action at many levels, from individual consumers to scientific organizations, special interest groups, and political leaders. In a complex society it is unrealistic to think that these entities would function together in a unified and cohesive fashion, but the more people are aware of the need to manage resources for sustainability the more likely our actions will lead in that direction.

Figure 15.5 U.S. Secretary of State Colin Powell was heckled and jeered as he spoke at the 2002 Johannesburg Summit. Powell argued that "The United States is taking action to meet environmental challenges, including global climate change," but the audience was skeptical.

Looking Forward

In this chapter we have seen that sustainability, or at least significant progress toward sustainability, is technically and economically feasible. The technology necessary to substantially improve efficiency of resource use either is available today or could be developed within a decade or two. The resources we depend upon and which are being depleted can be replaced. We don't need to feed two-thirds of our grain to livestock as is done in the United States—diets with much less meat can provide all the nutrition we need, and we would probably be much healthier if we changed our diets (Motovalli 2002). We don't need to cut old-growth forests to provide an adequate timber supply—we can make excellent products with smaller trees produced with intensive forest management, often at a lower cost than cutting forests in rugged mountains. We don't need to drive automobiles that get 19 miles to the gallon to get to school or work—autos with twice that efficiency are readily available and usually cost less.

The drive to sustainability is limited most by a cultural factor: the will to change. Personal habits such as a diet rich in meat or the enjoyment of driving a fast car, the social traditions of communities that are dependent on cutting trees in a remote area, and the self-interests of businesses and their shareholders demanding short-term profits are inertial forces. They keep us from changing in ways that we know we should, and they perpetuate the unsustainable practices that are occurring around the world today.

The recent history of the environmental movement has shown that when and where public sentiment for environmental protection and resource conservation is strong, our governmental, corporate, and individual behaviors respond. Maintaining that commitment is the best and perhaps the only way for human resource-use systems to become truly sustainable.

References and Additional Reading

Abramowitz, J. 1997. Valuing nature's services. In L. R. Brown, ed. *State of the World, 1997.* New York: W. W. Norton.

_____ and A. T. Mattoon. *Paper Cuts: Recovering the Paper Landscape.* Worldwatch Paper 149. Washington, D.C.: Worldwatch Institute.

Annan, K. A. 2002. Toward a sustainable future. *Environment* 44(7):10–15.

Anonymous. 2002. Data points: waste for money. *Scientific American* July:32.

Boerner, C. and K. Chilton. 1994. False economy: the folly of demand-side recycling. *Environment* 36(1):6–15, 32.

Borgstrom, G. 1965. *The Hungry Planet: The Modern World at the Edge of Famine.* New York: Collier Books.

Costanza, R. and H. E. Daly. 1992. Natural capital and sustainable development. *Conservation Biology* 6:37–46.

Costanza, R., R. d'Arge, R. de Groot, S. Farber, M. Grasso, B. Hannon, K. Limburg, S. Naeem, R. O'Neill, J. Paruelo, R. Raskin, P. Sutton, and M. van den Belt. 1997. The value of the world's ecosystem services and natural capital. *Nature* 387(6630): 253–260.

Council on Environmental Quality. 1981. *The Global 2000 Report to the President.* New York: Viking Penguin.

Cutter, S. L. 1993. *Living with Risk.* London: Edward Arnold.

_____. 1995. The forgotten casualties: women, children, and environmental change. *Global Environmental Change* 5(3):181–194.

Daly, H. E. and J. B. Cobb. 1989. *For the Common Good: Redirecting the Economy Toward Community, the Environment, and a Sustainable Future.* Boston: Beacon Press.

Dodds, F., ed. 2001. *Earth Summit 2002: A New Deal, Revised Edition.* London: Earthscan.

The Economist. 1995. More puff, less smoke. *The Economist,* September 2, p. 53.

Ehrlich, P. 1968. *The Population Bomb.* New York: Ballantine Books.

Flavin, C. 1994. Storm warnings: Climate change hits the insurance industry. *WorldWatch,* December, pp. 10–20.

Frosch, R. A. 1995. Industrial ecology: adapting technology for a sustainable world. *Environment* 37(10):16–24, 34–37.

Goodland, R. and H. Daly. 1994. Environmental sustainability: universal and non-negotiable. Paper presented to Ecological Society of America, Knoxville, TN, August 1994.

Graedel, T. E. and B. R. Allenby. 1995. *Industrial Ecology.* Upper Saddle River, NJ: Prentice-Hall.

Hamann, R., Z. Patel, and M. Pressend. 2002. Competing visions and conflicting priorities: a southern African perspective on the World Summit. *Environment* 44(6):8–21.

Kane, Hal. 1996. Shifting to sustainable industries. In L. R. Brown et al., *State of the World 1996.* New York: W. W. Norton, pp. 152–167.

Kates, R. W. 2000. Population and consumption: what we know, what we need to know. *Environment* 42(3):10–19.

———, W. C. Clark, R. Corell, J. M. Hall, C. C. Jaeger, I. Lowe, J. J. McCarthy, H. H. Schellnhuber, B. Bolin, N. M. Dickson, S. Faucheux, G. C. Gallopin, A. Grübler, B. Huntley, J. Jäger, N. S. Jodha, R. E. Kasperson, A. Mabogunje, P. Matson, H. Mooney, B. Moore III, T. O'Riordan, and U. Svedlin, 2001. Sustainability science. *Science* 292:641–642.

Management Institute for Environment and Business. n.d. Bayerische Motoren Werke, AG: A proactive approach to vehicle recycling. Management Institute for Environment and Business.

Meadows, D. H., et al. 1972. *The Limits to Growth.* New York: Universe Books.

Motavalli, J. 2002. The case against meat. *E Magazine* 13(1) Jan–Feb:26–32.

——— and J. Harkinson 2002. Buying green. *E Magazine* 13(5) Sept–Oct:26–33.

Myers, N. 2000. Sustainable consumption. *Science* 287:2419.

Najam, A., J. M. Poling, N. Yamagishi, D. G. Straub, J. Sarno, S. M. DeRitter, and E. M. Kim. 2002. From Rio to Johannesburg: progress and prospects. *Environment* 44(7):26–38.

Nash, J. and J. Ehrenfeld. 1996. Code green: business adopts voluntary environmental standards. *Environment* 38(1):16–20, 36–45.

National Research Council (NRC), Board on Sustainable Development. 1999. *Our Common Journey: A Transition toward Sustainability.* Washington, D.C.: National Academy of Sciences.

Network for Science and Technology for Sustainability. 2002. www.sustsci.harvard.edu.

Paddock, W. and P. Paddock, 1967. *Famine 1975!* Boston: Little, Brown.

President's Council on Sustainable Development, 1996. *Sustainable America: A New Consensus for Prosperity, Opportunity, and a Healthy Environment for the Future.* Washington, D.C.: U.S. Government Printing Office.

Probst, K. M., D. M. Konisky, R. Hersh, M. B. Batz, and K. D. Walker. 2001. *Superfund's Future? What Will it Cost?* Washington, D.C.: Resources for the Future.

Raven, P. H. 2002. Science, sustainability, and the human prospect. *Science* 297:954–958.

Serageldin, I. 1993. Making development sustainable. *Finance and Development* 30(4):6–10.

Shafik, N. 1994. Economic development and environmental quality: an econometric analysis. *Oxford Economic Papers* 46:757–773.

Steel Recycling Institute. 1996. *Recycling Scrapped Automobiles.* http://www.recycle-steel.org/Auto.html.

United Nations Economic and Social Council. 2001. *Implementing Agenda 21: Report of the Secretary General.* New York: United Nations.

United Nations Environment Program, 2002. *Global Environment Outlook 3.* New York: Oxford University Press.

U.S. Congress, Office of Technology Assessment. 1982. *Global Models, World Futures and Public Policy: A Critique.* Washington, D.C.: Superintendent of Documents.

U.S. Environmental Protection Agency. 2002. *Superfund 20th Anniversary Report.* www.epa.gov/superfund/action/20years/preface.html.

Waste Management of North America. 1992. *Recycling in the '90s: A Shared Responsibility.* Oak Brook, IL: Waste Management of North America.

Weinberg, A. M. 1984. Review of *The Resourceful Earth. Environment* 26:7, 25–27.

World Bank. 1992. *World Development Report 1992: Development and the Environment.* New York: Oxford University Press.

———. 1995. *Monitoring Environmental Progress: A Report on Work in Progress.* Washington, D.C.: Environment Department, World Bank.

World Commission on Environment and Development. 1987. *Our Common Future.* Oxford: Oxford University Press.

Worldwatch Institute. 1996. *WorldWatch Data Diskette.* Washington, D.C.: Worldwatch Institute.

Young, J. E. 1995. The sudden new strength of recycling. *WorldWatch* July/August:20–25.

Young, J. E. and A. Sachs. 1994. *The Next Efficiency Revolution: Creating a Sustainable Materials Economy.* Worldwatch Paper no. 121. Washington, D.C.: Worldwatch Institute.

For more information, check our web site at ***http://www.wiley.com/college/cutter.***

STUDY QUESTIONS

1. Do you think global models are better than science fiction or crystal balls in predicting the future?

2. What are the most important resource issues in your region? How do these compare to national issues? Global issues?
3. What can you do to address the local environmental issues facing your community? What can you do to stimulate others to take action as well?
4. How are resource issues different today than when your grandparents were your age? How might they differ when you become a grandparent?
5. Is sustainablity achievable in your lifetime? Why or why not?
6. In what specific ways can your college or university become more sustainable and join the ranks of other sustainable universities?

Glossary

Absolute scarcity A condition in which there is not enough of a resource to satisfy demand for it.

Acid mine drainage Water leaving a surface or underground mine enriched in acid, usually sulfuric acid.

Acid deposition The accumulation of acids, either in precipitation or through dry dustfall, on the land surface.

Active solar power Solar energy gathered by a device that collects this energy and mechanically distributes it to where it is needed.

Advection inversion A temperature inversion caused by warm air passing over a cool surface.

Age structure The relative proportions of a population in different age classes.

Agribusiness Large-scale, organized production of food, farm machinery, and supplies as well as the storage, sale, and distribution of farm commodities, for profit.

Agricultural runoff Water leaving areas of agricultural land use, usually enriched in nutrients, sediment, and agricultural chemicals.

Air quality index An index used to communicate health risks of air pollution to the public. The index ranges from 0–500 and uses a color-coded scheme to ease interpretation, with green meaning good air quality and shades of red indicating poor to hazardous air quality.

Ambient air quality The chemical characteristics of air as it exists in the environment; a measure of pollutant concentrations in the air.

Amenity resource A resource valued for nonmonetary characteristics, such as its beauty or uniqueness.

Amenity value The nonmonetary, intangible value of a good or service.

Anadromous fish Fish that breed in fresh water but spend most of their adult lives in salt water. Examples are salmon and striped bass.

Animal unit month The amount of forage needed to support a certain number of grazing animals for one month.

Anoxic Water without dissolved oxygen.

Anthracite The highest rank of coal, most modified from its original plant form.

Anthropogenic Of human origin, such as carbon dioxide emitted by fossil fuel combustion.

Aquiclude An impermeable layer that confines an aquifer, preventing the water in it from moving upward or downward into adjacent strata. Shale and some igneous rocks often form aquicludes.

Aquifer A geologic unit containing groundwater; an underground reservoir made up of porous material capable of holding substantial quantities of water.

Arable land Land that can be cultivated and can support agricultural production.

Arroyo A deep, steep-sided gully found in semiarid areas, particularly in the southwestern United States.

Augur mining A coal-mining technique using a screw that extracts coal as it is drilled into a deposit.

Baby boom A period from 1945 to the mid-1960s in which the average fertility rate in the United States was over 3 children per woman.

Benefit-cost analysis A process of quantitatively evaluating all the positive and negative aspects of a particular action in order to reach a rational decision regarding that action.

Bioaccumulation The tendency for a pollutant to accumulate in the tissues of plants or animals.

Biochemical decay Breakdown of pollutants in water through the action of bacteria.

Biochemical oxygen demand (BOD) The amount of oxygen used in oxidation of substances in a given water sample. Measured in milligrams per liter over a specific time period.

Biocide Willful destruction of living things.

Biogeochemical cycle The movement of a particular material through an ecosystem over long periods of time.

Biological diversity The range or number of species or subspecies found in a particular area.

Biomagnification An increase in the concentration of a pollutant as it is passed up the food chain, caused by a tendency for animals to accumulate the pollutant in their tissues.

Biomass The total amount of living or formerly living matter in a given area, measured as dry weight.

Biomass harvesting A forest harvest technique in which whole trees are chipped and used as fuel.

Biome A major ecological region within which plant and animal communities are similar in general characteristics and in their relations to the physical environment.

Bioregion A geographic area defined by ecological characteristics. A bioregion includes an area of relatively homogeneous ecological characteristics or a specific assemblage of ecological communities. It is similar to a biome but may refer to a smaller area with more specific characteristics.

Biosphere The worldwide system within which all life functions; composed of smaller systems, including the atmosphere, hydrosphere, and lithosphere.

Biosphere resources Resources associated with living organisms.

Biotic potential The maximum rate of population growth resulting if all females in a population breed as often as possible and all individuals survive past their reproductive periods.

Birth rate The number of babies born per year per 1000 population.

Bituminous coal A rank of coal below anthracite, characterized by a high degree of conversion from the original plant matter and a high heat content per unit weight.

Boreal forest A biome dominated by coniferous forests and found in relatively high altitudes or latitudes, almost exclusively in the Northern Hemisphere.

British thermal unit (Btu) The amount of energy required to raise the temperature of one pound of water one degree Fahrenheit at or near 39.2°F.

Bubble approach An approach to air pollution emissions control that allows a plant to consider emissions from several sources as combined emissions from the plant.

Bureau of Land Management (BLM) The U.S. Bureau of Land Management, part of the Department of the Interior, established in 1946 to administer federal lands not reserved for military, park, national forest, or other special uses.

Bycatch Non-target species caught by indiscriminate commercial fishing operations that are then discarded after the desired species are harvested.

Carbon budget An accounting of all flows of carbon between the atmosphere, biosphere, oceans, and the solid earth.

Carcinogen A substance that causes cancer.

Carrying capacity The maximum number of organisms in one species that can be supported in a particular environmental setting.

Cartel A consortium of producers of a single product who agree to coordinated actions to manipulate a market.

Catadromous fish Fish that breed in salt water but live most of their adult lives in fresh water. The American eel is an example.

Centralized energy An energy conversion technology in which the key conversion (such as combustion of coal to create electricity) is made at a large scale at a single site (such as a power plant).

Chaparral A subtropical drought-resistant and fire-prone shrubby vegetation associated with Mediterranean-type climates.

Chlorofluorocarbons (CFCs) A group of substances that are compounds of chlorine, fluorine, and carbon. They are widely used in refrigeration and many industrial processes and contribute to deterioration of stratospheric ozone.

Clean Air Act The name given to a series of air-quality improvement laws and their amendments passed in the United States beginning in 1963.

Clean Water Act The name given to a series of water-quality improvement laws and their amendments passed in the United States beginning in 1964.

Clear-cutting A forest harvest technique in which all trees in a particular area are cut, regardless of species or size.

Coal gasification A chemical process converting coal to a gas that can then be used in place of natural gas.

Cohort A group of individuals of similar age.

Coliform bacteria Bacteria of the species *Escherichia coli*, commonly occurring in the digestive tracts of animals; used as an indicator of the potential for disease-causing organisms in water.

Common property resource Resource such as air, oceans, or sunshine that is in theory owned by everyone but in practice utilized by a few. The question of regulation arises to prevent or lessen resource abuse.

Community A collection of organisms occupying a specific geographic area.

Concentration In the context of air or water quality, the amount of a substance per unit (weight or volume) of air or water.

Conservation The wise use or careful management of resources to attain the maximum possible social benefits from them.

Conservation tillage An agricultural system using tillage techniques designed to reduce soil erosion and overland flow. Most conservation tillage techniques involve less manipulation of the soil than conventional techniques, leaving more plant matter on the soil surface.

Consumer theory of value An approach to valuing commodities based on how much a consumer is willing to pay for them.

Consumptive use or **Consumption** Water use that results in water being evaporated rather than returned to surface water or groundwater after use.

Contingent valuation method A method for determining the value of a resource by asking how much people would be willing to pay for it under certain circumstances.

Continental shelf Area of the seafloor averaging less than 650 feet (200 m) deep, which generally was exposed at times of lower sea level in the past.

Contour plowing A soil conservation technique involving plowing parallel to the contour, across a slope rather than up and down it.

Cost-effectiveness analysis An analysis of all the costs involved in taking a specified action to determine the most efficient way to carry out the chosen action.

Cost theory of value An approach to valuing commodities based on the cost of production.

Crisis management A decision-making strategy that is reactive in nature. Once a resource issue becomes critical, then policy is determined to cope with the immediate problem without any consideration of long-term implications of such a policy.

Criteria pollutants Air pollutants, including carbon monoxide, hydrocarbons, lead, oxidants, particulates, nitrogen oxides, and sulfur oxides, for which maximum permissible concentrations in ambient air are established.

Critical mineral A mineral necessary for defense of the United States and available partly in America or partly from friendly nations.

Criticality A term used to describe environmental regions that are so degraded that economic activity and human habitation are impossible in the short term.

Crop rotation A soil conservation technique involving changing crops grown on a given parcel of land from year to year. Crop rotations may include fallow periods.

Cropland Land in which crops are regularly planted and harvested. It includes land in fallow or pasture as part of a regular rotation system.

Crown fire An intense forest fire that consumes the tops of trees as well as lower strata of vegetation.

Crude oil Unrefined petroleum as it is extracted from the ground; it is liquid at normal ambient temperatures.

Death rate The number of deaths per year per 1000 population.

Decentralized energy source An energy conversion system characterized by numerous small-scale facilities located at or near the end-use site. Photovoltaic cells are an example.

Decreaser A plant species in a range community that declines in importance as a result of grazing pressure. Usually, decreasers are the most palatable to the grazing animals.

Deep ocean Ocean areas seaward of the continental shelf.

Deforestation Any process of replacement of forest vegetation with other types.

Demographic transition The process by which a human population goes through a growth pattern, including an early phase of high birth and death rates, an intermediate phase of high birth rates but low death rates, and a later phase of low birth and death rates.

Deregulation The reduction or elimination of governmental market controls, widespread in commercial economies since the 1970s.

Desalination Artificial removal of salt from water, such as by distillation or reverse osmosis.

Desert A biome characterized by plants and animals adapted to extreme moisture scarcity.

Desertification A process of land becoming more desertlike as a result of human-induced devegetation and related soil deterioration, sometimes aggravated by drought.

Dilution In water quality, a reduction in pollutant concentration caused by mixing with water, resulting in a lower concentration of the substance.

Dissolved oxygen Oxygen found in dissolved form in water.

Dissolved solids Substances normally solid at ambient temperatures but dissolved into ionic form in water.

Diversification The trend in many large corporations toward ownership of a wide array of companies producing unrelated goods and services.

Domesticate A species that has been bred for specific characteristics that humans value, thereby rendering the species dependent on humans for its continued survival.

Doubling time The length of time needed for a population to double in size. It is a function of the growth rate.

Drainage basin An area bounded by drainage divides and defined with respect to a point along a stream. All the runoff generated within the area must pass the point along the stream; runoff generated outside the basin will not pass that point.

Drip irrigation An irrigation method involving small pipes placed at the base of plants, delivering water slowly to the plant roots.

Drought A period of time with unusually low precipitation.

Dry farming Agricultural production in climatically marginal lands without the use of irrigation.

Dry rock geothermal energy A method of extracting heat from the earth by pumping water through hot rocks.

Earth Summit The popular name given to the 1992 United Nations Conference on Environment and Development held in Rio de Janiero.

Earth systems science The study of earth's key biogeochemical processes from a holistic perspective.

Ecological economics An integrated study of ecological and economic perspectives on environmental issues.

Ecology The study of the interrelationships between living organisms and the living and nonliving components and processes that make up their environment.

Ecosystem The collection of all living organisms in a geographic area, together with all living and nonliving things they interact with.

Ecotone A transitional zone between two adjacent ecosystems.

Ecotourism Tourism focused on appreciation of nature rather than on built environments.

El Niño/La Niña A transient, periodic warming of the equatorial eastern Pacific Ocean, associated with fisheries depletion and large-scale climatic fluctuations.

Emissions Pollutants released to the atmosphere, measured at the point of release rather than after dilution in the atmosphere.

Emissions trading A procedure in air-quality regulation by which one polluter can acquire permission to discharge pollutants formerly discharged by another discharger that has ceased emitting pollutants.

Endangered species A species in imminent danger of extinction.

End-of-pipe pollution control Pollution control that focuses on removing pollutants from water or gases before they are discharged into the environment through controlling the point of discharge such as a tailpipe on an automobile or a smokestack at an industrial facility.

Energy budget An accounting of all energy inputs and outputs for a system.

Energy conservation Using energy resources in such a way as to minimize energy consumption in relation to benefits gained.

Energy intensity or efficiency The amount of utility, either work performed or income generated, gained per unit of an energy resource.

Environmental cognition The mental process of making sense of the world that each of us inhabits.

Environmental ethics A philosophical position regarding the relation between humans and nature.

Environmental justice The differential burden or disproportionate impact of natural resource use on people and places.

Environmental lapse rate The average rate at which temperature declines with increasing altitude in the troposphere.

Environmental refugee A person fleeing a natural or human-caused environmental disaster.

Environmental resistance Factors such as food supply, weather, disease, and predators that keep a population below its biotic potential.

Equity Fairness in the use and allocation of resources.

Erosion Removal of soil by running water or wind.

Erosivity The ability of rainfall to cause erosion. Erosivity is a function of rainfall intensity and drop size.

Estuary A semi-enclosed water body, open to the sea, in which seawater is significantly diluted by fresh water from the land.

Euphotic zone The upper portion of the sea, in which sunlight is intense enough to allow plant growth.

Eutrophication The process by which lakes become increasingly nutrient-rich and shallow. It is a natural process that is accelerated by water pollution.

Evapotranspiration The process by which liquid water is conveyed to the atmosphere as water vapor, including water use by plants.

Exclusive economic zone A zone of the oceans over which a particular nation has claims or exclusive control of certain economic activities, such as fishing.

Exploitation Use of a resource at the maximum profitable short-term rate, without regard for long-term resource quality or availability.

Externality A nonmarket exchange, in which at least one party to the exchange is not compensated and may have little choice in the exchange.

Extinction The process by which a species ceases to exist.

Farmland Land that is part of farm units, including cropland, pasture, small woodlots, and areas used for small farm roads and buildings.

Federal Land Policy and Management Act (FLPMA) Passed in 1976, this law consolidated diverse regulations of public land management and strengthened the power of the BLM to manage public lands.

Feedback An information transmission that produces a circular flow of data in a system.

Fertility rate The average number of children that women in a given population bear in their reproductive years.

Fertilizer A substance added to the soil to improve plant growth. The most commonly used fertilizers are those containing large amounts of nitrogen, potassium, and phosphorus.

Fire frequency The average number of fires per unit of time at a given location.

First law of thermodynamics The law of conservation of energy, which states that energy is neither created nor destroyed, but merely transformed from one state to another or converted to or from matter.

Fission A process of splitting heavy atoms of uranium or plutonium into lighter elements, thereby releasing energy.

Fixed costs Costs of operating a business that do not vary with the rate of output of goods and services.

Flood irrigation A means of irrigation whereby entire fields are occasionally inundated.

Flow resource A resource that is simultaneously used and replaced. Perpetual and renewable resources are flow resources.

Food chain A linear path that food energy takes in passing from producer to consumers to decomposers in an ecosystem.

Food security The condition of having both physical and economic access to the basic food that people need to function normally.

Food web A complex, interlocking set of pathways that food energy takes in passing from producer to consumers to decomposers in an ecosystem.

Furrow irrigation A type of irrigation in which water is allowed to flow along the furrows (troughs) between rows of crops.

Fusion The combination of two hydrogen atoms to create a helium atom, yielding energy.

Gaia hypothesis A view of earth history that emphasizes the earth's tendency to maintain a balance or equilibrium of natural systems.

GAP analysis The use of remote sensing and GIS techniques to identify holes or gaps in land ownership for species protection.

General circulation model A computerized representation of the earth's atmospheric and oceanic circulation system used to simulate weather and climate.

General Systems Theory A way of looking at the world or any part of it as an interacting set of parts.

Generational equity The fairness doctrine applied to subsequent generations so they receive the environment in the same or better condition than the generation before them.

Genetic damage Damage to individual cell tissues resulting in changes that are passed along to offspring in chromosomes.

Genetically modified (GM) crops Crops created through the detection, change, or moving of genes within an organism; the transfer of genes from one organism to another; the modification of existing genes; or the construction of new genes and their incorporation into an organism.

Geographic Information System (GIS) A computer database and data-manipulation system designed to use geographically organized data.

Geologic estimate of resource An estimate of the amount of a mineral resource in the earth based on information about the concentration and distribution of that mineral in rocks, without regard for the economics of extraction.

Geothermal energy Energy extracted from heat contained in rocks near the earth's surface.

Geopressurized resource A geothermal resource in which hot groundwater is pressurized by natural forces.

Global warming The warming of earth's atmosphere resulting from a human-caused increase in greenhouse gas concentrations.

Grassland A biome dominated by grasses. Most grasslands have semiarid climates.

Green Revolution A variety of agricultural systems developed for application in developing countries, involving the introduction of improved seed varieties, fertilizers, and irrigation systems.

Greenhouse effect The tendency of the atmosphere to be transparent to shortwave solar radiation but opaque to longwave terrestrial radiation, leading to a warming of the atmosphere.

Greenhouse gases Substances that are transparent to shortwave (solar) radiation but absorb longwave (terrestrial) radiation and thus contribute to warming of the atmosphere. Carbon dioxide, ozone, chlorofluorocarbons, methane, and water vapor are important greenhouse gases.

Ground fire A forest fire that burns only at ground level, consuming litter and downed trees but not live standing trees.

Groundwater mining. See *Overdraft*.

Groundwater Water below the ground surface, derived from the percolation of rainfall and seepage from surface water.

Guest worker A person allowed in a country on a temporary basis in order to increase the available labor force in that country.

Gully A steep-walled stream channel incised in the soil by accelerated erosion.

Gyres A circular flow pattern in the ocean.

Habitat Land that provides living space and sustenance for plants and animals.

Halocline A marked change in salinity at a particular depth in the ocean or an estuary; it signals the boundary between two layers of water.

Hardwoods Trees with particularly dense wood; primarily broad-leafed trees.

Heavy-water reactor A nuclear fission reactor using deuterium-enriched water to moderate the fission reaction.

High seas Areas of the oceans beyond the legal control of any nation.

High-temperature gas-cooled reactor A nuclear fission reactor using helium gas to transfer heat from the core to a steam generator.

Homestead Act A law passed in 1862 providing 160 acres of federal land free to settlers.

Homocentric A view of nature that considers only human, rather than plant or animal, needs.

Homosphere The lower portion of the earth's atmosphere, characterized by relatively uniform gaseous composition. Consists of the troposphere, the stratosphere, and the mesosphere.

Horizon A layer in the soil with distinctive textural, mineralogical, chemical, or structural characteristics.

Hydraulic mining Extraction of minerals from sediments using water to wash away less-dense materials.

Hydroelectric power Electricity generated by passage of runoff-derived water through a turbine, usually at a dam.

Hydrothermal mineralization A process of concentration of metallic ores caused by high-temperature geochemical processes in underground waters.

Hypoxic A condition in which lake or ocean water is significantly depleted in dissolved oxygen.

Illegal immigrant A person who enters and lives in a country in violation of that country's laws.

Incommensurables Effects of a given action that can, with some effort, be given monetary value.

Increaser A range plant species that is present in a range ecosystem prior to grazing and that increases in numbers or coverage as a result of grazing.

Incrementalism A type of decision-making strategy that reacts to short-term imperfections in existing policies rather than establishing long-term future goals. Decisions are made on a sequential basis and do not radically depart from existing policy.

Industrial ecology An approach to designing and operating industrial systems that function interdependently with natural systems.

Infiltration capacity The maximum rate at which a soil can absorb water.

Inorganic Describes a chemical substance that does not contain carbon.

Input Energy, matter, or information entering a system.

In-stream uses Uses of water that do not require it to be removed from a stream or lake. They include such things as shipping, swimming, and waste disposal.

Intangible A good, service, or effect of an action that cannot be assigned monetary value.

Integrated pest management A pest control technique that relies on combinations of crop rotation, biological controls, and pesticides.

Interbasin transfer A movement of water from one drainage basin to another, such as from the east side of the Rocky Mountains into the west-flowing Colorado River.

Internal waters Waters under the exclusive control of a coastal nation, including bays, estuaries, and rivers.

International Whaling Commission (IWC) An organization set up under the International Convention for the Regulation of Whaling in 1946, to regulate the whaling industry.

Invader A range plant species not present in a given area before grazing but entering the area as a result of the ecological changes caused by grazing.

IPAT An acronym for environmental impacts (I) equals population (P) times affluence (A) times technology (T).

Irrigation The artificial application of water to a crop or pasture beyond that supplied by direct precipitation.

Irrigation efficiency A measure of the volume of water applied in the root zone that is used by the crop expressed as a percentage. Indicates which irrigation method is the most efficient in terms of delivering water to the plants.

Kerogen A waxy hydrocarbon found in oil shale.

Labor theory of value An approach to valuing commodities based on the amount of human labor required to produce them.

Land Capability Classification System A scheme used by the U.S. Natural Resource Conservation Service for assessing and classifying the productivity of land units.

Landfill A land-based disposal method, in which waste is deposited in layers and covered with earth.

Law of conservation of energy *see First law of thermodynamics*.

Law of entropy The second law of thermodynamics. Entropy is a measure of disorder in a system.

Law of the Sea Treaty A treaty establishing jurisdiction over marine resources in coastal and deep-sea areas.

Leachate Water seeping from the bottom of a layer of ground and containing substances derived from that layer. Usually applied to landfills and other contamination sources.

Light-water reactor A type of nuclear power plant that uses ordinary water as the cooling medium.

Lignite A rank of coal characterized by a relatively low degree of modification of plant matter.

Limits to Growth A world model developed in the 1970s by a group called the Club of Rome; it predicted resource scarcity if world population and resource use growth continued.

Liquefaction Conversion of coal into a liquid hydrocarbon that can be transported by pipeline and burned as a liquid.

Liquid-metal fast-breeder reactor A nuclear fission reactor moderated and cooled by liquid sodium and used to convert nonfissionable material such as uranium-238 to fissionable material such as plutonium-239.

Liquid natural gas (LNG) Natural gaseous hydrocarbons that are pressurized and cooled in order to be stored and/or transported in liquid form.

Low level wastes Radioactive wastes that are neither spent nuclear fuel, waste from nuclear fuel reprocessing, nor high in concentration of elements with atomic numbers greater than 92.

Malthus British economist who wrote (1798) that populations increase geometrically while food supplies increase arithmetically.

Materials balance principle An approach to production systems that focuses on accounting and balancing inputs and outputs.

Maximum sustainable yield The largest average harvest of a species that can be indefinitely sustained under existing environmental conditions.

Megacities Urban places with more than 10 million people.

Mesosphere Layer of the atmosphere between 30 and 50 miles (50 and 80 km) in altitude, characterized by decreasing temperatures with increasing altitude.

Migration The movement of people from one area to another in response to warfare, environmental degradation, or perceived better opportunities.

Mined-land reclamation The return of land disturbed by mining to a more productive condition, usually a use similar to that existing before mining took place.

Minimum tillage A soil and water conservation technique that leaves the crop residue or stubble on the surface rather than plowing it under to minimize the number of times a field is tilled. Weeds are controlled by herbicides.

Mining Act An act passed in 1872 providing free access to minerals on federal lands.

Mixed cropping An agricultural system in which several different crops are grown in close proximity, in a rotation system, or both.

Mobile sources Sources of air pollution that move, such as automobiles, boats, trains, and aircraft.

Monkey wrenching Destructive vandalism directed against environmentally destructive activities. From Edward Abbey's famous novel, *The Monkey Wrench Gang*.

Monoculture An agricultural system in which a single crop is grown repeatedly over a large area.

Monopoly Control of access to a good or service by a single entity.

Montreal Protocol An agreement signed in Montreal in 1987 in which signatory nations consented to limit production and consumption of ozone-damaging chemicals.

Multinational corporation A business entity that operates in many nations and is not wholly subject to the laws of any one nation.

Multiple use The use of lands for as many different purposes as possible in order to gain maximum benefit from them.

Multiple Use Sustained Yield Act A law passed in 1960 establishing the principles of multiple use and sustained yield as guidelines for management of the national forests.

Municipal solid waste Mixed solid waste derived primarily from residential and commercial sources.

National Forest Management Act An act, passed in 1976, establishing operating principles and administrative divisions for the U.S. Forest Service.

National sovereignty The right of individual nations to look after their own interests first and foremost and to manage resources within their territorial borders any way they see fit.

Natural capital The stock of natural resources that provide or can be used to make benefical things.

Natural gas Gaseous hydrocarbons extracted from subterranean reservoirs that hold gas at normal ambient temperatures.

Natural increase In demography, the net change in population without regard to migration. It is the birth rate less the death rate, and it can be positive or negative.

Natural resource Something that is useful to humans and exists independent of human activity.

Natural resource accounting The inclusion of the full and often hidden costs of damages to natural resources in traditional benefit/cost analysis.

Neo-Malthusianism Modern advocates of Thomas Malthus's ideas; those who advocate birth control to avert overpopulation and who see overpopulation as ultimately leading to widespread malnourishment and famine.

NEPA The National Environmental Policy Act, signed on January 1, 1970, which established nationwide environmental goals for the United States and provided for the preparation of Environmental Impact Statements (EIS) to ensure compliance with those goals.

Net primary production The net amount of biomass created by plants in an ecosystem once the respiration by those plants is deducted.

Neutral stuff Something that exists but at present meets no known human material or nonmaterial needs.

Nonpoint source A pollution source that is diffuse, such as urban runoff.

Nonrenewable or **stock resources** Resources that exist in finite quantity and are not replaced in nature.

Off-stream use Use of water that requires that it be removed or withdrawn from surface or groundwater.

Oil Hydrocarbons found in the earth, liquid at normal ambient temperatures.

Oligopolistic competition A process in which a small group controls access to a good or service by agreeing on a single price or by restricting access to these commodities.

Oligotrophic Describes lakes that are relatively deep and nutrient-poor; opposite of eutrophic.

Organic Refers to substances containing carbon.

Output Energy, matter, or information leaving a system.

Overburden Rock and soil that lie above coal or other mineral deposits and that must be removed to strip-mine the coal.

Overdraft or **groundwater mining** Withdrawal of groundwater in excess of the replacement rate over a long period of time.

Overgrazing Grazing by a number of animals exceeding the carrying capacity of a given parcel of land.

Overland flow Water flowing on the soil surface and unchannelized, usually derived from precipitation that has not infiltrated.

Oxidants A group of air pollutants that are strong oxidizing agents. Ozone and peroxyacetylnitrate are among the more important oxidants. Also known as ground-level ozones.

Ozone depleting chemicals (ODCs) Substances that cause a reduction in stratospheric ozone.

Ozone hole A semipermanent depletion in stratospheric ozone concentration over a polar region. Most prominent over the South Pole.

Parent material The mineral matter from which soil is formed.

Particulate matter In refercnce to air quality, solid or liquid particles with diameters from 0.03 to 100 microns.

Passive solar power The collection of solar energy as heat at the end-use site, without any mechanical redistribution or storage of the energy.

Pastoral nomad A person who herds animals and has no permanent place of residence.

Pastoralist A person whose livelihood is based on grazing animals.

Pasture In U.S. terminology, land on which the natural vegetation is not grass but which is used primarily for grazing.

Peat The accumulated remains of plants, found in swampy or cool, humid areas. It is the initial material from which coal may be formed; may be dried and used for fuel.

Performance-based resource estimate An estimate of the quantity of a mineral deposit available in the earth, based primarily on the ability of prevailing technology to extract the mineral under existing and probable future economic conditions.

Permafrost Ground below 32°F (0°C) all year round.

Permeability A measure of the rate at which water will flow into or through soil or rocks.

Perpetual resources Resources that exist in continual supply, no matter how much they are used. Solar energy is an example.

Pesticide A general term used to refer to a chemical used to control harmful organisms such as insects, fungi, rodents, worms, and bacteria. Insecticides, fungicides, and rodenticides are kinds of pesticides.

Petajoule A unit of energy equal to 10^{15} (1,000,000,000,000,000) joules, or 947,800,000,000 British thermal units.

Photosynthesis The formation of carbohydrates from carbon dioxide and water, utilizing light as energy.

Photovoltaic cell A semiconductor-based device used to convert sunlight directly to electricity.

Placer deposit A deposit of a mineral formed by a concentration of heavy minerals in flowing water, such as by a stream or waves.

Point source A pollution source that has a precise, identifiable location, such as a pipe or smokestack.

Polluter-pays principle, or **residuals tax** A means of shifting the cost of pollution from the community to the polluter, usually in the form of a tax.

Pollution Human additions of undesirable substances to the environment.

Pollution prevention The elimination of potential pollutants at their source such as in a manufacturing process, rather than at the point of discharge to the environment (end-of-pipe) or later.

Pollution Standards Index (PSI) An index of air quality that is a combined measure of the health effects of several pollutants. It ranges from 0 (healthy) to 500 (extremely unhealthy).

Population dynamics The study of the rapidity and causes of population change.

Population pyramid A graphic representation of the number or portion of males and females in each of several age categories in a population.

Potential evapotranspiration The amount of water that could be evaporated or transpired if it were available.

Potential resource A portion of the natural or human environment that is not today considered of value, but that one day may gain value as a result of technological, cognitive, or economic developments.

Preservation The nonuse of resources; limited resource development for the purpose of saving resources for the future.

Primary standards Air pollution standards designed to protect human health.

Primary treatment Sewage treatment consisting of removal of solids by sedimentation, flocculation, screening, or similar methods.

Principle of limiting factors Whatever factor (nutrient, water, sunlight, etc.) is in shortest supply will limit the growth and development of an organism or a community.

Prior appropriation A doctrine of water ownership in which the first productive user of water establishes the right to the water indefinitely; the primary water-ownership doctrine in the western United States.

Privatization The transfer of government-owned resources, such as national forests, to private ownership or management.

Procedural equity A situation in which there is a differential application of environmental regulations, laws, or treaties such that some areas bear more pollution burdens than others.

Production theory of value An approach to valuing commodities based on the inputs of some critical commodity needed to make it.

Proxy value A price applied to a commodity that has no established market value.

Quad A measure of energy use, equal to one quadrillion (1,000,000,000,000,000) British thermal units.

Radiation inversion A temperature inversion caused by radiational cooling of air close to the ground.

Radioactivity The emission of particles by the decay of atoms of certain substances.

Railroad Acts A series of acts passed in the United States in the 1850s and 1860s granting large amounts of land to railroad companies as a subsidy to railway construction and stimulant to settlement of western lands.

Range condition As defined by the U.S. Forest Service, an estimate of the degree to which the present vegetation and ground cover depart from that which is presumed to be the natural potential (or climax) for the site.

Rangeland Land that provides or is capable of providing forage for grazing animals.

Recycling Reprocessing of a used product for reuse in a similar or different form.

Relative scarcity Short supply of a resource in one or more areas due to inadequate or disrupted distribution.

Renewable energy Energy resources that are produced naturally as fast as they are consumed, such as solar, wind, and hydroelectric power.

Renewable resource A resource that can be depleted but will be replenished by natural processes. Forests and fisheries are examples.

Replacement cost The cost of replacing a resource that is used.

Replacement level The number of births that will replace a population at the same size, without reduction or rise; also called Zero Population Growth.

Reserve In the context of mineral resources, a deposit of known location and quality that is economically extractable at the present time.

Residual Waste products of production. Air pollution is an example of residuals from manufacturing.

Residual tax A tax on production based on the amount of waste produced.

Residuals management An approach to production of goods and services that includes accounting of waste products associated with both production and consumption phases.

Resource Something that is useful to humans.

Resource recovery Separation of waste into recyclable components such as metal, glass, and heat from incineration.

Respiration Oxidation of food that releases oxygen, water, and energy, which are dissipated in the biosphere.

Reuse Repeated use of a product without reprocessing or remanufacture.

Rill A small channel created by soil erosion and small enough to be obliterated by plowing.

Riparian areas Lands adjacent to and subject to flooding by streams.

Riparian rights A doctrine of water ownership in which those whose land adjoins a stream have the right to use the water in the stream. It is the primary water-ownership doctrine in the eastern United States.

Ruminant One of a group of grazing animals, including cattle, bison, sheep, and goats, which have digestive systems particularly adapted to grasses.

Sahel A semiarid east–west swath across Africa, environmentally transitional between the Sahara Desert (to the north) and equatorial rainforests (to the south), in which recent desertification and drought have been particularly severe.

Salinity The concentration of mineral salts in water. The average salinity of the oceans is about 35 parts per thousand.

Saltwater intrusion Movement of salt water into aquifers formerly occupied by fresh water as a result of groundwater withdrawal in coastal areas.

Satisficing A decision-making strategy that seeks a course of action that is good enough but not necessarily perfect. A few alternatives are compared, and the best course of action is chosen from this limited range of options.

Savanna Tropical or subtropical semiarid grassland with scattered trees.

Second-growth forests Forests that have re-grown after harvest or other disruption such as fire.

Secondary standard An air-quality standard designed to protect human welfare (property, vegetation, etc.) as opposed to health.

Secondary treatment Sewage treatment that removes organic matter and nutrients by biological decomposition using such methods as trickling filters, aeration, and activated sludge.

Sedentarization Permanent settlement of once-nomadic people.

Sedimentation Deposition of solid particles by settling in a water body.

Selective cutting A timber-harvesting technique in which only trees of specified size or species are taken, leaving other trees.

Separate impacts Effects of a system's activity that can be measured separately.

Shadow price An artificial monetary value applied to those resources for which a simple price tag is not easy to calculate, for example, wilderness or habitat.

Shale oil see also *kerogen*. A petroleum-like substance found in high concentrations in some shale rocks.

Shelterwood cutting A two-phase timber-harvesting technique in which not all trees are taken in the first phase so that some trees may provide shelter for young seedlings; when these are established, the remaining older trees are cut.

Sievert The amount of radiation absorbed by the human body; measures the amount of human exposure to ionizing radiation where one unit is equal to 1 joule per kilogram or 100 rems.

Silviculture Intensive management of forest lands for increased production of trees.

Smog A term used to describe air pollution.

Social cost The cost of producing a good or service, plus its cost to humans in terms of pollution and other negative socio-environmental effects.

Social equity Situations in which there are different pollution burdens or access to resources based on race, ethnicity, age, or gender.

Softwoods Timber species with relatively low-density wood; primarily needle-leaf trees.

Soil A porous layer of mineral and organic matter at the earth's surface, formed as a result of the action of chemical and biological processes on rocks over a period of time.

Soil erodibility A measure of the inherent susceptibility of a soil to erosion, without regard to topography, vegetation cover, management, or weather conditions.

Soil fertility The ability of a soil to support plant growth through providing water, nutrients, and a growth medium.

Soil structure The way in which individual soil particles form aggregates, particularly the shapes and arrangement of such aggregates; especially important to soil hydrologic characteristics.

Soil texture The mix of different sizes of particles in a soil.

Solid waste Refuse materials composed primarily of solids at normal ambient temperatures.

Somatic damage Nonhereditary damage to individual cell tissues from radiation.

Species A group of organisms with similar genetic and morphologic characteristics that are capable of interbreeding.

Spent fuel Nuclear material that is no longer capable of sustaining the fission process.

Sprinkler irrigation Irrigation by pumping water under pressure through nozzles and spraying it over the land.

Stakeholders Individuals or groups who have a vested interest in a particular issue or policy or could be adversely affected.

Stationary source A pollution source that does not move, such as a smokestack.

Steady state When a system has inputs that equal outputs.

Stock resource See *Nonrenewable or stock resources*.

Stockpiling Amassing amounts of some substance well beyond present needs in anticipation of a shortage of that substance.

Strategic mineral A mineral necessary for defense purposes for which the United States is totally dependent on foreign sources.

Stratification A layering of a water body caused by differences in water density. It is commonly caused by temperature or salinity differences.

Stratified estuary An arm of the sea in which fresh water from the land overlies denser salt water.

Stratosphere Layer of the atmosphere between 3 and 30 miles (5 and 50 km) in altitude, characterized by increasing temperature with altitude.

Strip cropping A soil conservation technique in which parallel strips of land are planted in different crops.

Strip-mining or **surface mining** Extraction of a mineral from the ground by excavation at the ground surface.

Stubble mulch A soil covering composed of the unused stalks of crop plants.

Subbituminous coal A rank of coal intermediate between lignite and bituminous coal.

Subeconomic resource A resource that at present is unavailable for use because of the high cost of extraction.

Subsidence inversion A temperature inversion caused by differential warming of a sinking air mass. Upper portions of the mass are warmed more than lower portions, causing the inversion.

Subsidence Sinking of the land surface caused by removal of water, oil, or minerals from beneath the surface.

Substitutability The degree to which one material can be substituted for another in end uses.

Sulfur content The amount of sulfur found in coal. Combustion of coal with a high sulfur content results in emissions of sulfur oxides, which contribute to acid precipitation.

Surface fire A moderate-intensity forest fire in which low-level vegetation, such as shrubs, is consumed along with some of the surfaces (bark) of trees, but the crowns of trees are not consumed and trees survive.

Surface water Water and ice found in rivers, lakes, swamps, and other above-ground water bodies.

Suspended particulates In reference to air quality, solid or liquid particles with diameters from 0.03 to 100 microns.

Sustainability Economic growth with environmental responsibility; economic activity that could be carried on indefinitely without resource depletion.

Sustainable agriculture An agricultural system that is dependent solely on renewable resources and that maintains the soil in such a condition so that it will continue to be productive indefinitely.

Sustained yield Management of renewable resources conducted in such a way as to allow a constant rate of harvest indefinitely.

Synergistic impacts Effects of a system's activity that are different from the individual effects of component parts of the system.

Synfuel A contraction of synthetic fuel; liquid or gaseous fossil fuel manufactured from other fuels that are less useful as found in nature.

System An entity consisting of a set of parts that work together to form a whole. The human body, a transportation network, and the earth are all systems.

Tailings Solid waste products derived from mineral extraction or refinement.

Tar sand Sandy deposits containing heavy oil or tar. The sand must be heated to extract the oil.

Taylor Grazing Act An act passed in 1934 closing most United States public lands to homesteading and establishing controls on grazing use of federal lands.

Temperate forest A biome characterized primarily by deciduous broad-leaved trees.

Temperature inversion A condition in the atmosphere in which warm air overlies cool air. Inversions restrict vertical air circulation.

Terracing A soil and water conservation technique consisting of ridges on the contour, or level areas constructed on a slope.

Territorial sea A band of open ocean adjacent to the coast, over which the coastal nation has control. It is generally either 3 or 12 nautical miles (5.6 or 22 km) wide.

Tertiary treatment Any of a wide range of advanced sewage treatment processes aimed at removing substances not eliminated by primary or secondary treatment.

Thermal pollution Heat added by humans to a water body or to the air.

Thermocline A zone in a water body in which temperature declines rapidly with increasing depth. Vertical water circulation is limited by the presence of a thermocline.

Threatened species A species that is not endangered but has a rapidly declining population.

Throughput tax, or **disposal charge** A fee paid by a producer on materials that go into the production of polluting products. The fee reflects the social cost of the pollution.

Tidal power Energy generated by using tidal water-level differences to drive a turbine

Timber Culture Act An act passed in 1873 providing free access to timber on federal lands.

Toxic substance A substance that causes disease or death when organisms are exposed to it in very low quantities.

Total dissolved solids The total amount of dissolved solid matter found in a sample of water.

Total maximum daily loads A calculation of the maximum amount of a pollutant that a water body can receive and still meet water quality standards.

Transboundary pollution Transport of pollutants (particularly air pollutants) across national or state boundaries.

Trophic level One of the steps in a food chain.

Tropical rainforest A biome composed primarily of evergreen broad-leaved trees growing in tropical areas of high rainfall throughout most of the year.

Troposphere The lowest layer of the atmosphere, below about 9 miles (15 km) in altitude, characterized by decreasing temperature with increasing altitude.

Tundra A biome found in arctic and subarctic regions consisting of a dense growth of lichens, mosses, and herbs.

Underground mining A mineral extraction technique consisting of subsurface excavation with minimal disturbance of the ground surface.

Unidentified resource A mineral resource assumed to be present within known geologic districts, but not yet specifically located or characterized in detail.

Universal Soil Loss Equation A statistical technique developed by the U.S. Department of Agriculture for predicting the average erosion rate by rainfall under a variety of climatic, soil, topographic, and management conditions.

Upwelling An upward movement of seawater that usually occurs near the margins of oceans.

Uranium An element, of which two isotopes (^{235}U and ^{238}U) are important in atomic energy production.

Urban runoff Runoff derived from urban areas, usually containing relatively high concentrations of pollutants; also called urban stormwater.

Variable costs Costs of production that vary with the rate of output.

Visual blight Modification of a landscape that is visually undesirable.

Waste reduction Strategies for waste management that focus on reducing the amount of waste generated at its source.

Wastewater reclamation Any process in which wastewater is put to use, such as for cooling or irrigation, with or without treatment.

Water harvesting Any of several techniques for increasing the amount of runoff derived from a land area.

Water-holding capacity The ability of the soil to retain or store water.

Watershed management Integrative, holistic management of water resources organized at the watershed (drainage basin) scale.

Water table The upper limit of groundwater or of the saturated zone.

Weathering The breakdown of rocks into smaller particles or new chemical substances as a result of exposure to water and air at the earth's surface.

Willingness to pay A method of determining the proxy value of a resource by asking how much users of that resource would be willing to pay to use or not use it.

Windbreak A line of trees or shrubs planted perpendicular to the prevailing winds, designed to reduce wind velocities and thus reduce wind erosion.

Withdrawal The removal of water from surface water or groundwater.

Zero population growth A term applied to the fertility rate needed to attain a stable population over a long period of time.

Zoning A system of land-use management in which land is classified according to permitted uses.

INDEX

Absolute scarcity, 369
Accounting, 362
Acid deposition, 25, 26, 56, 271–279, 305, 369
Acid drainage, 325
Acid mine drainage, 369
Acid precipitation, 79, 237, 252, 253, 256, see Acid deposition
Acid rain, see Acid deposition
Active solar power, 338, 369
Adirondack Forest Reserve, 40
Adirondacks
  acid deposition in, 275
Advection inversion, 251, 369
Africa, AIDS in, 92–93
Age structure, 369
Age structure of populations, 95–98
Agencies
  International, See Food and Agricultural Organization, International Monetary Fund, International Union for the Conservation of Nature and Natural Resources, World Bank, United Nations Environment Programme, World Health Organization, Organization of Petroleum Exporting Countries (OPEC), European Environment Agency, European Environment Information and Observation Network, European Union World Bank
  U.S. Government, 54
Agenda 21, 47, 352, 365
Agribusiness, 113, 369
Agricultural land, distribution of, 109
Agricultural runoff, 369
Agricultural Stabilization and Conservation Service, 43
Agricultural technology, 182
Agriculture
  and water pollution, 233
  fertilizers in, 70
  impact of global warming on, 290
  mechanization of, 39
  origins of, 37, 107–140
Agricultural Adjustment Act, 43
AIDS, 92–93

Air pollution
  indoor, 8
  trends in, 55, 57, 248–269
Air quality index, 263, 369
Air Quality Management District (Los Angeles), 264–265
Alaska, 45, 52–53
Alaska Lands Bill, 52–53
Alaskan National Wildlife Refuge, 52–53
Alliance of Small Island States, 291
Aluminum ore, 296
Aluminum recycling, 308
Amazon forest, 151, 157–159
Ambient air quality, 254
  in megacities, 254–256
  United States, 261–262, 369
Amboseli Park, 187
Amenity resource, 369
Amenity value, 369
Amoco Cadiz, 207
Anadromous fish, 198, 369
Angola, diamonds in, 301
Animal rights, 7–8, 170–171
Animal unit month, 369
Animals, in agriculture, 125–128
Anoxic, 369
Antarctic Treaty and Convention, 47
Antarctica, 46, 281
Anthracite, 323, 325, 369
Anthropogenic, 369
Antiquities Act, 41
Aquaculture, 200
Aquiclude, 219, 369
Aquifer, 219, 222, 369
Arab-Israeli War, 317
Arable land, 66–67, 108, 369
Aral Sea, 176
Argentina, population, 90
Arroyo, 369
Asbestos Hazard Emergency Response Act, 268
Asian financial crisis of 1997, 317
Atmosphere, structure of, 248–250
Augur mining, 369
Auguring, 325
Australia, population, 90

Automobiles
  emissions from, 266
  fuel efficiency, 266, 342–343, 345
  recycling of, 358–359
  use of, 359

Baby boom, 94, 98, 369
Bacillus thuringiensis, 123
Bangladesh, population, 90
Basel Convention, 32
Basel Convention on the Control of Transboundary Movements of Hazardous Wastes and their Disposal, see Basel Convention
Bauxite, 296, 298
Belarus, 332
Benefit-cost analysis, 20–21, 369
Bennett, Hugh H., 43
Bhopal, 267
Bicycles, 360
Bioaccumulation, 79–80, 237, 369
Biochemical decay, 369
Biochemical oxygen demand (BOD), 233, 235, 239, 369
Biocide, 370
Biodiversity, 153, 157–159, 169–190, 198
  in forests, 142
Biogeochemical cycles, 70, 154–155, 370
Biological diversity, 370
Biological Weapons Convention, 47
Biomagnification, 80, 237, 370
Biomass, 370
Biomass harvesting, 146, 370
Biome, 63, 370
Biomes, map, 64
Bioregions, 62–63, 370
Biosphere, 370
Biosphere Reserve, 49
Biosphere Reserve Program, 183
Biosphere resources, 370
Biotechnology, see Genetically modified organisms
Biotic potential, 76, 370
Birth control, 87, 102
Birth rate, 89–95, 370
Bison, 177

382

Bituminous coal, 323, 325, 370
Black markets, 284
Black Sea, 196
BMW and recycling, 358–359
Boom and bust cycles, 309
Boreal forest biome, 65
Boreal forests, 159–160, 370
Boserup, Esther, 87
Bottle deposits, 355
Boundaries, maritime, 205
Brazil, 157–159, 174
    energy use, 341, 343
    population, 90, 96, 97
Breeder reactor, 329
British Thermal Unit (Btu), 316, 370
Brundtland Report, 47, 351, 353
Bubble approach, 370
Buffalo gourd, 169
Bureau of Land Management (BLM), 43, 139, 153, 185, 228, 370
Bureau of Reclamation, 228
Bush, George W., 45
Business
    and environment, 56
    and sustainability, 357–358
    and water pollution, 241
Bycatch, 201, 370

California, 346–347
    air quality, 57
    aqueduct, 223
    condor, 180
Canada
    and acid precipitation, 25
    hydroelectric power in, 334
    natural capital, 31
    population, 90
Canadian shield, 298
Cancer, 279, 332
Capital, human, 352
Carbon budget, 71–72, 370
Carbon cycle, 70
Carbon dioxide, 67, 70, 84, 118–119, 154–155, 159, 271, 285, 343, 359
    and coal combustion, 327
    atmospheric concentrations, 286
    emissions, 286–287, 290–291
Carbon monoxide, 253
    emissions of, 261
Carbon sequestration, 154–155
Carbon tetrachloride, 280
Carcinogen, 370
Carrying capacity, 37, 76, 82–83, 138, 370
Carson, Rachel, 43, 71, 168
Cartel, 18, 301, 370
Catadromous fish, 198, 370
Catalytic converters, 310
Cattle, 112
Center for Plant Conservation, 182
Central Arizona Project, 229
Centralized energy, 370
Centralized renewable energy, 333
Centrally planned economies, 17
Cereal production, 110, 113
Chaparral, 65, 370
Chernobyl, 45, 331, 332
Chickens, 112

China, 336–337
    acid deposition, 275
    air pollution, 254
    carbon emissions, 292
    coal resources, 325
    energy use, 341
    energy use in, 344
    forests, 151
    pollution in, 354
    population, 90
    population growth, 92
Chipko, 146–147
Chipping, 146
Chlorinated hydrocarbons, 79
Chlorofluorocarbons (CFCS), 279–284, 370
    atmospheric concentrations, 286
Chromium, 296, 303
Civilian Conservation Corps, 42
Clean Air Act, 44, 256, 266, 267, 278, 291, 308, 370
Clean Water Act, 209, 240, 242–243, 370
Clear-cutting, 146, 151, 370
Cleveland, Grover, 40
Climate and forests, 151
Clinton, Bill, 41, 56, 188
Coal, 4, 45, 292, 315, 323–327, 343
    carbon content, 285
    reserves of, 319
Coal gasification, 327, 370
Coastal Zone Management Act, 44, 210
Cobalt, 296, 303
Cod, 200
Coffee prices, 16
Cognition, 1–2
Cohort, 95, 370
Coliform bacteria, 233, 370
Colonialism, legacy of, 136
Commercial economies, 17
Commercial forest, 161
Common property resource, 25, 193, 203–204, 212, 370
Commoner, Barry, 43
Community, 371
Commuting, 359
Competition, 18
Comprehensive Environmental Response, Compensation and Liability Act, 44, 357
Computers, 311, 359
Concentration, 371
Congo (Dem. Rep.), population, 90, 95–96
Conservation, 371
    defined, 5–6, 9
    of energy, 320, 340–343
    vs. preservation, 41, 44
Conservation of energy, 69
Conservation Reserve Program, 133
Conservation tillage, 129–131, 131, 138, 371
Consultative Group on International Agricultural Research, 170
Consumer theory of value, 17, 371
Consumption, 104, 359
    of energy, 318
Consumptive use, 224, 227, 231, 371

Continental shelf, 198, 202, 212, 371
Contingent valuation, 21–22, 371
Contour plowing, 129, 131, 371
Contraceptives, 103–104
Convention on Biological Diversity, 47, 181, 188, 189
Convention on Conservation of Migratory Species of Wild Animals, 47
Convention on International Trade in Endangered Species, 47, 180
Convention on the Conservation of Migratory Species of Wild Animals, 181
Convention on Wetlands of International Importance, 181, see also Ramsar
Conversion factors, energy, 316
Cooperative Extension Service, 40
Copper, 296, 305
Corn, production of, 114
Cost theory of value, 18, 371
Cost-effectiveness analysis, 20, 371
Costs of production, 29
Cotton, 16
Crisis management, 55, 371
Criteria pollutants, 256, 371
Critical environmental zones, 176
Critical mineral, 303, 371
Criticality, 371
Crop rotation, 129, 371
Cropland, 112, 371
Crops, genetically modified, 123–124
Crown fires, 156, 371
Crude oil, 319, 371
Cuba, population, 90
Culture, 2
Currents, ocean, 195
Cyanide, 306
Cyanide fishing, 201

Dams, 179, 223, 224, 333–335, 336–337
DDT, 71, 79, 80, 181
Dead Zone, 195
Death rate, 89–95, 371
DeBeers cartel, 301
Decentralized energy source, 371
Decentralized renewable energy, 333
Decision-making, 20, 50–58
Decreaser, 371
Decreasers (range species), 126
Deep oceans, 198, 371
Deforestation, 67–68, 70, 71, 72, 371, see also Forests, clearance
    and global warming, 285
Demand, 18
Demographic transition, 99, 371
Department of the Interior, 186
Depletion of water, 225
Deregulation, 33, 346–347, 371
    of agriculture, 114
Desalination, 371
Desert, 371
Desert biome, 63
Desertification, 133–135, 371
Diamonds, 301
Diets, 108

Dilution, 371
Disease-causing organisms, 233, 234
Dissolved oxygen, 195–196, 232, 235, 237, 371
Dissolved solids, 234, 236, 371
Diversification, 29–30, 372
Dolphins, 201
Domestic use of water, 229
Domesticate, 372
Doubling time, 89, 372
Drainage basin, 222, 372
Dredging, 208
Drinking water, 79, 243
 access to, 244–246
Drip irrigation, 118, 372
Drought, 222, 372
Dry farming, 372
Dry rock geothermal energy, 372
Dusky sparrow, 181
Dust Bowl, 42–43

Eagle, bald, 181, 188
Earth Day, 50
Earth Summit, 47, 48, 352, 372
Earth system science, 11, 372
EarthFirst!, 153
Ecological economics, 15, 372
Ecology, 61, 169, 372
Economic conditions and mineral resources, 297
Economic development and air pollution, 254–256
Economic sustainability, 352
Economic systems, 17
Economics
 of natural resources, 15–35
 of soil erosion, 131–133
Eco-protectionism, 20
Ecosystem, 61–62, 372
Ecotone, 62, 198, 372
Ecotourism, 372
Education and birth control, 103
Efficiency, 353
Egypt, population, 90
Ehrlich, Paul, 104
El Niño, 194, 372
Elasticity, 15
Electric generation, 33, 69, 196–197, 231, 273, 318, 327–331, 333–335, 336–337, 343, 344, 346–347
Elephants, 180
Emerson, Ralph W., 40
Emigration, 98–99
Emission standards, 260
Emissions, 372
 nitrogen oxide, 273–275
 of air pollutants, 254, 259–261
 of carbon, see also Carbon dioxide
 sulfur dioxide, 273–275
Emissions trading, 26, 278–279, 292, 372
Endangered species, 21, 174, 372
Endangered Species Act, 181, 182, 186, 189, 213
Endocrine disruptors, 71
End-of-pipe pollution control, 357, 372
Energy budget, 372
Energy conservation, 69, 340–343, 344, 354, 372

Energy crises of 1970s, 317, 341
Energy crisis, California, 2000, 33, 346–347
Energy efficiency, 372
Energy intensity, 341
Energy resources, 314–348
 marine, 200–201
Entropy, 15, 69
Environmental cognition, 372
Environmental equity, 48
Environmental ethics, 372
Environmental justice, 6, 8, 372
Environmental lapse rate, 249, 372
Environmental movement, 43–44
Environmental policy International, 45–49
 U.S., 37–45
Environmental Protection Agency, U.S., 51, 242–243
Environmental refugee, 372
Environmental resistance, 372
Environmental sustainability, 352
Equity, 8, 48, 372
 social, 353
Erlich, Paul, 43
Erosion, 116–117, 151, 236, 372
 soil, 128–129
Erosivity, 372
Estuarine Research Reserves, 184, 186
Estuary, 194, 198, 372
Ethics
 environmental, 6, 8, 352
Ethiopia, population, 90
Euphotic zone, 195, 372
Europe, 332, 345, 361
 acid deposition, 275
 agricultural policy, 136–137
 fuel costs, 364
 population growth, 88–89, 99–100
European Environment Agency, 19
European Environment Information and Observation Network, 19
European Union, 19, 51, 281
Eutrophication, 195, 233, 372
Evaporation, 69
Evapotranspiration, 74, 151, 155, 219, 372
Exclusive economic zone, 204, 205, 373
Exotic species, see Introduced species
Exploitation, 9, 373
Exploration, for oil, 320
Externalities, 27, 362, 373
Extinction, 123, 168, 172, 373
Exxon Valdez, 16, 207, 209

Family planning, 103–104
Farm Service Agency, 43
Farmland, 373
 and Three Gorges Dam, 337
Federal Energy Policy Act, 347
Federal Energy Regulatory Commission, 347
Federal Land Policy and Management Act, 139, 373
Federal Water Pollution Control Act, 44, see also Clean Water Act
Feedback, 14, 76, 373

Fertility rate, 89–95, 373, see also Total fertility rate
Fertilizer, 70, 73, 83, 111, 115, 120, 235, 373
Fiber production, 144–150
Fire, 25, 36, 160
 in forests, 155–157
Fire frequency, 373
Fire, use of, 63
First law of thermodynamics, 373
Fish and Wildlife Service, 49
Fisheries, 199–200, 231–232
Fishery Conservation and Management Act, 181, 211
Fission, 329, 373
Fixed costs, 29, 373
Flood control, 336–337
Flood irrigation, 118, 373
Flow resource, 373
Food
 kinds consumed, 108
 production of, 110
 sources of, 108
Food and Agricultural Organization, 46
Food chain, 74–76, 80, 82, 198, 373
Food crops, 170, 182
Food production, 107–140, 199–200
Food Quality Protection Act, 71
Food security, 373
Food Security Act, 137
Food web, 169, 373
Forest, 15
Forest land
 distribution of, 143
 use of, 66
Forest product technology, 148–150
Forest Service, US, 156–157
Forestry
 sustained yield, 41
Forests
 clearance in U.S., 38, 39
 conservation of, 40
 impacts of mining on, 305
Fossil fuel reserves, 319
Fossil fuel use, 292
Fossil fuels, 71
Framework Convention on Climate Change, 47, 291
France
 nuclear power in, 329, 331
 population, 90
Freedom to Farm Bill, 33, 137, see also Food Security Act of 1996
Freon, 280, 281, 284
Freshwater mussels, 178–179
Fuel cycle
 nuclear, 329–331
Fuel efficiency vehicle, 342–343
Fuelwood, 144, 149, see also Wood
Fugitive sources of air pollutants, 260
Furrow irrigation, 373
Fusion, 373

Gaia hypothesis, 63, 373
Galapagos Islands, 207
Gandhi, Indira, 147
GAP analysis, 373

Garbage, *see* Solid waste
Gas, natural, 315
 carbon content, 285
Gasification, of coal, 327
Gasoline, price of, 314, 321
General Agreement on Tariffs and Trade, 46
General circulation model, 373
General Systems Theory, 10, 373
Generational equity, 48, 373
Genetic damage, 373
Genetically modified crops, 373
Genetically modified organisms, 123–124
Geographic Information System (GIS), 373
Geologic estimate of resource, 373
Geologic information and mineral resources, 297
Geology of minerals, 298–299
Geopressurized energy, 335, 373
Georges Bank, 198, 200
Geothermal energy, 335, 373
Germany
 population, 90
 recycling in, 356, 358–359
Glen Canyon Dam, 335
Global 2000, 351
Global Environmental Monitoring System, 254
Global positioning systems, 125
Global warming, 45, 72, 157, 175, 194, 284–293, 362, 373
 costs of, 290–291
Goats, 112
God Committee, 187
Gold, 296, 306
Gore, Albert, 45, 47, 56
Government agencies, U. S., 54
Grain
 land used for, 111
 supplies, 67
Grand Canyon, 41, 259, 335
Grassland, 169, 373
Grassland biome, 65
Grazing, 126–128
Great Barrier Reef Marine Park, 211
Great Lakes, 179, 224
Great Leap Forward, 92, 94
Green Dot program, 356
Green Revolution, 110, 123, 374
Greenhouse effect, 284–285, 374
Greenhouse gases, 285–286, 374
Greenpeace, 3
Gross National Product (GNP), 340
Ground fire, 156, 374
Ground-level ozone, *see* Oxidants
Groundwater, 5, 74, 218, 224, 374
Groundwater mining, 225
Groundwater pollution, 238–239
Groups, public interest, 56
Growth
 economic, 352
 limits to, 350
Guatemala, population, 90
Guest worker, 98, 374
Gulf of Mexico, 195
Gulf War, *see* Persian Gulf War

Gully, 374
Gyres, 374

Habitat, 175–178, 186, 231–232, 374
 forests as, 150–151
 marine, 197–198
Halocline, 194, 374
Halons, 280
Hardin, Garrett, 9, 25
Hardwood, 161–164, 374
Harrison, Benjamin, 40
Harvesting techniques, forest, 146–148
Hayes, Rutherford, 40
Hazardous Materials, 207–208, 267
Hazardous waste, 32, 44
Headwaters Forest, 152–153
Health and pollution, 49
Heat
 in water, 234, 237
Heavy-water reactor, 329, 374
Hetch-Hetchy, 41
High seas, 204, 374
High-temperature gas-cooled reactor, 329, 374
HIV, *see* AIDS
Hogs, 112
Homestead Act, 38, 139, 374
Homesteading, 43
Homocentric, 374
Homosphere, 248, 374
Hoover Dam, 42
Horizon, 374
Hot dry rock energy, 335
Hubbert, M. K., 322
Hunting, 175
Hurricane Andrew, 290, 361
Hurwitz, Charles, 152
Hydraulic mining, 306, 374
Hydroelectric power, 196–197, 224, 227, 231, 333–335, 344, 374
Hydrologic cycle, 73–74, 81, 159, 218
Hydrothermal energy, 335
Hydrothermal mineralization, 298, 374
Hypoxia, 196
Hypoxic, 374

Illegal immigrant, 374
Immigration, 90, 98–99
Incommensurables, 21, 374
Increasers (range species), 127, 374
Incrementalism, 55, 374
Index of Sustainable Economic Welfare, 363
India, 267
 energy use, 343
 forest resources, 146–147
 population, 90
 population growth, 95
 population control, 104
Indigenous cultures, 159, 170
Indonesia, population, 90
Indoor air pollution, 8, 254, 268
Industrial ecology, 358, 374
Industrial revolution, 100, 286, 288, 315
Infiltration capacity, 74, 128, 374
Information and mineral resources, 297
Inorganic, 374

Input, 374
In-stream use, 224, 230, 374
Insurance, 290, 361
Intangibles, 20, 21, 374
Integrated pest management, 84, 374
Intellectual property, 123
Interbasin transfer, 374
Interest rate, 21
Internal combustion engine, 315–316
Internal waters, 204, 205, 374
International Atomic Energy Agency, 46
International Convention for the Prevention of Pollution from Ships, 209
International Convention for the Regulation of Whaling, 46, 212
International Maritime Organization, 46
International Monetary Fund, 46
International Union for the Conservation of Nature and Natural Resources, 46
International Whaling Commission (IWC), 212, 374
Introduced species, 177–178
Invaders (range species), 127, 375
Invasive species, 127
IPAT, 104, 375
Iran, 317
 population, 90
Iron ore, 296, 297
Irrigation, 81, 111, 115, 117–119, 227–228, 375
Irrigation efficiency, 375
ISO 14000, 56
Israel, population, 90
Israel-Arab War, 317
Italy, population, 90, 96
Ivory, 180
Izaac Walton League, 184

Japan, 213
 acid deposition, 275
 population, 90
Johannesburg Summit, 365, *see also* United Nations World Summit on Sustainable Development
Justice, social, 353

Kennedy, John F., 43
Kenya, 187
Kerogen, 327, 375
Kesterson National Wildlife Refuge, 184
Kuwait, 314
Kuznets curve, 354
Kyoto Protocol, 45, 292, 365
Kyoto Protocol to the Framework Convention on Climate Change, *see* Kyoto Protocol

Labor theory of value, 17, 375
Labor, in agriculture, 124–125
Lacey Act, 181
Lagoons, 239
Lakes, 233–235
Land Capability Classification System, 375

Land degradation, 133–135, 176–177
Land Grant Colleges, 40
Land ownership, 144
Land use, 66–69
Landfills, 28, 155, 239, 297, 355–356, 375
Law of entropy, 375
Law of the Sea Treaty, 204–207, 375
Laws, U. S., *see* Agricultural Adjustment Act, Alaska Lands Bill, Antiquities Act, Asbestos Hazard Emergency Response Act, Clean Air Act, Clean Water Act, Coastal Zone Management Act, Comprehensive Environmental Response, Compensation and Liability Act, Endangered Species Act, Federal Energy Policy Act, Federal Land Policy and Management Act (FLPMA), Federal Water Pollution Control Act, Fisheries Management and Conservation Act, Fishery Conservation and Management Act, Food Quality Protection Act, Food Security Act, Freedom to Farm Bill, Homestead Act, Lacey Act, Marine Mammal Protection Act, Marine Protection, Research and Sanctuaries Act, Migratory Bird Conservation Act, Migratory Bird Hunting Stamp Act, Migratory Bird Treaty Act, Mining Act, Multiple Use and Sustained Yield Act, Multiple Use Sustained Yield Act, National Environmental Policy Act (NEPA), National Forest Management Act, NEPA, Ocean Dumping Act, Oil Pollution Act, Pittman-Robertson Act, Railroad Acts, Reclamation Act, Resource Conservation and Recovery Act, Safe Drinking Water Act, Superfund Amendments and Reauthorization Act, Surface Mine Reclamation and Control Act, Timber Culture Act, Toxic Substances Control Act, Weeks Act, Wilderness Act,
Leachate, 239, 375
Lead, 79, 277
  ambient concentrations, 262
  emissions of, 261
  in air, 253
  recycling of, 308, 312
Liability, 357
Life expectancy, 92
Light-water reactor, 329, 375
Lignite, 323, 325, 375
Limiting factors, 83
Limits to Growth, 350, 375
Liquefaction, 375
  of coal, 327
Liquid metal fast-breeder reactor, 329, 375
Liquid natural gas (LNG), 323, 375
Little Ice Age, 288

Livestock, production of, 111
Logging methods, 146–148
London Dumping Convention, 209
Los Alamos, NM, 156–157
Los Angeles, 61, 62, 229
  air pollution, 251, 264–265
Low-level wastes, 330, 375
Lumber, 21, 144–150

Machines, agricultural, 124–125
Magnesium, 296
Malaria, 290
Malthus, 375
Malthus, Thomas, 87
Mammals, marine, 212
Man and Nature, *see* Marsh, George P.
Manganese, 203, 303
Manufacturing, energy use in, 343
Manure, 121
Maquiladoras, 343
Marine Mammal Protection Act, 181, 213
Marine mammals, 212–214
Marine Protection, Research and Sanctuaries Act, 209, 210, 211
Marine Sanctuaries, 185, 186
Maritime boundaries, 205
Marpol, *see* International Convention for the Prevention of Pollution from Ships
Marriage and fertility, 103
Marsh, George P., 40, 43
Materials balance, 15
Materials balance principle, 375
Maximum sustainable yield, 375
Maxxam Corporation, 152–153
Meat consumption, 112
Meat production, 125–128
Mechanization of agriculture, 39
Medicines, 170
Mediterranean biome, 65
Megacities, 102
  pollution in, 254–256
Mesosphere, 250, 375
Mesquite, 2, 169
Metals as pollutants, 79
Methane, 285
  atmospheric concentrations, 286
Mexico City air pollution, 254
Mexico, population, 91
Midlatitude forest Biome, *see* Temperate forest Biome
Migration, 98–99, 375
Migratory Bird Conservation Act, 181
Migratory Bird Hunting Stamp Act, 181
Migratory Bird Treaty Act, 181
Milken, Michael, 152
Mined-land reclamation, 375
Mineral reserves
  geologic estimates of, 322
  performance-based estimates of, 322
Mineral resources, 201–204, 295–312
  Marine, 200–201
Minerals, 9, 16
Minimum tillage, 129–131, 375
Mining
  impacts of, 304–309
  ocean, 201

Mining Act, 38, 39, 308, 375
Mining techniques, 325
Missouri River, 222
Mixed cropping, 115, 375
Mobile sources, 252, 375
Monitoring, air pollution, 256
Monoculture, 115, 375
Monopoly, 18, 375
Monsanto, 123
Montreal Protocol, 281–282, 365, 375
Montreal Protocols on Substances that Deplete the Ozone Layer, *see* Montreal Protocol
Muir, John, 41
Multinational corporations, 29–32, 376
Multiple use, 49, 139, 144, 376
Multiple Use and Sustained Yield Act, 49, 376
Municipal solid waste, 376
Mussels, 178–179
Myanmar, population, 91

National Ambient Air Quality Standards, 256
National Emissions Standards for Hazardous Air Pollutants, 267
National Environmental Policy Act (NEPA), 44, 49
National Estuarine Research Reserves, 211
National Forest Management Act, 376
National Forests, U. S., 41, 68, 139, 156
National Marine Sanctuary Program, 211
National Park Service, 187
National Parks, 185, 186, 258, 259
National Parks, U. S., 41, 43, 49, 68, 156–157
National Plant Germplasm System, 182
National Pollutant Discharge Elimination System, 242–243
National Resources Inventory, 117
National sovereignty, 376
National Wilderness Areas, 186, 258
National wildlife refuges, 183, 184, 185, 186
Native Americans, 7, 24, 68, 134
  lands, 185
Natural capital, 363, 376
Natural Gas, 318–323, 376
  reserves of, 319
Natural hazards, 361
Natural increase, 90, 376
Natural resource, 376
Natural resource accounting, 376
Natural Resource Conservation Service, 43, 49, 117, 131
Natural Resources Defense Council, 56
Natural Resources Planning Board, 43
Nature
  value of, 30–31
  view of, 2–3, 6–7
Nature Conservancy, 184
Nature-centered view of resources, 61
Navigation, 231
Neoliberalism, 30, 33
Neo-Malthusianism, 87, 376
NEPA, 376

Net primary production, 82, 376
Neutral stuff, 1, 376
New Deal, 41–43, 136
Nickel, 296
Nigeria, population, 91
Nitrate, 235
Nitrogen, 233
Nitrogen cycle, 70, 72
Nitrogen fertilizers, 120
Nitrous oxide, 72, 253, 271, 285
 ambient concentrations, 262
 atmospheric concentrations, 292
 emissions of, 260
Nomad, pastoral, 111
Nonattainment areas, 261
Nondegradation of air quality, 258
Nonpoint source, 209, 232, 240–241, 376
Nonrenewable or stock resources, 376
North American Waterfowl Management Plan, 181
Northern coniferous forest biome, *see* Boreal forest biome
No-till farming, *see* Conservation tillage
NPDES, *see* National Pollutant Discharge Elimination System
Nuclear power, 315, 327–331, 344
Nuclear Test Ban Treaty, 46, 47
Nutrients, 209, 233, 234
 in forests, 150

Ocean currents, 195
Ocean dumping, 208
Ocean Dumping Act, 210
Ocean Dumping Convention, 47
Off-stream use, 376
Off-stream water use, 227–230
Ogallala aquifer, 226
Oil, 318–323, 343, 376
 Alaskan, 52–53
 marine resources, 200–202
 reserves of, 319
 uses of, 315–316
Oil Pollution Act, 210
Oil shale, 24, 327
Oil Spill Liability Trust Fund, 210
Oil Spills, 24, 207, 208
Old-growth forest, 150, 165, 175, 188
Oligopolistic competition, 376
Oligopoly, 18
Oligotrophic, 376
Oligotrophic lakes, 233
OPEC, 321
Open pit mining, 304
Ore production, U. S., 300
Organic, 376
Organic fertilizers, 121
Organization of Petroleum Exporting Countries (OPEC), 18
Organizations
 international, 46
 nongovernmental, 54–55, 56
Organochlorines, 122, *see also* Chlorinated hydrocarbons
Original forest, *see* Old-growth forest
Oslo Convention, 209

Our Common Future, 47, 351
Our Stolen Future, 71
Output, 376
Overburden, 325, 376
Overdraft, 225
Overdraft or groundwater mining, 376
Overgrazing, 40, 126–128, 133, 376
Overland flow, 128, 376
Ownership
 of forests, 161–164
 of land, 144
 of resources, 203–204
Oxygen-demanding wastes, 234, 235–236
Oxidants, 376
 emissions of, 260
Oxygen, dissolved, 195–196
Ozone, 249, 253, *see also* Oxidants
 ambient concentrations, 262
Ozone depleting chemicals, 376
Ozone depletion
 stratospheric, 279–284
Ozone hole, 45, 376

Pacific Lumber Company, 152–153
Pacific Northwest, 196–197
Pakistan, population, 91
Paper, 149–150
 recycling of, 356
Parent material, 116, 376
Particle board, 149–150
Particulate matter, 252, 376
 ambient concentrations, 261
 emissions of, 260
Passenger pigeon, 179
Passive solar power, 338, 376
Pastoral nomad, 376
Pastoralists, 111, 376
Pasture, 376
Peat, 323, 377
Performance-based resource estimate, 377
Permafrost, 159, 377
Permeability, 219, 377
Perpetual resources, 4–5, 377
Persian Gulf, 206, 317, 320
Persian Gulf War, 207, 314
Persistent pesticides, 79, 80
Peru, population, 91
Pest management, 84
Pesticides, 70, 78, 115, 122, 138, 236, 377
Pests, 115, 150
Petajoule, 316, 377
Petroleum, *see* Oil
Pfisteria piscicida, 200
Philippines, population, 91
Phosphorus, 235
Phosphorus cycle, 73
Phosphorus fertilizers, 120
Photochemical smog, 251
Photosynthesis, 69, 70, 82, 154, 195, 285, 377
Photovoltaic cells, 336, 338, 364, 377
Pigs, 112
Pinchot, Gifford, 41, 49
Pinelands National Reserve, 183
Pittman-Robertson Act, 181

Placer deposits, 298, 377
Plant Conservation, center for, 182
Plant nutrients, 233
Platinum group metals, 296, 303
 recycling of, 310
Plutonium, 329
Plywood, 149
Point source, 209, 232, 377
Poland, population, 91
Policy
 agricultural, 136
 natural resource, 49–50
Politics and resource use, 360
Pollutant decay, 79
Pollutant Standards Index, 262–264
Pollutants
 air, 252–253
 water, 234
Polluter pays principle, 28, 377
Pollution, 15, 175, 377
 and health, 49
 and tourism, 362
 credits, 26
 effects on biodiversity, 179
 extent of, 77
 marine, 207–209
 trading, 26
 water, 70, 224, 232–242, 304
Pollution control, 357
Pollution prevention, 31–33, 241, 357, 377
Pollution Standards Index (PSI), 377
Pollution trading, 265
Population, 76, 87–105
 control, 102–104
 dynamics, 89–95, 377
 environmental impacts of, 104
 pyramids, 95–97, 377
Population growth, history of, 88–89
Potential evapotranspiration, 63, 377
Potential resources, 169–170, 377
Pottassium fertilizers, 120
Powell, Colin, 365
Precautionary principle, 180
Precipitation, 159, 219
Precision farming, 125
Prescribed fire, 156–157
Preservation, 9, 377
 vs. conservation, 41, 44
Prevention of significant deterioration of air quality, 258
Price
 and mineral resources, 297, 299, 300
 of electricity, 334
 of fossil fuels, 343
 of gasoline, 314, 321, 342
 of lead, 312
 of oil, 317, 320
 of photovoltaic cells, 338
 variability of, 16
Pricing, 17
Primary production, 82
Primary standards, 256, 377
Primary treatment, 239, 377
Prime farmland, 117
Principle of limiting factors, 377
Prior appropriation, 228, 377
Privatization, 377

Procedural equity, 48, 377
Production theory of value, 18, 377
Protocol on the Prevention of Pollution from Ships, 47
Proxy value, 22, 377
Public Interest and Environmental Groups, 56
Public opinion and environment, 56
Pulp, 149–150

Quad, 377
Quiet Crisis, 43

Radiation inversion, 251, 377
Radioactivity, 234, 237–238, 330, 377
Radon, 268–269
Railroad Acts, 38, 377
Ramsar, see Convention on Wetlands of International Importance
Ramsar Convention, 211
Range condition, 377
Range management, 138–139
Rangeland, 126–128, 378
Rank, coal, 323
Reactors, nuclear, 329–331
Reagan, Ronald, 33, 44, 56, 58, 187, 243, 256, 267, 351
Reclamation Act, 39
Recreation, 232
  and forests, 154
Recycling, 17, 299, 302, 308, 309–311, 355, 358–359, 378
Reforestation, 68, 72
Refugees, 98
Regional Seas Program, 51, 210
Relative scarcity, 378
Renewable energy, 331–339, 344, 378
Renewable resource, 70, 378
Replacement cost, 24, 378
Replacement level, 95, 378
Reserve/production ratio, 299, 323
Reserves, 295–298
  of coal, 323–325
  of fossil fuels, 319
  of minerals, 299–301, 378
Reserves, mineral
  geologic estimates of, 322
  performance-based estimates of, 322
Reservoirs, 68
Residuals, 27, 378
Residuals management, 15, 378
Residuals tax, 28, 378
Resource, 378
Resource Conservation and Recovery Act, 44
Resource managers, 50–51
Resource recovery, 335, 378
Resources
  common property, 25
  defined, 1
  flow, 5
  in context of minerals, 295–298
  nonrenewable, 5
  of minerals, 299–301
  ownership of, 24–27
  perpetual, 4–5
  potential, 5
  renewable, 5
  stock, 5
  value of, 15, 17
Respiration, 69, 70, 154, 378
Restoration, forest, 150
Restructuring, economic, 345
Retorting, 327
Reuse, 358–359, 378
Rhinoceros, 179
Right of innocent passage, 204
Right of transit passage, 206
Rill, 378
Riparian areas, 378
Riparian doctrine, 228
Riparian rights, 378
Risk, 80
Romania, population, 91
Roosevelt, Franklin, 41–43
Roosevelt, Theodore, 41, 183
Rubber Tappers, 147
Rubber, natural, 170
Ruckelshaus, William D., 15
Ruminants, 112, 378
Runoff, 82, 151, 155, 219, 233, 240–241
Russia
  diamonds in, 301
  population, 91

Safe Drinking Water Act, 71, 242
Sahel, 378
Salinity, 193–194, 378
Salinization, 118
Salmon, 196–197, 236
Saltwater intrusion, 225, 378
Sanger, Margaret, 102
Sanitation, 244–246
Satisficing, 55, 378
Saudi Arabia, 319, 321
Savanna, 378
Savanna biome, 63
Scarcity
  absolute, 3
  relative, 3
Schurz, Carl, 40
Science and sustainability, 360
Scientific Revolution, 89
Seawater, properties of, 193
Secondary standard, 256, 378
Secondary treatment, 239, 378
Second-growth forest, 378
Sedentarization, 111, 378
Sediment, 236
Sedimentation, 337, 378
Seed, 122–124, 171
Selective cutting, 148, 378
Separate impacts, 378
September 11 attacks, 45, 318, 321
Sewage, 233, 239–240, see also Wastewater
Shadow price, 21, 378
Shale oil, 378
Sheep, 112
Shelf, continental, see Continental shelf
Shelterwood cutting, 147, 378
Shields, 298
Siberian forest, 159–160
Sierra Club, 41, 56
Sievert, 378
Silent Spring, 43, 71, 77, 168
Silviculture, 148, 378
Simon, Julian, 87
Smelting, 305
Smog, 248, 264, 379
Social agents, 50–51
Social capital, 363
Social costs, 27–28, 364, 379
  of mining, 305–309
Social equity, 48, 379
Social sustainability, 353
Softwood, 161–164, 379
Soil, 116–117, 379
  forest, 151
Soil conservation, 129–133
Soil Conservation Service, see also Natural Resources Conservation Service
  established, 42–43
Soil degradation, 133–135
  and global warming, 285
Soil erodibility, 379
Soil erosion, 38, 39, 128–129
  economics of, 131–133
Soil fertility, 379
Soil structure, 116, 379
Soil texture, 116, 379
Solar energy, 337–339, 344
Solar thermal power, 336
Solid waste, 57, 208, 311, 335, 355, 379
Somatic damage, 379
South Africa, population, 91
South Korea, population, 91
Sovereignty, 36
Soviet Union, 67, 160, 317, 332
Spaceship Earth, 10, 14, 43, 45
Spain, population, 91
Species, 379
  numbers of, 172–174
Species introductions, 177–178
Spent fuel, 379
Spotted owl, 151
Sprawl, 168
Sprinkler irrigation, 118, 379
Stakeholders, 50, 379
States, U.S., environmental regulation in, 56
Stationary source, 252, 379
Steady-state systems, 15, 379
Steel, 315
Stock resource, 322, 379
Stockholm Conference, 46
Stockpiling, 302–304, 379
Strategic mineral, 302–304, 379
Stratification, 233, 379
Stratified estuary, 194, 379
Stratosphere, 249, 279–284, 379
Strip cropping, 129, 131, 379
Strip-mining or surface mining, 379
Stubble mulch, 379
Sub-bituminous coal, 323, 325, 379
Subeconomic resources, 295, 379
Subsidence, 226, 325, 379
Subsidence inversion, 250, 379
Subsidies, 115, 136–137, 364
Subsistence economies, 17

Substitutability, 16, 299, 379
Sulfur
  emissions, 353–354
  emissions trading, 26
  in coal, 323, 325–326
  in fuels, 266
  in synfuels, 327
Sulfur content, 379
Sulfur dioxide, 252, 271
  ambient concentrations, 262
  emissions of, 260
Sulfur oxides, 288
Superfund, 44, 357, 363
Superfund Amendments and Reauthorization Act, 268
Supply, 18
Supply and demand, 27
Surface fires, 156, 379
Surface Mine Reclamation and Control Act, 326
Surface mining, 325–326, *see also* Strip mining
Surface water, 218, 379
Suspended particulates, 234, 379
Sustainability, 15, 61, 350, 352, 379
Sustainable agriculture, 138, 380
Sustainable development, 352–353
  United Nations World Summit on, 48
Sustained yield, 49, 139, 144, 380
Sustained yield forestry, 41
Sustainable Development, World Summit on *see* United Nations World Summit on Sustainable Development
Synergistic impacts, 380
Synfuels, 327, 343, 380
System, 380
Systems approach, 10

Tailings, 304, 330, 380
Tanzania, population, 91
Tar sands, 327, 380
Tax
  residuals, 28
  throughput, 28
Taxes, 364
Taylor Grazing Act, 43, 139, 380
Technology
  agricultural, 124–125, 182
  and mineral resources, 297
  forest products, 148–150
Tellico Dam, 186
Temperate forest, 380
Temperate forest Biome, 65
Temperature inversion, 250, 380
Tennessee Valley Authority (TVA), 27, 42
Terms of trade, 136
Terracing, 129, 380
Territorial sea, 203, 204, 205, 380
Tertiary treatment, 239, 380
Thailand, population, 91
Thermal pollution, 237, 380
Thermocline, 194, 380
Thermodynamics, 69
Thoreau, Henry D., 40, 160
Threatened species, 174–175, 380

Three Gorges Dam, 336–337
Three Mile Island, Pennsylvania, 330, 331
Throughput tax, or disposal charge, 28, 380
Tidal power, 336, 380
Timber, *see* Forests
Timber Culture Act, 38, 39, 380
Torrey Canyon, 207
Total dissolved solids, 233, 380
Total maximum daily load, 241, 380
Tourism, 362
Toxic Release Inventory, 77–78, 363
Toxic substance, 44, 77–79, 232, 234, 236–237, 239, 256, 306, 357, 380, *see also* Toxic wastes
Toxic Substances Control Act, 44
Toxic wastes, 208–209, *see also* Toxic substances
Trade, 136–137
  endangered species, 180
  in minerals, 295
Trade agreements, 18
Trading, emissions, 278–279, *see also* emissions trading
Tragedy of the Commons, 25
Transboundary pollution, 271, 380
Transnational Corporations, *see* Multinational Corporations
Trawling, 201
Treaties, 36, 47, 180, 281–282
Tree farming, 148
Tree spiking, 153
Trophic level, 75, 380
Tropical rainforest, 380
Tropical rainforest biome, 63
Troposphere, 249, 279, 380
Trust for Public Lands, 184
Tuna, 201
Tundra, 380
Tundra biome, 65
Turkey, 306
  population, 91

U. S. Forest Service, 364
Udall, Morris, 53
Udall, Stewart, 43
Ukraine, 332
  population, 91
Ultraviolet radiation, 279
UN Children's Fund, 46
UN Educational, Scientific and Cultural Organization, 46
UN Environment Program, 46
Uncertainty
  in resource management, 80
Underground mining, 325, 380
Unidentified resource, 295, 380
United Kingdom
  population, 91
United Nations, 45–49
United Nations Conference on Environment and Development, 45, 47, 188
United Nations Conference on the Human Environment, 47
United Nations Environment Programme, 47, 209

United Nations World Summit on Sustainable Development, 48
United States
  coal resources, 324
  land use, 68–69
  population, 91
Universal Soil Loss Equation, 380
Upwelling, 198, 199, 380
Uranium, 4, 329, 380
Urban land use, 68
Urban runoff, 380
Urban sprawl, 168
Urbanization, 101–102, 254

Valuation, 30–31
  Resource, 20–24
Value
  of biodiversity, 168–172
  of human life, 22–23
  of minerals, 295
  theories of, 17
Variable costs, 29, 380
Vehicle fuel efficiency, 342–343
Venezuela, 306
Vienna Convention for the Protection of the Ozone Layer, 47
Vietnam, population, 91
Virtual office, 359
Visual blight, 380
Volatile Organic Compounds, 253

Waste dilution, 230
Waste reduction, 356–358, 380
Waste, hazardous, 32
Waste, radioactive, 330
Wastes, high-level, 330
Wastes, low-level, 330
Wastewater reclamation, 381
Wastewater treatment, 239–240
Water
  in agriculture, 117–119, *see also* Irrigation
  renewable supply, 219–221
  withdrawals, 81
Water budgets, 74
Water cycle, *see* Hydrologic cycle
Water harvesting, 381
Water pollution, 70, 196–197
Water rights, 228
Water supply system, 223
Water table, 74, 218, 381
Water use, 227–232
Water-holding capacity of soil, 117, 381
Watershed management, 241, 381
Weathering, 298, 381
Weeks Act, 41
Wetlands, 211
Whales, 212–214
Whaling, 2, 46
White House Conference on Conservation, 41
Wild and Scenic Rivers, 184
Wilderness, 68
Wilderness Act, 44
Wildlife refuges, 68
Willingness to pay, 21, 22, 381
Wind energy, 339
Wind erosion, 128

Windbreak, 129, 131, 381
Withdrawal, 224, 227, 230, 381
Withdrawals of water, 81, 119
Women
   and birth control, 103
Wood, *see also* Forests
   as fuel, 315
Wood, uses of, 38
Woodland, *see* Forest

World Bank 30, 46, 337
World Health Organization, 46, 49

Yellowcake, 329
Yellowstone National Park, 41, 177
Yemen, population, 91
Yield
   of crops, 114
Yosemite, 41

Yosemite National Park, 187
Yucca Mountain, Nevada, 330

Zebra mussels, 178, 179
Zero population growth, 95, 381
Zinc, 16
Zoning, 69, 381